Cleaning Our Environment
A Chemical Perspective

A Report by the Committee on
Environmental Improvement
American Chemical Society
Washington, D.C.
Second Edition

Published October 1978

Price: $9.50

Order from American Chemical Society, Special Issues Sales,
1155 Sixteenth St., N.W., Washington, D.C. 20036

Library of Congress Catalog Card 78-73104

Contents

Preface

In April 1972 the American Chemical Society's Committee on Environmental Improvement recommended to the Society's Board of Directors that plans be made for a new edition of "Cleaning Our Environment" (1). The first edition had appeared two and a half years earlier; requests for it were still approaching 1000 monthly, and it was already the largest-selling single publication the ACS had ever produced. Also, the President's Science Advisory Committee had encouraged the Society to publish the supplement (2) that appeared in January 1971. These signs of sustained interest, and the rapid evolution of environmental science and technology, were good evidence that an updated edition would be needed in a few years.

In June 1972 the Board approved the Committee's recommendation. The Committee on Environmental Improvement appointed five of its members to a Steering Committee to oversee the project. The Steering Committee in turn recruited seven panels of experts to draft the report, and work began.

This new edition of "Cleaning Our Environment" follows the general scheme of its predecessor. The subtitle has been changed, however, from "The Chemical Basis for Action" to "A Chemical Perspective." In addition, the coverage has been expanded to include analytical chemistry, toxicology, and radiation as well as the four original topics: air, water, solid wastes, and pesticides. The chapters on analytical chemistry and toxicology outline principles and techniques that apply throughout environmental science and technology and that, accordingly, recur in relation to specific problems at many points in the report. In the new chapter on radiation, in contrast, the subject is treated in depth, as are the four topics from the first report.

The premise of "Cleaning Our Environment" has been that the ACS, as a scientific and educational body, should serve the public by producing an environmental report designed primarily for the involved and educated layman. In so doing we have focused on chemistry, chemical engineering, and the related disciplines. The expertise of the American Chemical Society and its 115,000 members lies, of course, in those areas. And we believe that progress in them is essential to a long-term, rational approach to understanding and conserving our environment.

Both editions have had two purposes. The first was to describe the status of the chemical science and technology of environmental improvement: what is known and how it is being used; what must be learned and how it might be used. The second purpose was to recommend measures that should help to accelerate the sound development and use of that science and technology.

---

1. "Cleaning Our Environment: The Chemical Basis for Action," American Chemical Society, 249 pp., Washington, D.C., September 1969.
2. "A Supplement to 'Cleaning Our Environment: The Chemical Basis for Action' — Priority Recommendations with Supporting Discussion," American Chemical Society, 20 pp., Washington, D.C., January 1971.

We have hoped in this way to foster the technical awareness of legislators, administrators, and others who must deal with environmental problems at one or more steps removed from direct involvement in the pertinent science and technology. If at the same time this report can refresh the specialist, so much the better. Also, we hope that it will attract the interest of scientists and engineers not now working on environmental problems who may have useful ideas to contribute to their solution. Finally, it appears that the first edition was used widely in the classroom, a function not really anticipated at the outset. Of the 75,000 copies distributed by the end of 1977, a large number had gone to educational institutions. We trust that this revision will prove equally useful to that audience.

In addition to the scientists and engineers who worked directly on this report, a number of authorities, both members and nonmembers of the American Chemical Society, were kind enough to provide comment and information. The authors have drawn also on the writings of many others. The Society and its Committee on Environmental Improvement wish to thank all of these people for their contributions. We wish also to thank Kenneth M. Reese, the project editor and writer, and the staff of the ACS Department of Chemistry and Public Affairs, especially Dr. Stephen T. Quigley, for their services.

T.E. Larson

ACS Committee on Environmental Improvement
Task Force for "Cleaning Our Environment"

Steering Committee

Dr. Thurston E. Larson, *Chairman*
Assistant Chief and Head of Chemistry Section
Illinois State Water Survey

Dr. Julian B. Andelman
Professor of Water Chemistry
Graduate School of Public Health
University of Pittsburgh

Dr. James P. Lodge, Jr.
Consultant in Atmospheric Chemistry
Boulder, Colo.

Dr. Daniel MacDougall
General Manager
Dow Pharmaceuticals — Canada

Dr. Robert E. Sievers
Professor and Co-Chairman
Department of Chemistry
University of Colorado

Chemical Analysis and Monitoring

Dr. Henry Freiser, *Chairman*
Professor of Chemistry
College of Liberal Arts
University of Arizona

Dr. Warren B. Crummett
Technical Manager, Analytical Laboratories
Michigan Division
Dow Chemical U.S.A.

Dr. Alfred H. Ellison
Deputy Director, Environmental Sciences Research Laboratory
U.S. Environmental Protection Agency

Prof. H.A. Laitinen
Graduate Research Professor
Department of Chemistry
University of Florida

Dr. Khalil H. Mancy
Professor of Environmental Chemistry
Department of Environmental and Industrial Health
School of Public Health
University of Michigan

Dr. L.B. Rogers
Graham Perdue Professor of Chemistry
University of Georgia

Dr. Rodney K. Skogerboe
Professor of Chemistry
Colorado State University

Dr. Vernon Stenger
Research Scientist, Analytical Laboratories
Michigan Division
Dow Chemical U.S.A.

Dr. Robert K. Stevens
Chief, Atmospheric Instrumentation Branch, Atmospheric Chemistry and Physics Division, Environmental Sciences Research Laboratory
U.S. Environmental Protection Agency

Dr. John K. Taylor
Chief, Gas and Particulate Science Division, Center for Analytical Chemistry
National Bureau of Standards

Toxicology

Dr. John Doull, *Chairman*
Professor of Pharmacology and Toxicology
Department of Pharmacology
University of Kansas Medical Center

Dr. J.C. Calandra
Professor of Pathology
Medical School
Northwestern University

Dr. Daniel MacDougall
General Manager
Dow Pharmaceuticals — Canada

Dr. Carrol S. Weil
Senior Fellow
Carnegie-Mellon Institute of Research
Carnegie-Mellon University

The Air Environment

Dr. James P. Lodge, Jr., *Chairman*
Consultant in Atmospheric Chemistry
Boulder, Colo.

Dr. A.P. Altshuller
Director, Environmental Sciences Research Laboratory
U.S. Environmental Protection Agency

Dr. C. Stafford Brandt
Private Consultant
Bellevue, Wash.

Dr. David S. Ensor
Manager, Aerosol Sciences Department
Meteorology Research, Inc.

Dr. Frances L. Estes
Staff Scientist
International Research
and Development Corporation

Dr. Bernard Weinstock
Manager, Chemistry Department
Scientific Research Laboratory
Ford Motor Company

The Water Environment

Dr. Russell F. Christman, *Chairman*
Professor and Chairman
Department of Environmental Sciences
and Engineering
School of Public Health
University of North Carolina

Dr. D.J. Baumgartner
Chief, Marine and Freshwater Ecology
Branch, Corvallis Environmental
Research Laboratory
U.S. Environmental Protection Agency

Dr. Brian W. Mar
Associate Dean and Professor
Department of Civil Engineering
University of Washington

Dr. R.T. Oglesby
Professor of Aquatic Science
Department of Natural Resources
Cornell University

Dr. Ranard J. Pickering
Chief, Quality of Water Branch
Water Resources Division
U.S. Geological Survey

Dr. Gerard Rohlich
C.W. Cook Professor
of Environmental Engineering
Department of Civil Engineering
University of Texas at Austin

Dr. Walter J. Weber, Jr.
Professor of Environmental and Water
Resources Engineering and Chairman,
University Program in Water Resources
University of Michigan

Solid Wastes

Dr. Virgil H. Freed, *Chairman*
Head, Department of
Agricultural Chemistry
Oregon State University

Mr. James F. Barbour
Research Assistant, Department of
Agricultural Chemistry
Oregon State University

Dr. Dirk R. Brunner
Sanitary Engineer, Solid and
Hazardous Waste Research Division
Municipal Environmental
Research Laboratory
U.S. Environmental Protection Agency

Dr. Joseph B. Farrell
Chief, Disposal Branch
Wastewater Research Division
Municipal Environmental
Research Laboratory
U.S. Environmental Protection Agency

Mr. Robert R. Groner
Research Assistant, Department
of Agricultural Chemistry
Oregon State University

Dr. Charles J. Rogers
Senior Research Chemist, Solid and
Hazardous Waste Research Division
Municipal Environmental
Research Laboratory
U.S. Environmental Protection Agency

Pesticides in the Environment

Dr. Donald G. Crosby, *Chairman*
Professor, Department of
Environmental Toxicology
University of California, Davis

Dr. Etcyl H. Blair
Director of Health
and Environmental Research
Dow Chemical U.S.A.

Mr. Roger C. Blinn
Manager, Registration and Technical
Information, International Plant
Industry, Research and Development
American Cyanamid Company

Dr. Cleve A.I. Goring
Director, Agricultural Products Research
Agricultural Products Division
Dow Chemical U.S.A.

Dr. Robert E. Hanson
Manager, Product Safety and Compliance
Shell Oil Company

Dr. P.C. Kearney
Chief, Pesticide Degradation Laboratory,
Agricultural Environmental Quality
Institute
U.S. Department of Agriculture

Dr. J.R. Plimmer
Chief, Organic Chemical Synthesis
Laboratory, Agricultural Environmental
Quality Institute
U.S. Department of Agriculture

Dr. William Upholt†
Director of Health Effects
and Science Policy
Office of Toxic Substances
U.S. Environmental Protection Agency

Dr. George W. Ware
Professor and Head
Department of Entomology
College of Agriculture
University of Arizona

Radiation in the Environment

Mr. Paul R. Fields, *Chairman*
Director, Chemistry Division
Argonne National Laboratory
U.S. Department of Energy

Dr. Cyril L. Comar
Director, Environmental Assessment
Department
Electric Power Research Institute

†Retired.

Dr. John C. Golden
Staff Radioecologist
Commonwealth Edison Company

Dr. Philip F. Gustafson
Director, Division of Environmental
Impact Statements
Argonne National Laboratory
U.S. Department of Energy

Dr. John H. Harley
Director, Environmental Measurements
Laboratory
U.S. Department of Energy

Dr. Jack Schubert
Professor of Environmental Health Science
Hope College

Dr. Donald C. Stewart†
Associate Director, Chemistry Division
Argonne National Laboratory
U.S. Department of Energy

Dr. G. Hoyt Whipple
Professor of Radiological Health
Department of Environmental
and Industrial Health
School of Public Health
University of Michigan

ACS Committee on Environmental Improvement, 1972–78

Dr. Thurston E. Larson, *Chairman*
Assistant Chief and Head of
Chemistry Section
Illinois State Water Survey

Dr. Francis J. Amore
Associate Chemist
Illinois State Water Survey

Dr. Julian B. Andelman
Professor of Water Chemistry
Graduate School of Public Health
University of Pittsburgh

Dr. Bernard B. Blaustein
Program Coordinator
Office of the Director,
Pittsburgh Research Center
U.S. Department of Energy

Dr. Donald G. Crosby
Professor, Department of
Environmental Toxicology
University of California, Davis

Dr. Virgil H. Freed
Head, Department of
Agricultural Chemistry
Oregon State University

Dr. Glenn E. Gordon
Professor of Chemistry
University of Maryland

Dr. Eugene V. Kleber
Executive Advisor, Atomics
International Division
Rockwell International

Dr. Ralph R. Langner
Director, Industrial Hygiene Laboratory
Dow Chemical U.S.A.

Dr. James P. Lodge, Jr.
Consultant in Atmospheric Chemistry
Boulder, Colo.

Dr. Daniel MacDougall
General Manager
Dow Pharmaceuticals — Canada

Dr. Nina I. McClelland
Vice-President, Technical Services
National Sanitation Foundation

Dr. James J. Morgan*
Professor of Environmental
Engineering Sciences
California Institute of Technology

Dr. Charles E. Moser**
Assistant to the Vice-President
and Senior Coordinator
Environmental Protection Department
Texaco, Inc.

Mr. Wendell F. Phillips
Director — Residue Research
Technological Resources, Inc.
Subsidiary of Campbell Soup Company

Dr. Robert C. Robbins*
Physical Chemist
Stanford Research Institute

Mr. Martin D. Schlesinger*
President
Wallingford Group Limited

Dr. Robert E. Sievers
Professor and Co-Chairman
Department of Chemistry
University of Colorado

Miss H. Gladys Swope*
Consulting Chemist, Waste Management
and Pollution Control
Madison, Wis.

Dr. Walter J. Weber, Jr.*
Professor of Environmental and Water
Resources Engineering and Chairman,
University Program in Water Resources
University of Michigan

Mr. J. Fred Wilkes
Consulting Chemical Engineer
La Grange, Ill.

Dr. John W. Winchester*
Professor, Department of Oceanography
Florida State University

Dr. Teh Fu Yen*
Associate Professor of Biochemistry,
Department of Medicine; Associate
Professor, Department of Chemical
Engineering and Environmental
Engineering Program
University of Southern California

Committee Staff

Dr. Stephen T. Quigley
Director, Department of Chemistry
and Public Affairs
American Chemical Society

Mr. David C. Wimert
Committee Liaison
Department of Chemistry
and Public Affairs
American Chemical Society

*Term expired before 1978.
**Deceased.

Chapter 1

Summary and Recommendations

1 / Summary and Recommendations

Introduction

This report assesses from a chemical point of view what we know of the environment and of how to conserve it and recommends actions designed to expand that knowledge and its use. The report is technically far-ranging, but selectively so. Omissions reflect the authors' choices of reasonable stopping points, not their belief that all problems not mentioned are well under control. The problems identified are in general well recognized by specialists. The Recommendations that follow, therefore, are aimed largely at areas in which deeper understanding, greater effort, or both are necessary to continued progress.

The nation's effort to achieve a cleaner environment has grown markedly during the past two decades. Public support has remained generally firm. The advent of the Environmental Protection Agency in 1970 and of various laws and regulations before and since have provided the legal and administrative tools, imperfect though they may be. Huge sums, both public and private, have been spent on facilities, equipment, research and development, and numerous ancillary activities.

This ambitious endeavor has produced clear-cut results. Our understanding of the environment and of man's impact on it, though far from adequate, have improved greatly in the past 20 years. Industrial pollution, especially of water, has declined significantly. Inroads are being made on automotive emissions as new, more stringently controlled cars gradually replace older vehicles. Municipal wastewater treatment is being upgraded, if relatively slowly. Data on solid waste management practices are uncertain, but resource recovery facilities, though few in absolute numbers, have been multiplying. Pesticides and radiation continue to arouse controversy, but for some years have been subject to more extensive regulation than any other environmental hazard.

Where environmental contaminants have been monitored sufficiently to reveal trends, the data suggest that emission controls are taking hold, but slowly and unevenly. The measured levels of airborne sulfur dioxide and particles, for example, have declined in many cities, but the ambient air quality standard for photochemical oxidants is being violated in almost every urban area in the country. The levels of organochlorine pesticide residues in water and wildlife appear to be falling, and improvements have been seen in a number of bodies of water in terms of traditional parameters like biochemical oxygen demand. Nevertheless, violations of water quality standards remain common.

On balance such findings are encouraging, but they must be viewed in context. Relatively few contaminants have been monitored to the point where trends can be detected, and new contaminants continue to be identified. We can describe with reasonable confidence the environmental behavior of very few pollutants. In particular it should be recognized that the links between human health and most environmental contaminants, at the levels at which they occur, are understood only tenuously at best. These and related deficiencies have not been neglected, but much more work will be needed. Among the basic requirements are these:

- Continued improvement in the analytical chemical methods needed to monitor, control, and study the environment.

- Better statistical tools for relating minimum monitoring requirements to the data desired (e.g., minimum number of stations, sampling frequency, etc., to determine the mean concentration of an air pollutant over an entire city for an entire year with an accuracy of ±10% at the 90% confidence level).
- More reliable data that can be used to define normal and deviant trends in the concentrations of environmental contaminants and to design valid models of subsystems of the environment.
- Sharply upgraded fundamental knowledge of how the environment and its inhabitants are affected by contaminants, especially by long-term, low-level exposure to them.

Advances in these areas will not come quickly or cheaply. However, the methods and knowledge sought are essential to the design of environmental programs that achieve the desired goals at minimum social and economic cost.

Many of the difficulties of conserving the environment are inherent, of course, in the size and complexity of the task. They can be resolved, given the time, the will, and the money. Numerous technical problems are accessible to existing science and technology, with economic factors the major constraint. Where programs lag or seem to lag, the reason may be unrealistic support or unrealistic prognoses. Countless man-hours are consumed by lawsuits and hearings, often involving scientific argument. However, legal and administrative contention are healthy to the extent that they lead to sounder environmental decisions, which cannot always be based solely on scientific data. Similarly, scientific argument will always exist and in fact is essential to the self-correcting process by which science moves ahead.

The social and economic aspects of environmental control are not the main concern of this report, but they cannot ultimately be considered apart from the technical aspects. A few examples bear mention. One is the need to assess carefully the costs and benefits of all programs, difficult though this may be in some cases. The costs of removing contaminants from municipal and industrial effluents, for instance, tend to rise exponentially with the fraction removed; an effort should be made to strike a sensible balance between the two. There is also a need for environmental measures that apply not simply to effluents, but at points much earlier in the processes that generate them. An example would be mechanisms for building social costs into the market system by giving manufacturers an economic incentive to design an appropriate degree of recyclability/disposability into their products. In this vein it would appear that the need to reduce pollution and conserve resources warrants higher quality and greater durability in practically everything that we use. Finally, we must minimize needlessly rigid and burdensome administration of environmental programs. To this end, the decisions of public agencies must regularly be tested and shaped by rigorous external challenge, both social and professional.

Recommendations

The following summary of this report contains in its discussion section 51 Recommendations. A number of them are composites of several of the 110 Recommendations in the report; at the end of each, the numbers of the corresponding Recommendations in the full text are given parenthetically. The summary concludes with the 27 Recommendations that are not covered in the discussion section but that appear with supporting remarks in the full text.

Chemical Analysis and Monitoring

Quantitative measurement, typified by chemical analysis and monitoring, is indispensable to progress in environmental science and technology. Quantitative data are needed to study the sources, behavior, and effects of contaminants; to set standards of quality; to evaluate abatement technologies; to build predictive models; and to prove violations for regulatory action. Environmental measurements range from determinations of specific contaminants in waste discharges, to analyses of receiving waters and air masses, to baseline probes of pristine environmental samples. The substances in question often are at very low concentrations and may be accompanied by interfering substances that obscure the analysis. Interactions among air, water, soil, and biological organisms may call for measurements in each, often by quite different techniques, greatly complicating the development of data that can be correlated and interpreted.

Analysts have responded remarkably well to the need to refine existing techniques and to develop new and sensitive procedures for identifying and quantifying, sometimes approximately, unanticipated or previously unrecognized constituents of the environment. Overall, however, progress in environmental measurement has been slow, in part because of the persistent shortage of specialists in the field. Analysis too often is treated as a secondary activity and assigned to unqualified personnel. One result has been masses of environmental data that frequently are useless for the intended purposes.

Today, we face explosive growth in the number of substances known or suspected to affect environmental quality. Analytical methods for such substances and their degradation products must be devised or adapted from existing procedures and subjected to the scientific scrutiny required to assess their reliability. An intensive, coordinated effort is essential if the tools and application of environmental analysis are to be upgraded to a level consistent with the need.

Recommendation AM I: *Analytical scientists must take the initiative in developing and promulgating standards for individual proficiency and for the determination of precision and accuracy, limits of detection, proof of presence of specific substances in environmental samples, and related aspects of quality assurance. (AM 1)*

Recommendation AM II: *Sharply increased financial support should be given to research and development on new and improved analytical methods with the sensitivity, specificity, and response time needed to pursue environmental programs effectively. Support should be strengthened correspondingly for more fundamental studies of new approaches to the measurement of pollutants. (AM 7)*

Recommendation AM III: *Training in analytical chemistry, both academic and on the job, must be funded at the much higher levels dictated by the fact that progress in environmental control has long suffered severely from an acute shortage of well-trained, innovative, and experienced analytical chemists. (AM 4)*

Analytical methods and instruments must serve a variety of purposes. Often, for example, the analyst will wish to measure a single compound, such as nitrogen dioxide in air. Many of the methods now used widely for such purposes, however, are not sufficiently specific to the compound of interest and do not provide the required information. Complementary to the measurement of single substances is measurement of classes of substances, such as total organic chlorine in water. Class analyses can provide useful information and, again, more reliable and accurate methods are an important need. Highly sophisticated equipment is being applied nowadays to environmental problems; an example is the combined use of the gas chromatograph, mass spectrometer, and computer to identify organic compounds at very low concentrations in water. Equipment of this kind is extremely useful, but it is costly and must be used in the laboratory. Much more suitable for many purposes would be simple, low-cost, portable instruments that could be employed in large numbers and would not necessarily require professional analytical chemists as operators.

Recommendation AM IV: *Balanced attention should be devoted to the development and improvement of techniques of two complementary types: methods for determining specific compounds and their concentrations in environmental samples; and methods for determining classes of substances. Paralleling these efforts should be the development of simple, low-cost methods for mass application. (AM 9, 10, 12)*

Toxicology

Toxicology is a central element of virtually all of the problems associated with improving our environment. The toxicologist is concerned broadly with "toxicity" and "hazard." Toxicity refers to the inherent ability of a substance to produce adverse symptoms or death in an organism; hazard refers to the probability that an organism will be harmed by exposure to a substance in a given quantity and manner.

The toxicologist assumes that well-designed studies of the effects of a substance on animals will allow him to predict whether it might pose a hazard to man. However, he is aware that such predictions at the current level of knowledge may suffer from significant uncertainties. One of these is the difficulty of extrapolating results in a relatively few test animals to thousands or even millions of humans. Biological variability is a second problem; seemingly duplicate studies frequently produce conflicting results. Further, humans are exposed to various pollutants in very low concentrations and in complex mixtures, conditions that are difficult or impossible to simulate realistically. Where a specific cause-and-effect relationship is suspected, epidemiological studies can help, but more than one cause may correlate well statistically with the effect of interest.

Work in environmental toxicology, involving nonhuman organisms in all phases of the environment, has expanded steadily in recent years. The rationale is that a direct toxic impact on one of the many species that occupy the typical ecosystem may indirectly affect one or more other species, including man. The methods of human toxicology can provide useful information on nonhuman species, but the environmental toxicologist must use combinations of additional procedures adapted to the more complex conditions of even the smallest ecosystems.

Toxicologic studies of all kinds are expensive in both time and money. The need to upgrade their effectiveness is urgent. It can be met to an appreciable degree only by pursuing the research necessary to improve the validity and relevance of current procedures and to replace them, where feasible, with new and more powerful methods.

Recommendation T I: *The significance of human exposure to environmental pollutants should be evaluated in terms of the contributions of all sources of the substances of interest, including food, air, and water, rather than by focusing separately on each such medium. Similarly, evaluation of the hazard of a given substance should span its lifetime, from the time it enters the environment until it is finally disposed of or converted to innocuous form. (T 1)*

Recommendation T II: *Support for the improvement of toxicologic testing methods should be supplied as rapidly as possible consistent with the ability of the available facilities and manpower to use such support effectively. The major improvement required is tests that singly or in combination have greater predictive relevance for humans than do current methods. Other goals include more rapid means of screening chemical agents for toxic potential and less costly protocols for animal studies. (T 2)*

Recommendation T III: *Toxicologic studies of nonhuman organisms, including plant life, should be pursued at a level and in a manner consistent with the fact that, ultimately, the well-being of humans is inseparable from the well-being of other forms of life. (T 3)*

A major goal of toxicology is to provide information needed to support regulatory decisions within the Federal Government. Such decisions are made as a rule without major difficulty, by balancing risks against benefits for the substance of interest. Problems arise, however, when the toxicologic evidence can lead to conflicting opinions or when its predictive relevance is uncertain. In these instances the Government traditionally has asked various groups of experts, public or private, to evaluate the data. A problematical element of the decision-making process is the so-called Delaney clause, which bans the use of any food additive that induces cancer when ingested in any amount by man or animal. The clause, in other words, recognizes risk but not benefit. Sensible alternatives to the Delaney clause are available and should be evaluated along with other means of reaching sound regulatory decisions involving toxicologic information.

Recommendation T IV: *Better mechanisms should be developed for incorporating toxicologic knowledge into the formulation of legislative and regulatory requirements. Initiatives in this respect should come from both the federal agencies involved and scientific bodies such as the National Academy of Sciences and the Society of Toxicology. (T 4)*

The demand for toxicologic services in this country is very high. Virtually all chemicals of commerce are subject by law to toxicologic evaluation, and resources are needed also for research on contaminants in the environment, including the work environment. In addition, our understanding of the basic mechanisms of toxicologic action must be expanded and put to practical use. Well-trained people are needed to discharge these tasks and, more broadly, to exercise the scientific judgment needed to translate evaluations of hazard into sound public policy. The necessary manpower is not now available and is unlikely to become available in the short term.

Recommendation T V: *Training programs in toxicology, at both the technician and professional levels, should be expanded to meet the manpower needs created by federal legislation, with support to come from federal funding arrangements like that in the National Institute of Environmental Health Sciences and from other public or private sources as necessary. (T 5)*

The Air Environment

The science and technology of air pollution and its control have advanced to a state that, though less than optimum, is nonetheless impressive. As one means of applying this knowledge, EPA in 1971 established ambient air quality standards for particulate matter, sulfur dioxide, carbon monoxide, nitrogen dioxide, oxidants/ozone, and hydrocarbons. The standards were based in part on the assumption that for each pollutant and each of its effects there is a threshold level below which no adverse effect occurs. This approach may lead to the conclusion that the only safe level is below the natural background. A second complication arises where pollutants are interrelated in behavior and the standards for them are not coupled accordingly. Further, various premises of the standards have been outdated by subsequent scientific findings. By law, a review of the standards must be completed by the end of 1980 and fresh reviews must be made every five years thereafter.

Recommendation A I: *Reviews of the National Ambient Air Quality Standards should include evaluation of the degree of improvement in air quality to be expected with existing standards and with alternative standards based on a variety of control schemes. (A 1)*

Flow, dispersion, transformation

Efficient management of the air environment depends on our understanding of how contaminants flow, disperse, and are transformed into other physical and chemical forms as they move from source to sink or receptor.

Since most man-made pollutants also occur naturally, or closely resemble naturally occurring substances, natural routes exist for converting them to forms not dangerous to life and removing them from the atmosphere. Such substances become pollutants when man overloads the natural disposal capacity, as with sulfur dioxide locally and regionally and carbon dioxide globally. In addition, some man-made substances differ from natural products enough not to fit easily into natural disposal schemes. Because these substances resist breakdown, they can become global pollutants; examples include the chlorofluoromethanes and, possibly, carbon tetrachloride.

In probing natural pollutant-removal processes and man's impact on them, we must know among other things what substances are present in the atmosphere. However, many substances exist in "clean air" or in the global atmosphere at concentrations that are at the extreme limit of sensitivity of the available analytical methods; others presumed to be present cannot be detected at all with current methods. The problem is illustrated by carbon monoxide. Fewer than 10 years ago, the gas was believed to be almost entirely man-made. Then it was discovered that most of the carbon monoxide in the atmosphere is in fact natural.

Chemists hypothesized that carbon monoxide could be produced naturally by oxidation of methane by hydroxyl radical and removed naturally by being itself oxidized by hydroxyl radical. These hypotheses could not be confirmed until very recently, however, when a means was developed of measuring the tiny amounts of hydroxyl radical involved.

Besides knowing what substances are present, we must know how they behave chemically and physically. Measures taken to reduce emissions of hydrocarbons and nitrogen oxides by mobile and stationary sources, for example, are based in part on knowledge of the processes that produce photochemical smog. These processes still are not well understood in detail; this is particularly true of the formation and role of particles or aerosols. Much remains to be learned also of chemical processes in the regional and global atmospheres. There is, for example, some evidence for a process that would very slowly remove carbon tetrachloride and chlorofluorocarbons from the lower atmosphere. The evidence is not unequivocal, however, and the question cannot be resolved in part because of inability to measure the pertinent substances accurately at concentrations of a fraction of a part per billion.

Meteorological processes also play a role in the behavior of airborne substances. Because of differential mixing rates, pronounced north-south variations exist in the global distribution of substances with atmospheric lifetimes of a few days to a few years. Thermal and aerodynamic effects contribute to the difficulty of constructing mathematical models for the distribution and concentrations of pollutants in urban and regional areas. Sound models should make it possible to predict emission limits for various pollutant sources so as to achieve the desired air quality at minimum social cost.

Recommendation A II: *A coordinated effort should be focused on certain interrelated problems: techniques for measuring contaminants at levels characteristic of "clean air" and for measuring the global background concentrations of airborne substances of all kinds; laboratory and field studies of the fundamental chemical processes that occur in "real" atmospheres, such as the formation of aerosols in urban air and the reactions of trace substances in the stratosphere; and the coupling of chemical behavior with meteorological phenomena like diffusion and dispersion in terms of the effects of contaminants on the urban, regional, and global atmospheres. (A 2–8)*

Recommendation A III: *Evaluations of the effects of contaminants on the urban, regional, and global atmospheres and of the changes to be expected from regulatory actions should be reviewed continuously on the basis of new scientific understanding of such problems. (A 9)*

Recommendation A IV: *Air-monitoring programs and field experiments must always include quality assurance of the resulting atmospheric data as an integral part of the effort. (A 10)*

Sources and control

Virtually every human activity — even physical exertion — causes some degree of air pollution, but the leading sources are agriculture, industry, generation of electric power, transportation, and space heating. This classification is not clear-cut, in that agriculture and transportation, for example, require products from industry, which generates pollutants when making them. Also, common to each major source is the consumption of energy generated by combustion, which in terms of processes is the largest single source of air contaminants.

Control of industrial air pollutants is based on emission limits established by the Environmental Protection Agency (EPA). The limits place ceilings on releases of specific pollutants from specific classes of industrial sources; they do not normally specify the means of compliance, such as process modification and stack-gas cleaning. Emissions from a given industrial process commonly are estimated in terms of an "emission factor:" the weight of pollutant produced per unit weight of the desired product. EPA has published extensive tabulations of emission factors, along with some data on how emissions are modified by various pollution-control devices. The growing use of control equipment, however, is creating a need for new sets of tables, and these have been appearing only slowly. Existing tabulations, moreover, contain only limited data on trace emissions (as opposed to major emissions like sulfur oxides), and in general they do not consider the effects of altitude.

The problem with trace emissions is exemplified by copper smelting. The amount of sulfur dioxide discharged per ton of copper is known reasonably well, but far fewer data exist on the traces of selenium dioxide discharged simultaneously. Concern for the trace composition of stack gases has come only recently; much more data and systematic compilation are needed to translate the concern into effective action. The need is important because many trace substances may be more active biologically than the primary contaminants.

Less critical, perhaps, are the effects of altitude on emission factors. Still, there is reason to expect that in some cases the chemistry of the discharge depends on altitude. Some combustion processes, for example, may yield less carbon dioxide and more carbon monoxide at high altitude than at low. These effects seem not to have been investigated in the context of emission factors.

The need is growing for more precise evaluation of emission factors for use in long-range planning, making land-use decisions, and modeling pollutant behavior when developing control strategies. As the need grows, the uncertainties become more important, and a sustained effort will be needed to clarify them.

Of the several means of controlling air pollution, the addition of a scrubber or other trapping device at the end of the process should normally be considered only where a valuable by-product can be recovered or where other approaches, such as process modification, have been tried and found wanting. However, add-on devices are used widely because often they offer the most expedient means of avoiding excessive pollution. Of the two broad classes of pollutants — particles, and gases and vapors — the former are the more tractable, but even so they present difficulties.

In general, the technology exists to remove particulate matter from stack gases with any degree of efficiency desired. Cost rises with efficiency, however, and removal becomes more costly and also more difficult as particle size decreases. Large particles like those that soil surfaces are removed fairly easily. Less readily trapped are particles below about 5 micrometers (5 millionths of a meter) in diameter, which are the ones most easily taken into the lungs and are the most potent biologically. The most difficult are particles of about 0.2 micrometer in diameter, which exert the maximum effect on visibility; they have been ignored in large part because collection efficiency is calculated as a rule on a mass basis, and in many processes the submicrometer particles comprise a relatively small fraction of the total mass of emissions. More should be learned of the size distribution of particles from all processes. A need exists also to evaluate separately the control of particles in the respirable and optically significant size ranges and to design equipment to collect them more efficiently.

Recommendation A V: *Further data must be obtained on emission factors for air pollution sources, especially with regard to emissions from control devices, emissions of less common and trace elements, the composition of emitted particles as a function of particle size, and the effects of altitude on emissions. Studies are needed on particle-control devices to determine differential efficiencies as a function of particle size and the effects of these differential efficiencies on effluent composition. (A 12)*

The generation of electrical power is responsible for major shares of the sulfur dioxide, nitrogen oxides, and particles emitted into the atmosphere. About half of the power is obtained by burning coal, and the fraction is certain to rise. Expansion of nuclear power generation is hampered by questions about safety, by social resistance, and by economic factors. Solar and other new forms of energy will contribute little for decades. Residual fuel oil and natural gas suffer from costs and short supply. Domestic reserves of coal, on the other hand, are plentiful, and that fuel is destined to become the predominant source of electrical power in the foreseeable future.

Efforts to control emissions from coal include development of more efficient combustion systems and of processes for converting the fuel to cleaner liquid and gaseous forms, but their impact will be minor until at least 1985. Such conversions and nearly all alternatives consume energy, so that valid assessment of alternatives requires honest energy accounting. Most coal-burning power plants, for example, have an energy efficiency of around 40% or less, and older coal-burning domestic furnaces have efficiencies of less than 20%, not including the energy used to deliver the coal and remove the ash. Therefore, although electrical-resistance heating sounds appallingly inefficient (typically about 30%, starting with coal at the generating plant), it could become a sound alternative as more efficient, less-polluting generating plants are developed.

EPA standards for power plants limit only their emissions of particles (fly ash), sulfur dioxide, and nitrogen oxides. The technology of fly ash and sulfur dioxide collection has improved greatly in the past decade. Electrostatic precipitators are used widely to collect fly ash, but they still face problems with low-sulfur coals because of the high electrical resistivity of the ash produced. A second problem is that toxic trace elements in coal appear to concentrate preferentially in the smaller, more readily respirable particles, which are collected less efficiently by gas cleaning equipment.

Where a major reduction in sulfur dioxide emissions is required, it can be accomplished by lime or limestone scrubbing. These processes are costly, however, and generate solid wastes that must be disposed of. More advanced systems recover sulfur as the element or as sulfuric acid; these products in principle could be sold, but would move utilities into an unfamiliar market where success is by no means certain. On the whole, the relative advantages of the various sulfur oxides control techniques remain difficult to assess. Large-scale operating experience is scanty, and the technology is changing relatively rapidly.

The only practical means of limiting nitrogen oxides emissions from power plants at present is to minimize their formation by modifying the combustion process. This approach at best can reduce emissions by about 50%. However, processes for removing nitrogen oxides from flue gases are being evaluated in this country, and one such process is being used commercially in Japan.

With the spread of air pollution control devices and cleaner-burning fuels, many cities have reached the point where gasoline-powered vehicles account for at least 90% of the total weight of air pollutants emitted. Federal emission standards for automobiles have grown increasingly stringent, notwithstanding periodic delays in their imposition. Control equipment installed in new cars has in fact produced measurable results, although the maximum impact of an emission standard does not come until about eight years after it is imposed because of the average life and growth in numbers of automobiles. In 1975, the automobile industry moved to a second-generation device, the catalytic converter, to control carbon monoxide and hydrocarbons. Despite early doubts the converters seem to be working, although relatively few long-term data are in hand on their performance in consumer hands. Catalytic control of nitrogen oxides simultaneously with carbon monoxide and hydrocarbons is well advanced experimentally and is expected to be introduced in some 1979 and 1980 models.

The apparent success of catalytic control postponed a highly capital-intensive conversion to new, less-polluting types of engines in this country. The small, stratified-charge engine being mass-produced abroad, for example, shows definite advantages in emission reduction, and the diesel has advantages as well. Longer-range possibilities are the gas turbine and external combustion engines; both are inherently low-polluting, but neither can yet match the blend of performance, durability, weight, and cost of the traditional engine. For the much longer run it can legitimately be questioned whether any western nation can continue to dedicate a large share of its resources to personal vehicles. Public transportation would seem to be the answer in the U.S., but real innovation will be needed to induce a significant fraction of the population to break away from the present dependence on the automobile.

Air pollution and energy

It is clear that air pollution and energy cannot be considered separately. Pollution controls do consume energy. But in many cases the penalty can be far offset by gains in efficiency that reduce both the consumption of energy resources and the generation of pollutants. Mounting costs, for example, make the use of hitherto wasted heat increasingly attractive. There seems no theoretical reason why even rather low-level process heat from industry could not be put to work in some way. This has been done sporadically; today, the need to obtain maximum energy from fuel that must be burned anyway puts a premium on developing the necessary concepts systematically and applying them as rapidly as possible.

At the heart of the energy crisis is technological inflexibility. Sources of energy are plentiful at home and abroad. The frequent problem is that we are not technologically flexible enough — and in some cases not technologically advanced enough — to harness them efficiently in the forms in which they occur. A power plant designed to burn gas or fuel oil does not work well on coal. The automobile requires a liquid fuel. A coal furnace will not burn wood efficiently. Millions of homes require gaseous fuel for heating space and water and a smaller number for cooking as well. Thus, for several decades at least, liquid and gaseous fuels must be available. For this reason, work on the liquefaction and gasification of coal is necessary despite the energy costs of such processes. Some critics suggest, for example, that oil from shale may supply less energy than is required to extract it. But even if this should prove correct, the process may well become advantageous if it yields a liquid fuel at less net energy loss than is incurred by liquefying coal.

Energy is available from natural sources like the sun and wind, but not continuously. To harness such sources effectively, surplus energy must be accumulated when it is available and stored for use during periods of darkness, calm, or other interruption of supply. Current methods for accumulating energy, such as the storage battery, are inefficient and unwieldy or have other serious drawbacks. A potentially useful approach is to store energy in the form of hydrogen liberated from water by a relatively efficient chemical or thermal cycle. The combustion of hydrogen in air yields nitrogen oxides, but little else in the way of by-product air pollutants. Hydrogen presents problems of its own, but is sufficiently promising to warrant careful attention. This does not mean that solar energy should be set to one side until the energy-storage problem is solved. Even with present technology, for example, solar heating of homes could reduce gas consumption significantly. The major problem is that there are no economical ways to retrofit existing houses.

Recommendation A VI: *The close ties between energy and air pollution warrant persistent emphasis on activities that include: development of more efficient, less inherently polluting vehicular engines; selective development of new concepts in public transportation; recomputation of cost-risk figures for nuclear and other forms of power to insure the explicit inclusion of true costs and hazards; studies of coal purification, liquefaction, and gasification with proper stress on the cleanliness of both processes and products; studies on potentially clean but intermittent energy sources (wind, waves, sun, etc.), coordinated with work on energy accumulation systems such as hydrogen generation; and development of federal economic incentives to encourage systematic use of waste heat. (A13–17)*

Effects of air pollutants

The acute toxicological effects of many air contaminants on man and other animals are understood reasonably well, but the effects of exposure to mixtures of gases and particles at very low concentrations are only just beginning to be comprehended. Numerous epidemiological studies associate the combination of sulfur dioxide, sulfates, and particles with health effects of varying severity, but the mechanism of action is obscure. The results of an epidemiological study of 1970–71, for example, suggested that adverse effects on health were associated more consistently with exposure to suspended sulfate than with exposure to sulfur dioxide or to total suspended particles. The significance of the observation, however, remains uncertain for two reasons: questions about the atmospheric chemistry involved — specifically, how and to what extent and in what compounds sulfate is formed in the air; and doubts about the validity of the data gathered by the air monitoring instruments used.

Observations of animals exposed to gas-aerosol systems show that particles may serve as respiratory irritants and thus enhance the respiratory action of other contaminants. Also, many specific substances in respirable particles are considered potential hazards to health. Vanadium inhibits a number of enzyme systems and, when inhaled, has been correlated with low cholesterol levels and upset of ascorbic acid (vitamin C) metabolism in man. A number of metals and, especially, particulate polycyclic organic matter have been indicted as environmental carcinogens.

Research on the effects of mixed pollutants, including mixed gas-aerosol systems, on animals has shown in general that mixed pollutants may act in several ways. The effects may be additive, amounting to the sum of the effects of each pollutant acting alone; they may also be greater or less than additive or may differ from the simple additive in some other way. Work with whole animals is complicated further by protective effects and cross tolerances. Thus the effects of contaminants are not independent of each other; the animal's total response involves its past history and its overall response to its environment. Physiological stress could be a factor in the changes in alkaline phosphatase activity found in the livers of rats that had inhaled acrolein, ozone, nitrogen dioxide, formaldehyde, or sulfur dioxide. Similarly, increased susceptibility to bacterial and viral infections follows exposure to auto exhaust, ozone, and nitrogen dioxide.

The complexities of research on whole animals have stimulated work on simpler biochemical systems designed to explain the basic mechanisms of the responses of cells to toxic or irritating substances. Such work has included research on various bacteria, individual tissues, and isolated enzyme systems. Ozone, nitrogen dioxide, peroxyacyl nitrate, and formaldehyde, for example, have been found to inhibit the activity of a number of enzymes in isolated systems, but with differences in rate, extent, and mode of action. Despite the progress made in experiments like these, the study of mechanisms of action of pollutants is hardly out of its infancy. Research on the biochemistry of the lung, which is crucial to true understanding of many mechanisms, is at about the same point. Evaluation of biological response, moreover, has been impeded by limitations in the precision, accuracy, and specificity of the available analytical chemical methods.

Recommendation A VII: *Better understanding of the reactions of sulfur oxides and nitrogen oxides in the atmosphere must be developed to make it possible to appraise accurately the potential physiological impact of those contaminants. (A 18)*

Recommendation A VIII: *Further research should be done on the effects of inhalation of metals at low and, where known, ambient concentrations; on the effects of combinations of air pollutants, both* in vitro *and* in vivo, *in carefully defined systems at known humidities and temperatures; and on the effects of contaminants on tissue oxidation, along with the uptake and destructive metabolism of contaminants by the tissues. (A 19, 20, 21)*

Recommendation A IX: *Epidemiological programs, such as EPA's now-terminated CHESS Program, ought to be supported by stronger overall analytical capabilities, firmer computer-based models, the development of measurement techniques for subclinical responses, and the determination of additional pollutants, some of which may actually account for effects that are being correlated with other, known contaminants. (A 22)*

Recommendation A X: *Studies of the biochemistry and biophysics of the lung should be pressed to provide the knowledge needed to examine more specifically the* in vivo *effects of exposure to ambient concentrations of air pollutants. (A 24)*

For more than a century, scientists have been studying the effects of air pollution on vegetation. During most of this period they were concerned primarily with visible effects on plants and the associated declines in growth and yield. The past decade, however, has seen clear recognition of the long-suspected reductions in growth and yield that can occur unaccompanied by visible symptoms. Effects of this kind caused by the photochemical smog complex are especially clear-cut. If the reductions in yield found experimentally with smog can be extrapolated only partly to other crops and conditions, the potential impact of more smog is rather frightening.

A major shortcoming of our knowledge of the effects of air pollution is the sparseness of information on the sensitivities of plants and other nonhuman organisms that do not clearly have economic value. Resources have been inadequate to study everything, and priority has been given to effects that can be related directly to health or economic loss. In the long run, however, all forms of life are interrelated, so that the ultimate survival of man may depend as much on the welfare of lichens and foxes as on that of wheat and hogs.

Recommendation A XI: *Studies of the effects of air pollutants, especially in combination, on plants must be continued, and studies of generalized losses in productivity should be expanded. The species studied should include those important in wild ecosystems as well as in agriculture and forestry; research on both plant and animal members of wild ecosystems should be integrated carefully with other research on the biological effects of air pollutants. (A 26, 27)*

Analysis of particles

The costs and effectiveness of air pollution control programs depend to a marked degree on our ability to detect and measure contaminants. A special problem in this respect is the analysis of particulate matter. It has become clear in recent years that the effects of particles are very much contingent on their size spectrum and on the distribution of chemical species within that spectrum. But the only particle-monitoring system used routinely is the very questionable high-volume sampler, which measures only the mass of the particles collected, irrespective of size distribution and chemical composition. A few collectors exist that separate particles by size, but all are subject to substantial uncertainties. Although a promising start has been made, the subject needs a great deal more emphasis.

In the absence of suitable instruments, special interest has been shown in optical techniques. One optical device, the integrating nephelometer, has proved especially useful in recent years. Basically the instrument measures the scattering of light by suspended particles; the results can be correlated with visibility (visual range), particle size and concentration, and even the color of the aerosol and objects viewed through it. The integrating nephelometer is a good means of measuring visibility-reducing aerosol (0.1 to 1.0 micrometer) and much of the respirable aerosol (under 5 micrometers). It is also useful for studying the chemistry of visibility-reducing aerosol. The understanding and control of visibility-reducing particles has been the goal of many air pollution control programs. Only in the past few years, however, has instrumentation, including the nephelometer, become available for studying visibility in a manner compatible with other aerosol measurements.

Recommendation A XII: *The development of inexpensive methods for measuring atmospheric aerosol in both fine and coarse regions in real time should be stimulated, along with development of methods for measuring sulfuric-acid aerosol. (A 28, 29)*

Recommendation A XIII: *A national data base should be established systematically for visibility measured instrumentally, so that trends in visibility (and air pollution) can be established more precisely and reproducibly than is possible with the visual estimates used largely today. The relationships among visibility and the sources of air pollution in an urban area should be defined; a by-product would be regional visibility models analogous to those used to relate ambient concentrations of pollutants to emissions from their sources. (A 30, 31)*

The Water Environment

A body of water is a dynamic system, steadily assimilating solids, liquids, and gases, both natural and man-made. Natural surface waters, moreover, normally teem with living organisms, which can powerfully affect the course of events in a given water system. All of these substances, living and nonliving, may flow, disperse, and interact chemically and physically before they reach a sink such as the ocean or a receptor such as a fish. Enroute they may assume a variety of chemical and physical forms.

Flow, dispersion, degradation

Sound management of the water environment depends in the long run on a thorough grasp of the flow, dispersion, and degradation of contaminants. This overall transport process determines in large measure the effects of waterborne substances on quality throughout a water system. Clear understanding of organic degradation phenomena is particularly important. Encouraging progress has been made in developing useful mathematical models of the physical aspects of pollutant transport. The major difficulties now are our incomplete perception of the factors that control the physical and chemical interactions of solids, gases, and liquids and of the nature and amounts of the substances that are interacting.

For almost no pollutant do we have adequate balances of the amounts that enter particular water systems and the amounts that leave. Useful data may emerge in time from the National Pollutant Discharge Elimination System, established in 1972. The system requires applicants for discharge permits to specify the physical and chemical characteristics of their discharges. Useful data should come also from the National Eutrophication Survey of 1972–75.

The main inorganic constituents of most wastewaters are known reasonably well. The specific organic compounds were largely unidentified until about 1970, when work began to detect and measure organic pollutants using sophisticated analytical technology. These analyses since have detected many individual organic compounds in the effluents from municipal and industrial waste treatment plants and in natural waters and public water supplies. An estimated 80% of the compounds identified are man-made.

The organic compounds being identified are present as a rule at very low levels — on the order of micrograms (millionths of a gram) per liter — but nonetheless must be viewed as a potential threat to public water supplies. Drinking water standards traditionally have been based on the premise that public water supplies would be drawn from natural waters, but this premise is growing progressively less tenable. The number of wastewater discharges in the nation is increasing, and the distances between discharge outlets and water supply intakes are decreasing. Sources of public water supply as a result are becoming more liable to pollution not only by untreated wastes but by the normal residual contaminants in treated wastewaters. Water supply treatment plants must cope with a growing load of pollutants, including some that most of them are not designed to remove.

Biological aspects

The physical and chemical fate of matter dissolved or suspended in water is influenced strongly by living organisms. Particularly important to water quality are the bacteria that oxidize organic compounds to obtain energy and carbon for synthesizing cellular material. These organisms degrade organic contaminants, but in the process they deplete the dissolved oxygen that is critical to the survival of most forms of aquatic life (the pollutant potential of organic material is commonly described in terms of "biochemical oxygen demand" — the amount of oxygen consumed in the degradation process). Organisms in water also take up nitrogen, phosphorus, and other cellular building blocks in addition to carbon. They do so at a rate determined by the proportions of the substances in the organisms and their abundances in the water. The uptake, release, and transformation of materials in solution is governed also by biochemical reactions related to the flow of energy through living communities.

At our present state of knowledge, the fates of a few materials in certain natural waters can be predicted with reasonable accuracy. But on the whole, the approach to such questions remains necessarily empirical. The difficulties are illustrated by the unexpected discovery in 1970 that bacteria in bottom environments can convert metallic mercury to soluble methyl mercury, which in turn can be taken up by organisms including fish. A second illustration is the extensive studies that have been needed to trace the movement of radionuclides through aquatic communities.

Particles

Particles are ubiquitous in natural waters. They can adsorb and otherwise bind various substances, can serve as sites for bacterial growth, and can act in other ways that may strongly affect the transport of pollutants. Particles in water vary widely in concentration and in their chemical, biological, and physical properties, but detailed data on these characteristics are not extensive. The transport of a few contaminants, especially pesticides, by waterborne particles has been studied to a limited degree; the subject deserves much greater effort.

Soil and groundwater

Knowledge of the transport of substances in soil and groundwater is growing more important for two reasons: the trend toward treating wastewater by spreading it on land; and the increased use of underground space for emplacing especially troublesome wastes. A related problem is pollution of groundwater, which normally is found in highly permeable strata of rock, sand, or gravel. Several decades of intensive effort have provided enough knowledge to permit water, including dilute wastewater, to be applied experimentally on land to recharge fresh groundwater lying not too far beneath the surface. Much of what is known of groundwater recharge is empirical, however, and large gaps exist in our understanding of the processes involved. Similarly, the use of underground space to store useful fluids and especially to dispose of fluid wastes is growing rapidly. It is important that we learn to monitor and predict the fate of underground wastes, whose disposition, composition, and movement are often unknown under present practice. The U.S. Geological Survey has documented many examples of groundwater pollution. As the country turns more to groundwater as a primary or supplementary source of water, its protection grows ever more necessary.

Pollutant sinks

Research on the sources, transport, and effects of water pollutants generally has taken precedence over research on the sinks: oceans, lakes, their sediments, and deep underground formations. In recent years, however, the sinks have begun to receive more attention; this is particularly true of the oceans, which cover some 70% of the earth's surface and contain all of the naturally occurring chemical elements. The effects of contaminants on the seas and other sinks depend on an intricate web of factors, including currents, mixing processes, and geochemical and biological events. The sheer size of the marine environment makes it potentially the biggest water pollution problem of all. Natural and man-made inputs may change oceanic processes in ways that impair the utility of the water to man; some such changes could be irreversible and have consequences unpredictable today. The seas and other sinks are a part of the cycle followed by waterborne substances and at present perhaps the least understood part of that cycle.

Recommendation W I: *The flow, degradation, and dispersion of water contaminants should be investigated in a sustained and coordinated program stressing: compilation of regional inventories of pollutants, preferably in terms of specific compounds, with emphasis on threats to sources of public water supplies; fundamental research on the action of natural mixed populations of bacteria and other organisms on specific compounds; comprehensive investigations of naturally occurring and pollutant particles to determine such characteristics as size, charge, composition, and adsorptive properties; improved mathematical descriptions of natural water systems subject to pollution; systematic studies of the flow and reactions of forms of phosphorus and nitrogen and of various organic substances in soil and groundwater; and transport and long-term deposition of pollutants in oceans, lakes, and deep underground formations. (W 1–7)*

Municipal wastewater treatment

The basic principles and technology of the processes used in treating municipal wastewater have been well established through years of practice. Refinements in design and improved operation are needed, however, to meet stringent modern standards. Many municipal facilities provide only primary treatment, but the law requires most to be providing a minimum of secondary treatment by 1983. Besides domestic sewage, municipal wastewater usually contains stormwater drainage, commercial wastes, and frequently industrial wastes. The amounts of commercial and industrial wastes may be large enough to warrant special measures, such as the addition of physical-chemical means of removing dissolved inorganic compounds.

Handling and disposing of sludges is the single most troublesome aspect of municipal treatment, accounting for up to 50% of capital and operating costs. Some sludge results from primary treatment, but most of it consists of excess microorganisms from secondary treatment, which involves biological oxidation by the activated sludge, trickling filter, or similar processes. Production of sludges is rising with the move toward secondary treatment as a minimum nationwide. Restrictions on ocean disposal and the rising costs of labor and land for land disposal are making the situation even more difficult.

Both primary and secondary sludges contain pathogens, numerous organic compounds, and many of the chemical elements, including toxic elements. The organisms in biological sludges flocculate into a structure from which it is extremely difficult to remove water, which must be done to reduce the volume of sludge to be handled. Sludge handling and disposal has four goals: to stabilize the organic matter; to reduce the volume by removing liquids; to destroy or control harmful organisms; and to obtain by-products whose use or sale reduces the overall cost of processing. A number of methods, and combinations of them, are used to achieve these goals, but the most effective means of sludge disposal are determined by local conditions. Accurate evaluation of alternative schemes requires thorough investigation of such questions as the value of liquid, dried, or composted sludge as a fertilizer or soil conditioner; underground disposal as in abandoned mines; and land reclamation. Evaluation of disposal on land further requires consideration of the potentially adverse effects of leachate or runoff on ground and surface waters, uptake of toxic elements by crops, and similar points.

Some sewerage systems carry storm water and sanitary sewage in the same conduit; when the combined flows exceed the capacity of the treatment plant, the excess is diverted untreated to the receiving stream. In addition, storm water discharged directly from separate storm drains contains significant amounts of pollutants, as does urban runoff that bypasses sewerage systems altogether. Various means of flow control, detention, and treatment are being tried to combat storm-water loads, whose impact can be so severe that pollution control programs may fail unless they are accounted for properly.

Recommendation W II: *A comprehensive effort should be made to optimize conventional treatment processes for municipal wastewaters, and to assess their potential relative to alternative processes, through expanded investigations of: new methods of handling sludges, with emphasis on reducing costs; the quantity, quality, and alternative treatments for combined sewage, wastewater from nonpoint sources, and urban storm waters; cost-effectiveness of physical-chemical treatment alternatives; and innovations based on fundamental understanding of microbiological processes. (W 8–11)*

Physicochemical and advanced treatment

Advanced waste treatment includes any process or combination of processes that will remove more contaminants from municipal and industrial wastewaters than will conventional biological treatments. Advanced treatment, therefore, generally involves chemical or physicochemical processes. These may be used to upgrade the effluent from conventional treatment or to modify or replace conventional treatment. Advanced treatment has two general goals: to maximize pollution control; and to expand water supply through water renovation and reuse.

Physicochemical processes that have proved particularly feasible for advanced waste treatment include coagulation, precipitation, filtration, adsorption, ion exchange, and certain membrane separation processes and chemical oxidations. These processes individually tend to be well suited to separating some classes or types of pollutants but not to separating others. Typically, therefore, various combinations of them must be used.

The development of physicochemical processes centered at first on "tertiary" systems designed to follow conventional primary and secondary treatment. This approach remains valid. However, it can have certain technical and economic disadvantages, and these have led to the concept of applying independent physicochemical treatment (IPCT) directly to raw wastewater rather than to secondary effluent. IPCT has been shown to be an attractive alternative to conventional biological treatment of municipal wastewater (and of industrial wastewater as well). Of the processes that make up IPCT systems, coagulation and adsorption on activated carbon are particularly important in advanced treatment schemes.

Coagulation involves the reaction of chemical coagulants with constituents of wastewater to form insoluble precipitates and to destabilize colloidal particles suspended in the water. Common coagulants are lime, iron and aluminum salts, and synthetic polymers. Primary sedimentation of screened, degritted, municipal wastewater rarely removes more than 50% of the suspended solids. The addition of coagulants, controlled mixing, and precipitation, however, can remove up to 99% of them as well as most of the soluble phosphorus and the metal ions found normally in wastewater. Pilot plant studies have shown that a well designed chemical coagulation-precipitation system can achieve at least the equivalent of conventional secondary treatment.

IPCT systems rely on activated carbon to remove dissolved organic materials from wastewater. The carbon effectively adsorbs most organic substances that are degraded by conventional biological processes as well as many that resist biodegradation. The process may follow conventional biological treatment or, with coagulation, may replace it. Because of the relatively large amounts of organic materials in wastewaters, treatment by carbon adsorption must be continuous and requires large amounts of carbon. The high initial cost of the carbon, moreover, requires that it be regenerated by controlled heating and reused.

Bacteria, viruses

Reliable removal or inactivation of disease-causing bacteria or viruses is growing more important for two reasons: waters that receive treatment plant effluent are being used increasingly for recreation; and direct reuse of treated water by humans will require means of preventing the numbers of organisms from building up in recycle systems. Coagulation and precipitation can achieve high removals of bacteria. Chlorination, the standard disinfection process in the U.S., can destroy more than 99% of the bacteria. High levels of chlorine (breakpoint chlorination) evidently inactivate viruses effectively, and coagulation-precipitation may be the next best weapon against them. However, further work is needed on disinfection by chlorine or other oxidants; since relatively high levels of them might be required, more should be known of their reactions with compounds in effluent streams to yield chlorinated organics or other undesirable compounds. Further work is needed also on removal of viruses by processes such as adsorption, ion exchange, and filtration.

Recommendation W III: *The capabilities of independent physicochemical treatment (IPCT) should be exploited by considering the process as an alternative to conventional biological treatment wherever new wastewater treatment facilities are being designed, and as a replacement for existing facilities that otherwise must meet higher effluent standards by adding tertiary processes; by developing less costly physicochemical processes for ammonia removal; and by developing comprehensive design models to optimize design and selection of combinations of physicochemical processes for complex IPCT systems. (W 12, 13, 15, 16)*

Recommendation W IV: *Chemical and biological characterizations should be made of advanced treatment systems operating at the large pilot-plant or demonstration-plant level, including the identification of specific chemical compounds and studies of the effects of chlorine or other oxidants on organic residues. Such studies should be comprehensive research and development efforts, not simply demonstrations that the systems operate properly in terms of the traditional parameters. (W 14)*

Recommendation W V: *Research should be maintained at a high level on removal or inactivation of viruses by wastewater treatment processes and on other means of preventing virus buildup in recycle systems. (W 17)*

Industrial facilities

It is difficult to generalize on the treatment of industrial wastewaters because their characteristics can differ markedly both within and among industries. For the same reason industry treats its wastewaters by many methods, including all of those applied to municipal wastewaters. A significant fraction of industrial wastewaters is treated in municipal facilities. Such wastes may contain substances that upset or pass through conventional municipal treatment processes.

In 1973, industry accounted for an estimated 43% of the biochemical oxygen demand discharged to receiving waters nationwide by point sources and 99% of the total suspended solids. The impact of industrial discharges, however, depends not only on these collective characteristics but on their content of specific inorganic and organic substances. Data on discharges of specific materials were inadequate by 1978, but they were beginning to accumulate. Such data could influence the nature of effluent limitation guidelines and performance standards for point sources of pollutants and pretreatment standards for industrial wastes to be discharged to public sewers. And as noted earlier, the data are essential to efficient management of specific water systems.

It has been estimated that perhaps 15% of the major industrial dischargers, and many smaller ones, did not meet the original (mid-1977) deadline for best practicable treatment established by the 1972 amendments to the Federal Water Pollution Control Act (the Clean Water Act of 1977 extended the deadline to mid-1984). It appears also that in some industrial categories the technologies on which EPA based its effluent limitation guidelines and discharge permits does not uniformly provide the anticipated level of control. But despite such difficulties it is apparent that the 1972 amendments spurred diverse technological efforts to control industrial water pollution.

Recommendation W VI: *Inventories should be compiled as quickly as possible of the specific substances in industrial wastewaters that are important or potentially important pollutants, and information should be made widely available on the technology of joint municipal-industrial treatment of wastewaters, particularly physical-chemical treatment. (W 18, 19)*

Agricultural pollution

Agriculture, including commercial livestock and poultry operations, is the source of many organic and inorganic pollutants in surface waters and groundwater. These contaminants include sediment from the erosion of crop land and compounds of phosphorus and nitrogen that originate in part in animal wastes and commercial fertilizers.

Animal wastes are high in biochemical oxygen demand, nitrogen, and phosphorus and often harbor pathogenic organisms like *Salmonella*. Commercial feeders as a rule contain their wastes and dispose of them on land, so that the main threat to natural waters is runoff and leaching as a result of precipitation. Land disposal is still the only practical method, but even the best land-disposal schemes face difficulties created by stricter regulation and large feeding operations. Uncertainties exist, for example, on the effects of high waste-loading rates on crops, soils, and groundwater. Better disposal methods are needed to reduce costs, including those of land and labor, to minimize nuisance, and to improve sanitation. One possibility is to process animal wastes for reuse. Manure, for example, could be converted to liquid or gaseous fuels; undigested nutritive material in manure can be converted to a feed supplement and recycled. Practical processes, however, remain to be demonstrated.

Runoff and leaching from barnyards, feedlots, and land treated with chemical fertilizers carry varying amounts of compounds of phosphorus and nitrogen into natural waters. The significance of these inputs, however, is not clear relative to those from other sources, such as raw or treated sewage and nitrate present naturally in soils. The concern over phosphorus is due to its role as a nutrient for plants whose excessive growth degrades surface waters. Nitrogen also is an important nutrient; in addition, nitrite formed from nitrate in groundwater can cause methemoglobinemia, which has resulted in sickness and sometimes death in both infants and ruminant livestock.

Recommendation W VII: *Strong support should be given to research and development on economic means of treating and disposing of animal wastes in agriculture, and basic data should be gathered systematically to delineate the relative importance of agricultural and rural sources of compounds of nitrogen and phosphorus in surface waters and groundwaters. (W 20, 21)*

Effects of water pollutants

The effects of water pollutants depend on many interrelated variables: the acidity or alkalinity of the water; its temperature; the degree of dilution and mixing; chemical and biochemical processes; the rate of flow; and reinforcing or counteracting actions of contaminants on one another. Further complexity arises from the economic need to relate water quality criteria to the generally recognized classes of water use: public supplies; freshwater aquatic life and wildlife habitat; marine aquatic life and wildlife habitat; industrial; recreation and aesthetic enjoyment; and agricultural. In this context, important gaps exist in our knowledge of the consequences of water pollution; notable instances include the effects involved in eutrophication, tastes and odors, and, especially, in human health.

At the levels typical today, inanimate water pollutants seem unlikely to exert significant acute effects on human health. Vigilance is warranted against the odd exception like nitrate in drinking water or cadmium in sludge-derived fertilizer. But the hazard of acute toxicity is less worrisome than that of chronic effects from long-term exposure to low levels of contaminants.

The problem is typified by the many organic compounds detected in drinking water supplies in the past few years. A few such compounds, at high concentrations, are carcinogenic in animals and thus may be carcinogenic in man, but the long-term, low-level effects of most are unknown. As water reuse grows more common, it becomes more important to remove potentially harmful substances at the source or by water treatment. The reuse need not be intentional; refractory chemicals discharged by one community need only enter the water supply treatment plant of a community downstream. Determination of those substances that might pose a threat in chronic exposure requires sharply improved knowledge of the specific compounds that are entering receiving waters. It also requires considerable work in areas like epidemiology and long-term, low-level exposure of laboratory animals to a variety of water pollutants.

The prospect of increased reuse of water, controlled or uncontrolled, has created growing concern over the hazard of bacterial and viral diseases such as salmonellosis and infectious hepatitis. Current bacteriological standards for both potable and recreational waters rely on counts of total coliform bacteria, which are taken to indicate a proportional level of fecal coliforms and viruses. Waterborne outbreaks of salmonellosis have occurred, however, where public water supplies met the standards, and salmonellae have been isolated from waters relatively low in both total and fecal coliforms. These and other findings suggest that research is badly needed on the correlation of currently used or new indicator organisms with waterborne disease. Key questions include the rates of dispersion and inactivation of viruses in recreational waters.

Concern over waterborne viral diseases has intensified the need to know more of the enteric viruses, those that occur in the gastrointestinal tract and feces of man and many lower animals. Of the enteric-virus diseases, only infectious hepatitis has been shown to be transmittable by the water route, and in the U.S. no large outbreaks of that disease have originated in water. It has been suggested, however, that evidence for viral diseases may be more apparent in secondary than in primary contacts, where the search thus far has concentrated. Furthermore, the ability of water treatment processes to remove or inactivate viruses remains uncertain, and relatively little is known of the levels of enteric viruses likely to be present in drinking water that meets current bacteriological standards. It might be possible to use groundwater basins for water treatment and dilution, but the opportunity cannot be assessed intelligently because remarkably few studies have been made on viruses in reuse systems for protracted periods. Data on the movement of viruses in groundwater are also extremely limited.

Recommendation W VIII: *Laboratory and epidemiological studies of the health effects of known water pollutants in long-term, low-level exposure should be strongly encouraged, and efforts should be made to define the risk of exposure to toxic substances in numerical terms, based on known statistics, as a means of establishing perspective in setting standards and regulations. (W 22, 23)*

Recommendation W IX: *Intensive effort should be devoted to studies of enteric viruses and their movement in soil and groundwater, and to correlations of bacterial indicator organisms with waterborne disease, particularly in recreational waters. (W 24, 25)*

Recommendation W X: *Public water supply treatment methods should be upgraded through research on removal and destruction of potentially harmful substances not removed by waste treatment practices or bypassed with insufficient dilution during plant outage periods or in times of disaster such as power failures. (W 26)*

Recommendation W XI: *Investigations should be pursued of the fundamental chemical and biological parameters of eutrophication and its effects. Development of effective and economic long-term controls will depend on considerably improved knowledge of factors such as mass balances for significant nutrients; the forms in which those nutrients exist in water; natural population dynamics; potentially limiting nutrients in specific situations; and algal, bacterial, and plant physiology in general. Investigations of a variety of lakes to document responses to changes in nutrient (especially phosphorus) inputs are especially needed if effective, economic, long-term strategies for controlling eutrophication are to be developed. Also critical to this question is a better understanding of nutrient sources, the transport of nutrients to standing bodies of water, and the biological reactivity of various chemical and physical forms of nutrients. (W 27)*

Alternative management strategies

The scope of this country's effort to conserve its waters makes institutional problems unavoidable. One such problem is the absence of a clear-cut, overall strategy for managing water resources. The development of a national water policy that stresses central administration of water programs based on watershed boundaries promises benefits that are unlikely to result from existing policies and institutions. Regionalization and centralization of water management on a river-basin basis has been practiced in England and Wales for several years. While the development was stimulated by a particular set of conditions, its success warrants close examination in terms of future directions for water management in the U.S.

Not the least of the advantages offered by regionalization is technical flexibility — the capacity to provide technical services to facilities that otherwise could not afford them, to select the best alternatives based on all the needs of all members of a river basin community and not a special group, to use a given water resource for the purposes to which it is most suited. Water management regions no doubt would be difficult to establish in this country, in part because hydrologic and political boundaries rarely coincide. But purposeful movement toward more regionalized operations under federally established policy seems essential to sound, long-term conservation of our water resources.

Recommendation W XII: *High priority should be given to the development of innovative alternative strategies for the management of our water resources. In particular, the strategy of regionalization, including the consolidation of management of river basins through authorities responsible for water supply, wastewater disposal, and overall water resources management and planning,* *should be carefully explored with its advantages and disadvantages enumerated. Recent British experience with consolidation of management warrants careful examination to determine its applicability to U.S. water management programs. (W 28)*

Solid Wastes

The basic science of solid waste handling and disposal remains in relatively primitive condition, but industry and government are bringing advanced technology from other fields increasingly to bear on the many problems that must be solved. Whatever the course of solid waste technology, the ultimate solution to the problem undoubtedly will be multifaceted. Important elements of that solution will be resource recovery and recycling, reuse by chemical transformation, and design of consumer products for prolonged use or at least for ease of materials recovery and recycling.

Estimates of the rates of generation of various types of solid wastes vary considerably. The absence of firm data stems partly from the relative lateness of interest in solid waste management and the resulting failure of public and private agencies to keep accurate records of wastes handled or, indeed, to keep records at all. At any rate, one estimate has this country generating about 10 pounds per person per day of household, commercial, and industrial solid wastes. Earlier data indicate that mineral, animal, and crop wastes would add perhaps another 100 pounds per person per day.

Hazardous wastes

Among types of solid wastes, special attention has been paid in recent years to hazardous materials — any waste or combination of wastes that poses a substantial present or potential hazard to human health or living organisms. A waste is regarded as hazardous if it is lethal, nondegradable, persistent in the environment, can be biologically magnified (as in food chains), or otherwise causes or tends to cause detrimental cumulative effects. General categories of hazardous wastes are toxic chemicals and flammable, radioactive, or biological substances, which include pathological and warfare agents. The wastes can be solids, sludges, liquids, or gases, and their sources are numerous and widely scattered. Prime producers are industry, the Federal Government, agriculture, and institutions such as hospitals and laboratories.

Treatment processes for hazardous wastes ideally should achieve four general ends: reduce the volume of the waste; separate the components; detoxify the waste; and recover useful materials. Treatment technology is generally available for most hazardous wastes, but no single process can perform all four functions; adequate treatment usually involves several processes linked in series. Residues from these processes, as well as untreated hazardous wastes, require ultimate disposal, and it is here that improved methods are needed.

Recommendation S I: *Increased disposal problems associated with disposal of sludges and hazardous wastes on land should receive the additional research and development required in the areas of health effects, environmental effects, control technology, management systems, and economic/legislative efforts. (S 3)*

Of the methods used in this country to dispose of solid wastes, disposal on land is by far the most common. Landfill probably accounts for more than 90% of the municipal refuse and incineration for most of the remainder. These traditional techniques very often are not applied properly, but modern practice is slowly taking hold. In addition, a strong trend has developed toward recovery of heat and materials from refuse.

Land disposal

An important aspect of disposing of solid wastes on land is the protection of groundwater from leachate. It has been estimated that fewer than one third of the nation's known land disposal sites meet state standards for sanitary landfills, and it is not certain that the standards are uniformly high enough to protect groundwater. Current research on sanitary landfilling is concerned mostly with municipal wastes. The work is directed primarily at identifying the potential leachate problem and providing appropriate solutions to it and to extending the life of the fill by pretreating the solid wastes. Experimental approaches to the leachate problem have included biological treatment of leachate as well as the use of aerobic stabilization or leachate recirculation to accelerate decomposition in the landfill. Pretreatment of wastes by shredding or baling offers advantages in handling and in greater disposal capacity per acre of landfill. Both techniques may ease the leachate problem. All of the foregoing approaches would, of course, increase the costs of sanitary landfilling.

Incineration

Incinerator and related thermal technology has been developing rapidly in the U.S. in recent years. The emphasis has changed, however, from simple reduction of wastes to ash and slag for disposal, to the recovery of thermal energy from solid wastes. Stringent particle emission limits have caused many conventional municipal incinerators to close down. The trend in particle emission control is toward electrostatic precipitators, which are well adapted to water-wall incinerators designed to generate steam. Water-wall units of this kind are operating in half a dozen U.S. municipalities, and others are in the planning or construction stage. A number of companies now burn in-plant wastes in conventional incinerators to produce steam. In development are numerous thermal processes for recovering energy from solid wastes. These include not only combustion processes, but also pyrolysis processes that yield gaseous or liquid fuels. The heterogeneous nature of municipal solid wastes creates technical problems for such energy recovery processes, and operating experience with them is not adequate as yet to support firm economic data and comparisons.

Recommendation S II: *The use of known science and technology peripheral to solid waste management in developing improved methods for sanitary landfill and incineration should be encouraged and supported. Specific efforts should be directed toward pollutant identification and the potential for pollutant migration into ground and surface waters from disposal of waste materials on land. (S 5)*

The upsurge of interest in recovering reusable materials from solid wastes has been focused primarily on municipal refuse. Hand sorting probably is the most common means of recovering salable materials from refuse, but mechanized systems are planned or operating at a number of sites, often as parts of energy-recovery schemes. In addition, attention is being devoted to problems and processes that affect the recyclability of individual materials.

Wastes and product design

Recyclability is governed in part by product design, including materials selection. In this light, a well designed product would be durable; relatively easy to disassemble, chemically or physically, for recycling; and readily degradable in the environment. The designer, however, must consider additional parameters, including consumer preferences, product safety, and the costs of materials and manufacture. Some design parameters, such as durability and degradability, may be in varying degree incompatible. More importantly, products are designed in an economic framework that does not reflect the design-related costs of solid waste handling and disposal, recovering or not recovering resources, and environmental impact. These costs might be built into the design process by various means, including a "product charge" imposed at the time of first sale. The creation of any such mechanism poses problems; among them are the difficulty of assigning equitable costs to waste disposal, and the perils of spawning yet another unwieldy bureaucracy and of channeling resources inadvertently into undesirable uses. Nevertheless, the prospective benefits of linking product design economically to its consequences for solid wastes are sufficient to warrant continued investigation of likely measures.

Recommendation S III: *Economic mechanisms that will encourage a more rational approach to solid waste management should continue to be developed and evaluated. Such measures should be directed at product design as well as at waste handling and disposal and resource recovery. (S 8)*

Effects of solid wastes

The primary purpose of controlling environmental pollution is to protect human health and the ecological integrity of the biosphere on which humans and other organisms depend. In terms of solid wastes, deeper understanding is needed in several areas: the effects of pollutants arising from the recovery of resources from wastes; the ultimate disposal of wastes; the mechanisms by which waste materials are transformed into pollutants not present in the original wastes; the mechanisms by which pollutants migrate from disposal sites; and the routes by which pollutants come into contact with the organisms they affect. Only within the framework of such knowledge can truly effective and economic solutions to solid waste recycle and disposal problems be developed.

Research on the effects of solid wastes pollution is a complex area, and some aspects of it take considerable time, often as much as decades. This is because the effects of pollutants on health often are subtle and may become apparent only when corrective measures are no longer possible. Indeed, the results of contamination often become known so belatedly that few recall any relationship between the cause and the effect. This slow response makes it difficult to determine and evaluate the impact of environmental degradation on receptor organisms, both individually and collectively.

Recommendation S IV: *Assessments of the environmental effects of pollutants must be made and documented as an integral part of any activity dealing with resource recovery and the ultimate disposal of wastes on land. Research to determine the levels at which effects occur and assessments to determine risks and cost/benefit levels are both required. (S 9)*

Pesticides in the Environment

Great concern continues to be expressed over the effects of pesticides in the environment. This concern has produced among other results increasingly rigid legal controls on the composition and use of the materials and on the amounts of residues permitted in environmental samples. Recent years also have seen careful and continuing examination of pest control practices and growing interest in a more balanced approach, including cultural and biological as well as chemical methods. If the concern over pesticides is to continue to yield positive results, it must be accompanied by full realization of the nature, uses, and value of pesticides as well as of the dangers they present.

Pesticide residues

A great deal of data on pesticides and their breakdown products in the environment have been gathered over the years. This has been true especially since the advent of highly selective and sensitive analytical techniques such as gas chromatography. Various agencies of the Federal Government have conducted nationwide monitoring programs for pesticides; at present, the Environmental Protection Agency, the Food and Drug Administration, and the Department of Agriculture are responsible for much of the pesticide-monitoring effort, although chemical manufacturers, food processors, and academic institutions contribute significantly.

Monitoring programs show that pesticide residues are widespread in soils and water, although the amounts detected usually are quite small. Minute amounts of pesticides have also been detected in air, and there is good evidence that residues of certain compounds are ubiquitous and travel great distances in the atmosphere. While pesticides most often are degraded to innocuous form in soil, air, and water, some may also move through these phases of the environment to living receptors, including plant and animal life, foods and feeds, and humans. Thus crops and foods are included in regular monitoring programs. The residues encountered most commonly in all phases of the environment have been those of the chlorinated hydrocarbon insecticides — the DDT group (including DDE and DDD), dieldrin, endrin, toxaphene, and chlordane. Recent data suggest that the amounts of these materials in the environment are declining. A decline would be expected to result from the shrinking usage of the materials overall because of technical factors and, especially, legal restrictions.

Recommendation P I: *Pesticide monitoring in all phases of the environment should be continued for specific substances recognized to be harmful, including the parent pesticides, their environmental transformation products, and their manufacturing by-products, and the identity of each substance should be confirmed by appropriate analytical methods. (P 1)*

Residues and health

Exposure of animals to pesticide residues, where it occurs, appears to be primarily through food; exposure through air and water appears to be minor, although aquatic animals can be exposed through water. Residues when ingested can have unexpected effects. A familiar example is the population decline in certain species of birds as a result of eggshell thinning caused by residues of chlorinated hydrocarbons. Human foods contain residues of pesticides, notably the chlorinated hydrocarbon and organophosphorus types, but at levels well below the acceptable maximums established by law and by international accord. Harmful effects from these residues have not been detected, but the relationship between pesticides and human health remains under continued study.

Humans have been poisoned fatally by many of the major pesticides; acute illness, whether caused by overexposure at work, accidental ingestion, or attempts at suicide, is all too well documented for these materials. The problem of acute intoxication has been approached by many means, including warning labels, use restrictions, and education of users and physicians. The success of these efforts can best be gauged by the fact that despite the vast increase in the availability and use of pesticides, the incidence of fatal pesticide poisoning in the U.S. — which held virtually constant at 1 per 1 million population per year over a 25-year period — has dropped continually during recent years although both total population and total accidental poisoning deaths have increased steadily.

Acute or subacute episodes are relatively clear-cut. It is much more difficult to evaluate the effects of long-term exposure of humans to comparatively low levels of pesticides. In the 1960's, humans in the U.S. and elsewhere were carrying a body burden of 3 to more than 20 parts per million of chlorinated hydrocarbon insecticides and their conversion products in fatty tissue. Measurements for 1970–74 show a general decline in the concentration of organochlorines in body fat, reflecting the decline in the use of the materials. There is no evidence at present that any group of pesticides other than the organochlorines is stored in the body or that the body contains metabolites resulting from the degradation of other pesticides to which the population is regularly exposed.

The chlorinated hydrocarbon insecticides, particularly DDT, have been in the human environment for more than 30 years; the older pesticides — lead arsenate, mercurials, copper compounds, and other inorganic materials — have been in the environment throughout this century and much of the 19th century. What effect have they had? Simply put, they have had no currently detectable effect on the population.

There is continuing fear that long-term exposure to pesticides causes cancer. Unfortunately there seems to be no sure test for carcinogenicity at present. Studies in animals often are conducted with levels of chemicals far exceeding actual environmental exposure, in special strains of laboratory animals, and under highly artificial conditions. They are conducted in this way for sound technical reasons, but large areas of doubt remain. An extensive review of test methods and the carcinogenicity of pesticides failed to produce agreement among experts on what tests to perform or how to interpret the results.

Certainly, research and testing must continue, and be augmented, to define the nature and causes of cancer and other ailments, and any part that pesticides might play. Carcinogenesis tests are required for registration of any new pesticide, and much emphasis also is being placed on study of existing ones. In fact, the cancellations by EPA of most of the uses of aldrin, dieldrin, chlordane, and heptachlor were based mainly on the fact that the compounds had been shown to be carcinogenic in laboratory animals. Carcinogenicity in test animals also was cited in the cancellation of most uses of DDT, but other environmental effects played a major role. In no instance, however, have pesticides tested in animals displayed carcinogenic effects even faintly comparable to those of natural agents to which people have been exposed — aflatoxin, cycasin, pyrrolizidine alkaloids, and many constituents of tobacco smoke, for example. Realistic prediction and evaluation of the carcinogenic potential of pesticides must include assessments of the contributions and interactive effects of simultaneous exposure not only to these agents but also to the host of other chemicals — natural and man-made — to which we are exposed daily. Evaluation of the mutagenic and teratogenic potential of pesticides and related compounds suffers from the same restrictions and uncertainties as the evaluation of carcinogenic effects.

Pesticides by design are biologically active and may be hazardous if misused. Nevertheless, the number of illnesses and deaths known to have been caused by accidental or deliberate misuse of the materials is far outweighed by the benefits these chemicals have brought in controlling disease-bearing pests and in increasing food and fiber production. In addition, there is no evidence to date that long-term, low-level exposure to pesticides at the concentrations found in our diet or in the environment in the U.S. has had a harmful effect on people, although extensive research is continuing.

Recommendation P II: *Efforts should be increased to interpret laboratory data on pesticide toxicity, including carcinogenicity, mutagenicity, and teratogenicity, in terms of the actual existence of hazard under conditions of practical use. (P 2)*

Pesticide regulation

A pesticide in this country is regulated from the first days of its initial discovery. Before a pesticide can be offered for sale it must be registered with the Federal Government. To register a new chemical with no previously established pesticidal use, the applicant must plan on three to five years to develop performance data representing the typical geographical distribution of the crop, residue data to establish a safe tolerance, and analytical methods for residues and for assay of the technical product and its formulations. He must compile a variety of other data on the compound: physical and chemical properties; manufacturing and purification processes; compositions of technical and formulated products; safety, including animal and plant metabolism and a substantial number of acute, subacute, chronic, and other toxicological trials; and environmental impact. If residues are found on crops or crop products that are fed to livestock, the law requires feeding studies on those animals. This product development and analytical work requires considerable expenditure for sophisticated equipment, computers, and instrumentation that further entail the skills of highly trained personnel. The entire registration process, up to the point where a new pesticide product is released for sale, requires (in the mid-1970's) as much as eight to 10 years and $8 million to $12 million in registration costs. In addition, when the product is on the market, its manufacture, packaging, handling, and use are regulated extensively under federal and state laws.

The complexity of the regulatory system in this country costs both applicant and government enormous amounts of time and money. To find an entirely new pesticide good enough to carry through registration and production usually requires the screening of thousands of potentially useful materials. The necessary research and development consume years that effectively reduce the period during which the product is protected by patents and thus the time available to the registrant to recover his costs and earn a profit. The costs involved are so high, moreover, that even the expense of registering a new use for an existing product may exceed the potential sales, especially on a minor crop. Improved regulation of pesticides certainly is providing a higher degree of safety to man and his environment. It is doing so, however, at a proportionally higher cost that must be borne in the long run by society in terms of less protection from pests, higher prices, or both.

Recommendation P III: *Realistic incentives should be provided by law to encourage the development of new, safer, and more effective pesticides and pest-control measures of other types. Such incentives might include extended patent protection, simplification of the registration process, and standardization of test procedures required for registration. (P 6)*

Control of environmental impact

In addition to legal restrictions, a variety of technical methods can help minimize the environmental impact of pesticides without sacrificing the benefits they provide. Promising results have come from grower contracts with professional crop-protection managers who undertake, for a fixed price, to guarantee effective pest control. The pest-control manager increases his profits by using the smallest possible amounts of chemicals in the most efficient way, and the grower receives his profit in any event. Crop management by continuous professional monitoring of developing pest populations and more accurate estimates of "economic threshold" — the amount of damage that can be tolerated — make it possible to reduce greatly the number of essential pesticide applications.

Marked reduction in the amount of pesticide used could be achieved in many cases by using improved application equipment. The development of better equipment and more efficient application methods also would be of major importance in reducing pesticide losses during spraying; such losses normally run 30 to 60%, but can amount to as much as 80%. Improved formulations of pesticides could reduce the amount of chemical required to control pests. Apparently, appropriate manipulation of pesticide formulations can greatly reduce the environmental mobility of the chemicals as well as reduce the total amount required for effective pest control. Presumably, formulations could be found that would increase the rate of degradation and so allow any desired degree of persistence to be designed into a pesticide product without having to find and register a totally new active compound.

Recommendation P IV: *Major improvements should be sought in methods of pesticide application, including increased application efficiency, control of volatilization and other kinds of movement, and storage stability. (P 7)*

Integrated control

Although the amounts of pesticides required could often be reduced, chemicals remain man's major weapon against pests for the foreseeable future. Nevertheless, the need for toxic chemicals can be reduced even further by alternative methods of pest control that now are practical as well as beneficial. Major alternative measures can be grouped under broad headings that include biological control, cultural control, genetic resistance in crops, sex attractants (pheromones), and integrated control.

Integrated control — pest management — combines maximum reliance on natural controls with appropriate use of other techniques: cultural control, pest pathogens, resistant crops, pheromones and other attractants, parasite or predator releases, and chemical pesticides if and when needed. Integrated control is not biological control nor the use of any single technique; it does not imply nonchemical farming, nor should it be identified with "organic gardening," which uses neither synthetic fertilizers nor pesticides. Integrated control, rather, is a system that embraces the best elements of all methods. For the long run, the integrated approach promises improved pest management with minimum harm to the environment and lower cost to the farmer.

Recommendation P V: *Systematic laboratory and field research on integrated pest management should receive increasing emphasis aimed at the earliest possible practical implementation. Chemists should be strongly encouraged to increase their direct collaboration with scientists of other disciplines in solving the chemical problems of integrated control, including those related to pheromones, natural repellents, and other control substances in addition to conventional pesticides. (P 8)*

Radiation in the Environment

Humans and other organisms have lived from the beginning in an environment containing low levels of radiation from naturally occurring radioactive sources. The fact went undetected until less than a century ago and for some years interested no one but a few specialists. The discovery of nuclear fission and its use in military weapons and in the reactors of a burgeoning electrical power industry have changed this picture completely. The problem of realizing the great benefits to be gained from radioactivity while at the same time controlling its hazard has become an international concern. A significant part of the problem is x-rays emitted by man-made devices.

The standards of safety and measurement relating to radioactivity are the business of a number of groups whose recommendations are in a continuing state of evolution. The most influential groups internationally are the International Commission on Radiation Units and Measurements (ICRU) and the International Commission on Radiological Protection (ICRP). In the United States, the National Council on Radiation Protection and Measurements (NCRP) was established by the National Bureau of Standards in 1929. With the growing awareness of the possible adverse effects of radiation on the human "genetic pool," the permissible exposure doses recommended by NCRP and ICRU have been set at conservative levels, particularly for the population at large. A number of federal agencies, but primarily the Nuclear Regulatory Commission (NRC), become involved in establishing and enforcing the legal standards necessary to insure that the exposure levels recommended by ICRU and NCRP are not exceeded. The tendency is to set the standards even more conservatively than do the international commissions.

Sources of exposure

Calculation of the average dose from environmental radioactivity for members of a large population is a difficult problem, complicated by the fact that many of the data required are not available. Estimates of past and present exposures are thought to range from about half to about twice the true values; estimates of future exposures probably are within 1/10th to 10 times the true values. It is not surprising, therefore, that different groups of estimators derive different values, although all agree in general on the relative importance of the exposures generated by individual sources of radiation.

Figures for average exposures, taken over the entire population of the United States, are dominated by the natural and diagnostic radiology exposures. Fallout is a smaller but significant third source. While the dosage associated with nuclear power is only a small fraction of the total, the estimates for that source are based on accident-free operations. It is true that, over the past 30 years, atomic energy facilities have had a remarkably good safety record — far better than in coal mining, for example. It is also true that strong arguments can be made that reactors are inherently incapable of behaving like an atomic bomb. Nevertheless, the large inventory of lethal materials in the core of an operating reactor is a potential hazard that cannot be ignored.

Natural exposure to radiation is estimated still to be the dominant factor in the year 2000. This assumes tacitly that man will be able rigorously to contain all radioactivity arising from his use of the fission process. The amounts of radioactive materials to be produced are sobering. Nevertheless, there is no apparent technological reason why the activity cannot be contained, although the problem requires maximum attention. The primary sources of man-made radionuclides are nuclear reactors, nuclear detonations, and, to a much lesser extent, charged ion accelerators. Secondary sources of radioactivity in the environment are more numerous. They include reactor fuel fabrication and processing plants, research laboratories, transportation and waste storage facilities, hospitals, industrial users, and the release of natural radioactivity through mining and ore refining. The radioactive products formed in fission power reactors will be the major problem for some decades.

The nuclear fuel cycle

The nuclear fuel cycle comprises the various steps involved in using uranium as the basic fuel for producing electrical power. Each step involves the possibility of producing radioactive waste in gaseous, liquid, or solid form. However, the waste problems are relatively less serious until after the enrichment stage, in which the concentration of the fissionable uranium-235 in the uranium is raised above its natural level of 0.72% in gaseous diffusion plants. The enriched material goes to any of a number of private or governmental facilities for fabrication into reactor fuel elements. Under present circumstances, the steps in the nuclear fuel cycle up to this point involve materials of relatively low radiation intensity, such as natural or slightly enriched uranium; they do not involve potential environmental impacts of widespread significance. This picture would change somewhat for the fabrication plants if plutonium assumed a larger role in reactor fuels. Elaborate arrays of glove boxes and complicated ventilation systems would be required, particularly since the material is a powder during much of the fabrication process.

Nuclear power plants in the U.S. today are predominantly light-water reactors (LWR's) of two types: the pressurized water reactor (PWR), which accounts for about 60% of the total, and the boiling water reactor (BWR). Radioactive effluents from a reactor can be either gaseous or liquid. Small quantities of radioactive and nonradioactive gases as well as soluble and insoluble solids are formed in the primary coolant by neutron activation of corrosion products, water, and traces of dissolved air; fission products also may enter the primary system from leaks in the fuel cladding. Leakage from fuels with failed cladding can occur during refueling or during storage of spent fuels under water in canals. The mixtures of nuclides in liquid or gaseous effluents will vary with the reactor type, and the basic cleanup techniques will vary correspondingly.

To obtain a license to operate a reactor, an applicant must demonstrate his ability to establish an adequate monitoring and surveillance system for radioactivity releases. Operators are required also to report semiannually the quantities of discharged nuclides and the environmental levels of radiation and radioactivity that result from plant operation. EPA has studied intensively the effluents and their effect on the surrounding environment at the Yankee Nuclear Power Station, which is an older PWR, and at the Dresden Nuclear Power Station, an older BWR. Similar studies are under way or planned at newer reactors of both types. The Yankee and Dresden studies both concluded that the impact of the plants on the surrounding environment was minimal; radioactivity above the natural background was detected only with great difficulty. While there are some dissenters, such surveys have led most observers to decide that LWR's in normal operation do not greatly threaten the environment or public safety. The main concern is the possibility of accidents.

Fuel reprocessing

The buildup of fission products and loss of reactivity permits the "burning" of only about 50% of the uranium-235 in a typical LWR fuel before it must be replaced. The great economic value of the remaining uranium-235, and of the fissionable plutonium-239 produced in the reactor, is the chief argument for reprocessing the fuel to recover the two isotopes. The Carter Administration's position (in 1978) is that the value of plutonium recycle does not warrant the risks of nuclear proliferation arising from the weapons potential of separated plutonium. In fact, very little spent fuel from commercial power reactors in the U.S. has ever been reprocessed; virtually all spent fuel assemblies from such reactors remain in interim storage.

Government-owned fuel reprocessing plants have been built at Hanford, Wash.; Oak Ridge, Tenn.; Savannah River, S.C.; and the National Reactor Testing Station in Idaho. In these processes, the transplutonium elements remain with the fission products, as does about 0.5% of the plutonium. The separated uranium is returned to the enrichment plants, and the plutonium at present is converted to a solid and stored. The fission-product fraction is the "high-level waste" that is of major concern in terms of disposal or long-term storage. Since the fuel is dissolved completely, all of the radioactive fission-product gases are released. Current practice, however, is to store the used fuel elements in fuel-pool storage facilities at the reactor for "cooling," so that all of the short-lived species have decayed to stable daughters by the time fuel is shipped to the reprocessing plant. The most important gaseous radioactive nuclides remaining are tritium, iodine-129, krypton-85, and perhaps traces of iodine-131. The amounts of them emitted are quite small, but research and development is under way on means of trapping all four.

Recommendation R I: *Considerable progress has been made in devising methods for removing radioactive species from gaseous and liquid effluents. Efforts to improve such techniques should continue with particular attention paid to the tritium, iodine, and xenon-krypton release problems and with even more emphasis on removal of trace transuranics from emissions and effluents. (R 5)*

Recommendation R II: *Strong emphasis should be placed on developing more efficient methods for complete removal of actinides and other long-lived species in reprocessing light-water and breeder reactor fuels in the event that circumstances call for the country to reconsider its present position on the reprocessing question. Innovative processes having the effect of simplifying the final waste disposal problem should receive particular attention. (R 6)*

Radioactive waste management

The bulk of the waste material that is either inherently radioactive or is contaminated with radioactivity is generated either in the nuclear power cycle or in government-owned production and research facilities. Smaller amounts can arise in departments of nuclear medicine in hospitals, in university research laboratories, in industrial plants using radioisotopes for various applications, etc. All users of significant amounts of radioactivity must be licensed by NRC or by a "license-agreement state" — a state that has assumed from NRC the responsibility of regulating the distribution, handling, and use of radioactive materials. Establishment of a waste-disposal system is part of the licensing procedure; the ground rules for disposal are established by NRC. Wastes are classified as low-, intermediate-, or high-level, depending on their concentration of radioactivity and their potential hazard.

Low-level wastes are those that can be discharged to the environment under controlled conditions after reasonable dilution or simple processing. By far the bulk of the low-level waste is disposed of by near-surface burial on land. Materials at the intermediate level of activity require more than simple processing. They may be treated by methods like evaporation, ion exchange resin column, and flocculation to split them into a high-level concentrate and a low-level liquid. As the quantities generated increase, however, it is probable that intermediate-level material will simply be combined and solidified with high-level wastes. A type of solid waste that might be considered in the intermediate class is contaminated obsolete equipment from fuel fabrication plants, reactors, etc. Normally such items can be decontaminated and buried at licensed sites. However, highly alpha-contaminated units of this type that cannot be cleaned sufficiently are a disposal problem that presumably will become more urgent as first-generation facilities in the nuclear fuel cycle, including reactors, eventually are decommissioned and if plutonium is used to a greater extent in fuel fabrication.

Recommendation R III: *The problem of decontamination or disposal of alpha-contaminated equipment should receive a high level of attention. (R 7)*

Process solutions from fuel reprocessing plants constitute essentially all of the high-level wastes. They present one of two separate problems, depending on whether or not they contain long-lived alpha emitters. If these are not present, methods must be devised to isolate the wastes for 600 to 700 years to allow them to decay to an innocuous level. If substantial amounts of alpha emitters are present, the time of isolation must be of the order of several hundred thousand years. Guaranteeing this isolation is an exceptionally difficult problem, but not necessarily an insoluble one.

Past and current practice has been to store the liquid wastes in tanks, which everyone understands to be purely an interim measure. In the U.S. the tanks are encased in concrete and buried underground. Tank farms, comprising a total of about 200 tanks, are located at Hanford, Savannah River, and near Idaho Falls. The tank farm system has not been free of problems. During 1958–73, the Atomic Energy Commission reported 17 leaks involving a total of 447,000 gallons of waste. The liquid soaks into the ground, but the radioactive species appear to be held sufficiently by the ion exchange action of the soil so that the activity remains within 10 or 15 feet of the tank and well above the water table. The fact that the leaks occur at all, however, is a matter of considerable concern. As a result, each new generation of tanks has incorporated additional safeguards and more sophisticated instrumentation.

Converting the liquid wastes to solids offers many advantages for long-term storage or ultimate disposal. A very considerable amount of research has been undertaken to find methods leading to maximum immobilization of the active species. Four processes have been developed in the U.S. and demonstrated on an engineering scale. These are the pot, spray, phosphate glass, and fluidized bed techniques.

Most waste management plans call for interim storage of solidified materials for as long as 30 years before final disposal. This is primarily because the heat released by the fission products can raise initial temperatures at the centerline of a 6-inch-diameter pot filled with solidified waste to almost 1000°C. Concepts for interim storage accordingly have envisaged water-filled canals, air-cooled annular bins, or air-cooled concrete vaults. An essential safeguard is a thoroughly reliable backup method for removing heat should the primary cooling system fail. The rate of heating by the fission products drops sharply during the first few years as the shorter-lived species decay to stable forms. These latter are different elements than those present originally, a fact that produces changes in the properties of the calcined or melted materials. For most of the solidification processes, clear definition of the effect of these changes remains to be determined.

Recommendation R IV: *Research on waste solidification methods should be maintained at a high level because of the obvious advantages to be gained in terms of the final disposal problem. Particular attention should be paid to the effects of chemical changes with time upon the water solubility and physical properties of the solid. (R 8)*

The time will arrive eventually when the problem of disposing of the solidified materials permanently must be faced. Various alternatives have been considered. They include disposal in abandoned salt mines, in deep vaults mined in bedrock, in Antarctic rocks, in deep holes drilled in long-lasting ice sheets like those of Greenland or Antarctica, and in large cavities blasted into deep, impervious rock strata by nuclear or conventional explosives. Carefully engineered, near-surface underground vaults have been proposed for storing solidified wastes in retrievable form until a completely satisfactory method of final disposal can be developed. This plan would entail the costs of very long maintenance and surveillance to be certain the materials were contained safely.

The ultimate disposal of high-level wastes would be simplified considerably if plutonium and other actinide elements could be removed from them essentially completely and either handled separately or destroyed by long irradiation in a reactor. The latter would be a very long process at the neutron fluxes of most present day reactors, although the seriousness of the actinide disposal problem conceivably could justify the design and construction of special, high-flux reactors dedicated only to this purpose. Ultimate disposal of high-level wastes might be eased also if the cesium and strontium could be separated from the other fission products, since these are the two chief heat-producing fractions. The cesium and strontium presumably would be stored separately in mausolea, although the problem of their ultimate disposal eventually would have to be faced.

Recommendation R V: *A major effort should be made to simplify the radioactive waste disposal problem by separating the actinides, and the possibility of destroying the separated actinides in nuclear reactors should be thoroughly investigated. (R 9)*

Recommendation R VI: *Research should continue to find economical means of isolating cesium and strontium from radioactive wastes for separate management, although these elements are not as critical a problem as are the actinides. (R 10)*

Detection and monitoring

A number of radioactivity monitoring networks have been established by federal, state, and foreign agencies. EPA and many more localized agencies operate extensive water quality sampling and analysis programs. The Radiation Alert Network gathers atmospheric samples at 68 stations throughout the 50 states as well as in Puerto Rico, Panama, and Guam. An elaborate air and water sampling network operates in connection with the Nevada nuclear testing site. A Pasteurized Milk Network comprises 61 sampling stations in the U.S., one in Puerto Rico, and one in the Canal Zone. The Canadians have an air and precipitation monitoring network, and a Pan American Air Sampling Program exists at 12 locations throughout South America. The Nuclear Regulatory Commission operates a network for determining monthly deposition rates for strontium-90 at 33 sites in this country and 90 sites abroad. Sampling and measurement techniques are not completely standardized among all the surveillance groups, so intercomparisons are sometimes difficult. A tremendous mass of data is nevertheless available.

EPA has produced a guide for establishing surveillance networks around various nuclear facilities in an effort to insure that the data obtained are uniform and can be used to calculate the exposure dose to any of the populations at risk. A discussion of monitoring techniques and philosophy similar to the EPA guide has also appeared. In spite of such guides, however, it is not uncommon to find considerable disagreement in data based presumably on samplings taken under similar circumstances and in similar locations by different groups. This problem has been intensified as private companies have entered the field of providing monitoring services for a fee.

Recommendation R VII: *Federal or state certification of all laboratories offering monitoring services should be considered as a means of standardizing the methods used and improving quality control for the data produced. (R 12)*

Recommendation R VIII: *Existing radioactivity monitoring programs around possible emitters of radioactivity to the environment should be maintained and in some cases expanded. (R 13)*

Despite the problems with standardization, the detection and monitoring of radionuclides in the environment have improved enormously since the period of intense weapons-testing fallout of the 1950's and early 1960's. At that time, careful and tedious chemical separation was required, followed by counting of the alpha or beta-gamma rays emitted by the separated element. Now that we have much better alpha and gamma spectrometers, simpler separations and in some cases direct measurements are adequate. This improvement is obtained only at considerable cost in equipment and counting time. A number of excellent laboratory instrumental systems are now available for measuring environmental levels of radionuclides. Because of its extremely soft (low-energy) beta radiation, tritium is a particularly difficult radioisotope to measure accurately. Flow-through instruments have been developed for tritium measurements in the field, but they are subject to a number of interferences that make the data obtained of questionable value in low-level activity situations. Field measurements in the environment are limited largely to measurements of gamma emitters; a gamma level of a millionth of a roentgen per hour can be measured to better than ±25%. Some additional information is available from *in situ* spectrometry.

Recommendation R IX: *Efforts should continue to develop highly sensitive, reasonably accurate instruments for measuring low-level activity in the field. (R 14)*

Additional Recommendations

The following Recommendations appear with supporting discussion in the indicated chapters of this report.

Chemical Analysis and Monitoring

Recommendation AM2: *Public information programs in government and industry should stress the critical function and limitations of chemical analysis in environmental programs as well as the significance of "limits of detection" and similar analytical concepts.*

Recommendation AM3: *Support must be maintained at an adequate level for the programs of standardization, collaborative testing, and quality assurance necessary to demonstrate the equivalency and reliability of analytical methods and instruments. This support should extend to the development of calibrating materials for analytical instruments and of appropriate standard reference materials.*

Recommendation AM5: *Research on remote methods of air analysis and on methods for analyzing actual samples (grab samples) should proceed in parallel in recognition of the fact that each approach has intrinsic strengths and weaknesses and that remote techniques must be validated by field and laboratory analyses of actual samples.*

Recommendation AM6: *Methods for removing and concentrating organics in water (i.e., collection on sorbents or solvent extraction) for subsequent separation and analysis should be developed further, especially for quantitative measurements.*

Recommendation AM8: *The adequacy of sampling procedures, including positional and temporal factors, must be evaluated critically in all analytical and monitoring work related to the environment. Representative and valid samples, and the maintenance of their integrity, are essential to the development of analytical data that accurately portray the condition and dynamics of all phases of the environment.*

Recommendation AM11: *More research should be conducted on the identity and concentration of naturally-occurring compounds in air and water from pristine areas to improve assessments of the impact of contamination by man.*

The Air Environment

Recommendation A11: *Increased attention must be given to the energy cost of agriculture, and means must be sought to decrease it.*

Recommendation A23: *Further research should be pursued on the responses of air contaminants to photoirradiation and the importance of these responses in the development of adverse effects on health.*

Recommendation A25: *More precise and specific analytical methods should be developed for determining specific biological responses to individual air contaminants, such as the conversion of hemoglobin to carboxyhemoglobin by carbon monoxide.*

Recommendation A32: *National regulations should be established for ambient visibility, which is an obvious effect of air pollution noticed by most people, but priority should remain on health-oriented regulations.*

Solid Wastes

Recommendation S1: *The appropriate federal, state, and local government agencies should press their efforts to define the nature and magnitude of the solid waste problem both now and in the future. Augmented programs are necessary in education, research, development and demonstration, and local and regional planning for solid waste management, utilization, and disposal.*

Recommendation S2: *Research and development on processes for recovering various minerals from mining and processing wastes should be maintained at an adequate level against the day when changing economics warrant the recovery of such minerals. Work on other means of utilizing or disposing of these wastes should be maintained at a steady level.*

Recommendation S4: *Continuing attention should be paid to collection and transportation of municipal refuse, both in the development of improved technology and in mechanisms for promoting the application of such technology by local agencies who are responsible for handling and disposing of municipal refuse.*

Recommendation S6: *Research and development on utilization and recycle of components of solid wastes, including their energy content, should be main-*

tained at a level that will insure that radically new approaches are not overlooked or inadequately investigated.

Recommendation S7: *Efforts by private industry to improve the economics of the automobile scrap processing industry should be stimulated. The improvement of present scrapping methods or the development of radically new scrapping methods that would permit the use of less costly equipment should be pursued at all levels.*

Pesticides in the Environment

Recommendation P3: *Research and education emphasizing human safety in the manufacture, transportation, and application of pesticides should receive high priority, especially in relation to nonagricultural uses.*

Recommendation P4: *Regulatory practices should continue to be based on human and environmental safety, and on good agronomic practice consistent with safety, rather than on the prevailing sensitivity of analytical chemical methods.*

Recommendation P5: *The uses, application rates, and environmental fates of pesticides in nonagricultural applications should be identified and given increased research and regulatory attention.*

Recommendation P9: *More attention should be devoted to the role of containers in environmental pollution by pesticides, including container decontamination, reusable containers, possible limitations on package size, and the safety aspects of containers in home use.*

Recommendation P10: *Methods should be developed for the practical chemical or biological disposal of every registered pesticide; this might reasonably be required of registrants before their pesticides are registered.*

Radiation in the Environment

Recommendation R1: *In cases of significant accidental exposure, excretion of the bone-seeking strontium radioisotopes can be accelerated by prompt medical treatment. While some progress has been made, similar treatments are much less available for the transuranics and rare earths. Research in search of suitable techniques should continue for all of the "nuclides of special concern."*

Recommendation R2: *Radioisotopes unquestionably have furnished a powerful new diagnostic tool to medical science. Efforts should continue to develop techniques for minimizing patient exposure by optimizing the chemical forms used in introducing the material into the body and by working out methods utilizing short-lived isotopes to minimize overall retention times. The problem of adequate regulatory control of the use and disposal of radioisotopes used in medicine should receive constant attention in view of the rapid expansion in the use of such materials.*

Recommendation R3: *Relatively little is known of the possible synergistic or antagonistic effects of radiation combined with chemical pollutants on humans and ecological systems. This area of research should be vigorously pursued.*

Recommendation R4: *One of the important sources of occupational radiation doses within nuclear power stations results from the chronic buildup of radioactive materials, especially cobalt-58 and -60, which accumulate in the primary system. Reduction of this source of radioactivity by off-line or preferably on-line chemical cleaning of the primary system should be given greater attention.*

Recommendation R11: *Further research on the geochemical behavior of plutonium and the other actinides in soils and rocks, as well as on their chemistry in the other portions of the environment, is required with high priority because of the toxicity of this man-made group of very heavy elements.*

Chapter **2**

Chemical Analysis and Monitoring

(Continued)

(Continued from page 41)

2 / Chemical Analysis and Monitoring

INTRODUCTION

Quantitative measurement, typified by chemical analysis and monitoring, is indispensable to progress in environmental science and technology. Subjective criteria — odors, tastes, smog, turbidity — can provide useful benchmarks of environmental quality. But quantitative physical and chemical data are needed to study the sources of contaminants, their behavior en route from source to sink or receptor, and their effects; to set standards of quality; to evaluate abatement technologies; to build predictive models; and to prove violations for regulatory action. Our ability to measure chemical parameters precisely rests on a solid scientific base (Table 1) (1) whose growth has accelerated in recent years. Concern for the environment and the consequent legislation, however, have accelerated even more rapidly. Thus the need to acquire sound environmental data clearly has outstripped the means of doing so.

The environment embodies severe analytical problems. The necessary measurements range from determinations of specific contaminants in waste discharges, to analyses of receiving waters or air masses, to base-line probes of pristine environmental samples. The substances in question often are at very low concentrations — parts per million or less by weight or volume.* They may be accompanied by other substances that interfere with the analysis. The possible interactions of air, water, soil, and biological organisms may call for measurements in each, often by quite different techniques, greatly complicating the development of data that can be correlated and interpreted. The discussion in this chapter is restricted largely to determinations of substances in air and water. However, the principles and techniques involved apply to soil or any other medium.

Analysts have responded remarkably well to the need to refine existing techniques and to develop new and sensitive procedures for identifying and quantifying, sometimes approximately, unanticipated or previously unrecognized constituents of the environment. Overall, however, progress in environmental measurement has been slow. One problem is the persistent shortage of specialists in the field. Another is that analysis too often is considered a secondary activity; unwarranted assumptions are made of the simplicity of the measurement and the reliability of the method. Such difficulties make the measurement of all but the simplest parameters, and the interpretation of the results, by other than qualified analytical chemists a dubious procedure. The result in part has been masses of data that frequently are useless for the intended purposes. An intensive, coordinated effort is essential if environmental analysis is to be upgraded to a level consistent with the need.

Analytical Concepts and Procedures

Chemists over the years have worked out certain criteria for the acceptability of analytical methods when applied to specific substances in specific condi-

*Modern practice is to report analytical data for air in terms of milligrams (mg) or micrograms (μg) per cubic meter (m^3). Similarly, data for water are given in mg, μg, or nanograms (ng) per liter (l). However, data sometimes are still reported in parts per million (ppm), per billion (ppb), or per trillion (ppt) by volume for air and by weight for water. Data in this chapter generally will be given in both forms to show their equivalency.

tions. Numerous techniques have met these criteria, typically through months and years of use by many analysts in many laboratories. Today, we face explosive growth in the number of substances newly realized or suspected to affect environmental quality. Analytical methods for such substances and their degradation products must be devised or adapted from existing procedures and subjected to the scientific scrutiny required to assess their reliability.

Table 1

Classification of common analytical methods

General classification	Subclassification	Quantity measured
Gravimetric	Direct method	Weight of compound containing species
	Indirect method	Loss of weight by volatilization of species, leaching, etc.
Volumetric	Titration methods	Volume of solution that is chemically equivalent to the wanted species
	Gas analysis	Volume of a gaseous species produced or consumed
Optical	Emission spectroscopy	Radiation emitted by species
	Absorption spectroscopy including colorimetry	Radiation absorbed by species
	Polarimetry	Rotation of polarized light by species
	Refractometry	Refractive index of solution of species
	Turbidimetry, nephelometry	Scattering of radiation by species
Electroanalytical	Potentiometry	Potential of an electrode in equilibrium with the species
	Conductimetry	Conductance of a solution of the species
		Quantity of current equivalent to the species
	Polarography	Current associated with reaction of a species at a polarizable electrode
	High-frequency methods	Capacitance of solution of species
Chromatography	Gas, liquid, thin layer, ion exchange	Separation of species in a multistage, countercurrent manner based on differences in distribution ratios of components of mixtures between a mobile bulk phase and an immobile dispersed phase.
Miscellaneous	Mass spectroscopy	Mass-to-charge ratio of decomposition products of species
	Radiochemical	Radioactive decay of species
	Thermal-conductivity methods	Thermal conductance of species
	Enthalpy titrations	Heat of reaction of species

Source: American Chemical Society.

Skoog, D.A., West, D.M., "Analytical Chemistry: An Introduction," Holt, Rinehart and Winston, New York, N.Y., 1965.

The ideal analytical method responds only to a single substance, such as sulfur dioxide in air or nitrate in water; it reveals the exact amount of that substance in the medium at hand over the pertinent range of concentrations. Analytical methods nearly always vary from the ideal, however, in characteristics that include specificity and selectivity, sensitivity, and detection limit. Specificity and selectivity relate to the number of substances to which a method will respond. Few methods are specific to a single substance, but many are selective — that is they will respond preferentially to any of several related substances in the presence of certain foreign substances (interferences). Sensitivity describes the smallest amount of a substance to which a method will respond; detection limit is the smallest amount to which it will respond detectably. In principle the two are equivalent, but with an instrumental method, for example, electronic "noise" may obscure small responses, so that the detection limit will be somewhat larger than the sensitivity. It should be noted that if a method shows no response to a given sample the analyst cannot say that the substance of interest is not present, but only that it is not present in an amount that the method will detect.

Precision and accuracy

The inherent shortcomings of an analytical method can be overcome in varying degree by properly selecting the conditions of use and by manipulations such as concentrating the sample and chemically removing interferences. In addition, moreover, the results obtained with a method tend to be sensitive to the skills and environment of the analyst. This being so, the acceptability of a method for general use is best evaluated on the basis of the "precision" and "accuracy" (2) of the results obtained with it by a number of analysts in different laboratories. When the method is applied in this way to a series of samples containing the same concentrations of the substance to be measured,

- **Precision** describes the variation or scatter among the results, irrespective of the true composition of the sample.
- **Accuracy** describes the difference between the average result and the correct result.

Accuracy in particular is not easy to achieve, especially among different laboratories and techniques. Although the analyst always tries for high accuracy, he relies mainly as a rule on the precision of his data, at least during the early stages of the particular problem. When the correct result of an analysis is unknown, agreement of the results obtained by two methods, whose chemical principles are as unrelated as possible, is used as a measure of probable accuracy. Such an approach is always desirable, but is certainly needed when a Standard Reference Material is not available for crosschecking, as with a remote sensing method, for example. The approach is also desirable when the analysis may be subject to unknown effects, such as those caused by interfering substances. Accuracy is especially difficult to achieve with authentic environmental samples, whose complexity can lead to insidious errors that may defy anticipation.

Standardization of methods

Standardization of methods is a general practice in analytical chemistry. The "standard method" for a given substance, when used in the intended circumstances by different analysts in different laboratories, is expected consistently to produce results that are accurate and precise within the same known and acceptable limits. Standard methods are essential for substances that must be determined routinely, as in commerce, government regulation, or medicine — the phosphorus content of steels, the concentrations of pesticide residues on foods, the level of albumin in urine. In consequence, chemists in such areas have developed many standard methods as the needs arose. The same has been true of environmental analysis, although progress has not fully kept pace with the need. Without an appropriate arsenal of standard methods, routine analytical data are less reliable, data from different sources often cannot be compared with confidence, and environmental quality control programs cannot function equitably and rationally.

The standardization process requires as a rule not only that a method be selected, but also that it be appropriately modified and improved. Certain limitations may have to be faced as well. The method should not require excessive time. It should not require equipment that will be too costly for those who must use it. In addition to the analytical technique involved, a standard method generally will cover steps such as sample collection, storage (if necessary), and preparation. The standardization procedure is extremely critical and must not be compromised. If a method must be modified for "field use," the expected sacrifice in performance must be recognized and noted in the evaluation of results. In environmental measurement in particular, the pressure to accelerate methods development and the growing reliance on analytical instrumentation have sharply increased the emphasis on two procedures — collaborative testing and calibration.

Collaborative tests

Accepted methods of chemical analysis have evolved traditionally through the time-consuming process of research, documentation in the scientific literature, and evaluation by peers. Before it is accepted as standard, a method should additionally be subjected to collaborative testing, a procedure that is being employed increasingly to fill environmental needs. Two organizations that have been especially active in collaborative testing are the American Society for Testing and Materials and the Association of Official Analytical Chemists (3–5). In one form of collaborative test, "identical" samples are sent to selected collaborators who determine the substance of interest by any familiar method. The results of such a test, properly designed and executed, make it possible to select the most consistent of the methods used. In a second approach, each collaborator applies the same analytical method or procedure to his sample. Here the results indicate the reliability or performance characteristics of that method; the combined results may be analyzed for accuracy (mean error) and precision (standard deviation), and these values may be used to compute the so-called "total error," a measure of the acceptability of the method (6). Although a standard method, ideally, should have been subjected to collaborative testing, some have not been so tested. This is particularly true of methods of air analysis, where it has become possible only lately to generate realistic test atmospheres for the collaborators to analyze.

Calibration

A recent trend in environmental analysis is toward heavy reliance on analytical instruments, as opposed to the traditional "wet chemical" techniques. In the typical wet chemical procedure, chemicals are added to the test solution and measurements are made after suitable reactions have occurred. Such methods as a rule determine a substance directly or can be calibrated against a readily available primary substance such as pure iodine. Analytical instruments, however, determine a substance in the context of the environmental system in which it is measured; they must be calibrated, therefore, against a standard that closely resembles that system. An instrument for measuring ambient carbon monoxide, for example, would be checked against a calibrant containing a precisely known concentration — say, 10 mg/m^3 (9 ppm) of carbon monoxide in air; a conductivity meter would be calibrated in part against a 0.001 normal solution of potassium chloride in water. Analytical instruments in general include a sensing system, which responds to the substance to be measured, and a readout system, which displays the measured value. The readout system usually can be adjusted to display, for a given measured value, any of a number of values in a range characteristic of the instrument. Calibration consists of adjusting the readout to display precisely the value measured. Most instruments must be calibrated at regular intervals, since the performance characteristics of their electronic and mechanical components tend to drift with time.

Calibrating materials often are difficult to prepare in such a way that their precise composition is certain. Analysts can ease this problem by crosschecking their calibrants systematically both within and among laboratories. Calibrants also can be checked in some cases against the Standard Reference Materials prepared and distributed by the National Bureau of Standards (7, 8). SRM's serve basically as primary standards of known composition and homogeneity. They allow a body of analytical data to be validated (or invalidated) by tracing the measurements to a common basis. SRM's are not simply devised, and relatively few are available thus far that apply directly to environmental analysis. Most measurements related to current standards of ambient air quality can be traced indirectly to SRM's, however. Measurements of many inorganic substances in water can be traced to SRM's as well. Analyses for organic substances in water, on the other hand, cannot in general be validated in this way. A major problem here is the very large number of organics that may have to be monitored. SRM's cannot practically be created for each; those to be developed, therefore, must be selected with care to achieve any significant degree of traceability. An approach to quantitative analysis that has been used in the absence of SRM's is to prepare standard aqueous solutions by dissolving known amounts [at the ppb (μg/l) level] of various organic compounds in carefully purified water (9).

Quality assurance

We noted earlier that the results obtained with a given analytical method tend to be sensitive to the skills and environment of the analyst. More broadly, it must be recognized that no amount of methods development, collaborative testing, and related practices can assure the soundness of environmental measurements unless analysts perform with proficiency and judgment. Poor data can be obtained with a standard method as readily as with any other. Conversely, although standard methods have distinct advantages in routine use, data obtained with different methods can be compared and evaluated satisfactorily, providing proper emphasis is placed on the precision and accuracy of the data and on quality control protocols. When evaluating data, analysts also must be aware of the limits of detection and specificities of their methods and, where appropriate, of the need to confirm the presence of specific substances in environmental samples by the use of combinations of methods. The cause of improving our environment cannot be advanced rationally unless the results of environmental measurements are correct and legally defensible.

Recommendation AM1: *Analytical scientists must take the initiative in developing and promulgating standards for individual proficiency and for the determination of precision and accuracy, limits of detection, proof of presence of specific substances in environmental samples, and related aspects of quality assurance.*

Recommendation AM2: *Public information programs in government and industry should stress the critical function and limitations of chemical analysis in environmental programs as well as the significance of "limits of detection" and similar analytical concepts.*

Recommendation AM3: *Support must be maintained at an adequate level for the programs of standardization, collaborative testing, and quality assurance necessary to demonstrate the equivalency and reliability of analytical methods and instruments. This support should extend to the development of calibrating materials for analytical instruments and of appropriate standard reference materials.*

Recommendation AM4: *Training in analytical chemistry, both academic and on the job, must be funded at the much higher levels dictated by the fact that progress in environmental control has long suffered severely from an acute shortage of well-trained, innovative, and experienced analytical chemists.*

MEASUREMENTS IN AIR

The need to monitor contaminants in ambient air — a need that became apparent only in the mid-1950's — brought with it a new spectrum of analytical tasks. To handle them, techniques in use in the chemical process industries were modified and pressed into service. These first-generation methods, with some improvements, still are used widely. The past decade, however, has seen steady progress (10) toward the better analytical technology that is essential for many purposes: to monitor ambient air for compliance with quality standards; to detect hazardous air pollutants; to test stationary and mobile sources for compliance with pollutant-emission standards; to do research on the transport, behavior, effects, and ultimate fate of air contaminants; to develop predictive mathematical models of polluted air masses; and to fill other, related needs.

The first-generation methods, with certain exceptions, are based on colorimetry. The gas to be determined is collected in a water solution, where it reacts with a color-forming reagent; the light absorbance of the colored compound, which is measured spectrophotometrically, is proportional to its concentration. The exceptions to the colorimetric approach include, for example, carbon monoxide, which is measured by nondispersive infrared spectroscopy. These methods are useful, in the hands of capable analysts, but often they suffer from lack of both specificity and sensitivity. A further disadvantage is that they do not permit real-time monitoring — that is, they do not provide data quickly enough to permit corrective measures to be applied, if indicated, to the source(s) of the offending emissions. Such problems have spurred a considerable effort, both governmental and private, to devise better methods and equipment. One result has been a rash of new instruments for monitoring the five major gaseous air pollutants: sulfur dioxide, oxides of nitrogen, oxidants, carbon monoxide, and hydrocarbons. By 1973, about 100 manufacturers were marketing more than 300 such instruments based on more than 20 different operating principles (11). Numbers like these can be deceptive, however. A major difficulty is that relatively little has been done systematically to establish the equivalency of one method or instrument with another. Thus data on a contaminant acquired with one instrument often cannot be compared meaningfully with data on the same contaminant acquired with a different instrument.

Analysis and Air Quality Standards

The Clean Air Amendments of 1970 authorized the Environmental Protection Agency to promulgate national standards for ambient air quality, for new stationary sources of air pollutants, for hazardous air pollutants, for mobile sources of air pollutants, and for fuels and fuel additives. The amendments also called for the states to develop formal plans to meet the federal standards for ambient air quality. These implementation plans were to be based on EPA guidelines (12) that were to include air quality monitoring techniques.

To develop the measurement methods and instrumentation needed to support the 1970 amendments, EPA undertook a research and development program planned under contract by a large industrial laboratory (13). The plan included a review of the state of the art of measurement of each of some 30 air pollutants; it considered each pollutant in terms of the measurement methodology that would be necessary to develop and implement the type of air pollution standard that probably would be needed. The plan also considered sampling procedures and supporting measurements such as those needed to convert the concentration of a contaminant in a source emission to a mass emission rate. EPA for the most part has followed this measurement research and development plan. Funding limitations have extended the original timetable, however, and needs that have emerged since the plan was prepared are being added from time to time.

Air pollution control legislation before 1970 provided for the achievement of adequate air quality by state and local agencies on the basis of federal criteria (14–19); it also authorized the establishment of emission standards for passenger cars and other light-duty vehicles. Thus measurement research and development related to the pre-1970 legislation dealt mostly with air quality and light-duty vehicle emissions, and great progress has been made in these areas (20, 21). The 1970 amendments, however, called for the control of emissions from stationary sources as rapidly as possible and authorized EPA to set standards based on the best control technology available. To support these new control activities, EPA mounted a major effort to provide means of measuring contaminants in emissions from stationary sources (22–24).

EPA analytical methods

EPA has promulgated a number of standards under the Clean Air Amendments of 1970 and has proposed others (25–45). By early 1978, the promulgated standards included six for ambient air quality and some three dozen others for emissions from mobile sources and from new stationary sources in 24 categories of industry. The standards relate mainly to particulate matter, sulfur dioxide, oxides of nitrogen, photochemical oxidant, hydrocarbons, carbon monoxide, hydrogen sulfide, and the hazardous air pollutants mercury, beryllium, asbestos, and vinyl chloride.

For most of these standards, EPA specified a detailed measurement procedure — the reference method — for the contaminant involved. (Some standards for stationary sources rely on the estimated effectiveness of the required control measures, such as floating roofs on storage vessels for petroleum liquids, rather than on monitoring of emissions by specified techniques.) The reference methods are proven techniques or instruments of reasonable cost that are comparatively easy to operate and maintain. Legal compliance with a standard can be determined only by an EPA-specified reference method or "equivalent method." An equivalent method is one that meets certain performance requirements and gives results that bear a consistent relationship to those obtained by the reference method.

New approaches to the application of reference and equivalent methods took effect in February 1975 (46, 47). The new concepts apply to continuous analyzers for contaminants in ambient air. The reference method for carbon monoxide, for example, had been nondispersive infrared (NDIR) spectrometry. Under the new approach, this reference method was replaced by a specified "measurement principle" (NDIR) and calibration procedure. No single reference method is specified; there can be as many as there are models of NDIR analyzer that meet the requirements. A non-NDIR measurement principle, manual or continuous, that meets the requirements may be designated an equivalent method; if it is clearly superior to NDIR it may even replace that technique as the legally acceptable measurement principle. Decisions on superiority, however, are to be based on costs and benefits, as well as on analytical capability, so that measurement principles are expected to be replaced only infrequently. (Among other effects, this new approach to legally acceptable methods should accelerate to some extent the demonstration of degrees of equivalency among the many instruments now available commercially.) EPA-specified analytical methods for pollutants in ambient air appear in Table 2.

EPA reference methods for monitoring contaminants emitted by stationary and mobile sources were unaffected by the foregoing new approach to methods for contaminants in ambient air. Effective in October 1975, however, continuous monitoring was required for some emissions from stationary sources. At that time, procedures were established for calibrating the required continuous systems against the specified reference methods (48). Selected EPA reference methods for emissions from stationary and mobile sources appear in Table 3.

Standardization of methods for air

Many methods of air analysis are used in addition to those specified by EPA. More advanced and costly techniques are used regularly in research by both EPA and others. For routine use, a company or other organization usually will prefer methods best suited to its own analytical needs and constraints, which will vary with the processes involved, the purposes of the analyses, and other factors. Standardization of these as well as of the EPA methods has been under way for some years by groups such as the American Society for Testing and Materials (49) and the Intersociety Committee for a Manual of Methods of Air Sampling and Analysis (50). Collaborative testing of such methods, however, presents unique difficulties and, in fact, has become generally feasible only in the past few years.

Table 2

EPA analytical methods for pollutants in ambient air

Pollutant	Measurement principle or method	Reference method	Equivalent method
Total suspended particles	High-volume sampler (manual method) (gravimetric)	High-volume sampler	None possible (a)
Sulfur dioxide	Pararosaniline (manual method) (colorimetric)	Pararosaniline (b)	Manual or continuous (c)
Carbon monoxide	Nondispersive infrared spectrometry	d	Manual or continuous (c)
Ozone (photochemical oxidant)	Chemiluminescence (O_3-ethylene reaction)	d	Manual or continuous (c)
Nitrogen dioxide	Chemiluminescence (NO-O_3 reaction)	d	Manual or continuous (c)
Total hydrocarbons	Gas chromatography (flame ionization)	Gas chromatography (flame ionization)	None

a. No equivalent method is possible because the pollutant is defined as the material collected by the high-volume sampler during the sample-collection phase of the reference method.

b. Since the reference method is manual, it is defined by a series of explicit manual operations, so that no other reference method can exist.

c. A manual equivalent method must show a consistent relation to a reference method; an equivalent continuous analyzer must show a consistent relation to a reference method and meet performance specifications.

d. None specified. Any continuous analyzer is a reference method if it meets performance specifications.

Source: Environmental Protection Agency.

The normal procedure in round-robin tests is to make up a standard sample of the material to be analyzed — whether steel, or contaminated water, or detergent powder — divide it into a number of fractions whose compositions can be presumed to be identical, and send the fractions to the collaborating laboratories for analysis. It is not possible, though, to divide samples of contaminated atmospheres into identical fractions. One alternative is to send the collaborators to the same location, where they can sample a real atmosphere. The contaminant of interest must be present at the right time and in the desired range of concentration, however, and the statistical validity of the results may be open to question. A second alternative is to make up a synthetic test atmosphere and send samples of it (in small cylinders, for example) to the collaborating laboratories. This approach has long suffered, however, from the difficulty of generating known amounts of a pollutant gas at the very low levels found in the atmosphere. The same difficulty has hampered the development of primary reference standards of known composition for calibrating instruments for air analysis.

Recent years have seen steady progress with these problems. For a few gases, at least, the solution has been the permeation tube. The tube is made of Teflon and is about the size of a lead pencil 5 to 10 cm long. It is filled with the liquefied contaminant gas and sealed. The gas escapes through the wall of the tube under its own pressure at a slow but constant rate that can be measured by weighing the tube periodically. The escaping gas is picked up in a stream of metered air to produce with great accuracy concentrations well below 1 ppm. The first permea-

Table 3

Selected EPA reference methods for emissions from stationary and mobile sources

Pollutant	Reference method	Basis of detection
Particulate matter		
Stationary sources	EPA sampling train	Gravimetric
Sulfur dioxide		
Stationary sources	Barium-thorin titration	Volumetric
Carbon monoxide		
Stationary sources	Nondispersive infrared spectrometry	Infrared
Mobile sources	Nondispersive infrared spectrometry	Infrared
Hydrocarbons		
Mobile sources	Gas chromatography	Flame ionization
	Nondispersive infrared spectrometry	Infrared
Nitrogen oxides		
Stationary sources	Phenoldisulfonic acid	Colorimetric
Mobile sources	NO-O_3 reaction	Chemiluminescence
Hydrogen sulfide		
Petroleum refineries	CdS-iodine reaction	Volumetric

Source: Environmental Protection Agency.

tion tubes, for sulfur dioxide, became available as Standard Reference Materials from the National Bureau of Standards in 1971–72 (51); NBS later perfected a similar device for nitrogen dioxide (51, 52). Commercial suppliers offer permeation tubes for gases that include vinyl chloride and hydrogen sulfide, and work on additional gases is under way (53).

The permeation tube is not always the best choice in devising air pollutant reference samples, so other techniques are in use as well. Calibrated ozone generators are being used (51). Small cylinders containing known amounts of carbon monoxide (51) or hydrocarbons (54) are available as reference standards. A large batch of atmospheric particulate matter was being characterized physically and chemically in 1977 for use as a source of reference standards (51). Still another development involves a gas phase reaction (titration) in which nitric oxide and ozone react to form nitrogen dioxide. With this system, analytical techniques for nitric oxide, nitrogen dioxide, or ozone can be calibrated against a reference sample of any one of the three (55).

With the advent of such techniques has come accelerated, if not fully satisfactory progress in collaborative testing and calibration standards. EPA is pursuing collaborative tests of its reference methods, and the results began to appear in 1972 (56–60). The first to be tested was the high-volume method for suspended particulate matter. In this case, no means exists yet of generating a suitable test atmosphere, so that the collaborators sampled a real atmosphere at the same site. The American Society for Testing and Materials also has conducted collaborative tests of its analytical methods, using basically the on-site approach. By the end of 1974, when the program ended, ASTM had tested six of its methods for ambient air and five for stationary source emissions (61).

Measurement strategies

Various measurement strategies are used or are being considered in the air pollution control program. Air pollutant levels in an urban area or a small locality, for example, may be measured by one or several monitoring stations. Alternatively, if current research succeeds, the levels may be measured by long-path spectroscopic techniques. Stationary-source stack emissions may be measured by conventional analysis of a sample extracted from the effluent gas, by an across-the-stack optical technique, or by remote measurement of the plume from the stack. Epidemiological studies of the effects of air contaminants on humans in an urban area call for still other strategies. One approach is multistation monitoring of pollutant levels in the locality where the subjects live and work; a second approach is for the subjects to carry with them miniaturized instruments that record and integrate the levels of contaminants to which they are exposed. More broadly, some situations may call for instrumentation that measures contaminants continuously in ambient air or in a source emission; others may require that a central laboratory analyze samples collected on the scene in a bubbler, on a solid absorbent, on a filter, in an evacuated cylinder, or in an expandable plastic bag.

Monitoring Air Contaminants

Many organizations, both governmental and private, are at work on means of measuring specific air contaminants routinely, and a considerable body of technology is now in being (22, 62, 63) (Table 4). Much of EPA's program, in fact, has involved the evaluation and application of technology developed elsewhere, both under contract and independently. It is doubtful that any industry, for example, uses EPA methods exclusively; research in air pollution, moreover, demands the use of additional, sophisticated techniques not normally well suited to repetitive analyses. Nevertheless, the reference and equivalent methods established by EPA (Tables 2 and 3), despite their recognized imperfections, provide a basic context for existing techniques for monitoring specific substances in air. An important proviso is that all methods should be validated and the differences among them recognized and noted in any evaluation of data.

Particulate matter

To measure particulate matter in ambient air, EPA specifies the high-volume sampler (25). This device draws air rapidly through a filter, much like a vacuum cleaner does; the volume of the filtered air and the weight-gain of the filter are measured to determine the concentration of dust and particulate matter in the air being tested. A collaborative test has shown that the method can yield results of relatively high precision (56). Scientists agree in general, however, that much better means of measuring particles in ambient air are needed urgently.

The composition and effects of particles tend to vary with their size, so that much of the research on the monitoring problem is aimed at methods that, unlike the high-volume sampler, fractionate particles into at least two size ranges. (Particles between 0.1 and 10 μm contribute most of the mass of airborne particles; those between 0.1 and 1 μm are largely responsible for haze and turbidity; and those smaller than about 1 μm are the most likely to enter and be retained by the respiratory system.) In 1976, EPA was field-testing a dichotomous sampler that separates ambient particles into the respirable and nonrespir-

Table 4

Selected instrumental methods for monitoring ambient air

Method	Sulfur dioxide	Nitrogen oxides	Oxidant	Carbon monoxide	Hydrocarbons
Conductimetric	X				
Colorimetric	X	X	X		
Coulometric	X	X	X	X	
Chemical sensing electrode	X	X			
Chemiluminescence		X	X		
Gas chromatograph-flame photometric detector	X				
Gas chromatograph-flame ionization detector				X	X
Nondispersive infrared				X	
Nondispersive ultraviolet absorption			X		
Correlation spectroscopy	X	X			
Bioluminescence	X	X		X	

Source: Hollowell, C.D., McLaughlin, R.D., "Instrumentation for Air Pollution Monitoring," *Environ. Sci. Technol.*, **7,** 1011 (1973).

able size ranges. Interest is high also in continuous measurement of particles in ambient air. Work is under way, for example, on instrumentation based on the attenuation of beta radiation by particulate matter collected on a filter (64, 65), a technique demonstrated earlier (66).

Particulate matter in emissions from stationary sources is measured gravimetrically after collection with the EPA sampling train (26). One problem here is that, under certain conditions, particulate matter can form in the sampling system during collection, so that the method indicates a higher concentration of particles than actually is present in the air being sampled. Particles also can remain within the sampling probe, so that a procedure is required for removing and weighing this part of the collected sample. Continuous measurement of particulate matter in source emissions is a current need and, as with ambient air, the beta attenuation method is among those being studied for the purpose.

Particle composition. A pressing need in the characterization of particulate matter is sharply upgraded means of identifying the chemical elements and compounds that it contains. Such data are important in several areas: research on atmospheric chemistry; assessment of the effects of particulate matter on health; identification of sources of atmospheric particles; and support of air quality standards for specific elements or compounds that, in ambient air, may exist at least partly in particulate form. Of the several possible techniques for elemental analysis of particulate matter, x-ray fluorescence, both wavelength dispersive (67) and energy dispersive (68, 69), appears to be quite promising. These methods may supplement or replace previously used techniques like atomic absorption spectroscopy (70, 71), emission spectroscopy, and neutron activation analysis (72).

Elemental analysis by x-ray fluorescence has been shown to agree within 20% of neutron activation analysis for calcium, chromium, manganese, iron, nickel, copper, zinc, and lead (73). These methods do not measure specific compounds. Research has shown, however, that important knowledge of the compounds of sulfur that a particulate sample may contain can be deduced from the results of extensive elemental analysis by x-ray fluorescence of aerosols collected in two size ranges (74).

Hundreds of organic compounds occur in particulate matter or adsorbed on the surface of largely inorganic particles. Some of these compounds are known to be carcinogenic or toxic. The organic constituents are usually separated and identified by chromatography and mass spectrometry. The relatively nonvolatile compounds, such as polynuclear aromatic hydrocarbons, can be separated by high performance liquid chromatography and measured by ultraviolet absorption or fluorescence detectors. Extraction with solvents and other techniques can be used to prepare samples for analysis. The more volatile compounds in aerosols can be determined by gas chromatography.

Filter problems. Numerous problems have been encountered with the filters used to collect particles for both chemical and physical analysis. No single filter material is suited to all tasks. Elemental analysis by x-ray fluorescence, for example, requires a material with very low background levels of impurities (74). For maximum sensitivity in mass measurement by beta attenuation, the filter material must have relatively low mass per unit area; at the same time it must be strong enough to withstand normal use and handling. A major unresolved problem is the tendency of filter materials to adsorb or absorb nonparticulate matter such as water and organics from the air passing over the filter. Several research programs are under way on these problems.

The technology of aerosol characterization has been reviewed (75); the review includes discussion of techniques for collecting and analyzing the chemical properties of airborne particles.

Sulfur dioxide

For measuring sulfur dioxide in ambient air, EPA specifies the pararosaniline method (25). A collaborative test has shown that the method is capable of satisfactory precision (59). The pararosaniline technique is cumbersome, however, and an instrumental equivalent method would be desirable. Sulfur dioxide (SO_2) in ambient air can be measured continuously by a flame photometric detection system. The detector measures total sulfur; the method can be used, therefore, when the amount of sulfur-containing material present is known to be at least 95% sulfur dioxide (76) or after sulfur compounds have been separated from one another by gas-liquid chromatography (77). Ambient SO_2 also can be measured by instruments based on coulometric and ultraviolet methods. Other reference methods are specified for measuring SO_2 in stack emissions from fossil fuel-fired power plants and for measuring the gas (as well as sulfur trioxide and sulfuric acid mist) in emissions from sulfuric acid plants (26). Sulfur dioxide in power plant stack emissions can be measured continuously by nondispersive infrared spectroscopy after the sample is conditioned to remove particulate matter and water vapor and to cool the gas to near ambient temperature (78).

Nitrogen oxides

Ambient air contains both nitrogen dioxide (NO_2) and nitric oxide (NO), but the EPA standard is for NO_2. For some years, NO_2 in ambient air was measured by absorbing it in a sodium hydroxide solution to form sodium nitrite, which was determined colorimetrically following a color-forming reaction with another reagent. This reference method was revoked in mid-1972 after it was found to be inadequate (79). EPA worked intensively to develop and evaluate improved collection techniques (55). These efforts involved the evaluation of several collection solutions (80–82); the requirements included high specificity for NO_2 and a relatively constant efficiency of collection of the gas. In December 1976 EPA specified an instrumental method based on chemiluminescence as the measurement principle for NO_2 in ambient air (83). This continuous technique is based on the chemiluminescent reaction between nitric oxide and ozone (21); it requires in part the conversion of NO_2 to NO in a suitable sampling system.

For emissions from mobile and stationary sources, the EPA nitrogen oxides standard is for NO_x, the sum of NO_2 and NO. Mobile source emissions of NO_x are measured by the chemiluminescence procedure used for nitrogen dioxide in ambient air (84). NO_x in power plant emissions is absorbed in a special solution and measured colorimetrically (26). Continuous measurement of NO_x in power plant emissions has been proposed using nondispersive infrared, nondispersive ultraviolet, and an electrochemical technique (85).

Photochemical oxidant, hydrocarbons

EPA's ambient air standard for photochemical oxidant is based on a measurement principle for ozone, which comprises virtually all of the oxidant (after correction for NO_2). The technique involves a chemiluminescent reaction of ozone with ethylene and permits ozone to be measured continuously (25). A collaborative test of the method showed that satisfactory precision can be obtained (57, 60). Oxidant is not a source emission, but rather is formed in the atmosphere by photochemical reactions involving hydrocarbons and nitrogen oxides. No need exists, therefore, for source measurements of ozone. The control of photochemical oxidant is based instead on emission standards for hydrocarbons.

The reference method for hydrocarbons in ambient air is based on a flame ionization detector (25). The detector measures total hydrocarbons, but one hydrocarbon in ambient air, methane, does not take part in the photochemical reactions that form oxidant. Thus the analytical equipment is arranged to measure total hydrocarbon in the sample and, in a second step, to separate and measure methane alone. The value for oxidant-forming hydrocarbons is obtained by difference.

Hydrocarbons in mobile source emissions are measured by a flame ionization detector or by nondispersive infrared spectroscopy (29). Flame ionization detection is used for all mobile source emissions; nondispersive infrared may be used for heavy-duty vehicles. Both measurements are for total hydrocarbons. Methane is not subtracted, since it is only a minor constituent of hydrocarbon emissions by mobile sources. This may change, however, with the increasing use of catalytic exhaust systems on vehicles made after 1974. The catalytic devices tend to promote the oxidation of methane (to carbon dioxide and water) less effectively than do earlier exhaust systems, so that the fraction of methane in vehicle exhaust may rise in the future. Further, chemists now believe that some other hydrocarbons, such as ethane, propane, acetylene, and benzene, may be relatively unreactive in the formation of photochemical oxidant in the atmosphere. Measurement techniques for these hydrocarbons are being developed, therefore, should it be deemed advisable to distinguish them, along with methane, from the other hydrocarbons in mobile source emissions.

EPA has neither promulgated nor proposed stationary source emission standards for hydrocarbons that require the compounds to be measured.

Carbon monoxide

The measurement principle for carbon monoxide in ambient air (25) and in emissions from mobile (29) and stationary (27) sources is nondispersive infrared spectroscopy (NDIR). A collaborative test of the method for ambient air analysis showed satisfactory precision (58). While NDIR is adequate to support the ambient air quality standard for carbon monoxide (10mg/m^3 or 9 ppm), it is less effective at very low concentrations of the gas — in the range of 1 or 2 mg/m^3. At such levels, carbon monoxide can be converted to methane and measured in that form by gas chromatography with a flame ionization detector.

Hydrogen sulfide

To measure hydrogen sulfide in emissions from petroleum refineries by the EPA method, the gas is first collected in alkaline cadmium hydroxide solution to form cadmium sulfide, which precipitates. The solid cadmium sulfide is then dissolved in hydrochloric acid, and the hydrogen sulfide evolved is absorbed in a known volume of iodine solution. The iodine consumed is a measure of the hydrogen sulfide content of the original sample (27).

EPA has not established a reference method for hydrogen sulfide in ambient air, but others have reported methods for the compound (86, 87).

Hazardous air pollutants

The law requires EPA to set emission standards for "hazardous" air pollutants. Through 1977, such standards had been promulgated for mercury, beryllium, asbestos, and vinyl chloride.

Measurement of the hazardous air pollutant mercury, in both particulate and gaseous emissions from stationary sources, begins with an appropriate sampling procedure and collection of the sample in an iodine monochloride solution. The mercury collected is reduced to elemental form in basic solution by dihydroxylamine sulfate. The elemental mercury is then purged from solution with a stream of air and analyzed with an atomic absorption spectrophotometer in the flameless mode (28).

The hazardous air pollutant beryllium is measured by atomic absorption after appropriate sampling and digestion in an acid solution (28). Equivalent methods, such as gas chromatography of a volatile beryllium chelate derivative, may also be used (28, 88).

No suitable technique exists for measuring the hazardous air pollutant asbestos in emissions from sources, so that the current standards are based perforce on "visible" emissions. Asbestos in ambient air is sampled on filters and measured, while still on the filter, by electron microscope methods (89); other techniques have been reported (90). These techniques are not rapid enough, however, to support an emissions standard, and EPA is working currently to provide an adequate means of determining asbestos in source emissions.

The hazardous air pollutant vinyl chloride in source emissions is measured by gas chromatography with a flame ionization detector (91).

Automatic Monitoring

Automatic monitoring of airborne contaminants offers great advantages in air pollution control programs. Manpower requirements are reduced. Continuous recording of pollutant measurements reveals trends in ambient concentrations, which may vary with sunlight intensity, traffic density, and other factors. Comparisons of data for different contaminants may reveal relationships among them that help to clarify their chemical and physical behavior in air. And these and other kinds of data are necessary to build and verify models of urban atmospheres for use in arriving at effective pollution control strategies at minimum social cost.

The benefits to be gained have spurred considerable progress in automatic, continuous instrumentation. The results can be seen in part in the air quality monitoring networks that the states are required by law to establish. Of six ambient pollutants being monitored by these networks in 1974, all but one (suspended particulate matter) were being measured by instrumental methods that were either approved by EPA or had the potential to be approved (Table 5). And in 1975, as mentioned earlier, continuous monitoring technology for stationary source emissions had reached a point at which the agency felt justified in requiring its use for some such emissions (48). These include nitrogen oxides, sulfur dioxide, and oxygen or carbon dioxide emitted by fossil fuel-fired steam generators; nitrogen oxides from nitric acid plants; and sulfur dioxide from sulfuric acid plants. In addition, air emitted by fossil fuel-fired steam generators and by catalyst regenerators for fluid-bed catalytic cracking units must be monitored continuously for opacity.

Table 5

Pollutant method station summary for state air quality monitoring networks (1974)

Pollutant	Method or principle[a]	No. of stations	Percent of total	Ap-proved	Unap-proved[b]	Unac-ceptable
Total suspended particles	High-volume sampler (FRM)[a]	3683	100	X		
Carbon monoxide	Nondispersive infrared (FRM)	316	96	X		
	Coulometric	2	n[c]			X
	Flame ionization	10	4		X	
	Total	**328**				
Sulfur dioxide	Colorimetric	122	6		X	
	Conductimetric	93	4		X	
	Coulometric	223	10		X	
	Flame photometric	59	3		X	
	Sequential conductimetric	6	n		X	
	Pararosaniline (FRM)	1648	77	X		
	Total	**2151**	**100**			
Nitrogen dioxide	Colorimetric	139	11		X	
	Coulometric	5	1		X	
	Chemiluminescence (FRM)[d]	49	4	X[d]		
	Saltzman bubbler	5	1			X
	Sodium arsenite (orifice)	294	24		X	
	Sodium arsenite (frit)	730	59		X	
	Total	**1220**	**100**			
Oxidant	Alkaline potassium iodide	10	3			X
	Coulometric	34	9		X	
	Neutral potassium iodide	71	19		X	
	Phenolphthalein	1	n			X
	Alkaline potassium iodide bubbler	18	5			X
	Ferrous oxidation	20	5			X
Ozone	Chemiluminescence (FRM)	225	59	X		
	Coulometric	1	n		X	
	Ultraviolet				X	
	Total	**380**	**100**			

a. FRM, Federal reference method or principle.
b. "Unapproved" means the method has the potential to be approved by EPA.
c. n = negligible.
d. Designated as measurement principle in December 1976.

Source: Hoffman, A.J., et. al., "EPA's Role in Ambient Air Quality Monitoring," *Science*, **190**, 243 (1975). Copyright 1975 by the American Association for the Advancement of Science.

Research in Air Analysis

Research conducted to develop the information required to set air quality standards demands as a rule considerably greater analytical sophistication than is characteristic of the time- and cost-limited methods and equipment used in monitoring and related tasks. Research methods used commonly include spectroscopy, particularly infrared (92), as well as gas chromatography (93), electron microscopy, x-ray fluorescence, x-ray diffraction, mass spectrometry, emission spectroscopy, and neutron activation analysis. Instrumental techniques that show promise include the use of excitation bandwidths that stimulate fluorescence (94, 95) or fragment the target molecule into more readily detectable gaseous species (96). For the most part these methods employ laboratory instruments; special procedures must be devised for collecting samples in the field and transporting them, in unaltered form, to the laboratory for analysis. In some instances, however, it is possible to install equipment of this type in mobile laboratories so that measurements can be made in the field.

Currently under development are remote and long-path techniques for measuring pollutants in ambient air or in emissions from stationary sources (97, 98). A remote method measures the contaminant at a distance; a long-path method measures the total amount of the contaminant along a given path or column of air. Neither technique requires that a sample be collected. Ambient ozone concentrations have been measured along a path 0.67 km long, using a directly tunable carbon dioxide laser at one end of the path and a mirrored reflector at the other end (99). These measurements have been verified using a method for comparing the results obtained from long-path monitors with those from point monitors (100). Tunable semiconductor diode lasers have been used to make similar measurements of carbon monoxide (101). Long-path techniques of this type are the subjects of continuing research and development. So, too, are methods of determining spatial concentration contours of gases by means of pulsed-laser radiation backscattered from particles in the atmosphere. The results of initial development of these systems suggest that they can sensibly be projected to evolve toward greater coverage in time and space over distances on the order of kilometers. This evolution will be paralleled by increased spectral coverage — and thus the ability to detect more pollutants — and output power for tunable laser sources (102).

Recommendation AM5: *Research on remote methods of air analysis and on methods for analyzing actual samples (grab samples) should proceed in parallel in recognition of the fact that each approach has intrinsic strengths and weaknesses and that remote techniques must be validated by field and laboratory analyses of actual samples.*

Research is in progress also on the detection and measurement of hazardous organic and inorganic air pollutants like carcinogens and mutagens (103). Such compounds are present in polluted air in both gaseous and particulate form. Their concentrations, however, may be minute — in the range of micrograms per cubic meter (ppb) — so that investigations of them call for reliable sampling methods and sensitive analytical instrumentation. By 1977, scientists at EPA had devised an array of procedures capable of assaying 200 to 300 such substances. In general, these procedures include sampling with filters or sorbent media and analysis by various forms of chromatography, mass spectrometry with computer processing, spectrophotofluorimetry, and x-ray fluorescence or atomic absorption. EPA and others have used such methods to measure in urban air a variety of hazardous and potentially hazardous substances, including the carcinogenic benzo(a)pyrene and other polyaromatic hydrocarbons and the carcinogens vinyl chloride and dimethylnitrosamine (103–105).

The problem of preventing degradation of the air in nonurban areas (106) is complicated by the fact that the very low levels of contaminants involved cannot in general be detected by available monitoring techniques. More sensitive and accurate methods are required both to establish the necessary base-line concentrations and to detect the onset of degradation.

Methods designed to meet these requirements have been developed and used in the field, especially in research on the background levels of various constituents of the global atmosphere (107–108). Relatively little progress has been made, however, toward systematic monitoring of base-line concentrations on a global or regional basis.

Recommendations applicable to air analysis, but also to other phases of the environment, appear at the end of this chapter.

MEASUREMENTS IN WATER

The measurement technology of water, like that of air, serves diverse purposes: surveillance of ambient waters or public water supplies for compliance with standards; tests of raw water entering water treatment plants and industrial facilities; monitoring of pollutants in municipal and industrial wastewaters; research on problems like eutrophication and the behavior and ultimate fate of waterborne contaminants. The water analyst traditionally has used methods borrowed from other fields, where they have been found to be reliable, and scaled down to the required sensitivity. These techniques have been based largely on gravimetry, titrimetry, and spectrophotometry. Modern lowering of tolerances, however, has made some such methods obsolete; others take too much time. The trend, at any rate, is toward instrumental techniques. The state of the art of water analysis related to pollution control, as of about 1970, has been summarized (109); instrumentation and techniques available in 1974 for sampling, monitoring, and analyzing plant effluents or surface waters have been itemized and described (110). Reviews of progress in water analysis appear regularly (111).

Analysis and Water Quality Standards

The federal Water Quality Act of 1965 required the states and other jurisdictions to establish ambient quality standards for interstate waters; the Federal Water Pollution Control Act Amendments of 1972 extended the standards program to intrastate waters. As is the case with air, legal compliance with these standards in terms of specific contaminants is determined by analytical methods specified by EPA. Water analysis has a long history, and many standard (if not always ideal) methods have evolved. EPA recognizes "approved" methods drawn from the two primary compilations of standard methods (49, 112) and from the agency's own methods (113). EPA regulations provide also for "alternate test procedures." These approved and alternate methods are analogous to the agency's reference and equivalent methods of air analysis.

Besides extending the ambient quality standards program, the 1972 amendments established limits on the nature and amounts of pollutants in all point discharges to the nation's waters. They required EPA also to "promulgate guidelines establishing test procedures" for the analysis of waste effluents. The published guidelines (114) covered 71 parameters; the approved methods were drawn from the primary compilations and from EPA techniques. Later, EPA proposed changes in the guidelines including the addition of 15 parameters, bringing the total to 86 (115).

Standardization of methods for water

Standardization of methods has occupied water analysts increasingly in the 1970's. Most water analysis laboratories use the standard methods in the two primary compilations (49, 112). In some cases these methods are not sufficiently sensitive, however. And by the time a volume of analytical methods is published, some of them are likely to have been outdated by recent improvements. Thus water laboratories at all levels — federal, state, municipal, private — must pursue research on methods actively to stay abreast of analytical developments.

In the Federal Government, a program is under way to develop a handbook of recommended standard methods, including analytical methods, for use by federal agencies in collecting water data (116). In mid-1974 the work was in a second, expanded phase which was to include methods for making hydrometeorological observations, snow and ice surveys, and evapotranspiration measurements in addition to methods for determining the quantity and quality of water. In 1977, the U.S. Geological Survey published "Chemical and Physical Quality of Water and Sediment," one of 10 chapters that will make up the handbook.

Standardization is in progress worldwide as well. The International Standards Organization has established Technical Committee 147, Water Quality, to standardize all aspects of analysis, especially where internationally compatible data are necessary. Several international scientific bodies concerned with water analysis are collaborating on plans for an interlaboratory program (117). It calls for comparative analyses of standard reference samples in water-quality laboratories throughout the world.

Despite the activity in water measurement technology, both current analytical methods and environmental data often are proving to be questionable. Collaborative testing by the Environmental Protection Agency, the American Society for Testing and Materials, and others has shown that few methods for inorganic ions and relatively simple organic substances are "acceptable" (total error of no more than ±50%); for many important parameters no acceptable methods exist. Tests of methods for complex mixtures of organics produce even poorer results; at the same time, no well-accepted test substances are available to provide a basis of comparison for the tests.

EPA has estimated the precision of routine atomic absorption measurements at ±20%, with no assessment of their accuracy (113). The U.S. Geological Survey's round-robin testing of atomic absorption methods for 11 common metals showed, through its 20th cycle, standard deviations of 10 to 50% even when unusually divergent results were excluded (118). Such results, from competent laboratories often using standard methods, certainly justify the concern for the quality of routine environmental data on water.

General experience in trace analysis points to inadequate if not faulty calibration of instruments as the prime reason for these uncertainties. Intralaboratory measurements commonly are precise, while data from different laboratories may show significantly large biases. Accurately certified Standard Reference Materials have corrected some such situations by providing a common basis for measurement standardization. The development of appropriate Standard Reference Materials and the encouragement of their use should do much to improve the state of the art of water analysis. Certainly no data should be admitted to national repositories unless suitable standards are used as evidence of analytical quality control.

Measurement program design

The accelerating demand for environmental data on water has been partly responsible for a subtle, persistent problem — inadequate design of measurement programs. Too often the parameters measured have been inconsistent with the type of information sought. Measurements of emissions at the source, for example, are generally quite specific and indicate the need for or the efficiency of an abatement procedure. Measurements in a receiving medium away from the source of emission can characterize the medium itself, or some part of it, and indicate its unsuitability for an end use, such as human consumption. A well-designed measurement program is marked, first, by clearly defined goals. Other considerations include suitable analytical parameters and methods; carefully specified numbers and locations of measurement sites; and proper frequency of sampling (or interrogation of monitors). Also required is a data processing plan. By recycling the data through an appropriate scheme, it is possible regularly to reassess the soundness of the types, numbers, sites, and frequencies of measurement (119, 120).

Sampling techniques

Techniques for insuring that samples for analysis truly represent the body of water being monitored leave much to be desired. Few constituents yet can be monitored in place, as by probe-type sensors; for the foreseeable future most analyses will require sophisticated equipment generally unsuited to use in the field. Samples often must be collected at the scene, therefore, transported to the laboratory, and perhaps stored there for a time before they are analyzed. In the process, changes can result both from reactions within the sample and interactions between sample and container. The sampling operation itself is often faulty because of an indifferent operator or contamination by the sampling device or the sampling platform (such as a ship). More broadly, both sampling procedure and interpretation of analytical data may suffer from inattention to factors like the position of the sampling point in the body of water, the time of day, the season, and previous weather conditions.

Monitoring Water Contaminants

For many purposes, useful qualitative information on water can be derived from certain general or collective parameters determined by analytical methods of considerable historical background. Parameters of this kind include pH, conductance, total dissolved solids, total suspended solids, acidity or alkalinity, hardness, biochemical oxygen demand, and dissolved oxygen. Also well established are methods for many specific inorganic substances, such as various metals, chloride, and carbonate, that long have been prime targets of the water analyst. Of much more recent concern are the many specific organic compounds created by new technology or by degradation of wastes; methods for these materials are in an evolutionary stage, although some procedures have reached remarkable levels of sensitivity. Traditional analytical techniques for a number of water pollutants appear in Table 6; in addition, recent years have seen a strong trend toward instrumental methods of water analysis (Table 7).

Inorganics in water

For inorganics in water, instruments based on electrochemistry and on the optical techniques, emission and absorption spectroscopy, have been especially prominent. Examples include ion-selective electrodes and automated spectrophotometric procedures for phosphate and nitrogen. Inorganic instruments often have been limited to determining a single chemical element in each sample of water, but the modern trend is toward determining several elements in each sample simultaneously. Atomic absorption analysis for metals is moving in this direction (121); the multielement capability of x-ray fluorescence spectroscopy and neutron activation analysis compensate for their relatively high cost. Environmental programs often may require more than 100 samples daily to be run on a single instrument, and multielement analysis thus offers significant gains in efficiency.

Organics in water

For organics in water, few classical techniques have been available. A notable exception is the phenols, which long have been known to cause odor and unpleasant taste in drinking water. Colorimetric methods have been devised that are sensitive to 5 μg/l (5 ppb) of phenol itself and to somewhat larger concentrations of phenols of higher molecular weight. For most organics in water, however, new methods perforce are being developed (122). Gas chromatography, infrared spectroscopy, and mass spectrometry have provided most of the essential tools; ultraviolet spectroscopy, fluorimetry, and Raman spectroscopy are seeing limited use. Especially useful has been the tandem use of the gas chromatograph, a sensitive means of separating volatile organic components, and the mass spectrometer, a means of identifying them. An unknown organic structure is identified in essence by comparing its mass spectrum with those of known structures until a match is found. The known spectra of many organics are being collected in data banks that allow the comparison to be made readily by computer (123). This approach is greatly extending the qualitative utility of the gas chromatograph–mass spectrometer procedure; there is an important need to obtain more qualitative data on the many organic compounds present in water. In the past few years, high-performance liquid chromatography has sharply improved the efficiency of analyses of organic compounds that do not volatilize readily. The methods used include ion-pair, ion-exchange, reversed-phase, and gel-permeation chromatography.

Oxygen demand. Historically, organic contaminants in water have been measured in terms of biochemical oxygen demand (BOD) or chemical oxygen demand (COD). Oxygen demand values are significant because they indicate the amount of oxygen potentially required in a natural body of water to oxidize organic matter completely to carbon dioxide, nitrogen, and water. The BOD and

COD techniques determine the amount of oxidant required to oxidize all of the organic material in the sample; they do not identify individual organic compounds or classes of compounds. The BOD test measures the amount of molecular oxygen consumed and customarily requires five days (112). The COD test measures the consumption of the oxidant chromic acid and requires one to two hours [49 (Part 31), 112]. (Variants of the COD test use permanganate as the

Table 6

Selected methods for monitoring water quality

Physical, demand, and nutrient parameters and testing methods

Parameter	Method
Solids	Gravimetric
Color	Visual comparison
Turbidity	90° scatter photometer
Radioactivity	Gross α, β, ν, tritium
Specific conductance	Wheatstone bridge
Biochemical oxygen demand	5-day—20°C
Chemical oxygen demand	Dichromate reflux
Ammonia	Distillation-nesslerization or Auto Analyzer
Kjeldahl nitrogen	Digestion-distillation or AutoAnalyzer
Nitrate	Brucine sulfate or AutoAnalyzer
Nitrite	Diazotization or AutoAnalyzer
Total phosphorus	Persulfate digestion or AutoAnalyzer
Phosphate-ortho	Single reagent, stannous chloride, or AutoAnalyzer

Testing methods for organic and general parameters

Parameter	Method
Phenols	Colorimetric
Oil and grease	Hexane Soxhlet extraction
Total organic carbon	Combustion-infrared
Surfactants	Methylene blue active substance
Pesticides	Gas chromatography
Alkalinity	Electrometric titration or AutoAnalyzer
Bromide	Colorimetric
Chloride	Mercuric nitrate titration or AutoAnalyzer
Cyanide	Silver nitrate titration or pyridine pyrazalone
Fluoride	SPADNS (a dye) with distillation or probe
Hardness	Ethylenediaminetetraacetic acid titration, AutoAnalyzer, or atomic absorption
Sulfate	Turbidimetric or AutoAnalyzer
Sulfide	Titrimetric or methylene blue colorimetric
Sulfite	Iodide-iodate titration

Source: Ballinger, D.G., "Instruments for Water Quality Monitoring," *Environ. Sci. Technol.*, **6**, 130 (1972).

oxidant in acid or alkaline conditions.) Within the past decade, the COD-chromic acid method has been automated and the reaction time shortened.

Also available are instruments that determine a quantity closely related to COD within about two minutes. One of these is based on combustion in an atmosphere limited in oxygen; it measures the oxygen remaining after the

Table 7

Instrumental methods for analysis of water

Instrumental method	Parameter measured
Continuous samplers	
Gas sensing electrode	O_2; NH_3; SO_2
Turbidimetric; nephelometric	Turbidity
Transmissometric	Turbidity
Wheatstone bridge	Conductivity
pH meter	H^+ activity
Thermistor; thermocouple; resistance change	Temperature
Reflectance	Oil and grease
Potentiometric	Oxidation-reduction potential; selective ions (e.g., F^-)
Semicontinuous samplers	
Atomic absorption	Hg
Molecular absorption	Metals; total alkalinity; Cl^-; F^-; total hardness; nitrogen (NH_3); nitrogen (total, Kjeldahl); nitrogen (nitrate-nitrite); nitrogen (organics + NH_3); phosphorus (all forms); silica (dissolved); SO_4^{2-}; PO_4^{2-}; Cl_2 (free; combined); CrO_4^{2-}; hydrazine; MnO_4^-; SO_3^{2-}; phenols; chemical oxygen demand; organic carbon
Electrochemical (amperometric)	Cu
Potentiometric	Selective ions (e.g., Cl^-, CN^-, F^-)
Laboratory analytical systems	
Differential pulse voltammetry	Electroactive constituents
Anodic stripping	Mercury-soluble metals
Atomic absorption	Metals
Molecular absorption	Gases (see semicontinuous samplers)
Emission spectroscopy	Metals
Neutron activation	Elemental analysis
X-ray fluorescence	Elemental analysis
Gas sensing electrode	Dissolved oxygen; biochemical oxygen demand
Ion-selective electrode	F^-; CN^-, NO_3^-; Cl^-; others
Gas chromatography	Oils; phenols; pesticides; sludge digester gas
Mass spectrometer	Metals; pesticides (in conjunction with a gas chromatograph)

Source: Phillips, S.L., Mack, D.A., MacLeod, W.D., "Instrumentation for Water Quality Monitoring," *Anal. Chem.*, **46**, 345A (1974).

organics have been burned and, by difference, yields a value called TOD (total oxygen demand) [49 (Part 31)]. The other rapid method employs combustion of the organics in a carbon dioxide atmosphere; it measures the amount of carbon monoxide produced and gives a value called CO_2D (chemical oxygen demand by combustion in carbon dioxide). TOD and CO_2D agree reasonably well with each other. Both methods, however, oxidize whatever nitrogen is present, in ammonia or amino groups, to free nitrogen or nitric oxide; both, therefore, give higher values than the COD method, which does not oxidize ammonia or amino nitrogen. The individual instruments are calibrated against standard solutions of known compounds. The instruments are sensitive over the range of concentrations of organics found normally in domestic wastewaters and down to a limit set by the uncertainty caused by differences in the amounts of dissolved oxygen in samples and standard solutions. The TOD instruments are designed for repetitive, on-stream analysis, but can be used also for laboratory analyses; the method is at least as good as BOD or COD for routine water-quality monitoring, and several well-tested instruments are available. The CO_2D instrument is for laboratory use only and therefore costs somewhat less than the TOD instrument.

Total carbon. A number of instruments can measure the organic content of water rapidly in terms of the element carbon. These devices convert the organic matter to carbon dioxide, which is measured in a nondispersive infrared analyzer; in one modification the carbon dioxide is reduced to methane and measured in that form with a flame ionization detector. Organics are converted to carbon dioxide by combustion, either with excess oxygen or an oxidizing catalyst, or by digestion with persulfate in a sealed tube. The combustion methods are the fastest. The results are expressed as total carbon (TC) [49 (Part 31)]; they may be corrected to total organic carbon (TOC) [49 (Part 31), 112] by deducting the carbon present in the sample as carbonate, which can be determined instrumentally in two minutes. (Alternatively, the organic carbon can be determined alone after removing the carbonate carbon by acidification and purging with air or oxygen.) Some such instruments are equipped to account for the carbonate carbon automatically. Certain instruments, in addition, have been modified to determine total carbon at concentrations as low as 0.1 mg/l (0.1 ppm).

The development of these rapid methods for determining oxygen demand and organic carbon has improved enormously the efficiency of water pollution monitoring efforts. The new methods are very useful also in studies of waste control processes. Regulatory agencies, moreover, now allow the techniques to be used, instead of the slower BOD and COD tests, to obtain analytical data for required periodic reports on effluent streams. For this purpose, however, factors must be established to correlate the results of the methods with BOD and COD measurements on samples of similar character.

Sample concentration. Analysis of water for specific organic compounds or groups of compounds at low concentrations requires as a rule that the organics be concentrated before analysis, whether by gas chromatography or one of the foregoing collective methods. Two widely used means of concentration are extraction of the organics from water with an organic solvent and adsorption of the organics on charcoal or other adsorbents.

One standard extraction procedure, aimed specifically at chlorinated hydrocarbon pesticides, involves extracting the pesticide from water with the organic solvent hexane (112). The hexane is then evaporated to concentrate the sample and analyzed by electron-capture gas chromatography (EC-GC). The method is reported, for example, to be sensitive to 10 ng/l (10 ppt) of the insecticide lindane in relatively unpolluted water. In a second procedure of this kind, 5 μg/l of an organic compound, bis(2-chloroisopropyl) ether, was extracted quantitatively from 5 l of water with 5% ethyl ether in hexane (124). The sample, again, was concentrated by evaporating the solvent and analyzed by EC-GC.

Two standard adsorption procedures for concentrating organic contaminants involve passing the water through charcoal and extracting the adsorbed organics from the charcoal with chloroform. One of the two methods is for use with drinking water or other clear water supplies (112); the other is for use with surface water other than municipal drinking water supplies (112). In both cases the amount of the contaminant present is determined ultimately by weight. A third charcoal adsorption procedure, which included analyses by GC-MS, infrared spectroscopy, and chemical methods, identified a number of chlorinated hydrocarbons in drinking water at concentrations that probably were below 1 μg/l (124). These carbon adsorption methods are relatively slow — they require from a few days to more than two weeks to complete and are subject to error caused by biological degradation during the adsorption period. They do, however, permit many organics to be detected in water at extremely low levels. Many organic compounds have also been detected in water by procedures involving adsorption on synthetic resins (125–127).

In still another procedure, the sample of water is drawn into a hypodermic syringe with an equal volume of helium or another pure, inert gas (128). A vaporizable organic in the water will move into the gas until its concentrations in the two phases reach an equilibrium. This principle, combined with gas chromatography and a flame ionization detector, has permitted the detection of 1 to 3 ng/l of certain low-molecular-weight aliphatic hydrocarbons (alkanes or cycloalkanes) in water. The method has been reported also to have detected 4 to 12 ng/l of aromatic hydrocarbons with no more than eight carbon atoms, and less than 1 μg/l of chloroform in city tap water. Other variations, such as sparging with nitrogen and collecting the volatiles on a porous polymer (e.g., Tenax GC), permit organic compounds to be measured at concentrations of less than 1 μg/l. The volatile organics are thermally desorbed into a cooled gas chromatography column for measurement.

Recommendation AM6: *Methods for removing and concentrating organics in water (i.e., collection on sorbents or solvent extraction) for subsequent separation and analysis should be developed further, especially for quantitative measurements.*

Sample cleanup. Regardless of the method used to concentrate organics for analysis, the determination of specific substances at low concentrations is likely to be complicated by interfering compounds. Selective cleanup procedures usually are necessary to remove troublesome interferences and ensure that the detecting device actually is measuring the substance sought and only that substance. Analysis can be extremely difficult, for example, when the substance of interest is biologically active at minute concentrations and must be distinguished not only from closely related compounds, but also from its own, less active isomers — compounds that have the same elemental composition but differ in structure and properties. One instance of this situation involved work on chlorinated dibenzo-p-dioxins and dibenzofurans (129). The compounds are possible impurities in the herbicide 2,4,5-T; of these various related compounds, only one is highly undesirable.

Common cleanup procedures include extractions, ion-exchange processes, liquid column chromatography, and thin-layer chromatography. All of these may usefully precede GC or GC-MS determinations. The more complex the cleanup procedure, however, the greater the possibility of error through loss or contamination of the sample. GC column preparation and conditioning and instrument-conditioning demand great care. The quality of the analytical results often depends largely on the skill and experience of the chemist and the attention he is willing to devote to each detail of the procedure.

Automatic Monitoring

Water monitoring data are still obtained widely by manually operated instruments in the laboratory using samples brought from the scene. Within the limits of current technology, however, automated procedures are growing steadily more common. Automated methods reduce costs by using manpower and equipment more efficiently. And they shorten the time between sampling and analysis, thus tending to minimize changes in the sample that make it nonrepresentative of the parent water and to reveal more quickly the need for abatement action, if indicated. Automated water monitoring systems usually are designed:

- To provide rapid acquisition of data, in part to help insure consistent compliance with standards of quality.
- To establish automatic control of treatment processes.
- To verify and provide data for predictive calculations of water quality, including computerized mathematical modeling to develop design and operational criteria for pollution control.

Most laboratory instruments can be automated to some degree, at least, but the two main approaches to automatic monitoring involve electrochemical transducers or sensors and wet chemical methods. In the first approach, the sensor is simply inserted into the sample (or, in effect, the body of water) without altering its physicochemical nature; in the second, chemicals are added to the sample and colorimetric measurements are made after the desired reactions have taken place. The principal restraints on these monitoring systems are the limited variety of sensors available and, in some cases, poor performance characteristics.

Transducer systems

Equipment based on tranducers or sensors usually can measure temperature, electrical conductivity, dissolved oxygen, turbidity, pH, sunlight intensity, chloride, oxidation-reduction potential, and alpha and beta radioactivity. Parameters that can be measured for specific purposes include fluoride, nitrate, cyanide, sulfide, and copper (119, 120, 130). Sensors also have been developed that combine immobilized enzymes with electrochemical sensors in the same unit; an instrument of this kind based on the enzyme cholinesterase, for example, can be used to monitor water for organophosphate and carbamate insecticides (131). Sensor-based systems can be designed to monitor rivers and streams continuously and unattended, except for periodic servicing, and to record and telemeter data to a central point, such as a computer center (Fig. 1). The sensors used include electrodes, thermistors, and photoelectric cells. In addition, the past decade has seen the development of selective-ion or chemical-sensing electrodes (132). They include sensors for calcium, mag-

FIGURE 1

Automatic water quality monitor installation[a]

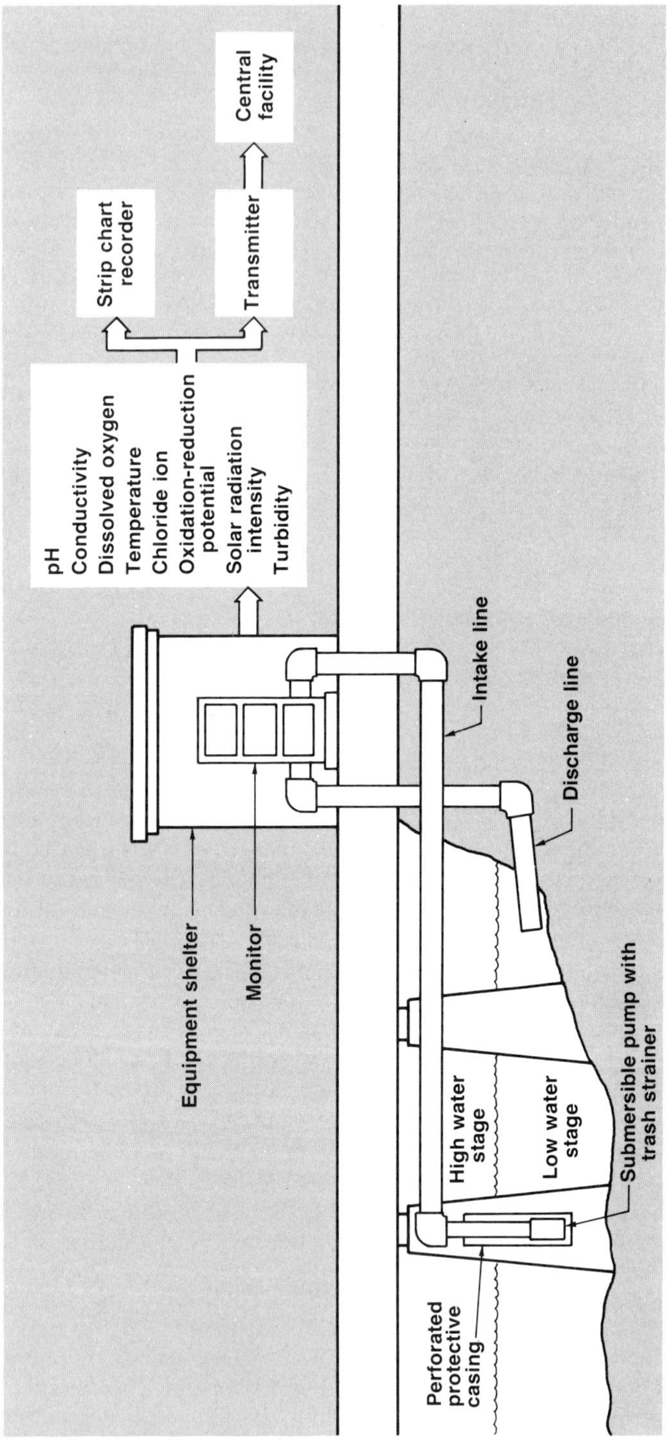

a. One of several possible configurations.

Source: "A Program Guide to Automated Instrumentation for Water Pollution Surveillance," U.S. Department of the Interior, Federal Water Pollution Control Administration, Cincinnati, Ohio, 1966.

nesium, chloride, cyanide, copper, fluoride, lead, nitrate, sulfide, and iodide. Besides the other uses of selective-ion electrodes, attempts have been made to incorporate them into water quality monitoring systems (130), but success thus far has been limited. The reliability of sensor-based equipment depends on factors that include proper calibration, service, and maintenance (133–143).

Wet chemical systems

The second main type of automatic monitoring system is based on automation of wet chemical analyses that traditionally have been performed manually. Such systems essentially can sample, filter, dilute, add reagents, mix, heat, and digest and, after waiting for color to develop from reactions in the sample, perform colorimetric measurements. All of these procedures are automated and applied to a succession of samples moved by a fixed-speed peristaltic pump. One of the main advantages of automated chemical systems is their ability to measure many different parameters. They are used most widely in laboratories that handle large numbers of samples of water daily (144–149). Attempts have been made to use such equipment, housed in a trailer on a river bank or on board ship, to monitor water quality, but this application faces limitations; turbidity in the water interferes with colorimetric measurements; more frequent servicing is required than for transducer-based equipment; and capital and operating costs tend to be higher than for transducer systems.

Combined systems

Certain sensor systems can be adapted to use in monitoring by adding chemicals to the sample of water, on-stream, to adjust the pH or mask troublesome interferences. The combined system can be fully automated. Typical applications include monitoring metals in water supplies in the range of parts per billion (130). These systems usually are tailored to a specific type of water and often are used to monitor certain industrial wastewaters.

Research in Water Analysis

A notably neglected area of water analysis is identification of the form or individual species in which an element or compound is present. Organic contaminants usually are identified by class and often as specific compounds. Inorganic pollutants, however, are measured typically as total element; little or nothing is learned of other important characteristics. An element may exist in water in ionic form, for example, in several oxidation states; it may be contained in a variety of complexes; it may be a constituent of a number of compounds suspended in particulate form. A case in point is phosphorus (150), which improved analytical techniques might distinguish in water in at least four different forms; in true solution as orthophosphate; as a component of suspended mineral material; adsorbed on suspended material; and in soluble organic material. More should be known particularly of the availability of adsorbed and organic phosphorus as a plant nutrient. Means of making such distinctions for various elements are still relatively rare, yet the environmental significance of the possible forms of an element in water may vary widely.

The identification and measurement of limiting nutrients in eutrophication would be eased by better methods not only for phosphorus but also for low levels of the possible forms of nitrogen (151) and for trace metals and trace organic growth factors in water. Improved procedures are required also for traces of organic compounds that may be important in three problem areas: toxicity to man and animals; color; and taste and odor (152, 153). Current methods of concentrating samples for organic analyses, such as carbon adsorption and solvent extraction, have limitations that need to be understood more fully; improved techniques should be sought directly for identifying and measuring organic contaminants in water at concentrations of from milligrams per liter down to micrograms per liter or lower. The past few years have seen steady progress toward these goals, especially with chromatography and mass spectrometry. Even these tools, however, are often not sensitive enough to determine organic contaminants at the levels at which they occur in natural waters and wastewaters. Preliminary concentration procedures are needed that alter neither the organic compounds to be measured nor their relative distributions in complex mixtures. Techniques used in biochemical analysis can perhaps be adapted to analysis of trace organics in water.

Among more specific requirements are a simple, reliable, specific test for hydrogen sulfide and its ionic forms in water; more sensitive and quantitative techniques for measuring turbidity; an organic nitrogen method to replace the one published by Kjeldahl in 1883; and a means of measuring oxygen demand in undisturbed lake bottoms (154).

Interest has grown rapidly in the condition of estuaries in this country, but efforts to determine the quality of estuarine waters have been hampered by lack of suitable analytical methods (155). Such methods must be applicable to turbid waters of high and variable salinity, containing many substances in far higher concentrations than are found in the open ocean.

There is a pressing need to develop sensor systems for a number of substances: phosphates; traces of metals such as lead, copper, cadmium, arsenic, zinc, and mercury; and traces of organic materials like pesticides, phenols, and carcinogens. Also lacking is a means of monitoring microorganisms (156). For automatic water monitoring in general, urgent requirements include much broader exchange of information on current practices and the establishment of a data bank on performance characteristics under various field conditions.

The following concluding Recommendations apply generally to the environment, not to water alone.

Recommendation AM7: *Sharply increased financial support should be given to research and development on new and improved analytical methods with the sensitivity, specificity, and response time needed to pursue environmental programs effectively. Support should be strengthened correspondingly for more fundamental studies of new approaches to the measurement of pollutants.*

Recommendation AM8: *The adequacy of sampling procedures, including positional and temporal factors, must be evaluated critically in all analytical and monitoring work related to the environment. Representative and valid samples, and the maintenance of their integrity, are essential to the development of analytical data that accurately portray the condition and dynamics of all phases of the environment.*

Recommendation AM9: *Increased attention should be given to the development of methods for determining the specific compounds and their levels of concentration in environmental samples. Many of the methods used widely at present* *are not sufficiently specific and do not generally provide adequate information on the particular species present in such samples.*

Recommendation AM10: *Continued development of accurate and reliable means of compound class analysis should be recognized as still an important need. Such methods — total organic chlorine, total aldehydes, total organic carbon, total polynuclear aromatic hydrocarbons, etc. —are complementary to more specific methods.*

Recommendation AM11: *More research should be conducted on the identity and concentration of naturally-occurring compounds in air and water from pristine areas to improve assessments of the impact of contamination by man.*

Recommendation AM12: *Development of simple, low-cost analytical instruments for large-volume use should proceed apace with work on more sophisticated research instrumentation.*

LITERATURE CITED

1. Griefer, B., Taylor, J.K., "Pollutant Analysis Cost Survey," EPA-650/2-74-125, Environmental Protection Agency, Washington, D.C., 1974.
2. "Guide for Measures of Precision and Accuracy," *Anal. Chem.*, **40**, 2271 (1968).
3. "Official Methods of Analysis,—A.O.A.C.," 12th ed., Association of Official Analytical Chemists, Washington, D.C., 1975.
4. "Manual for Conducting an Interlaboratory Study of a Test Method," ASTM Special Technical Publication No. 355, American Society for Testing and Materials, Philadelphia, Pa., 1963.
5. "Collaborative Study Procedures of the Association of Official Analytical Chemists," *Anal. Chem.*, **50**, 337A (1978).
6. McFarren, E.F., Lishka, R.J., Parker, J.H., "Criterion for Judging Acceptability of Analytical Methods," *Anal. Chem.*, **42**, 358 (1970).
7. Cali, J.P., et al., "Standard Reference Materials: The Role of SRM's in Measurement Systems," Special Publication 260-46, National Bureau of Standards, U.S. Department of Commerce, Washington, D.C., 1974.
8. Cali, J.P., "The NBS Standard Reference Materials Program: An Update," *Anal. Chem.*, **48**, 802A (1976).
9. Sievers, R.E., et al., "Environmental Trace Analysis of Organics in Water by Glass Capillary Column Chromatography and Ancillary Techniques," *J. Chromatog.*, **142**, 745 (1977).
10. Saltzman, B.E., Burg, W.R., "Air Pollution," *Anal. Chem.*, **49**, 5, 1R (1977).
11. Hollowell, C.D., McLaughlin, R.D., "Instrumentation for Air Pollution Monitoring," *Environ. Sci. Technol.*, **7**, 1011 (1973).
12. *Fed. Regist.*, **36**, 15486, Aug. 14, 1971.
13. "A Working Document for Air Pollution Measurement Techniques Development, Fiscal Years 1972–1977," Prepared under Contract No. 22-69-154 by ESSO Research and Engineering Company for the Environmental Protection Agency, December 1971, NTIS No. EPA-R4-73-015, National Technical Information Service, Springfield, Va.
14. "Air Quality Criteria for Sulfur Oxides," National Air Pollution Control Administration, Washington, D.C., Publication No. AP-50, April 1970.
15. "Air Quality Criteria for Particulate Matter," National Air Pollution Control Administration, Washington, D.C., Publication No. AP-49, January 1969.

16. "Air Quality Criteria for Carbon Monoxide," National Air Pollution Control Administration, Washington, D.C., Publication No. AP-62, March 1970.

17. "Air Quality Criteria for Photochemical Oxidant," National Air Pollution Control Administration, Washington, D.C., Publication No. AP-63, March 1970.

18. "Air Quality Criteria for Hydrocarbons," National Air Pollution Control Administration, Washington, D.C., Publication No. AP-64, March 1970.

19. "Air Quality Criteria for Nitrogen Oxides," EPA Air Pollution Control Office, Washington, D.C., Publication No. AP-84, January 1971.

20. Stevens, R.K., O'Keeffe, A.E., "Modern Aspects of Air Pollution Monitoring," *Anal. Chem.*, **42,** 143A (1970).

21. Stevens, R.K., Hodgeson, J.A., "Applications of Chemiluminescent Reactions to the Measurement of Air Pollutants," *Anal. Chem.*, **45,** 443A (1973).

22. "Developments in Sampling and Analysis Instrumentation for Stationary Sources," *J. Air Pollut. Control Assoc.*, **23,** 587 (1973).

23. "Continuous Monitoring of Stationary Air Pollution Sources," *J. Air Pollut. Control Assoc.*, **25,** 803 (1975).

24. Midgett, M.R., "How EPA Validates NSPS Methodology," *Environ. Sci. Technol.*, **11,** 655 (1977).

25. *Fed. Regist.*, **36,** 8186, April 30, 1971.

26. *Fed. Regist.*, **36,** 24876, Dec. 23, 1971.

27. *Fed. Regist.*, **39,** 9308, March 8, 1974.

28. *Fed. Regist.*, **38,** 8820, April 6, 1973.

29. *Fed. Regist.*, **37,** 24250, Nov. 15, 1972.

30. *Fed. Regist.*, **38,** 1254, Jan. 10, 1973.

31. *Fed. Regist.*, **38,** 1258, Jan. 10, 1973.

32. *Fed. Regist.*, **38,** 19088, July 17, 1973.

33. *Fed. Regist.*, **38,** 19050, July 17, 1973.

34. *Fed. Regist.*, **40,** 24350, June 5, 1975.

35. *Fed. Regist.*, **40,** 27574, June 30, 1975.

36. *Fed. Regist.*, **40,** 33152, Aug. 6, 1975.

37. *Fed. Regist.*, **40,** 43850, Sept. 23, 1975.

38. *Fed. Regist.*, **40,** 46240, Oct. 6, 1975.

39. *Fed. Regist.*, **40,** 48292, Oct. 14, 1975.

40. *Fed. Regist.*, **40,** 59532, Dec. 24, 1975.

41. *Fed. Regist.*, **41,** 2232, Jan. 10, 1976.

42. *Fed. Regist.*, **41,** 2332, Jan. 10, 1976.

43. *Fed. Regist.*, **41,** 3826, Jan. 26, 1976.

44. *Fed. Regist.*, **41,** 18498, May 4, 1976.

45. *Fed. Regist.*, **41,** 23059, June 8, 1976.

46. *Fed. Regist.*, **40,** 7042, Feb. 18, 1975.

47. Hoffman, A.J., et al., "EPA's Role in Ambient Air Quality Monitoring," *Science*, **190,** 243 (1975).

48. *Fed. Regist.*, **40,** 46240, Oct. 6, 1975.

49. "Annual Book of ASTM Standards," Part 26 (Gaseous Fuels; Coal and Coke; Atmospheric Analysis); Part 31 (Water), American Society for Testing and Materials, Philadelphia, Pa., 1976.

50. Intersociety Committee, "Methods of Air Sampling and Analysis," 2nd ed., American Public Health Association, Washington, D.C., 1977.

51. National Bureau of Standards, Tech. Note 585, Washington, D.C., January 1972.

52. Hughes, E.E., et al., "Performance of a Nitrogen Dioxide Permeation Device," *Anal. Chem.*, **49**, 1823 (1977).

53. Singh, H.B., et al., "Generation of Accurate Halocarbon Primary Standards with Permeation Tubes," *Environ. Sci. Technol.*, **11**, 511 (1977).

54. National Bureau of Standards, Tech. Note 505, Washington, D.C., October 1969.

55. *Fed. Regist.*, **38**, 15174, June 8, 1973.

56. "Collaborative Study of Reference Method for the Determination of Suspended Particulates in the Atmosphere (High Volume Method)," prepared for EPA by Southwest Research Institute under Contract No. CPA 70-40, June 1971, National Technical Information Service No. PB 205-892, Springfield, Va.

57. "Collaborative Study of Reference Method for the Measurement of Photochemical Oxidants Corrected for Interferences Due to Nitrogen Oxides and Sulfur Dioxide," prepared by Southwest Research Institute for EPA under Contract No. CPA 70-40, July 1973.

58. "Collaborative Study of Reference Method for the Continuous Measurement of Carbon Monoxide in the Atmosphere (Non-Dispersive Infrared Spectrometry)," prepared by Southwest Research Institute for EPA under Contract No. CPA 70-40, May 1972, National Technical Information Service No. PB 211-265, Springfield, Va.

59. "Collaborative Study of Reference Method for Determination of Sulfur Dioxide in the Atmosphere" (Pararosaniline Method), prepared by Southwest Research Institute for EPA under Contract No. CPA 70-40, September 1971, National Technical Information Service No. PB 205-893, Springfield, Va.

60. "Collaborative Study of Reference Method for Measurement of Photochemical Oxidants in the Atmosphere (Ozone-Ethylene Chemiluminescent Method)," prepared by Southwest Research Institute for EPA under Contract No. CPA 70-40, February 1975.

61. "1974 Annual Report of Committee D-22 on Sampling and Analysis of Atmospheres and Project Threshold," American Society for Testing and Materials, Philadelphia, Pa.

62. Chapman, R.L., "Continuous Stack Monitoring," *Environ. Sci. Technol.*, **8**, 520 (1974).

63. Hochheiser, S., Burmann, F.J., Morgan, G.B., "Atmospheric Surveillance: The Current State of Air Monitoring Technology," *Environ. Sci. Technol.*, **5**, 678 (1971).

64. "Development of a Nucleonic Particulate Emission Gauge," Final Report by Industrial Nucleonics for EPA Contract No. 68-02-0210, 1972, National Technical Information Service No. PB 209-954, Springfield, Va.

65. Macias, E.S., Husar, R.B., "Atmospheric Particulate Mass Measurement with Beta Attenuation Mass Monitor," *Environ. Sci. Technol.*, **10**, 904 (1976).

66. Nader, J.S., Allen, D.R., *Am. Ind. Hyg. Assoc. J.*, **21**, 300 (1960).

67. "Development of X-Ray Fluorescence Spectroscopy for Elemental Analysis of Particulate Matter in the Atmosphere and in Source Emissions," prepared by the Naval Research Laboratory under an Interagency Agreement with EPA, EPA-R2-72-063, October 1972.

68. Giauque, R.D., et al., "Trace Element Determination with Semiconductor Detector," *Anal. Chem.*, **45**, 671 (1973).

69. Goulding, F.S., Jaklevic, J.M., "X-Ray Fluorescence Spectrometer for Airborne Particulate Monitoring," EPA Report No. EPA-R2-73-182, April 1973.

70. Thompson, R.J., Morgan, G.B., Purdue, L.J., "Analysis of Selected Elements in Atmospheric Particulate Matter by Atomic Absorption," *At. Absorpt. Newsl.*, **9**, AA595 (1970).

71. Ranweiler, L.E., Moyers, J., "Atomic Absorption Procedure for Analysis of Metals in Atmospheric Particulate Matter," *Environ. Sci. Technol.*, **8,** 152 (1974).

72. Dams, R., et al., "Nondestructive Neutron Activation Analysis of Air Pollution Particles," *Anal. Chem.*, **42,** 861 (1970).

73. Hammerle, R.H., et al., "Test of X-Ray Fluorescence Spectrometry as a Method for Analysis of the Elemental Composition of Atmospheric Aerosols," *Anal. Chem.*, **45,** 1939 (1973).

74. Dzubay, T.G., Stevens, R.K., "Applications of X-Ray Fluorescence to Particulate Measurements," Proceedings of the Second Joint Conference on the Sensing of Environmental Pollutants, Washington, D.C., December 1973. Instrument Society of America, Pittsburgh, Pa., 1973, pp. 211–216.

75. Mueller, P.K., Kothny, E.L., "Air Pollution," (Analytical Reviews), *Anal. Chem.*, **45,** 5, 1R (1973).

76. Stevens, R.K., O'Keeffe, A.E., Ortman, G.C., "Absolute Calibration of a Flame Photometric Detector to Volatile Sulfur Compounds at Sub-Part-Per-Million Levels," *Environ. Sci. Technol.*, **3,** 652 (1969).

77. Stevens, R.K., et al., "Gas Chromatography of Reactive Sulfur Gases in Air at the Parts-Per-Billion Level," *Anal. Chem.*, **43,** 827 (1971).

78. "Monitoring Instrumentation for the Measurement of Sulfur Dioxide in Stationary Source Emissions," TRW, Inc., Contract Report for EPA Contract No. EHSD 71-23, EPA R2-73-163, 1973, National Technical Information Service No. PB 220-202, Springfield, Va..

79. Hauser, T.R., Shy, C.M., "Position Paper: NO_x Measurement," *Environ. Sci. Technol.*, **6,** 890 (1972).

80. Christie, A.A., Lidzey, R.G., Radford, D.W.F., *Analyst*, **95,** 519 (1970).

81. Salzman, B.E., "Colorimetric Micro Determination of Nitrogen Dioxide in the Atmosphere," *Anal. Chem.*, **26,** 1949 (1954).

82. Mulik, J., et al., "New Methods for the Collection and Analysis of Atmospheric NO_2," presented at the 165th National Meeting of the American Chemical Society, Dallas, Tex., April 1973.

83. *Fed. Regist.*, **41,** 52686, Dec. 1, 1976.

84. Sigsby, J.E., et al., "Chemiluminescent Method for Analysis of Nitrogen Compounds in Mobile Source Emissions (NO, NO_2, and NH_3)," *Environ. Sci. Technol.*, **7,** 51 (1973).

85. Snyder, A.D., et al., "Instrumentation for the Determination of Nitrogen Oxides Content of Stationary Source Emissions," Monsanto Research Corp., Contract Report, EPA Contract No. EHSD 71-30, National Technical Information Service PB Nos. 204-877 and 209-190, Springfield, Va.

86. Okita, T., Lodge, J.P., Jr., Axelrod, H.D., "Filter Method for the Measurement of Atmospheric Hydrogen Sulfide," *Environ. Sci. Technol.*, **5,** 532 (1971).

87. Natusch, D.F.S., et al., "Sensitive Method for Measurement of Atmospheric Hydrogen Sulfide," *Anal. Chem.*, **44,** 2067 (1972).

88. Ross, W.D., Pyle, J.L., Sievers, R.E., "Analysis for Beryllium in Ambient Air Particulates by Gas Chromatography," *Environ. Sci. Technol.*, **11,** 467 (1977).

89. "Development of a Method for the Determination of Asbestos in Ambient Air," prepared by Battelle Memorial Institute under Contract No. CPA 22-69-110 to EPA, August 1971.

90. Spurny, R.K., et al., "The Sampling and Electron Microscopy of Asbestos Aerosol in Ambient Air by Means of Nucleopore Filters," *J. Air Pollut. Control Assoc.*, **26,** 496 (1976).

91. *Fed. Regist.*, **41,** 46560, Oct. 21, 1976.

92. "Spectroscopic Methods for Air Pollution Measurement," P.L. Hanst, *Adv. in Environ. Sci. Technol.*, J.N. Pitts, R.L. Metcalf, eds., John Wiley & Sons, Inc., New York, N.Y., 1971.

93. Altshuller, A.P., "Gas Chromatography in Air Pollution Studies," *J. Gas Chromatogr.*, **1,** 6 (1963).

94. Tucker, A.W., Peterson, A., Birnbaum, M., "Fluorescence Determination of Atmospheric NO and NO_2," *Appl. Opt.*, **12,** 2036 (1973).

95. Okabe, H., Splitstone, P.L., Ball, J.J., "Ambient and Source SO_2 Detector Based on a Fluorescence Method," *J. Air Pollut. Control Assoc.*, **23,** 514 (1973).

96. McClenny, W.A., Hodgeson, J.A., Bell, J.P., "Photofragment Detection of Nitrogen Dioxide," *Anal. Chem.*, **45,** 1514 (1973).

97. Ludwig, C.B., Bartle, K., Griggs, M., "Study of Air Pollutant Detection by Remote Sensors," NASA Report, National Technical Information Service Publication No. N69-31961, Springfield, Va., 1971.

98. Herget, W.F., Conner, W.D., "Instrumental Sensing of Stationary Source Emissions," *Environ. Sci. Technol.*, **11,** 962 (1977).

99. "Development of a Gas Laser System to Measure Trace Gases by Long Path Absorption Techniques," prepared by General Electric Co., Ordnance Systems, for EPA under Contract No. CPA 68-02-0757, February 1972.

100. McClenny, W.A., et al., "Methodology for Comparison of Open-Path Monitors with Point Monitors," *J. Air Pollut. Control Assoc.*, **24,** 1044 (1974).

101. Hinkley, E.D., "Bistatic Monitoring of Gaseous Pollutants with Tunable Semiconductor Lasers," Symposium on Remote Sensing of Environmental Air Pollutants, Pittsburgh Conference on Analytical Chemistry and Applied Spectroscopy, Cleveland, Ohio, March 1974.

102. McClenny, W.A., Herget, W.F., Stevens, R.K., "A Comparative Review of Open-Path Spectroscopic Absorption Methods for Ambient Air Pollutants," in "Analytical Methods Applied to Air Pollution Measurement," Ann Arbor Science Publishers, Ann Arbor, Mich., 1974, Chap. 7.

103. Sawicki, E., "Analysis of Atmospheric Carcinogens and Their Cofactors," Symposium on Environmental Pollution and Carcinogenic Risks, Lyons, France, Nov. 3–5, 1975.

104. Pellizzari, E.D., et al., "Determination of Trace Hazardous Organic Vapor Pollutants in Ambient Atmospheres by Gas Chromatography/Mass Spectrometry/Computer," *Anal. Chem.*, **48,** 803 (1976).

105. Fine, D.H., et al., "Determination of Dimethylnitrosamine in Air, Water and Soil, by Thermal Energy Analysis: Measurements in Baltimore, Md.," *Environ. Sci. Technol.*, **11,** 581 (1977).

106. *Fed. Regist.*, **38,** 18985, July 16, 1973.

107. Axelrod, H.D., et al., "A System for Background and Regional Monitoring," in Special Environmental Report No. 3, Observation and Measurement of Atmospheric Pollution, World Meteorological Organization, Geneva, Switzerland, 1974.

108. Shendrikar, A.D., Lodge, J.P., Jr., "Microdetermination of Ammonia by the Ring Oven Technique and Its Application to Air Pollution Studies," *Atmos. Environ.*, **9,** 431 (1975).

109. Mancy, K.H., Chapter 6 in "Analytical Chemistry: Key to Progress on National Problems," National Bureau of Standards Special Publication 351 (1972).

110. Chermisinoff, P.N., *Pollut. Eng.*, **6,** 36 (1974).

111. Fishman, M.J., Erdmann, D.E., "Water Analysis," *Anal. Chem.*, **49,** 5, 139R (1977).

112. "Standard Methods for the Examination of Water and Wastewater," 14th ed., American Public Health Association, Washington, D.C., 1975.

113. "Manual of Methods for Chemical Analysis of Water and Wastes, 1974," EPA-625/6-74-003a, Environmental Protection Agency, National Environmental Research Center, Cincinnati, Ohio.

114. *Fed. Regist.*, **38,** 28758, Oct. 16, 1973.

115. *Fed. Regist.*, **40,** 24535, June 9, 1975.

116. "Recommended Methods for Water Data Acquisition — Preliminary Report," Federal Interagency Workgroup on Designation of Standards for Water Data Acquisition, Office of Water Data Coordination, U.S. Geological Survey, Washington, D.C., 1972.

117. "Standardization in Hydrology and Related Fields," Report on WMO/IHD Projects, No. 18, World Meteorological Organization, Geneva, Switzerland, 1973.

118. "Report on Analytical Evaluation Program," U.S. Geological Survey, Water Resources Division, Lakewood, Colo., 1972.

119. Mancy, K.H., "Proceedings of World Health Organization Seminar on the Design of Environmental Information Systems," World Health Organization, Geneva, Switzerland, 1973.

120. Mancy, K.H., "Proceedings of the World Health Organization Symposium on Automatic Water Quality Monitoring in Europe," P.A. Krenkel, ed., **28**, 141 (1971), Vanderbilt University Press, Nashville, Tenn.

121. Mitchell, D.G., Jackson, K.W., Aldous, K.M., "Application of a Silicon-Target Vidicon Detector to Simultaneous Multielement Flame Spectrometry," *Anal. Chem.*, **45**, 1215A (1973).

122. "Identification and Analysis of Organic Pollutants in Water," L.H. Keith, ed., Ann Arbor Science Publishers, Inc., Ann Arbor, Mich., 1976.

123. Hoyland, J.R., Neher, M.B., Battelle Columbus Laboratory Annual Report to the U.S. Environmental Protection Agency, Project No. 16020 HGD, Battelle Columbus Laboratory, Columbus, Ohio, March 1973.

124. Kleopfer, R.D., Fairless, B.J., "Characterization of Organic Components in a Municipal Water Supply," *Environ. Sci. Technol.*, **6**, 1037 (1972).

125. Burnham, A.K., et al., "Identification and Estimation of Neutral Organic Contaminants in Water," *Anal. Chem.*, **44**, 139 (1972).

126. Gesser, H.D., et al., *Anal. Lett.*, **4** (12), 883 (1971).

127. Uthe, J.F., et al., *Environ. Lett.*, **3** (2), 117 (1972).

128. McAuliffe, C., "GC Determination of Solutes by Multiple Phase Equilibration," *Chem. Tech.*, **1**, 46 (1971).

129. Crummett, W.B., Stehl, R.H., *Env. Health Perspectives*, September 1973, pp. 15–25.

130. McClelland, N.I., Mancy, K.H., *J. Am. Water Works Assoc.*, **64**, 795 (1972).

131. Gray, D.N., Keyes, M.H., Watson, B., "Immobilized Enzymes in Analytical Chemistry," *Anal. Chem.*, **49**, 1067A (1977).

132. Frant, M.S., "Detecting Pollutants with Chemical-Sensing Electrodes," *Environ. Sci. Technol.*, **8**, 224 (1974).

133. Durst, R.A., "Determination of Fluoride by Analate Additions Potentiometry," *Mikrochim. Acta.*, **3**, 611 (1969).

134. Riseman, J.M., "Specific Ion Electrodes as Transducers in Continuous Monitoring Applications," in "Water Quality Instrumentation," Vol. 1, J.W. Scales, ed., Instrument Society of America, Pittsburgh, Pa., 1972.

135. Oliver, R.T., Mannion, R.F., "Analysis Instrumentation," Vol. VIII-3, Plenum Press, New York, N.Y., 1971, pp. 1–7.

136. Bunton, N.G., Crosby, N.T., *Water Treat. Exam.*, **18**, 338 (1969).

137. Langmuir, D., Jacobson, R.L., "Specific-Ion Electrode Determination of Nitrate in Some Freshwaters and Sewage Effluents," *Environ. Sci. Technol.*, **4**, 834 (1970).

138. Mahendrappa, M.K., "Determination of Nitrate Nitrogen in Soil Extracts Using a Specific Ion Activity Electrode," *Soil Sci.*, **108**, 132 (1969).

139. Williams, L.M., *Tetracol.*, **8**, 41 (1969).

140. Collis, D.E., Diggens, A.A., *Water Treat. Exam.*, **18**, 192 (1969).

141. Crosby, N.T., "Equilibria of Fluorosilicate Solutions. Fluoridation of Public Water Supplies," *J. Appl. Chem.*, **19**, 100 (1969).

142. Light, T.S., in "Ion Selective Electrodes," R.A. Durst, ed., Spec. Publ. 314, National Bureau of Standards, U.S. Department of Commerce, Washington, D.C., 1969, Chap. 10.

143. Riseman, J.M., *Am. Lab.*, July 1969, p. 32.

144. O'Brien, J., Fiore, J., *Wastes Eng.*, March 1962, p. 2.

145. Smith, R.V., Ciaccio, L.L., Lipchus, R.L., "Determination of Potassium in Food Using the Technicon AutoAnalyzer Flame Photometer and Solid Prep Sampler," *J. Agr. Food Chem.*, **15**, 408 (1967).

146. "Advances in Automated Analysis," Vol. I, II, Halos, Miami, Fla., 1970.

147. Anderson, N.G., "Computer Interfaced Fast Analyzers," *Science*, **166**, 317 (1969).

148. Blaedel, W.J., Hicks, G.P., in "Advances in Analytical Chemistry and Instrumentation," C.N. Reilley, ed., Vol. 3, Wiley-Interscience, New York, N.Y., 1964, pp. 105–140.

149. DuCros, M.J.F., Salpeter, J., "Automated Methods for Assessing Water Quality Come of Age," *Environ. Sci. Technol.*, **9**, 929 (1975).

150. Lee, G.F., Clesceri, N.L., Fitzgerald, G.P., "Studies on the Analysis of Phosphates in Algal Cultures," *Int. J. Air Water Pollut.*, **9**, 715 (1965).

151. Bankert, S.F., Bloom, S.D., Dietrich, F.S., "Rapid Determination of Very Low Nitrogen Levels in Water," *Nature*, **242**, 270 (1973).

152. Baker, R.A., "Trace Organic Analyses by Aqueous Gas-Liquid Chromatography," *Int. J. Air Water Pollut.*, **10**, 591 (1966).

153. Baker, R.A., Malo, B.A., in "Proceedings, the National Symposium on Quality Standards for Natural Waters," Continued Education Series No. 161, School of Public Health, University of Michigan, Ann Arbor, Mich., 1966, p. 275.

154. Stein, J.E., Denison, J.G., "In Situ Benthic Oxygen Demand of Cellulosic Fibers," paper presented at 3rd International Conference, Munich, Germany, 1966.

155. Jenkins, D., "Analysis of Estuarine Waters," *J. Water Pollut. Control Fed.*, **39**, 159 (1967).

156. Oleniacz, W.S., et al., "Chemiluminescent Method for Detecting Microorganisms in Water," *Environ. Sci. Technol.*, **2**, 1030 (1968).

Chapter 3

Toxicology

3 / Toxicology

INTRODUCTION

Toxicology is a central element of virtually all of the problems associated nowadays with improving our environment. The science in a very real sense provides a bridge between the chemical and the biological worlds, and we must comprehend the strengths and weaknesses of that bridge before entrusting to it the weight of decisions that will affect not only our own lives, but those of future generations as well.

The traditional definition of toxicology as the science of poisons clearly is no longer adequate. All substances, even essential nutrients like vitamins and amino acids, are poisons when taken in too large a quantity or in the wrong way. The technical committee of the Society of Toxicology has defined the field as the science that deals with the adverse effects of useful and noxious chemicals. This definition is useful since it emphasizes the role of the toxicologist in studying the deleterious effects of chemical agents on biological systems. Toxicology deals also, however, with other aspects of useful and harmful substances: origin, properties, physiological actions and symptoms, lethal dose, antidotes, specific identification, and quantitative measurements and their evaluation and interpretation.

The toxicologist is concerned broadly with "toxicity" and with the evaluation of "hazard" or "risk." Toxicity refers to the inherent ability of a compound to produce adverse symptoms or death in a living organism. Hazard refers to the probability that an organism will be harmed by exposure to a substance in a given quantity and manner. Risk refers to the probability that a given use of a hazardous substance will result in harm. Any substance, as we have noted, can be toxic when administered in sufficient amount and by the appropriate route. But this does not mean that the amounts of the substance to which an organism normally may be exposed pose a significant hazard or, indeed, any hazard at all.

Much of the toxicologic research of recent years has been designed to determine the amounts of various chemical agents that can be added safely to the environment for some useful purpose. Among these agents are drugs, food additives, and pesticides. Additionally, however, the toxicologist's work encompasses agents like air and water pollutants that enter the environment inadvertently. Further, it involves the assessment of hazard not only to man, but also to other organisms: mammals, reptiles, birds, insects, bacteria, plants. Links between organisms may be important. The toxicologist may wish to know, for example, whether a given water pollutant is biologically magnified as it moves up a food chain, such as plankton-fish-bird, that may or may not end with man. More broadly, he must consider the toxicologic significance of the contributions of all sources of the substance of interest, as well as the hazard posed by the substance in the various forms that it may assume during its lifetime in the environment.

Recommendation T1: *The significance of human exposure to environmental pollutants should be evaluated in terms of the contributions of all sources of the substances of interest, including food, air, and water, rather than by focusing separately on each such medium. Similarly, evaluation of the hazard of a given substance should span its lifetime, from the time it enters the environment until it is finally disposed of or converted to innocuous form.*

Given the spectrum of interests in toxicology, it is not surprising that the field has evolved as a multidisciplinary science. Because of this multidisciplinarity, there probably exists no single satisfactory formal educational program in toxicology, broadly defined. By the same token, practitioners of the science often are classified in terms of their specialties. Some chemists, for example, can be classified as forensic toxicologists by virtue of their specialized training in analytical chemistry and the law, which enables them to work effectively in establishing agents responsible for injury and death for medical-legal purposes. Physicians who are concerned with the adverse effects of drugs and chemical agents and with acute poisoning may be classified as clinical toxicologists. Scientists who are involved in industrial hygiene and safety may be classified as occupational toxicologists. And a growing number of scientists involved in regulatory and other aspects of environmental control might be called environmental toxicologists.

The unifying theme that links all of these specialists is their concern with evaluating risk in some type of biological system. Prediction of risk is a fundamental responsibility of the toxicologist regardless of his primary orientation. He may be attempting to identify a hazard in food or water, to evaluate a new drug, to determine a cause of death or injury, but he is seeking in the end to assess the risk to a biological system exposed to a chemical agent.

Legislative background

The demands on toxicology have expanded rapidly during the past two decades with the passage of a succession of federal laws designed to upgrade the quality of many aspects of the environment. The Food, Drug, and Cosmetic Act of 1938 was broadened and strengthened by the Miller Pesticide Amendment of 1954, the Food Additives Amendment of 1958, the Color Additive Amendment of 1960, and the Kefauver-Harris Amendments of 1962. Other pertinent measures include the Clean Air Act of 1970, the Occupational Safety and Health Act of 1970, the Consumer Product Safety Act of 1972, the Environmental Pesticide Control Act of 1972, and the Water Pollution Control Act of 1972. Most recently have come the Toxic Substances Control Act of 1976 and various amendments to earlier legislation.

Such legislation, where it raises toxicologic issues, usually embodies the risk-benefit concept. That is, in making regulatory decisions, the responsible parties must properly balance the risks of using a pesticide, for example, against the benefits that its use might bring, or the risks of allowing a pollutant to enter the environment at a given level against the cost of reducing that level. The toxicologist's task is to provide the prediction of risk required to support the decision-making process.

We will not focus here on the biologic effects of specific pollutants. We will consider rather a more fundamental issue: the place of toxicology as a science in decision-making processes that involve links between chemicals and health. To do this, we must deal with certain questions:

- How valid are the tools that the toxicologist uses to formulate his predictions of risk?
- How are such predictions used in making regulatory decisions, and how can the process be facilitated or improved?
- What are the high-priority needs of toxicology itself in its role as a bridge between chemicals and biological systems, and how might they best be met?

THE TOOLS OF THE TOXICOLOGIST

Predictions of toxicologic hazard to humans are derived for the most part from controlled studies in animals. The few exceptions involve substances whose effects already are well understood. Prescription drugs, for example, must be tested in humans for safety and efficacy before marketing, but such tests are preceded by extensive testing in animals. Pesticides that act by inhibiting the enzyme cholinesterase, a reversible effect, may be tested at subacute levels in humans to confirm the results of animal studies. The behavioral effects of carbon monoxide inhalation, which also are reversible, may be studied in humans as well. There are other instances, but on the whole, toxicologic studies are not conducted with human subjects. Testing any chemical in humans poses a degree of risk; excepting drugs, where large potential benefits may exist, the risk usually is considered unreasonable. In addition, it is generally impractical to do highly controlled toxicologic experiments — lifetime feeding studies, for example — with human subjects. The only alternative is to work with animals that respond to the substance in question as much like humans as possible. In this vein it should be noted that toxicologic evidence can be obtained directly from all species but man; the problems of extrapolating from one species to another arise only when man is the subject of interest.

Basic to the process of evaluating risk is the concept of dose response. An animal's (and a human's) response to a chemical can vary significantly, both in and among individuals of the species. Because of the evidence from studies of many chemicals, however, the toxicologist assumes as a rule that the response to a new compound will intensify in a predictable way as the dose or level of exposure increases. For the same reason, he assumes that some range of gradually increasing doses of any chemical agent will produce no measurable adverse response in a given species, but that as the dose continues to rise a toxic level can be reached. The goal of an animal study is to determine in the laboratory the dosage and conditions of exposure that will produce the toxic effect on the species and to use this information to predict what may occur in real-life situations in the same or another species, including humans.

Toxicologic studies in animals include the determination of acute toxicity (single-dose studies), subchronic toxicity (repeated-dose studies), and chronic toxicity (long-term studies). They include also a variety of special studies designed to determine whether a chemical is mutagenic, teratogenic, or carcinogenic or can produce cataracts, decreased fertility, or other adverse effects.

The animals in these kinds of studies usually include "controls," animals that are not exposed to the test substance. Controls are employed to help confirm that effects seen in exposed animals really are due to that substance. Laboratory rats, for example, may develop tumors spontaneously. Thus the appearance of tumors in rats exposed to a given compound may be simply coincidental. But the appearance of significantly more tumors in an exposed group than in an unexposed but otherwise comparable group — the controls — provides much better evidence that the compound is tumorigenic. Controls may be required for more than one variable in an experiment. Where a substance is fed to animals in a formulation with other ingredients, one group of controls might receive those ingredients but not the test substance, while a second group receives neither. In any event, the basic diet, housing, and other general conditions of an experiment must be the same for all animals involved, including the controls, if the results are to be considered valid.

The intended application of a prospective commercial chemical agent, and the way in which it will be used in that application, determine in part the kinds of animal studies that are performed. A potential food additive or drug, for example, may be subjected to a more extensive battery of animal studies than a chemical to be used in a manufacturing process. The toxicity of a food additive when taken orally may be studied more thoroughly than its toxicity when inhaled, applied to the skin, injected intravenously, or administered by some other route. The same principles apply to studies of compounds that enter the environment inadvertently. It is evident, therefore, that toxicity testing cannot be a cookbook procedure if it is to establish the relative safety or hazard of differing substances under widely varying conditions. No single protocol or set of protocols can be universally applicable; professional judgment is an essential ingredient of every study.

Acute studies

The acute toxicity of a chemical agent is determined by first exposing each of a group of animals to a single dose and observing the types of effects produced and their durations. [Appropriate numbers of control (unexposed) animals may be included in the experiment.] The study is continued with a series of graded dosage levels to determine the minimum dose required to produce each of the toxic manifestations observed.

Because of variation within species, the minimum dose statistically most likely to produce a given symptom in random individuals is the one that produces it in 50% of an infinite number of the species. This median dose must be estimated by working with a practical number of animals — on the order of 5 to 10 per dose level and 50 to 100 total — and using statistical methods to interpret the results. One such method used commonly is probit (probability unit) transformation, which converts the test data into a dose-response curve (Fig. 1). If death is the response measured, the curve will show the median lethal dose (LD50), the amount that most likely will result in 50% mortality. Similarly, if the response measured is a toxic effect or symptom short of death, the curve will show the median effective dose (ED50). Usually, the LD50 or ED50 is expressed as grams or milligrams of toxicant administered per kilogram of body weight to the test animal (g/kg or mg/kg); also used is milligrams of toxicant per square meter of surface of the body. The LD50's of various organic compounds, administered orally to rats and through the skin to rabbits, appear in Table 1. The average values illustrate the variability of the lethal dose of different chemicals in animals of the same species; the ranges of values illustrate the variability of the lethal dose of a single chemical among animals of the same species.

Acute toxicity studies usually are performed first in mice or rats because they are relatively small and inexpensive. Later the studies may be extended to a larger species, such as dogs or monkeys, to obtain information on the variation of response between species. The results of such studies depend not only on variations within and between species, but also on many other factors: the test animals' age, sex, genetic background, nutritional state, and endocrine status (pregnancy, etc.); the volume and concentration of the formulation; the vehicle used to prepare the chemical agent for formulation; the route and rate of administration; and a variety of environmental factors, such as temperature, pressure, humidity, and previous exposure to other chemicals or environmental agents. The results depend also on how the animals are housed, the conditions of handling, the season of the year, and other factors. These factors cannot be duplicated exactly between labs or even in the same lab and so may affect the reliability of the studies significantly. Despite these difficulties, the LD50 and, to a lesser extent, the slope of the probit-transformation (dose-response) curve provide the single most reliable and commonly used indication of toxicity.

Subchronic studies

The subchronic toxicity test is similar to the acute test except that it involves multiple doses over periods of from several days to months. Rats and dogs are the species used most commonly for these studies, and usually both males and females are used. The range of dosage levels includes the vehicle only (for the

FIGURE 1

Portrayal of dose-response information

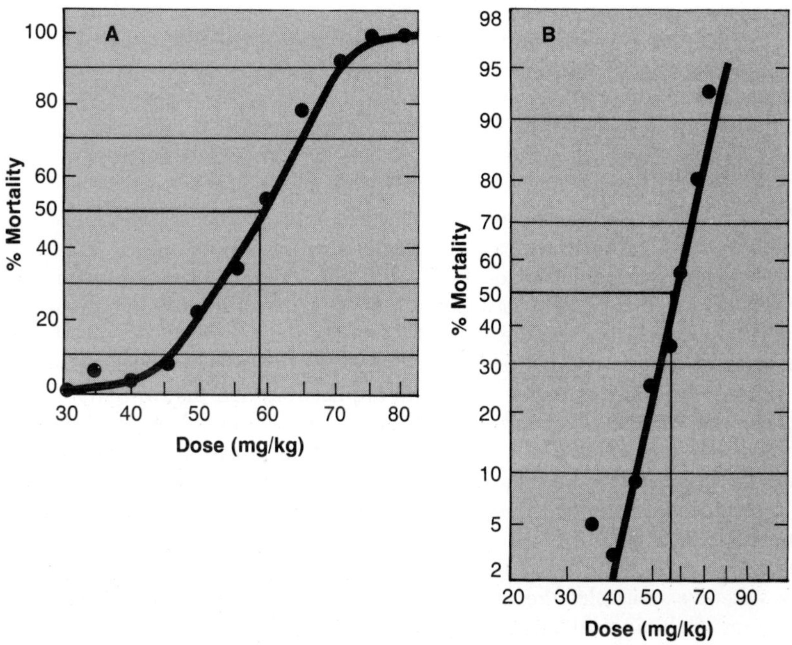

The two plots show different means of portraying dose-response data for a compound administered orally to mice. The response measured is death, and the LD50 is 58 mg/kg. In A, a smooth curve is obtained by plotting the data as a frequency distribution. In B, a straight line is obtained by plotting dose on a logarithmic scale and mortality on a probit scale. The purpose of these maneuvers is to convert the raw data into more understandable form.

control group), a no-observed-effect dose, a clearly toxic or lethal dose, and several intermediate doses. Various evaluative procedures are conducted frequently throughout the experimental period; they include observation of clinical symptoms and of hematological and biochemical parameters. All of the animals that die during the study, and those that are sacrified at its conclusion, customarily are examined thoroughly, both grossly and microscopically. In most such studies, the oral route of exposure is used because of the ease of adding the experimental agent to the diet. Inhalational, dermal, or parenteral exposures can be used, however, if they are the potentially hazardous routes.

Chronic studies

The third major type of study, for chronic toxicity, involves exposure periods of 1.5 to two years. For some strains of rats and mice a test that long may constitute essentially a lifetime study, which is appropriate for substances such as insulin or food additives, to which humans may be exposed for a major part of their lives. Because chronic toxicity tests are so costly and time-consuming, it is critical that the proper doses be selected for them. Chronic studies usually are not initiated, therefore, until all possible dosage-related information has been obtained from the acute and subchronic studies. A major goal of chronic studies is to detect potential carcinogenic effects of the experimental agent, but the basic purpose is to detect all chronic effects. The chronic test should focus not only on carcinogenicity, therefore, but should involve complete pathological evaluation with detailed observations of clinical symptomatology, hematology, and biochemistry.

Study guidelines

Toxicity testing, as mentioned previously, cannot be a cookbook procedure; it requires a high level of professional knowledge and judgment. Studies of the validity and reproducibility of the results of some tests of acute toxicity and irritation in animals (1, 2) have shown significant variability. The results of subchronic and chronic studies doubtless would display even greater variability unless the studies were designed and conducted properly. Experiments on the toxicity and hazard of a given material require detailed planning by the toxicologists, biochemists, pathologists, geneticists, hematologists, biostatisticians, and others who are to take part. The experiments with animals concluded, the team must assess the relative safety or hazard of the test material for man. Any experiment intended to support such assessments must be designed and interpreted in accord with guidelines (3, 4) generally agreed upon by a group of toxicologists. Briefly, these guidelines contemplate:

- Wherever feasible, the use of test species that handle the substance in question biologically as much like man as possible.
- The use of the same routes of administration in test animals as are to be used in man.
- The use of several dose levels, where practical, on the grounds that all toxicologic and pharmacologic actions in man are dose-related.
- The presumption that, for any material and adverse effect in man or animal, some dose level exists below which the adverse effect will not appear.
- The application of valid statistical tests for significance to results in the randomized experimental animals.

The study guidelines must take into account the fact that different animals metabolize some chemicals differently. The detoxification of the resulting metabolites, or the biochemical enhancement of their actions, cause some of the biological reactions of man or other animals that might appear to be caused directly by the original chemicals. One indication of the applicability of the results of acute tests on animals to the prediction of hazard for man is whether the median effective doses (LD50's, for example) in different species are similar. If the metabolic fate of the chemical is not known, and the LD50's differ considerably among the species tested, the margin of safety for man must be based on the LD50 determined for the most sensitive species.

Tests of chronic toxicity in animals are intended to simulate the exposure of man to low levels of the chemical in question during large fractions of a lifetime. For this purpose only a few laboratory animals — including the rat, the mouse, and the hamster — are sufficiently short-lived to be practical experimental subjects. In addition to chronic toxicity studies, the biological fate of each material studied should be determined whenever possible. A substance that can produce chronic effects in humans cannot be relied on to produce observable symptoms in short-lived animals before they die of natural causes if they are exposed to it at the low levels anticipated for humans. Typically, therefore, the animals are exposed to higher doses so that the nature of the hazard, if any, can be defined. The rationale is that chronic effects produced by high doses in a short period may also be produced by much lower doses in a much longer period; similarly, if the high doses produce no chronic effects, none will result from low doses. Scientific experience indicates that this approach is generally sound. Again, however, careful experimental design and the use of valid statistical methods are essential if the approach is to yield results that can be extended with reasonable confidence to man.

Table 1

Comparison of LD50's for different chemicals (grams per kilogram)

Compound	Single oral dose, rats Avge (range)	Single skin penetration,[a] rabbits Avge (range)
4,4′-Biphenyldiol	9.85 (6.11–15.9)	1.78 (0.84–3.79)
Oxydiacetic acid, disodium salt	2.44 (1.91–3.12)	1.27 (0.78–2.07)
Toluene diamine, 2,3- and 3,4-isomers, mixed	0.81 (0.59–1.12)	1.12 (0.62–2.04)
Benzyl ammonium tetrachloro-iodate	1.23 (0.94–1.62)	0.84 (0.60–1.18)
Nitrosophenylhydroxylamine	0.49 (0.29–0.83)	6.40 (0.50–81.7)
Di-2-chloroethyl maleate	0.071 (0.056–0.090)	0.14 (0.06–0.32)
4-Thiocyanoaniline	0.24 (0.17–0.33)	0.16 (0.12–0.21)
Bis(triphenylsilyl)chromate	3.36 (1.35–8.40)	0.71 (0.32–1.54)

a. Applied dermally.

Source: Carpenter, C.P., Weil, C.S., Smyth, H.F., Jr., "Range-Finding Toxicity Data: List VIII," *Toxicol. Appl. Pharmacol.*, **28**, 313 (1974).

Other toxicity studies

The results of acute, subchronic, and chronic toxicity studies may lead the toxicologist to use additional procedures in his efforts to evaluate the risk to humans associated with a given chemical (5, 6). Some of these procedures were mentioned earlier. They include further investigations of carcinogenic potential as well as of mutagenic and teratogenic potential; of effects on fertility, reproduction, and lactation; and of effects on isolated tissues and organs and on organ systems. They may also include studies of the compound's toxicologic mechanism of action and kinetics, including absorption, metabolism, distribution, and excretion; the development of antidotes for overdoses; and any of a variety of other special studies. Some of these procedures are relatively new and consequently may not be officially required by the testing protocol being used. New procedures, furthermore, may not have been applied to compounds tested several years previously, which complicates toxicologic comparisons of new agents with structurally related ones tested in the past.

Mechanism of action. The behavior of a compound in the body — its mechanism of action — depends in part on its structure, so that structurally related compounds tend to behave similarly. Results obtained with a particular chemical, therefore, can provide useful guidelines for investigating the toxicity of a new chemical of the same structural class. The structure-activity relationship may hold only within wide limits, however. More broadly, the ability to predict the toxic potential of chemical agents of any type requires much improved knowledge of mechanisms of action. One needs to know, for example, if a compound acts directly on one or another enzyme, if it is converted in the body to more or less toxic metabolites, and if it tends to bind to particular molecules in the body. Deeper understanding of mechanisms of action has a number of potential benefits, including improved tests for toxicity, antidotes for toxic chemicals, and more reliable extrapolations of dose-response relationships from laboratory animals to man.

Organ systems. Toxic agents typically affect various "target" organs, directly or indirectly. Thus toxicologic evaluations often call for examination of organs and organ systems — the liver, the kidney, the respiratory and cardiovascular systems. Such examinations have focused as a rule on physical and functional changes, like organ enlargement or inflammation and impaired metabolism or respiration. In recent years, however, the emphasis in research has moved beyond these physical and functional changes to their underlying causes. This effort has shown in part that the chemical that affects an organ may be a metabolite that, once formed in the body from the original substance, binds so rapidly to tissue components that it cannot be detected by the normal means, such as analyses of tissue fluids. It has been found also that the effect of exposure to a toxic substance may be controlled by the rate at which the cells of the target organ can "repair" the damage. That is, a toxic symptom may reflect the overwhelming of the organ's capacity to adapt to a change in its environment. Research on the detailed responses of organs and organ systems to specific substances, although a relatively new field in the 1970's, has proved a most promising approach to toxicologic problems. Among other benefits, it should effectively complement parallel work on mechanisms of action, especially in a multidisciplinary context.

Carcinogenic screens. The cost and time entailed by traditional methods of toxicologic testing have spurred considerable research on short-term tests or screens, especially for carcinogens (7, 8). By 1978, several such tests were being proposed for use in industry. The intent of the techniques is to predict the carcinogenic potential of chemicals cheaply, rapidly, and reliably. They are not considered likely replacements for long-term studies in animals, which probably will remain the best — if not fully satisfactory — measure of carcinogenic hazard to man. Instead, the idea is to use batteries of short-term tests, each compensating for shortcomings among the others, to screen chemicals that are candidates for long-term studies of carcinogenicity. The results might indicate, for example, that the carcinogenic potential of a prospective new drug is sufficiently low to warrant long-term tests or, conversely, that it is too high to justify further testing and expense. Short-term tests also could be used to determine the carcinogenic potential of specific chemicals among the large number in the environment for which long-term studies in animals are simply impractical.

The most sensitive of the existing short-term techniques assays the ability of a chemical to cause mutations in strains of *Salmonella* bacteria by damaging the genetic material deoxyribonucleic acid (DNA). The method, by nature, is an indicator of mutagenicity; its use as an indicator of carcinogenicity is based on the concept that damage to DNA is somehow involved in the induction of cancer. In one study, the *Salmonella* test responded correctly to almost 90% of a number of known mammalian carcinogens and noncarcinogens (9). These results demonstrate that some carcinogens are also mutagens, but they do not demonstrate unequivocally that a compound shown by the system to be mutagenic is necessarily carcinogenic in mammals. Nor do they demonstrate that a compound that passes the test is necessarily noncarcinogenic. Still, the validity of the test is considered relatively high, and by 1978 it was being considered for use by many companies. The test costs a few hundred dollars per chemical tested and can be completed in several days, as opposed to the $500,000 or more and two or more years required for long-term studies in animals.

Behavioral effects. A potentially useful but still embryonic means of evaluating toxicity is the observation of behavioral effects. Changes in behavior are known to result from toxic effects on physiological systems. Intoxication by heavy metals like mercury, for example, impairs alertness and orientation; inhalation of carbon monoxide slows the speed of reaction to stimuli. Where such behavioral symptoms occur before physiological changes can be detected, they should serve as indicators of otherwise unsuspected toxic effects. On the whole, however, the ability to monitor the behavioral symptoms of the physiological effects of environmental toxicants, especially at low levels of exposure, is primitive even in the controlled conditions of the laboratory; its improvement will require much progress in the relatively new sciences of behavioral pharmacology and behavioral toxicology.

Epidemiology. Humans are exposed to certain substances, such as air pollutants, in very low concentrations and in very complex mixtures. Animal studies may provide useful information on the effects of exposures like these, but often it is difficult or impossible to devise animal experiments that simulate the conditions of human exposure in anything like a realistic sense. With problems of this type, epidemiologic studies may be in order. Where the toxicologist starts with a cause — a chemical agent — and seeks to determine its effects, the epidemiologist starts with an effect — bronchitis, say — and seeks to determine its cause. The general approach is to attempt statistically to correlate the incidence of the effect in a large population with that population's exposure to a suspected or unknown causative agent. A basic limitation of epidemiology is that more than one cause may correlate rather well statistically with the effect of interest; in such instances the toxicologist, working from cause to effect, may be able to supply clarifying insights. Toxicology and epidemiology, in other words, are complementary approaches to the evaluation of the toxicologic hazard.

ENVIRONMENTAL TOXICOLOGY

Toxicologic studies of nonhuman organisms, including plant life, in all phases of the environment have multiplied steadily since about 1950. The trend does not mirror declining emphasis on human toxicology — the main theme of this chapter — but rather the realization that ultimately man is inseparable from his environment. Pollution has damaged commercial shellfish beds. Certain chemicals can accumulate biologically to toxic levels as they move through food chains that terminate with man or with organisms that he values. Other substances may exert subtle effects that can upset natural or man-made ecosystems important to human well-being.

The principles of environmental and human toxicology are essentially the same, but the focus is different (10). With nonhuman species, dose-response investigations are oriented toward populations; with humans, they are oriented toward individuals. Biologically, nonhuman species range from the relatively simple (microorganisms) to the complex (mammals). In addition, many species occupy the typical ecosystem, so that a direct toxic impact on one species may indirectly affect one or more other species. The methods of human toxicology can provide useful information on nonhuman species, but for reasons like the foregoing the environmental toxicologist must employ combinations of additional procedures adapted to the more complex and variable conditions found even in the smallest ecosystems.

The methods and species used to assess environmental impact depend in part on the substance in question — its chemical and physical properties and where and in what amounts it is likely to enter the environment. Controlled studies in the laboratory are used to detect acute, subchronic, and chronic effects, if any, in individual species expected to be exposed; the subjects might be a mammal and a bird, for example, or a fish and an alga. Such studies also may show whether the substance can be bioaccumulated and thus may affect other species via food chains. If further investigation is indicated, several species may be exposed together in the laboratory in "controlled microcosms" designed to represent a step closer to the real environment (11). Beyond this point come uncontrolled field studies of populations and ecosystems of varying extent. Investigations like these are rarely pursued to support the commercial development of specific chemicals, which normally are cleared or rejected earlier in the testing process. More often, field studies are designed to uncover or clarify subtle, unexpected effects of environmental stress or to explore the ecological impact of large, inadvertent releases of potentially harmful substances. It should be noted also that successive levels of testing for environmental effects bring correspondingly higher costs as well as progressively greater difficulty in interpreting the findings.

THE VALIDITY OF TOXICITY TESTS IN PREDICTING RISK

The toxicologist assumes, on the basis of scientific evidence on many compounds, that a series of critically designed studies of the physiological, pharmacological, and biochemical behavior of a substance in animals will allow him to predict whether it will constitute a potential hazard to man. He assumes on the same basis that the studies will allow him to predict to some extent, at least, the magnitude of that hazard. Under the best of circumstances, then, toxicity studies should provide both qualitative and quantitative information: qualitative in terms of the kinds of effects that are likely to occur and quantitative in terms of the amounts of the agent required to produce those effects. Even in the best circumstances, however, predictions based on animal tests at the current level of knowledge may involve significant uncertainties.

Well designed and conducted toxicity studies in animals may fail to predict adverse reactions in man for various reasons. For one thing, toxic symptoms in man are often subjective, whereas symptoms observed in animals are primarily objective. Compounds that produce headache, giddiness, or irritability in man, for example, may get a clean bill of health in animals. A second problem is that most animal tests involve a relatively small number of subjects — 50 to 100 animals per group. If the results of a test in 50 or 100 animals are negative, one might reasonably conclude that the chances that an adverse reaction will occur are less than one in 50 or one in 100. But when the data are extrapolated to man, the number of subjects may run into millions for a widely used compound; the adverse effect, if it occurs, may do so in an enormous number of people, even at an incidence much lower than one in 100. In principle, the problem can be eased by studying very large numbers of animals — the so-called "kilomouse" or "megamouse" experiments. In practice, however, logistical and experimental difficulties multiply rapidly as the number of animals increases, so that such experiments are by no means a panacea.

Another problem with animal tests arises when two compounds exhibit synergistic or potentiated toxicity. Most tests with animals are done with single compounds. Man, however, is exposed to complex mixtures of chemicals, sometimes when both the chemicals and their concentrations in the environment are changing, so that synergistic toxic effects can develop unexpectedly. Finally, we have the unavoidable problem of biologic variability in any type of animal study. Despite the strenuous efforts of toxicologists to develop uniformity in the protocols for similar tests, seemingly duplicate studies frequently produce conflicting results. Such an outcome complicates decision-making. Suppose, as a hypothetical example, that one observes an increased incidence of liver tumors (hepatomas) in mice fed a high dose of a compound and that duplicate studies with the same strain of mice and the same dose of the compound yield no evidence of increased tumor formation. What can one say about the safety of the compound? How many negative experiments are required to counteract the single positive experiment? Since no two animal studies are ever absolutely identical, it is not surprising that questions of this sort often involve philosophic and political rather than statistical arguments.

For reasons like the foregoing, several chemical agents that passed toxicity tests in animals with flying colors have been launched commercially but later withdrawn from the market because of unexpected adverse reactions. A chemical thus found to be "unsafe," whether it be a drug, pesticide, food additive, or other product, represents a poor investment for any company. Manufacturers, therefore, are at least as concerned as the general public about the ability of toxicity studies to prevent such disasters. It follows that they will continue to use the current toxicologic testing methods only as long as they are clearly better than anything else we have. It should be kept in mind also that toxicity tests can be guilty of overpredicting as well as of underpredicting hazard and that economic problems can result from overprediction in the same way that safety problems can result from underprediction.

Toxicity studies are expensive in both time and money, and it is not likely that controversy can be avoided — or test results improved — simply by adding stringent new requirements to current protocols. Instead, we need to improve the validity and relevance of the current procedures and to replace them with whatever newer ones are shown to have superior predictive ability.

Current developments in testing for carcinogenicity illustrate how procedures might gradually be upgraded. The present standard test for carcinogenicity involves long-term or lifetime exposure of groups of animals to the test substance, beginning as a rule when the animals are at the weanling stage. It has been shown, however, that some substances that do not produce tumors when tested in weanling animals will produce tumors of the bladder, for example, if exposure is initiated when the animals are in the uterus. For such compounds, then, exposure beginning in the uterus and continuing after birth and weaning would be a more rigorous test than exposure beginning at weaning. In a different series of investigations, on the other hand, the animals were exposed between birth and weaning by introducing the test substance orally through tubes. The results were criticized on the basis that "under these conditions, extraordinarily high peak intragastric dosages of the material would reach the target organs in contrast with the levels reaching these target organs when the compound was administered in the diet" (12).

This criticism illustrates one of the problems of upgrading animal tests. Toxicologists have long recognized that newborn and very young animals may be more susceptible than adults to the toxic effects of a chemical. One reason is the increase in liver-enzyme activity, and thus in detoxification ability, that occurs in animals as they mature. Sometimes it is quite difficult technically, however, to convert such knowledge into toxicologic studies that are both valid and relevant to real-world situations.

Trends in Toxicologic Methods

Toxicology is among the most traditional of disciplines. It is an inexact science, and so its practitioners must rely very much on professional experience. They are highly reluctant to abandon methods of evaluation that have provided a measure of assurance over the years in favor of newer procedures that might be more rational scientifically but are unproven by time. There is even a certain reluctance to devise and propose new procedures for fear that they may be added to the formidable battery of tests already required under various laws and regulations. The reluctance is due partly to the fact that while new tests may be added, older ones tend to remain past the point of clear-cut utility. In short, replacement of traditional methods by new ones is certain to proceed slowly. It seems probable that the traditional methods will be modified gradually by more rational selection of the levels and durations of exposures, the species employed, and the parameters determined.

In the past, toxicologic studies commonly have been conducted at levels of exposure that would provide certain arbitrarily established margins of safety relative to probable levels of exposure in use. The concept of margin of safety, however, has not proved acceptable for carcinogenic chemicals because the existence of no-ill-effect doses for carcinogens has not been demonstrated. The tendency in the 1970's, moreover, is to extend this philosophy to mutagenic and teratogenic chemicals. This emphasis on total lack of hazard from carcinogens, mutagens, and teratogens would seem to dictate that a toxicology program for a new compound be initiated by defining a general toxicologic profile for the compound without regard to the margin of safety.

One approach to defining such a profile might be to start a toxicologic evaluation with a study of effects on a target organ or organ system and separate studies to assess carcinogenic, mutagenic, and teratogenic potential. These tests would be complemented by metabolic studies of the chemical to determine its patterns of absorption, excretion, and distribution in the body as well as its ultimate fate. This combination of studies might permit the chemical's toxicologic acceptability to be judged — in terms of a "go" or "no-go" decision — earlier in the testing process than now is common. Should the decision be "go," the combination also would provide guidance to a rational design for chronic studies. The question of margin of safety would be considered at the conclusion of the testing program.

Recommendation T2: *Support for the improvement of toxicologic testing methods should be supplied as rapidly as possible consistent with the ability of the available facilities and manpower to use such support effectively. The major improvement required is tests that singly or in combination have greater predictive relevance for humans than do current methods. Other goals include more rapid means of screening chemical agents for toxic potential and less costly protocols for animal studies.*

Recommendation T3: *Toxicologic studies of nonhuman organisms, including plant life, should be pursued at a level and in a manner consistent with the fact that, ultimately, the well-being of humans is inseparable from the well-being of other forms of life.*

USE OF TOXICOLOGIC INFORMATION IN REGULATORY DECISIONS

A major goal of toxicology in our context here is to provide the information needed to support regulatory decisions. A report on chemicals and health (13) by a panel of the President's Science Advisory Committee has stated two key principles of the decision-making process. The first is that regulatory procedures should insure balanced consideration and balanced decision in regulatory actions. This view implies consideration of both the direct and indirect consequences of each of the possible actions. The second principle is that, where knowledge is so inadequate as to make the reality of a possible threat quite tenuous, the proper response is to seek more knowledge, not to take drastic action or to do nothing. The report pointed out also that those involved — but especially the Federal Government — must try to relate the strength of the response to the seriousness of the threat. It noted further that all interested segments of the public should be fully informed and should understand the kinds of knowledge and judgments that underlie regulatory actions.

Benefits and risks

An important aspect of the decision-making process is that it must take into account the benefits as well as the hazard that may be associated with the product at hand (14). Balancing benefits against risks sometimes is relatively easy: if a person is dying of a particular disease, administration of a drug that may produce undesirable side effects clearly is preferable to letting the person die. The decision is more difficult if the condition may or may not be fatal: if a patient has high blood-cholesterol, for example, the physician must decide whether the level is high enough to warrant the administration of a cholesterol-lowering but potentially hazardous drug. On the whole, however, exposure to a prescription drug is optional and aimed at a specific problem in an individual; responses of humans to the drug are known from clinical studies at the minimum. The benefit-hazard ratio may be markedly more difficult to assess for nontherapeutic agents. Food additives or pesticide residues, for example, tend to be less potent than drugs, but exposure to them is not optional as a rule, and their effects are known usually only from animal studies. The difficulty can be illustrated by a pesticide residue in a major food. Here, one must weigh the hazards of long-term, low-level exposure of a significant fraction of a total population against the higher quality and quantity of the food to be obtained by using the pesticide.

Delaney clause

A special example of the long-term, low-level exposure problem is provided by the Delaney clause in the Food Additives Amendment of 1958 to the Food, Drug, and Cosmetic Act. Under the Delaney clause, no food additive, direct or indirect, can be used ". . . if it is found to induce cancer when ingested by man or animal, or if it is found, after tests which are appropriate for the evaluation of the safety of food additives, to induce cancer in man or animal . . ." Direct food additives include substances like antioxidants and colors; indirect additives include substances, such as pesticides, that can find their way into food inadvertently as a result of use elsewhere in the growing-processing-packaging process. The clause seems logical on the surface, but it does create problems. In lifetime feeding studies on rats, a dose must be fed that will result in some significant biological effect. With compounds that have a high acute oral toxicity, the dose that will result in some biological alteration often is relatively low, but with relatively nontoxic materials, very high doses are used. If no carcinogenicity problems are encountered with the highly toxic material, a tolerance — the maximum amount permitted in food — can legally be established for it. If the relatively nontoxic material displays any carcinogenicity, however, even at exceedingly high doses, it must be banned from use under the terms of the Delaney clause. This requirement, in effect, leaves no room to evaluate the benefit-hazard ratio: the compound must be banned irrespective of its benefits to society and the degree of the carcinogenic hazard.

Implicit in the Delaney clause is the concept that threshold doses do not exist for carcinogenic compounds — that any dose, no matter how small, will exert a carcinogenic effect. This view is contrary to toxicologic experience, which indicates that, for any substance, a dose exists that will produce no response in a given species and that as the dose increases the response will appear and intensify. Reputable authorities can be found on both sides of the controversy, and the scientific evidence thus far has proved insufficient to resolve it. Toxicologists argue, however, that even if thresholds or "no-ill-effect" levels do not exist for carcinogens, the public benefits less from the Delaney clause than it would from the alternative procedure: weighing the benefits of a compound against the degree of hazard it poses and on that basis determining the amount of it to be permitted in food. Sound scientific judgment still might dictate that the compound not be used, but it also might dictate that the benefits warrant its use, subject to a safety factor consistent with the level of risk.

Making the decision

However such difficulties are resolved, the toxicologist's evaluation of risk provides only part of the risk-benefit information, and it is sometimes the easiest part to obtain. The social benefit of an artificial color in a food, for example, is more difficult to assess than that of a lifesaving drug or of a pesticide that brings definable economic gains. For any decision-making process to work, we must be able to consider both sides of the story. Otherwise, we will be unable to regulate wisely the chemicals in our environment. If we recognize that complete safety is not attainable — that always we must live with a degree of risk — we can then consider a specific chemical rationally in terms of its risks and contributions to human welfare.

Most regulatory decisions that derive from some form of toxicologic information are made without great difficulty. Where the indication of a specific toxic effect is clear-cut, and the experimental dose-response information is adequate, it is relatively simple to establish no-ill-effect levels and apply appropriate safety factors. If the toxicologic data are from studies in humans, the safety factor as a rule of thumb is set at 10 — that is, routine exposure of humans is limited to one tenth of the no-ill-effect level determined experimentally. If the data are from animal studies, the safety factor is set at 100, again as a rule of thumb. On these bases regulations can be devised that protect the public health satisfactorily. Problems in decision-making can occur, however, when the toxicity data may be inadequate, when the experimental evidence can lead to conflicting conclusions, and when the ability of the interspecies comparisons accurately to predict the situation for man is questionable. In this kind of situation, the Government traditionally has asked for assistance from the National Academy of Sciences, professional societies, intra- and extramural groups from the National Institutes of Health, and a variety of other outside agencies.

Typically, bodies asked for assistance in such cases form advisory panels of experts to evaluate the toxicologic information; increasingly, these panels include representatives of consumer groups. In making their recommendations, advisory panels must consider not only the toxicologic information, but also how the compound will be used (anticipated levels and patterns of exposure, etc.). Sometimes, different investigators may reach different conclusions from the same experimental evidence. Reports of advisory groups, therefore, may contain minority viewpoints or other evidence of conflicting opinions. In some cases, controversy has resulted from panel members' bias. Lack of objectivity, however, does not necessarily signify lack of integrity; conflict of interest cannot always be assumed for representatives of industry nor discounted for representatives of academe and government. In commenting on the cigarette–lung cancer controversy, one scientist observed (15) that "The reputable men are indeed reputable but they are also human so that their judgment is not a reflection of diabolical intent or subverted aims but rather is a reflection of clouded objectivity in the evaluation of findings or the application of uneven standards to the assessment of environmental hazards."

It has been suggested (16) that the utility of an advisory panel depends on whether its members are "good or bad choosers." Good choosers tend to have considerable experience in experimental design, the analysis of results, and the extension of the results of animal experiments to the prediction of risk for man. Decisions reached through a consensus of "disinterested scientists," or by majority vote, or by averaging the choices of the "good" and "bad" choosers are useless. This argument leads to the conclusion that panels composed of "good choosers" are those most likely to be able to weigh the results of toxicologic studies rationally in the light of risk and benefit.

Recommendation T4: *Better mechanisms should be developed for incorporating toxicologic knowledge into the formulation of legislative and regulatory requirements. Initiatives in this respect should come from both the federal agencies involved and scientific bodies such as the National Academy of Sciences and the Society of Toxicology.*

PERSPECTIVES IN TOXICOLOGY

The demand for toxicologic services in this country by 1978 was unprecedented and still growing. Virtually all chemicals of commerce are subject to toxicologic evaluation under federal law. Toxicologic resources also are needed for research on inadvertent contaminants in the environment, including the work environment. Progressive manufacturers traditionally have evaluated their products and operations toxicologically, but the proliferating uses of toxicology are bringing more rigorous requirements and more sophisticated methods for establishing probable safety. Behind these many activities stands a public that has become acutely aware of the potential hazard of exposure to environmental agents.

The current challenge to toxicologic resources is exemplified by the requirements of the Toxic Substances Control Act of 1976, which took effect Jan. 1, 1977. Earlier legislation in general calls for thorough toxicologic tests of drugs, food additives, and pesticides before they are put on the market. The new act authorizes the Environmental Protection Agency to require manufacturers and processers of all other potentially harmful chemical substances and mixtures to test them, as appropriate, for hazard. The tests required may cover, for example, acute toxicity and carcinogenic, mutagenic, teratogenic, behavioral, and synergistic effects, as well as environmental characteristics like persistence and effects on food chains. Where indicated, EPA may call for premarketing studies and for studies of chemicals already in use when the act took effect. The number of chemical substances in commercial use by the end of 1977 was problematical, but estimates ranged up to 75,000. In addition, new substances enter the market at rates estimated at from 200 to 1000 annually.

The numbers and costs of the toxicologic tests to be performed under the act remain uncertain, but clearly they will be significant. The acute toxic effects of many commercial chemicals are known reasonably well, but the chronic effects of most commercial chemicals are not. EPA has estimated that tests will be needed on some 200 existing chemicals and 150 new ones annually during the five years 1978–82. The agency envisioned limited tests for 315 of these 350 chemicals and major tests for 35 of them; by early 1978, a federal interagency panel had recommended health and environmental studies of some 50 large-volume chemicals. In 1975, EPA projected average costs of $10,000 to $20,000 per chemical for limited tests and $200,000 to $400,000 for major tests, which also would require at least three years to perform. These estimates were too low for chronic tests being planned or under way in mid-1977, when the cost of two-year studies of ethylene oxide was expected to exceed $750,000. Costs aside, however, the Toxic Substances Control Act represents without question a considerable new work load in toxicology.

The pressure on toxicologic resources is administrative as well as legislative. In 1976, the Food and Drug Administration concluded that it was finding too many deficiencies in the conduct and reporting of toxicologic studies, both by industry, which does most of the testing, and by government. By year's end, FDA had proposed regulations for good practice in nonclinical (animal) laboratory studies (17); as of July 1978, the final regulations had not appeared. In addition, the agency anticipated that its across-the-board food additive review programs, under way in 1977, might raise questions about some additives cleared for use and marketed in earlier years. New questions could call for new tests, using the newest available technology.

Research and training

The rapid growth in the toxicologic work load is both a deterrent and a stimulus to the development of the science. Those employed in toxicology are so occupied by required studies, usually under conventional protocols, that they have little time or financial support for research on means of obtaining better toxicologic information at less cost in time, talent, and funds. The growth in the field also is creating shortages of the trained manpower required to do both studies and fundamental research, and a task force of the National Institute of Environmental Health Sciences reported on the problem in 1976 (6). Some of the necessary research and training is funded by the Federal Government, the major supporter of both, and industry is increasing its support. Whatever the source, however, it appeared by 1978 that considerably more would be needed in both research and training. In addition, it would seem desirable not only to educate new or entry-level specialists in toxicology and related disciplines, but also to train scientists from other disciplines in these fields.

We noted earlier that probably there exists no single satisfactory formal training program for the toxicologist, defined as one who can evaluate the probable biological impact of a chemical agent on the total environment and can do so in terms of the practical considerations of benefit-versus-risk under probable conditions of exposure or use. A toxicologist must be a biologist-chemist-pathologist-medical scientist. Formal training may be in any of these areas, but the toxicologist must move beyond the comfortable confines of a specialty. Specialists are required for the multiple facets of a toxicologic evaluation, but a multidisciplinary grasp is needed to plan a program, monitor the studies, and evaluate the results. Scientific development tends toward specialization, but able specialists must be encouraged to develop knowledge and interest in the larger problem.

Recommendation T5: *Training programs in toxicology, at both the technician and professional levels, should be expanded to meet the manpower needs created by federal legislation, with support to come from federal funding arrangements like that in the National Institute of Environmental Health Sciences and from other public or private sources as necessary.*

Facilities

The toxicologic activity of the 1970's has pressed facilities as well as manpower very hard. A number of manufacturers have sharply expanded their toxicology laboratories, and many contract-toxicology laboratories have expanded also. The Chemical Industry Institute of Toxicology, formed in 1975, had some 30 corporate members by 1978 (18). The institute performs and supports research on all aspects of the toxicology of commodity chemicals and funds several postdoctoral fellowships. Also expanding their efforts are federal facilities such as EPA, FDA, and the National Institute for Environmental Health Sciences.

Information handling

The limitations on testing facilities and personnel place a premium on effective dissemination of information that will help to forestall unwarranted duplication of toxicologic studies. The most extensive effort in the collection, storage, and retrieval of toxicologic data is the Toxicology Information Program of the National Library of Medicine, which was created in 1967. By 1977, the program's TOXLINE computer-based system had on file more than 380,000 references to published toxicologic information; computer-access to the system was available at some 280 organizations throughout the nation. In mid-1976, the Toxicology Information Program launched a monthly "current-awareness" publication (*Tox-Tips*) to disseminate quick announcements in all areas of long-term toxicologic testing. A second federal source of information is the "Registry of Toxic Effects of Chemical Substances," published annually by the National Institute for Occupational Safety and Health. The first edition was compiled in 1971 (under the name "Toxic Substances List"). The 1977 edition (19) contained 98,993 listings of chemical substances: 26,478 were for different chemicals and 72,515 for synonymous names and codes for those chemicals. The type of information included is illustrated in Figure 2.

The flow of toxicologic information is hampered to a degree by strictures on the disclosure of proprietary data. Under the Toxic Substances Control Act, EPA has discretionary authority to release nominally confidential information when doing so is essential to protect health or the environment. Under the Food, Drug, and Cosmetic Act, however, certain information submitted by drug companies, for example, is held confidentially by FDA. This is true of that agency's very large holdings of the results of chronic toxicity studies of proprietary products. Means might be sought of exploiting this resource, and one possibility involves the optimum duration of toxicologic studies. Many toxicologists believe that studies of six months or one year, using appropriate dosages, should reveal nearly all toxic effects except carcinogenesis. They believe, therefore, that longer studies may tie up hard-pressed toxicologic resources unnecessarily, but there is relatively little hard evidence for this argument. It should be possible to resolve the question, without violating confidentiality, by using FDA's files to do comparative analyses of studies shorter and longer than, say, one year.

Conclusion

The overriding problem in toxicology in the 1970's is the gap between work load and manpower. Our understanding of the basic mechanisms of toxicologic action must be expanded sharply. The new knowledge must be used to upgrade

FIGURE 2

Example of a typical entry in the "Registry of Toxic Effects of Chemical Substances"

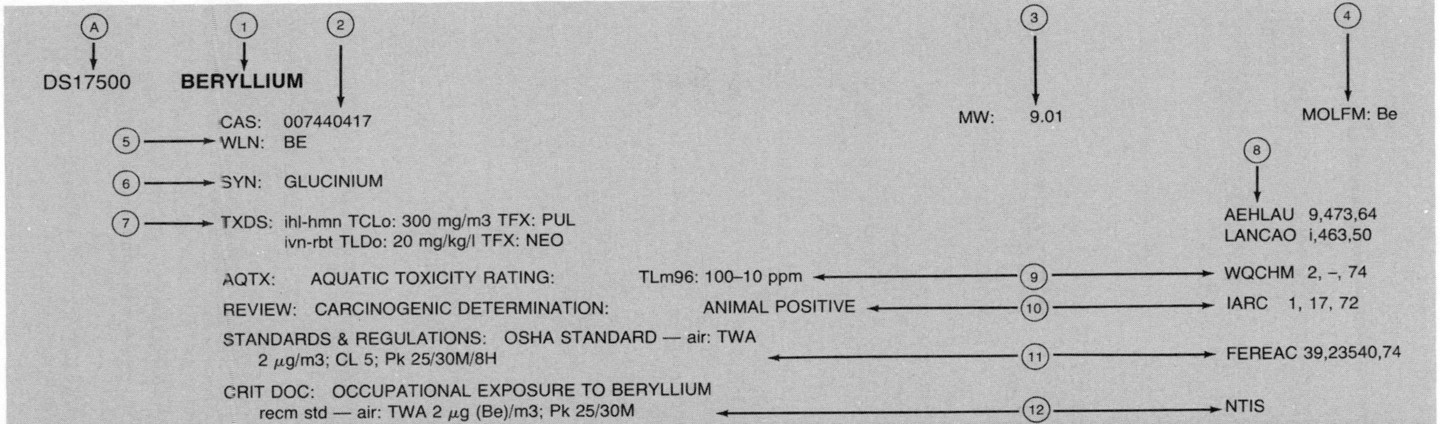

Key[a]:

A. Sequence number in this listing.
1. Prime name of compound.
2. Chemical Abstracts Register Number, which is a number assigned to this compound so that it may be uniquely identified.
3. Molecular Weight of this compound.
4. Molecular Formula or Elemental Formula of this compound.
5. Wiswesser Line Notation, which is a formula defining the structure of this compound.
6. Synonyms, common names, trade names, and other chemical names.
7. Toxic dose line.
8. This is the reference to the original article or source from which the toxic data were derived.
9. Aquatic toxicity rating is a relative toxicity to aquatic life.
10. Review of this compound.
11. Standards and regulations entries indicate a standard for this substance has been promulgated by a Federal Agency.
12. A Criteria Document supporting a recommended standard has been published by NIOSH, U.S. Department of Health, Education, and Welfare.

a. Entries have been abbreviated by omitting references to fuller explanations in text of document.

Source: "Registry of Toxic Effects of Chemical Substances," 1976 Edition, National Institute for Occupational Safety and Health, Washington, D.C., p. xi.

the utility of the classical toxicologic tests as predictive tools for man and other organisms and, with time, to develop more effective testing procedures. Well-trained people are needed to discharge these tasks. More broadly, they are needed to exercise scientific judgment, to formulate from test results the predictions of toxicologic hazard essential to sound decisions on public policy. The necessary manpower is not now available and is unlikely to become available in the short term. The rapidly growing demand for toxicologic services will not be met adequately without corresponding growth in support for academic training and research in the toxicologic sciences.

LITERATURE CITED

1. Weil, C.S., Wright, G.J., "Intra- and Interlaboratory Comparative Evaluation of Single Oral Test," *Toxicol. Appl. Pharmacol.*, **11**, 378 (1967).
2. Weil, C.S., Scala, R.A., "Study of Intra- and Interlaboratory Variability in the Results of Rabbit Eye and Skin Irritation Tests," *Toxicol. Appl. Pharmacol.*, **19**, 276 (1971).
3. Weil, C.S., "Guidelines for Experiments to Predict the Degree of Safety of a Material for Man," *Toxicol. Appl. Pharmacol.*, **21**, 194 (1972).
4. "Guidelines for Carcinogen Bioassay in Small Rodents," National Cancer Institute, NCI-CG-TR-1, DHEW Publication No. (NIH)76-801, U.S. Government Printing Office, Washington, D.C., February 1976.
5. "Man's Health and the Environment — Some Research Needs," Report of the Task Force on Research Planning in Environmental Health Science, U.S. Department of Health, Education, and Welfare, U.S. Government Printing Office, Washington, D.C., March 1970.
6. "Human Health and the Environment — Some Research Needs," Report of the Second Task Force for Research Planning in Environmental Health Science, U.S. Department of Health, Education, and Welfare, Publication No. NIH 77-1277, Government Printing Office, Washington, D.C., December 1976.
7. Bridges, B.A., "Short Term Screening Tests for Carcinogens," *Nature*, **261**, 195 (1976).
8. Fox, J.L., "Ames Test Success Paves Way for Short-Term Cancer Testing," *Chem. Eng. News*, **55**, 50, 34 (1977).
9. McCann, J., et al., "Detection of Carcinogens as Mutagens in the *Salmonella*/Microsome Test. Assay of 300 Chemicals," *Proc. Nat. Acad. Sci.*, **72**, 5135 (1975), McCann, J., Ames, B., "Assay of 300 Chemicals: Discussion," *Ibid.*, **73**, 950 (1976).
10. "Principles for Evaluating Chemicals in the Environment," National Academy of Sciences, Washington, D.C., 1975.
11. Metcalf, R.L., "A Laboratory Model Ecosystem to Evaluate Compounds Producing Biological Magnification," *Essays Toxicol.*, **5**, 17 (1974).
12. "Report of the Secretary's Commission on Pesticides and their Relationship to Environmental Health," U.S. Department of Health, Education, and Welfare, Washington, D.C., December 1969.
13. "Chemicals & Health," President's Science Advisory Committee, Science and Technology Policy Office, National Science Foundation, U.S. Government Printing Office, Washington, D.C., 1973.
14. "Decision Making for Regulating Chemicals in the Environment," National Academy of Sciences, Washington, D.C., July 1975.
15. Kotin, P., "The Cigarette-Lung Cancer Imbroglio and Public Opinion," *J. Am. Med. Assoc.*, **211**, 506 (1970).

16. Weil, C.S., "Experimental Design and Interpretation of Data from Prolonged Toxicity Studies," Proc. 5th Int. Congr. Pharmacology, San Francisco 1972, vol. 2, pp. 4–12, Karger, Basel, 1973.

17. *Fed. Regist.*, **41,** 51206, Nov. 19, 1976.

18. *Chem. Ind.* (London), June 17, 1978, page 393.

19. "Registry of Toxic Effects of Chemical Substances," 1977 Edition, National Institute for Occupational Safety and Health, U.S. Government Printing Office, Washington, D.C., September 1977, 2 vols.

Chapter 4

The Air Environment

4 / The Air Environment

INTRODUCTION

During the first half of the 1970's, the term "air pollution" certainly completed the transition from relative obscurity to everyday use. It is less certain that popular understanding of the problems and possibilities implied by the words made a comparable transition. Public officials find their mail divided almost equally between demands that a given source of contaminants, since it emits nothing visible, must be left alone, and demands that "the air pollution that is killing me" be abated at once. One official, bombarded by citizens' demands that he halt emissions from an electical utility, remarked, "I ask them if they are volunteering to sit in the cold and the dark without electricity until the technological problems are solved. I have yet to get a reply." The dialog typifies the common tendency to seek simplistic and minimal solutions to an extraordinarily complex set of problems.

Despite such difficulties, the science and technology of air pollution and its control have advanced by the late 1970's to a state that, though less than optimum, is nonetheless impressive. This knowledge reflects a substantial intellectual effort. Contaminants in the air must often be detected and measured at concentrations in the range of parts per million (ppm) or less by volume or milligrams (mg=a thousandth of a gram) or micrograms (μg=a millionth of a gram) per cubic meter (m^3) by weight. The physical, chemical, and physiological behavior of pollutants and combinations of pollutants at such levels must be clarified. And to cope with these and the many related problems a variety of analytical chemical techniques and instruments must be devised and perfected.

The scientific problems have been rivaled by those of building a sound legal-regulatory structure for managing the air environment. The structure in place by 1978 evolved in response to many, often conflicting pressures; its defects are not surprising, given the complexities of preserving the quality of the air in a modern industrial society. Beyond doubt, however, the creation and implementation of effective laws and regulations would be eased markedly by a sharply improved grasp of the pertinent science and technology, which will be treated here in four broad and interrelated areas:

- The flow and dispersion of air contaminants and their transformation into other chemical and physical forms in the local, regional, and global atmospheres.
- The means of avoiding the generation of air pollutants or of abating pollution if it cannot be avoided.
- The effects of air pollutants on plant and animal life and on inanimate objects and materials.
- The means of detecting and measuring air pollutants and their effects (see also Chapter 2).

Background

The atmosphere is a dynamic system. It steadily absorbs a variety of solids, liquids, and gases from both natural and man-made sources. These substances may travel through the air, disperse, and react among themselves and with other substances both chemically and physically. Eventually, whether or not in their original forms, they may reach a sink, such as the ocean, or a receptor, such as a man. Some, such as helium, escape from the earth's atmosphere. Others, such as carbon dioxide, may enter the atmosphere faster than they return to their sinks and thus gradually accumulate in the air.

Clean, dry air contains 78.09% nitrogen, by volume, and 20.94% oxygen (Table 1). The remaining 0.97% of the gaseous constituents of dry air includes small amounts of carbon dioxide, helium, argon, krypton, neon, and xenon, as well as very small amounts of other organic and inorganic gases whose concentrations may differ with time and place. Water vapor normally is present in air at concentrations of 1 to 3%. The air also contains aerosols — dispersed solid or liquid particles. They may range in size from clusters of a few molecules to diameters of a few tens of micrometers (1 μm is one millionth of a meter).

The local air mass and its contents are parts of the regional and global atmospheres and ultimately must be treated as such. Contaminants generated locally often leave the local atmosphere. Little is known of the fate of many such substances, but they may have important geochemical or geophysical effects. Carbon dioxide and fine particles, for example, are produced in large amounts by man; both may be affecting global temperature by upsetting the infrared radiation balance of the planet.

The five most common primary air pollutants in tons emitted annually nationwide are carbon monoxide, sulfur oxides, hydrocarbons, nitrogen oxides, and particulate matter (Tables 2, 3)*. Many others are emitted in smaller amounts, however, and weights emitted do not in general correlate with effects produced. A given receptor, for example, may be much more susceptible to damage by one primary contaminant than by another. Primary pollutants may react with other substances in the air to yield secondary pollutants that exert smaller or larger effects than their precursors. And the behavior and lifetime of a contaminant in air may vary with factors such as weather conditions and the other contaminants present.

Few data exist on the precise concentrations of pollutants in ambient air before the early 1950's. Carbon dioxide, carbon monoxide, and sulfur dioxide were measured in the U.S. and Europe as long ago as the 1920's and in some instances earlier. Such analyses were rare, however, and generally of questionable accuracy.

Since the early 1950's, systematic monitoring of ambient air has been undertaken on a slowly expanding scale by state and local agencies and by the Environmental Protection Agency and its predecessors. These efforts have involved primarily urban areas. They have produced considerable data, of varying reliability, on total suspended particles, sulfur dioxide, carbon monoxide, and oxidants (mainly ozone) and lesser amounts of data on nitrogen dioxide and hydrocarbons. The measured maximum levels of these six contaminants in 1976 were in the range of the permissible maximums set by the National Ambient Air Quality Standards (Table 4) and often exceeded those values (1,2). The data available, according to EPA, allow nationwide trends — generally downward — to be inferred for total suspended particles and sulfur dioxide in urban areas during 1970–76; data for carbon monoxide, nitrogen dioxide, and oxidants are insufficient to allow nationwide trends to be inferred. Pollutants other than

*Carbon dioxide produced by man's combustion processes is an air contaminant, strictly speaking, and its emissions far exceed those of the pollutants in Table 2. It is omitted from the table because it is not considered a contaminant that can be controlled except by replacing combustion with another source of energy, such as nuclear or solar energy.

these five are not as a rule monitored systematically in ambient air except in special, relatively short-range studies.

Legislative framework

The legislative framework for managing the air environment is based on the federal Clean Air Act of 1970 and Clean Air Act Amendments of 1977, whose principal forerunners were the Clean Air Act of 1963, the National Motor Vehicle Emissions Standards Act of 1965, and the Air Quality Act of 1967. The act of 1970 (Public Law 91-604) provides for uniform national standards of ambient air quality. It provides also for interstate pollution control regions, formal state plans for meeting air quality standards, pollutant emission standards for specified sources, and national emission standards for hazardous air pollutants. In addition to other requirements, the act of 1970 included deadlines for achieving certain of its goals. The act was modified and expanded, but not fundamentally altered, by the amendments of 1977 (Public Law 95-95).

In April 1971, EPA published National Ambient Air Quality Standards for six contaminants: particulate matter, sulfur dioxide, carbon monoxide, nitrogen dioxide, oxidants/ozone, and hydrocarbons (Table 4). The standards are of two types: primary standards are designed to protect human health; secondary standards are designed to protect property, plant life, and aesthetics and in general to promote human welfare. Through 1977, the six original national standards for ambient air quality remained in effect without major revision. No additional standards had been established, although in December 1977 EPA

Table 1

Composition of clean, dry air near sea level

COMPONENT	CONTENT % by volume	CONTENT ppm	COMPONENT	CONTENT % by volume	CONTENT ppm
Nitrogen	78.09%	780,900 ppm	Hydrogen	.00005%	0.5ppm
Oxygen	20.94	209,400	Methane	.00015	1.5
Argon	.93	9,300	Nitrogen dioxide	.0000001	0.001
Carbon dioxide	.0332	332	Ozone	.000002	0.02
Neon	.0018	18	Sulfur dioxide	.00000002	.0002
Helium	.00052	5.2	Carbon monoxide	.00001	0.1
Krypton	.0001	1	Ammonia	.000001	.01
Xenon	.000008	0.08			
Nitrous oxide	.000033	0.33			

Note: The concentrations of some of these gases may differ with time and place, and the data for some may be open to question. Single values for concentrations, instead of ranges of concentrations, are given above to indicate order of magnitude, not specific and universally accepted concentrations.

Source: "Air Chemistry and Radioactivity," Junge, C.E., Academic Press, New York, N.Y., 1963, p. 3.

"Air Pollution," Vol 1, 2nd ed., Stern, A.C., ed., Academic Press, New York, N.Y., 1968, p. 27.

"Sources, Abundance, and Fate of Gaseous Atmospheric Pollutants," Robinson, E., Robbins, R.C., prepared for American Petroleum Institute by Stanford Research Institute, Menlo Park, Calif., 1968.

"Energy and Climate," National Academy of Sciences — National Research Council, Washington, D.C., 1977 (for carbon dioxide).

Pierotti, D., Rasmussen, R.A., "The Atmospheric Distribution of Nitrous Oxide," *J. Geophys. Res.*, **82**, 5823 (1977).

Table 2

Nationwide air pollutant emissions estimates, 1975[a] (millions of tons and percent of total)

Origin	Carbon monoxide	Hydrocarbons	Nitrogen oxides	Sulfur oxides	Particles
Transportation	77.4 (80.4%)	11.7 (37.9%)	10.7 (44.2%)	0.8 (2.4%)	1.3 (7.2%)
Fuel combustion (nontransportation)	1.2 (1.3%)	1.4 (4.5%)	12.4 (51.3%)	26.3 (79.8%)	6.6 (36.8%)
Industrial processes	9.4 (9.8%)	3.5 (11.3%)	0.7 (2.9%)	5.7 (17.3%)	8.7 (48.3%)
Solid waste disposal	3.3 (3.4%)	0.9 (2.9%)	0.2 (0.8%)	<0.1 (0.3%)	0.6 (3.3%)
Miscellaneous	4.9 (5.1%)	13.4 (43.4%)	0.2 (0.8%)	0.1 (0.3%)	0.8 (4.4%)
Totals	**96.2 (100%)**	**30.9 (100%)**	**24.2 (100%)**	**32.9 (100%)**	**18.0 (100%)**

a. Preliminary

Source: "National Air Quality and Emissions Trends Report, 1975," EPA-450/1-76-002, Environmental Protection Agency, Research Triangle Park, N.C., 1976.

proposed a standard for particulate lead (3). The law required that most states meet the primary standards by May 31, 1975. In addition, a court decision of May 1972 determined that, in areas where the air is cleaner than required by the standards, "significant deterioration of air quality" must be prevented.

Ambient air quality standards had to be set to achieve control of various pollutants. The concentrations selected as standards, however, were based on the evaluation of many factors (4) and are not absolute dividing lines between healthy air and unhealthy air. In mandating standards the law assumes that for each pollutant and each of its effects there is a threshold concentration below which no adverse effect occurs. This approach may not prove valid in the long run; considering genetic variability and the occasional hypersensitive individual, we may arrive at a conclusion that the only safe level is below the natural background. A second complication arises where pollutants are interrelated in behavior and the standards for them are not coupled accordingly. Further, various premises of the initial standards have been outdated by subsequent scientific findings. For such reasons, the amendments of 1977 call for a review of the National Ambient Air Quality Standards to be completed by Dec. 31, 1980, and for fresh reviews every five years thereafter. The amendments authorize EPA to review the standards at shorter intervals and to revise them where indicated.

Recommendation A1: *Reviews of the National Ambient Air Quality Standards should include evaluation of the degree of improvement in air quality to be expected with existing standards and with alternative standards based on a variety of control schemes.*

Under the act of 1970, EPA established 247 Air Quality Control Regions (AQCR's) covering the 50 states, Guam, Puerto Rico, and the Virgin Islands. The law recognized that air pollution does not conform to state boundaries. Thus the AQCR's are laid out to reflect air pollution problems common to each and arising from factors such as meteorological trends, locations and sizes of pollutant sources, social and governmental structures, and patterns of urban growth. The authority to implement national air quality standards is retained by each state for the portions of AQCR's within its boundaries. The state exerts this authority in

Table 3

Nationwide air pollutant emissions estimates, 1970–75 (millions of tons)

Year	Particles	SO_x	NO_x	Hydrocarbons	CO
1970	26.8	34.2	22.7	33.9	113.7
1971	24.9	32.3	23.4	33.3	113.7
1972	23.4	36.7	24.6	34.1	115.8
1973	21.9	35.6	25.7	34.0	111.5
1974	20.3	34.1	25.0	32.9	103.3
1975[a]	18.0	32.9	24.2	30.9	96.2

a. Preliminary

Source: "National Air Quality and Emissions Trends Report, 1975," EPA-450/1-76-002, Environmental Protection Agency, Research Triangle Park, N.C., 1976.

accord with a State Implementation Plan (SIP), which must provide in part for emission limitations and timetables for compliance, air monitoring systems, inspections, and adequate staff and funding. The implementation plan is drawn up by the state, but must be approved by EPA. If a state fails to draft a satisfactory implementation plan, EPA must supply one.

Although the states are generally responsible for controlling contaminant emissions within their borders, the law authorizes EPA to set emission standards for new stationary (nonautomotive) sources, mobile (automotive) sources, and hazardous air pollutants. New Source Performance Standards for stationary sources (5) are set for categories of sources and, within each category, place maximums on emissions of specified contaminants. The standard for the cate-

Table 4

Summary of national ambient air quality standards[a]

Pollutant	Averaging time	Primary standards	Secondary standards	Comments
Particulate matter	Annual (geometric mean)	75 $\mu g/m^3$	60 $\mu g/m^3$	The secondary annual standard (60 $\mu g/m^3$) is a guide for assessing State Implementation Plans to achieve the 24-hour secondary standard.
	24-hour[b]	260 $\mu g/m^3$	150 $\mu g/m^3$	
Sulfur oxides	Annual (arithmetic mean)	80 $\mu g/m^3$ (0.03 ppm)	—	
	24-hour[b]	365 $\mu g/m^3$ (0.14 ppm)	—	
	3-hour[b]	—	1300 $\mu g/m^3$ (0.5 ppm)	
Carbon monoxide	8-hour[b]	10 mg/m^3 (9 ppm)		
	1-hour[b]	40 mg/m^3 (35 ppm)	(same as primary)	
Nitrogen dioxide	Annual (arithmetic mean)	100 $\mu g/m^3$ (0.05 ppm)	(same as primary)	
Photochemical oxidants	1-hour[b]	160 $\mu g/m^3$ (0.08 ppm)	(same as primary)	
Hydrocarbons (nonmethane)	3-hour (6 to 9 a.m.)	160 $\mu g/m^3$ (0.24 ppm)	(same as primary)	The hydrocarbon standard is a guide to devising State Implementation Plans to achieve the oxidant standard. The hydrocarbon standard need not be met if the oxidant standard is met.

a. In December 1977, EPA proposed an ambient air quality standard for particulate lead: 1.5 $\mu g/m^3$ monthly average, to be attained by 1982 and maintained thereafter.

b. Not to be exceeded more than once per year.

Source: Environmental Protection Agency.

gory "steam generators," for example, limits emissions of particles, sulfur oxides, and nitrogen oxides. The standard for sulfuric acid plants limits emissions of sulfur oxides and acid mist. The standard for phosphate fertilizer plants limits emissions of fluorides. Most of the standards for new stationary sources include limits on one or more of the "criteria pollutants" — those for which National Ambient Air Quality Standards exist. No emission limits are set for oxidants/ozone, however, because that criteria pollutant stems largely from photochemical reactions in the air; control is based on limiting the emissions of the nitrogen oxides and hydrocarbons involved in the reactions.

The 1970 act accelerated the earlier schedule for abating pollution by motor vehicles (cars, trucks, buses). Emissions of hydrocarbons and carbon monoxide by new vehicles were to be 90% lower in 1975 than in 1970; emissions of nitrogen oxides were to be 90% lower in 1976 than in 1971. Emission standards for new vehicles were established accordingly, but the schedule for compliance was extended at various times, most recently by the amendments of 1977.

The national standards for emissions of "hazardous" air pollutants by stationary sources cover contaminants that ". . . may cause, or contribute to, an increase in mortality or an increase in serious irreversible, or incapacitating reversible, illness." Through December 1977, EPA had designated asbestos, beryllium, mercury, and vinyl chloride as hazardous air pollutants. The agency had established emission standards for various categories of stationary sources of these substances.

Deadlines

The deadlines (attainment dates) embodied in the Clean Air Act of 1970 had not in general been met through December 1975 (6), and the amendments of 1977 extended them through 1982. EPA estimated from monitoring data that some 47% of the Air Quality Control Regions met the ambient air quality standard for particles by the statutory date, May 31, 1975, and that some 86% met the standard for sulfur dioxide. Monitoring data for carbon monoxide and oxidants were less broadly informative. An analysis in spring 1975, however, showed violations of the carbon monoxide standard in 28% of the AQCR's and of the oxidants standard in 32% of them. Oxidant levels as high as twice the national standard were found in rural areas during 1975. The picture for nitrogen dioxide was obscured by the discovery, in 1972, that the prescribed monitoring method (the reference method) gave concentrations that were higher than actually existed (7) (EPA adopted a new method in 1976). Reliable data available in spring 1975, meanwhile, indicated that nitrogen dioxide levels were at or above the national standard in only about 6% of the AQCR's. Data on criteria pollutants for 1976 showed no surprising new trends (1,2). Where carbon monoxide was measured, the trend was downward, owing mainly, apparently, to auto emission limits; oxidant remained a widespread problem.

The foregoing data do not mean that the carbon monoxide standard, for example, was being exceeded in 28% of the U.S. in spring 1975, nor are they necessarily an accurate guide to air quality. In the first place, if a standard is exceeded at a single measurement site, the entire AQCR is rated in violation. In the second place, an AQCR is not rated in violation, irrespective of air quality, for pollutants not being measured.

Delays occurred also in the scheduled rollback of automotive emissions, the major source of urban carbon monoxide and an important contributor to oxidant levels. The original new-vehicle standards for carbon monoxide and hydrocarbon emissions were suspended in favor of less-stringent, interim standards for model years 1975 and 1976; nitrogen oxides limits were revised similarly for model years 1976 and 1977. The amendments of 1977 extended the 1977 limits for all three contaminants through model year 1979, and delayed until 1981 the imposition of the most stringent standards in the law. These delays may well have an impact on compliance with attainment dates.

FLOW, DISPERSION, TRANSFORMATION

Efficient management of the air environment depends in the long run on our understanding of how contaminants flow, disperse, and are transformed chemically into other physical and chemical forms as they move from source to sink or receptor. An understanding of these natural pollutant-removal processes is especially important on a local and regional basis. It is less so on a global basis: many substances do not mix thoroughly into the global atmosphere because their lifetimes (residence times) there are shorter than the time required to do so. The exceptions include carbon dioxide and particles, whose global behavior and possible effects on climate are matters of some concern.

Most of man's technology consists really in controlling and, usually, accelerating natural processes, so that most of the pollutants he generates occur also in nature. The difference is in the amounts produced. Sulfur dioxide, for example, is always present in air as a result of naturally occurring oxidation. The amount of the gas in the air in a given geographic region, however, can be increased sharply by man's use of combustion — accelerated oxidation — of sulfur-containing fuels. This man-made excess of sulfur dioxide becomes "pollution" to the extent that it overloads the natural disposal capacity of the region.

Growing comprehension of pollutant-removal and similar processes led in the early 1970's to the development of the Gaia hypothesis (after the Greek earth goddess) (8–11). The hypothesis visualizes a homeostatic or equilibrium-seeking earth; it holds essentially that the biosphere, the entire biological system of earth, behaves like a single organism (Gaia) that acts to optimize the terrestrial environment in its own behalf. Quite good evidence indicates, for example, that while life has been on earth the temperature of the surface has changed very little although the energy arriving from the sun may have as much as doubled. Gaia apparently has compensated for rising incident energy by adjusting such parameters as the nature of the earth's surface and of the atmosphere. Growing plants over the centuries have produced enough hydrocarbons to coat the earth's land surface with tar to a depth of some meters. Complex sequences of chemical reactions, however, destroy these hydrocarbons and return most of the carbon in them to the atmosphere as carbon dioxide, which can be utilized by plants. Volcanic emissions and newly exposed igneous rock contain large quantities of chemically reduced substances that absorb oxygen during natural weathering. The resulting oxidizing reactions consume oxygen and would tend to deplete the atmosphere of the element if new oxygen were not produced. The key reaction in the production of new oxygen has been hypothesized to be the production of methane and its transport into the stratosphere (8); on this basis, the biosphere would appear to divert energy into the production of methane so as to balance the somewhat variable removal of oxygen.

The homeostatic earth of the Gaia hypothesis has certain consequences for air pollution:

- Since most man-made pollutants also occur naturally, or closely resemble naturally occurring substances, routes already exist for converting them to forms not dangerous to life and removing them from the atmosphere.
- Substances sufficiently refractory (resistant to breakdown) to become global pollutants, such as the chlorofluorocarbons, are refractory precisely because they differ from natural products enough not to fit easily into natural disposal schemes.
- Excepting these novel materials, pollution results not from the presence of wastes, but from their presence in amounts large enough to overload the natural disposal capacity.
- Since a homeostatic earth reflects the activities of the entire biosphere, the impact of pollution on the lowliest bacteria may well be fully as important as its impact on man in terms of the latter's long-range survival on earth.

The distinction between natural and man-made contaminants sometimes is not obvious. We tend to think of a sandstorm in the Sahara Desert as a natural phenomenon. But within historic times that area was reasonably well covered with vegetation. Some scientists believe that it was denuded as a result of overgrazing by domestic animals (12); there is evidence that overgrazing has nearly turned the Sahel into desert (13) and suspicion that a similar change could occur in South America (14). If this is true, then air pollution by flying sand in the Sahara (and elsewhere) may not be natural, but the result of human mismanagement of the land.

The homeostatic nature of earth is suggested by comparisons of air analyses. The approximate concentrations of four gases in unpolluted tropical atmospheres appear in Table 5 (15). These levels probably are representative also of the trace (background) composition of the atmosphere in temperate latitudes within perhaps an order of magnitude (from 0.1 to 10 times the true values). The background concentrations of three of these gases and of two others in the atmosphere of the central U.S. appear in Table 6 (16). The concentrations of Table 6 are not totally natural. They represent the natural background plus the sum of contributions from a number of distant man-made sources. Nevertheless it can be seen that, even in a generally populated area, at points not under the immediate influence of a major city, the atmospheric composition is shifting toward the background level as a result of the natural air-cleaning processes that are a part of the metabolism of the hypothetical Gaia. Specifically, the concentrations of nitrogen oxides and sulfur dioxide in Table 6 are well below typical urban values but higher than the background levels of Table 5. The concentration of ammonia, on the other hand, is distinctly lower in Table 6 than in Table 5. Ammonia is primarily of natural origin, and less of it is formed in most cities than in the countryside. It may be part of the natural mechanism for removing acidic gases.

The final consequence of the Gaia hypothesis, if it is correct, is that few kinds of pollution or other environmental disturbance produce effects as serious as would be predicted in the absence of this kind of biospheric feedback. On the other hand, a disturbance large enough really to overpower this feedback may have even larger consequences than would be predicted, since it may also affect other feedback loops.

A warning is necessary about the data presented here on the composition of unpolluted air. Almost without exception the measured values are at the extreme limit of sensitivity of the analytical methods used. Far more work must be done on more sensitive methods before such values can be considered reliable

Table 5

Concentrations of atmospheric trace gases in the American Tropics [Parts per billion by volume (micrograms per cubic meter by weight)][a]

	Ammonia (NH_3)	Nitric oxide (NO)	Nitrogen dioxide (NO_2)	Sulfur dioxide (SO_2)
Forest (under canopy)				
Ducke Forest Res. (Brazil)	9 (6.9)	0.1 (0.13)	0.2 (0.4)	0.3 (0.9)
Albrook Forest (Panama)	16 (12.0)	0.5 (0.57)	0.4 (0.8)	1.1 (3.1)
Other Panama	13 (9.9)	0.3 (0.40)	0.3 (0.6)	0.5 (1.4)
Forest (above canopy)				
Albrook Forest (Panama)	15 (11.4)	0.4 (0.53)	0.5 (1.0)	0.9 (2.6)
Other Panama	15 (11.4)	0.3 (0.40)	0.3 (0.6)	0.3 (0.9)
Interior river				
Rio Taruma/ Solimoes (Brazil)	25 (19.0)	0.3 (0.40)	0.3 (0.6)	0.3 (0.9)
Chagres R. (Panama)	5 (3.8)	0.7 (0.94)	0.6 (1.2)	0.3 (0.9)
Savannah/cleared area				
Panama	15 (11.4)	0.5 (0.57)	0.6 (1.2)	1.0 (2.9)
Maritime				
Caribbean				
Ft. Sherman, Panama	31 (23.6)	0.7 (0.94)	0.5 (1.0)	1.3 (3.7)
Other Panama	9 (6.9)	0.3 (0.40)	0.4 (0.8)	0.4 (1.1)
Barbados	15 (11.4)	0.3 (0.40)	0.3 (0.6)	1.1 (3.1)
Advance II	16 (12.0)			
Pacific				
Bay of Panama	18 (13.8)	0.6 (0.80)	0.7 (1.5)	1.3 (3.7)
Generalized tropical values	15 (11.4)	0.4 (0.53)	0.4 (0.8)	0.9 (2.6)

a. Values given in source in parts per billion only; converted to micrograms per cubic meter at 0° C and 760 torr.

Source: Lodge, J.P., Jr., *et al*, "Atmospheric Trace Chemistry in the American Humid Tropics," *Tellus*, **26**, 250 (1974).

for use in planning or for determining compliance with nondegradation regulations (those designed to prevent "serious deterioration of air quality" where quality exceeds the national standards).

Implicit in the data of Tables 5 and 6 is one additional point. North-south mixing in the atmosphere is generally slower than east-west mixing, and mixing across the equator is still slower. As a result, there are pronounced north-south variations in the global distribution of substances with atmospheric lifetimes in

Table 6

Concentrations of trace gases measured during 1971 near Athensville, Ill., and Cofman, Mo. [Parts per billion by volume (micrograms per cubic meter by weight)][a]

Time, CDT[b]	Hydrogen sulfide (H_2S)	Sulfur dioxide (SO_2)	Nitrogen dioxide (NO_2)	Ammonia (NH_3)	Aldehydes[c] (RCHO)
		Oct. 13, Athensville			
0700		d	1.7 (3.5)	2 (1.5)	
0830		d	2.0 (4.1)	10 (7.6)	
1030		d	1.2 (2.5)	5 (3.8)	
1230		d	1.2 (2.5)	2 (1.5)	
1430		d	1.1 (2.3)	6 (4.6)	
1630		d	1.5 (3.1)	5 (3.8)	
Avg		0.25 (0.71)	1.4 (2.9)	5 (3.8)	
		Oct. 26, Cofman			
1250		d	1.1 (2.3)	d	5.0 (6.7)
1505		d	1.0 (2.1)	d	5.3 (7.1)
1725		d	1.1 (2.3)	d	2.7 (3.6)
1920		3.2 (9.12)	1.2 (2.5)	d	1.5 (2.0)
Avg		1.0 (2.85)	1.1 (2.3)	2 (1.5)	3.6 (4.8)
		Oct. 31 to Nov. 1, Athensville			
1900			0.9 (1.8)	2 (1.5)	2.0 (2.7)
2000	0.08 (0.12)			d	
2100			1.5 (3.1)	d	3.6 (4.8)
2300			5.0 (10.3)	d	1.4 (1.9)
0030	0.08 (0.12)				
0100			2.2 (4.5)	d	2.2 (2.9)
0255			1.4 (2.9)	d	1.6 (2.1)
0400	0.15 (0.23)				
0500			1.6 (3.3)	d	1.1 (1.5)
0700			1.4 (2.9)	7 (1.5)	0.9 (1.2)
0800	0.09 (0.14)				
0900			2.9 (5.9)	5 (3.8)	0.8 (1.1)
Avg	0.10 (0.15)		2.1 (4.3)	3 (2.3)	1.7 (2.3)

a. Values given in source in parts per billion only; converted to micrograms per cubic meter at 0° C and 760 torr.

b. The time reported is the midpoint of the sampling period.

c. Calculated as formaldehyde, probably the predominant aldehyde present.

d. In constructing the averages, an entry of d was considered to be one half of the sensitivity of the method used.

Source: Breeding, R.J., et al, "Background Trace Gas Concentrations in the Central United States," *J. Geophys. Res.*, **78**, 7057 (1973).

the range from a few days to a few years. Products of human origin are concentrated in northern mid-latitudes where the population is the greatest. Products of the land surface or of coastal regions are more prevalent in the northern hemisphere, while products of the oceans are most concentrated in the southern hemisphere.

Recommendation A2: *Greater attention must be paid to developing and validating techniques for measuring contaminants at the levels characteristic of "clean air" and for measuring the global background concentrations of airborne substances of all kinds, irrespective of their known sources and effects.*

Transport, Diffusion

The flow and dispersion of matter are physical processes commonly called transport and diffusion by specialists. For air pollutants they are aptly illustrated by the behavior of the effluent or plume from a factory or power-plant stack. The effluent generally is warmer than the environment it enters. It cools by radiating heat and by diffusion processes and heat transfer to the cooler air, but until equilibrium is reached it remains warmer and thus less dense than the surrounding atmosphere and attempts to rise through it. With no wind, a stack plume may rise vertically to great height, or it may stop rising at some lower point and spread into a large, flat layer, depending on when its density matches that of the surrounding air. When there is wind, the faster it blows, the less the plume rises before its buoyancy is dissipated by diffusion processes. At high speed, the wind essentially shears off the effluent at the top of the stack. A plume may travel horizontally with the wind, or it may propagate in great loops that touch the ground a short distance from the stack, exposing that area momentarily to high concentrations of contaminants. This complex repertory of behaviors results from two atmospheric phenomena: thermal structure and aerodynamics.

The thermal structure of an air mass is related to its stability. Temperature in the lower atmosphere (the troposphere) tends to decline with increasing altitude at a rate called the "lapse rate." When the lapse rate is about 1°C per 100 m of altitude, a parcel of air moved up or down, and allowed to expand or contract to reach the density of the surrounding air at its new altitude, will also reach the temperature of the air at that altitude and will remain there. If the lapse rate is more than 1° C — as it would be near ground being warmed by intense sunlight, for example — a parcel of air given a slight upward push remains warmer than its surroundings and continues to rise. Such air is called "unstable." If the lapse rate is less than 1°C per 100 m, a parcel given a slight upward push remains colder than its surroundings and sinks to its original position. Such air is called "stable." The extreme case of stable air — where the temperature actually rises with increasing altitude — is called an "inversion." With no wind, a stack plume will rise to great height in unstable air. In stable air the plume will stop rising at a height determined by the interaction of heat loss, dilution, and its initial buoyancy.

With wind, aerodynamic effects arise. Wind of more than a certain velocity does not move as a solid block of air across even a perfectly smooth earth. Instead it breaks into eddies of all sizes, its speed and direction only averaging the nominal speed and direction of the air mass. The earth's surface, moreover, is by no means smooth, but is covered with roughness elements ranging from blades of grass through houses, ocean waves, and mountains. All of these cause eddies and fluctuations in wind speed and direction. Hence, a strong wind that might be expected to carry off smoke in a perfectly straight line may move it up, down, and sideways from the general direction of travel, causing the looping mentioned earlier. Simultaneously, smaller eddies may tend to spread the plume so that its cross section grows with distance traveled.

Mathematical models

Because of these thermal and aerodynamic effects, it can become extremely difficult to trace a particular parcel of polluted air to its source. The parcel may have traveled under an inversion for some time, perhaps, then risen as the sun heated the ground and dissipated the inversion, and then returned to the surface in a large, vertical eddy; at the same time, chemical transformations may have been occurring in the parcel. If, instead of a single source, the question concerns the impact of an additional source on a city with thousands of individual sources, the problem becomes even more complex. Still, it is clearly important to be able to correlate dangerously high concentrations of pollutants with their sources, to predict the effects of adding additional sources, and to regulate emissions in a manner that will achieve the proper control overall. A better understanding of the problem is badly needed, and the quest for it has involved three general approaches. Each results in one or more mathematical expressions called models.

The model for a given area is supplied data on source locations and emission rates, chemical transformations, meteorology, and other pertinent factors; it is then used to calculate the distribution and concentrations of ambient pollutants in the area. A sound model should make it possible to predict emission limits for various sources in consonance with changing conditions so as to achieve the desired air quality at minimum social cost.

The simplest of the three approaches to the problem is the "rollback" model, so named because it involves a rollback of local rates of contaminant emission (17). Briefly, the model assumes that if the rate today, when we have an air pollution problem, were reduced to the rate of an earlier year, when there was no problem, we would no longer have the problem today. This assumption leads to a simple mathematical proportionality and is certainly the simplest tack. On the other hand, the method is anything but rigorous. At best it would be strictly correct only under four conditions: the size of the source region (a city, for example) did not change during the period considered; the general climate did not change; the imposed controls decreased discharges from all sources proportionately; and the pollutants are mixed completely, so that the concentration of each is uniform throughout the area considered. These conditions are never met completely, and the results produced by the rollback model are, therefore, extremely approximate.

The second class of models, probably best called "semiempirical," embrace a much wider range of complexity. They vary from single, relatively simple initial equations to enormously complex expressions into which local topography, roughness introduced by buildings, and many other factors are worked according to reasonable relationships. Common to all semiempirical models are certain adjustable parameters. The output of the model is compared with events in the real world, and these parameters are adjusted until, for at least a few situations, the model yields the answer shown by experience to be correct. The parameters adjusted, it is assumed that the model, applied to new and different arrangements of pollutant sources, will yield accurate distributions of ambient pollutant concentrations (18). Even a relatively simple semiempirical model may require a rather large computer, not because of conceptual complexity, but because a city contains so many interacting pollutant sources.

The third approach is the "primitive equation" model. It starts with the fundamental equations that describe how energy is transferred from the sun to the earth's surface and thence to the atmosphere and those that describe on the most basic level the motions of air in response to this energy and to the earth's topography and rotation. Fundamental physical constants are introduced, but not adjustable parameters. Theoretically it should be possible to describe in this fashion the motion of the entire atmosphere of the earth or any portion thereof. In practice the equations become so complex that even the largest computers take longer to solve them than the phenomena themselves take to happen. Thus the pollution forecast for tomorrow would not be ready until the day after tomorrow.

As of the late 1970's, it appears in general that rollback models, though widely employed, are too inaccurate to be particularly useful, and that primitive-equation models will require advances in computers that are still some years in the future. Some of the simpler semiempirical models have proven remarkably useful and perhaps unexpectedly accurate. The many adjustable parameters of the larger and more complex models cannot be optimally adjusted without a large body of quite detailed meteorological and pollutant distribution data and other information. Erroneous data yield a bad model; for that reason, attempts thus far to adjust semiempirical models using existing data have had discouraging results. In consequence, it is probable that a simple model is preferable, on the grounds that added complexities do not improve the model's performance commensurately with the cost incurred.

It is certainly to be hoped that the improved data needed to validate models will begin to emerge during the late 1970's or soon thereafter. The generation of such a body of data was the goal of the Regional Air Pollution Study (RAPS), a five-year, $22 million effort in St. Louis funded by EPA (19). The RAPS monitoring network began to operate in mid-1974, and the field measurement program was concluded in mid-1977. The use of the resulting data to evaluate existing models and, to a lesser extent, to develop new ones was under way in mid-1978. The funds allocated to RAPS initially, however, were not increased in step with inflation and may well have been insufficient to accomplish the main mission of the study satisfactorily.

Partly related to the modeling problem is the use of statistical techniques to arrive at the monitoring scheme required to obtain a given body of data. Suppose that one wished to determine the mean concentration of an air pollutant over an entire city for a full year with an accuracy of ± 10% at the 90% confidence level. In the interest of efficiency it would be desirable to employ the minimum number of monitoring stations and the minimum rates of sampling necessary to achieve that goal. However, we do not really have the statistical tools needed to make such determinations with confidence.

Chemical Transformations

As pollutants move through the atmosphere, they undergo chemical transformations that proceed at a wide range of rates and yield a variety of products. The general trend of these transformations is toward atmospheric self-purification, but some of the intermediate substances produced along the way are biologically very active at the concentrations in which they occur in highly polluted air. Other contaminants, whether primary or secondary, may be capable of affecting the physical environment, sometimes in ways that extend in principle, at least, to global phenomena such as climate. The following discussion covers some of the more important transformations that have been studied to date.

Particles

Particles, both liquid and solid, are the most widespread of all the substances that usually are considered pollutants (20). Particles of natural origin include sea salt and those produced by erosion, volcanoes, and forest fires. Man-made particles larger than 10 μm in diameter come mainly from mechanical processes such as erosion, grinding, and spraying. Those between 1 and 10 μm stem partly from mechanical processes and also include industrial dusts and ash. Among particles in the size range 0.1 to 1 μm, sulfates and the products of combustion begin to predominate, along with aerosols formed by chemical reactions in the air. Relatively little is known of the chemical nature of particles in the size range below 0.1 μm. Over cities, however, their concentration characteristically is higher than the natural level, and the excess seems to be due to combustion. Particles larger than about 2 to 5 μm settle on surfaces and constitute "dustiness." However, they have little effect on visibility or human health, since they neither scatter light efficiently nor penetrate beyond the nose or pharynx.

The size distribution of airborne particles tends to be relatively constant. It covers a range of diameters of about 0.001 μm to a few tens of micrometers. Particles in the range of 0.1 to 10 μm account for most of the mass and a large fraction of the numbers of those present. The relatively constant size distribution has been hypothesized to result from the steady generation of small particles by condensation and of large ones by mechanical processes, combined with the nature of the mechanisms that remove particles from air (21). This explanation is qualitatively good, but quantitative support for it is weak. A second explanation is that the observed size distribution is simply the statistical result of mixing particles from many sources. It should be noted that the normally constant size distribution is not found in the immediate vicinity of sources of particles; in particular, a shift to much larger sizes is characteristic of arid areas, where the soil contributes a large portion of the particles in the air.

The small condensation aerosols and the larger, mechanically generated aerosols differ widely in size, so that the size distributions of atmospheric aerosols normally have at least two maxima. The large size range usually has a peak somewhere between 5 and 20 μm, and the small size range usually has one in the vicinity of 0.5 μm. Near sources of particles there may also be a much smaller maximum (22) unless the sources produce large particles initially as with the arid areas of the preceding paragraph.

Particle behavior. It is not possible to generalize on the chemical behavior of particles because they are so diverse and little understood. Chemical reactions may occur between particles, especially those in the size range below 0.1 μm, which collide relatively often with each other. Particles also may react with gases. Particle-particle and particle-gas reactions have not been studied extensively, however, although recent years have seen considerable work on the formation and role of particles in photochemical smog and on sulfate aerosol in general.

Water vapor plays an important and often-neglected role in the chemistry of particulate matter. One example is the reaction between gaseous ammonia and sulfuric acid mist to form the salt ammonium sulfate. At high humidity, the rate of reaction is controlled by the rate at which ammonia in the air can diffuse to the sulfuric acid droplets, where it reacts. At low humidity, the rate of reaction is controlled by the rate at which ammonium sulfate formed in the reaction diffuses away from the surface and into the acid droplets, thus exposing more surface for reaction with ammonia.

Gravitational settling is the main mechanism by which particles are removed ultimately from the air, but there are intervening mechanisms that vary with the size of the particle. Particles of less than 0.1 μm in diameter move randomly in air, collide often with other particles, and thus grow rapidly by coagulation; they would soon vanish from the air were they not replenished constantly by condensation processes. Particles in the next larger size range, 0.1 to 1.0 μm, probably are also removed from the air primarily by coagulation. They grow more slowly than the smaller particles, however, because they are somewhat less numerous, move less rapidly in air, and thus collide less often with other particles. At diameters larger than 1 μm, particles begin to develop appreciable settling velocities, and above 10 μm they begin to settle relatively rapidly, although particles as large as 10 μm can be kept airborne by turbulence for extended periods.

Particles may be removed from the air by impinging on buildings, trees, and other objects. Rain also removes them, but the effect is negligible at particle diameters of less than 2 μm. Particles smaller than 2 μm can occur in rainwater, however, if they originate in clouds, since cloud droplets are more effective than raindrops in collecting smaller particles.

Particles generated in an urban atmosphere normally remain airborne for only a few days, although, depending on their size, they may remain airborne for several weeks. Gravitational settling prevents larger particles, such as fly ash and soil, from traveling very far from their sources. The lifetime of a particle in air is a strong function of the height at which it is introduced, however, and large or intense sources, such as big metropolitan areas, erupting volcanoes, nuclear explosions, and forest fires, can produce particles that circulate globally.

Measurements of the concentrations of airborne particles at remote points are scanty, but the global level is believed to be rising (23). Intuitively it would appear that a layer of aerosols above the ground would invariably cool the ground, since less radiation would reach it. In fact, at least three effects are involved. The aerosols scatter light in all directions, although the majority of the energy still travels in nearly its original direction. The aerosols absorb energy and are thereby heated. And the aerosols prevent the radiation of heat from the ground by a blanketing effect. It is difficult to do unequivocal experiments in nature on these phenomena, and the mathematical treatments turn out to be extremely sensitive to the precise optical properties of the aerosols, which themselves are not well known. As a result, it is not yet clear whether aerosol layers actually are causing heating or cooling of the underlying surface (24–30) — that is, whether they are raising or lowering mean global temperature.

Sulfur dioxide: the sulfur cycle

Sulfur dioxide (SO_2) is one of several forms in which sulfur exists in air. The others include hydrogen sulfide (H_2S), sulfur trioxide (SO_3), sulfuric acid

(H_2SO_4), and sulfate salts such as ammonium sulfate [$(NH_4)_2SO_4$]. These compounds and derivatives of them are generated by man, but they also occur naturally in the global sulfur cycle, which extends not only through the atmosphere, but through soils, waters, plants, and animals as well (Fig. 1). The relative contributions of man and nature are uncertain, although man has modified the cycle enough so that it can no longer be studied in its natural state (31).

Compounds of sulfur come from many sources. An estimated 99% of the sulfur dioxide that enters the atmosphere is produced by man and 1% by volcanoes, the only natural source known to be significant. Some two thirds of the man-made sulfur dioxide results from combustion of sulfur-containing fuels and one third from industrial processes such as petroleum refining and the smelting of nonferrous metals; such sources also emit a small amount of sulfur trioxide. Hydrogen sulfide is emitted by industrial processes and volcanoes; it is released in possibly far greater amounts by decaying organic matter on land and in the oceans. Sulfate aerosol from sea salt is injected into the air in large amounts by the bursting of bubbles at the surface of the seas. Some evidence suggests that organisms in oceans and soils may produce organic sulfur compounds, such as dimethyl sulfide, which enter the sulfur cycle. These biogenic sources, in fact, have been estimated to account for at least half of all the sulfur in the global atmosphere. On the other hand, evidence based on the contents of sulfur isotopes in compounds of the element indicate that biogenic processes may not be important sources of atmospheric sulfur.

The contributions of the various sources of airborne sulfur compounds can be compared in terms of their sulfate equivalents. On this basis, man accounts for perhaps a quarter of the sulfur compounds in the atmosphere at any given time, sea salt for about a quarter, and biogenic and other processes for about a half (assuming that biogenic sources are in fact important contributors). These global estimates differ from those for the two hemispheres. Man-made sources, for example, are much more common in the northern hemisphere; they contribute some 40% of the airborne sulfate equivalent there, as opposed to 6% in the southern hemisphere. Similarly, man may account for considerably more than 25% of the sulfate equivalent in highly polluted air.

Sulfur and its compounds undergo a variety of chemical reactions in air. Sulfur itself and hydrogen sulfide are oxidized to sulfur dioxide, which in turn is oxidized to sulfur trioxide. The latter dissolves in water droplets to form sulfuric acid, which may react further to form ammonium sulfate or other sulfate salts. Both acid and salts exist in air as aerosols. Sulfur dioxide also can be oxidized in water droplets that contain ammonia; the end product, again, is ammonium sulfate. These reactions proceed at rates that vary widely with the circumstances. Oxidation in water droplets, for example, can be speeded by the catalytic effects of ions of metals like iron and manganese. In air that contains nitrogen dioxide and hydrocarbons, sulfur dioxide is oxidized very rapidly in a photochemical reaction process that yields aerosols containing sulfuric acid.

Aerosols containing sulfate are distributed broadly through large regions of the United States and elsewhere (32, 33). One of the fundamental questions concerning them is the speed with which sulfur dioxide is converted to sulfuric acid droplets and the rate and means by which these droplets are removed from the air. Research in the area (34–37) has been criticized on the grounds that its results cannot be extrapolated to conditions that actually exist in the air. Sulfuric acid in the air has been studied (38, 39), but the speed and mechanism of its formation there and the details of its adsorption on particles remain obscure.

FIGURE 1

Sources and sinks of atmospheric sulfur compounds

(10^6 tons of sulfate equivalent per year)

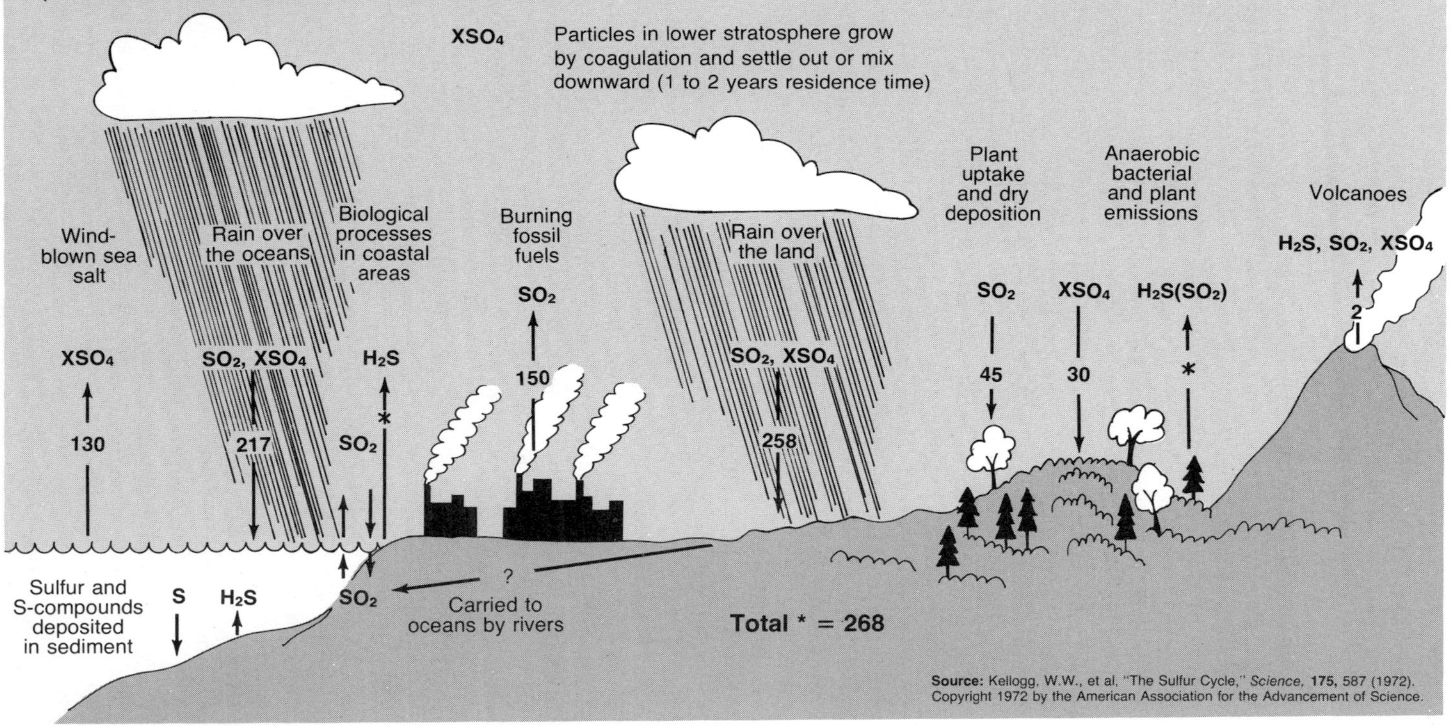

Source: Kellogg, W.W., et al, "The Sulfur Cycle," *Science*, **175**, 587 (1972). Copyright 1972 by the American Association for the Advancement of Science.

A number of mechanisms remove sulfur compounds from the air. Sulfate aerosol, including sulfuric acid, is washed out by rainfall. It is removed also by impingement on soil and vegetation and on buildings and other structures. Sulfur dioxide can diffuse through the atmosphere to the surface, where it is taken up in substantial amounts by soil, water surfaces, and vegetation.

Acid precipitation. The mechanisms that remove sulfur compounds from the air can produce significant local or regional effects. One of these is acid precipitation, a growing problem both in Europe and the eastern U.S. (40, 41). It results mainly from increased use of sulfur-containing fuels. When rain and snow absorb airborne sulfur compounds they become more acidic, a change reflected by lowered pH. A pH of 7.0 is considered neutral; values below 7.0 are increasingly acidic; values above 7.0 are increasingly alkaline. Absolutely pure rainwater has a pH of about 5.5 because of dissolved carbon dioxide, but a variety of processes normally tend to raise that figure once the rain has fallen. The pH of precipitation over parts of Europe began to decline in the 1960's, and a corresponding decline was observed in the pH of surface waters. The pH of Lake Vanern in Sweden fell from about 7.3 in 1965 to about 6.8 in mid-1967. By 1976, lake waters in parts of Norway had dropped below pH 5.5 and in some cases as low as 4.0. The pH of lake waters depends on factors such as temperature, biological activity, and mineral composition as well as on precipitation, but acid precipitation had become a significant influence by the 1970's. Sulfur dioxide over western Europe tends to lie in layers 1 to 2 km above the surface and may travel many kilometers before being precipitated in rain or snow. An estimated three quarters of the sulfur deposited in precipitation in Norway and Sweden originates in industrial areas in the United Kingdom and Europe.

Declining pH threatens aquatic life, since most organisms cannot live at less than pH 4.0. Studies in Norway show few or no fish in lakes below pH 5.5; they show also that declining pH markedly reduces the numbers and abundance of species of phytoplankton, zooplankton, and bottom-dwelling (benthic) organisms. Acid precipitation also has been implicated in damage to fish populations in lakes in Ontario, Canada, and in the Adirondack Mountains of New York. The phenomenon may affect vegetation as well. Research in southern Sweden suggests that it may have reduced forest growth about 4% between 1950 and 1965.

Lifetimes, global effects. The lifetimes of sulfur compounds in air reflect the competing chemical and physical processes that modify and remove them. Estimated lifetimes in the lower troposphere are less than a day for hydrogen sulfide, several days for sulfur dioxide, and several days to one week for sulfate aerosol. Within these lifetimes, such substances may travel up to several thousand kilometers from their sources. It has been shown that under fairly typical conditions about two thirds of a quantity of sulfur dioxide confined to the layer of air next to the ground would be removed in the first 500 km of travel (42). Lifetimes tend to increase with altitude; a compound must be long-lived to reach high altitude, and once there it diffuses only slowly to the lower altitudes where it is more likely to be transformed and removed. A layer of sulfate aerosol of uncertain origin, for example, exists in the stratosphere, 16 to 18 km above the earth's surface. The lifetime of these particles is estimated at one to two years. The global significance of man-made emissions of sulfur compounds is obscure. Sulfate aerosol is widespread, as noted above. The fraction of it originating with man is quite uncertain, but nevertheless could be contributing to a possible increase in the global concentration of airborne particles with its potential effects on mean global temperature. In the long run, therefore, man-made emissions of sulfur cannot be excluded as a factor in geochemical and geophysical phenomena.

Nitrogen oxides

Of the eight possible oxides of nitrogen, only three — nitrous oxide (N_2O), nitric oxide (NO), and nitrogen dioxide (NO_2) — are known to be important constituents of the atmosphere. The most plentiful of the three is nitrous oxide, whose global concentration is about 0.33 ppm (6.6×10^2 $\mu g/m^3$) (43). Nitric oxide has been detected in nonurban air in the American tropics at levels of up to 0.7 ppb (0.94 $\mu g/m^3$) (Table 5); estimates of its global concentration range from less than 0.2 ppb (0.26 $\mu g/m^3$) to 10 times that level. Nitrogen dioxide has been measured in the same area at levels of up to 0.7 ppb (1.5 $\mu g/m^3$); estimates of its global level range from less than 0.5 ppb (1.0 $\mu g/m^3$) to 4 ppb (8 $\mu g/m^3$). More recent measurements, in the western United States, showed that the upper limit for nitrogen dioxide in the global troposphere is often 0.1 ppb (0.2 $\mu g/m^3$) (44).

Nitrous oxide arises naturally from biological action in soils, and this emission is increased by the application of artificial nitrogen fertilizers. Small amounts are contributed by automobiles using catalytic converters to reduce hydrocarbon emissions and by combustion processes generally, and the oceans have been proposed as a source. Man's overall contribution of nitrous oxide, though uncertain, would be expected to be relatively small. There is evidence, however, that the global level of the gas increased by perhaps 1.5% during the period 1964–74 (45). Nitric oxide and nitrogen dioxide are emitted by some chemical processes, but the major source is combustion, which converts nitrogen in the air chiefly to nitric oxide. The latter is oxidized slowly by oxygen, and quite rapidly by ozone, to nitrogen dioxide. Man-made sources cannot account for all of the nitric oxide and nitrogen dioxide that enter the air globally or even for a major part of it. Biological processes in soil may be a very large contributor, but little is known of such processes.

Nitrous oxide is relatively inert, and its lifetime in the lower atmosphere is uncertain and a subject of considerable interest. A sizable fraction of the tropospheric nitrous oxide mixes into the upper stratosphere, where part of it is oxidized to nitric oxide. The nitric oxide takes part in a process that is thought to be important in maintaining the level of ozone in the stratosphere. Long-term changes in the concentration of nitrous oxide, therefore, are cause for concern, although the entire chain of reactions is far from being understood. The stratosphere is the only identified sink for tropospheric nitrous oxide, but recent measurements indicate that the ground surface is a major sink and may even be the largest one (46). Estimates based on these measurements put the lifetime of nitrous oxide very approximately at about eight years.

Nitrogen dioxide is a strong absorber of ultraviolet light from the sun and for this reason is the trigger for photochemical reactions that produce smog in polluted air. The gas also can combine with water vapor in the air to form nitric acid, which in turn can react with ammonia or particles to form nitrates such as ammonium nitrate. Nitrogen dioxide that does not react photochemically ultimately becomes nitrate salt aerosol, mainly in the particle size fraction larger than 1 μm or so, which settles from the air or is removed by rain. Nitrogen dioxide has been thought to have a lifetime of about three days in urban air. It has been found recently, however, that nitrogen-containing gases can be converted to unexpectedly large amounts of nitrate on the filters of equipment used to sample ambient air for nitrate aerosol (47). The finding casts considerable doubt on earlier measurements of nitrate and calls for reassessment of the lifetimes of nitrogen dioxide and related gases in urban air. On a global basis, the lifetimes of nitric oxide and nitrogen dioxide are about five days.

Nitrogen cycle. The nitrogen oxides in the atmosphere are among many nitrogen compounds that take part in the global nitrogen cycle. Nitrogen itself makes up about 80% of the earth's atmosphere by volume, and its compounds are essential to plant and animal growth. Organic nitrogen compounds break down to yield the nitrogen compound ammonia (NH_3) and possibly some of the nitrogen oxides. On the whole, however, the global nitrogen cycle is not well understood, although it seems certain that man contributes a relatively insignificant fraction of the global content of airborne nitrogen compounds.

Hydrocarbons

Man produces only an estimated 15% of the total global emissions of hydrocarbons to the air. He is the leading contributor, however, in urban areas, where these compounds of carbon and hydrogen exert their chief pollutant effect by taking part in the chemical reactions that cause photochemical smog. Man-made sources of hydrocarbons include incineration, evaporation of industrial solvents, and combustion of coal and wood, but the major one is the processing and use of petroleum and its products, especially gasoline. Natural sources include forests and vegetation, which emit large amounts of hydrocarbons of the terpene class, and bacterial decomposition of organic matter, which produces very large amounts of methane (CH_4).

Man-made and naturally occurring reactive hydrocarbons are both removed from the air by photochemical reactions involving nitrogen oxides and ozone. These are the same reactions that create smog in highly polluted urban air. Methane, a nonreactive hydrocarbon, appears to be removed largely by oxidation by hydroxyl radical to form carbon monoxide. The global concentration of methane is about 1.55 ppm (11 x 10^2 $\mu g/m^3$), and its estimated lifetime in air is about two years. Data on nonmethane hydrocarbons (two to 10 carbon atoms) in other than urban areas are quite limited. A typical midcontinent concentration, in the absence of local man-made sources, is 25 to 30 $\mu g/m^3$; the global level has been estimated at 6 to 7 $\mu g/m^3$ (48).

Carbon monoxide

Our view of the global balance of carbon monoxide (CO) has changed substantially since 1969, when the gas was believed to be almost exclusively a man-made pollutant (49). Combustion was considered its only significant source, with the automobile accounting for 80% of global emissions. The residence time or turnover rate of carbon monoxide in the atmosphere was estimated to be three years, but the mechanism of its removal was uncertain. By 1975, a number of analyses had shown that nature produces some 10 times more carbon monoxide than man and that the residence time may be as short as one month. Also, the proposed major removal mechanism, reaction with hydroxyl radical (HO), had been placed on much sounder ground.

The new conclusions on natural sources and residence time stemmed from three independent research efforts (50). One relied on the fact that the content of radioactive carbon-14 in a sample of carbon monoxide will vary with the sample's origin and history. A second relied on the variation in the content of the stable isotopes carbon-13 and oxygen-18 in carbon monoxide with the atmospheric concentration of the gas. In the third approach, the importance of natural sources of carbon monoxide and its atmospheric residence time were deduced from a photochemical model of the lower atmosphere.

The proposed removal mechanism for carbon monoxide involved its oxidation by hydroxyl radical in the atmosphere to form carbon dioxide (51). Later it was pointed out that the large natural source of carbon monoxide could be the oxidation of atmospheric methane by hydroxyl radical (52). For these reactions to occur at the required rates the atmospheric concentration of the hydroxyl radical during daylight would have to average only about 4.7×10^6 molecules per cubic centimeter (53) or two ten millionths of a part per million (2×10^{-7} ppm). [A recent reevaluation of the rate of reaction of carbon monoxide with hydroxyl radical (54) would cut this estimate in half.] A technique using a tunable ultraviolet laser beam has been developed to measure this very small concentration of hydroxyl radical in air, and preliminary measurements are consistent with the predicted concentration (55).

Ambient levels. Carbon monoxide in ambient air has been measured extensively in remote places in both hemispheres. The average concentration in the northern hemisphere is 0.14 ppm (.17 mg/m^3); the average in the southern hemisphere is significantly lower, falling to 0.04 ppm (.05 mg/m^3) near Antarctica. These observations are consistent with a residence time for carbon monoxide that is much shorter than the exchange time between the two hemispheres, which is about one year. Removal processes in addition to oxidation by hydroxyl radical have been identified for carbon monoxide. They include biological activity in a number of soils and transport into the stratosphere with rapid conversion to carbon dioxide. Natural sources of carbon monoxide besides the oxidation of methane by hydroxyl radical have also been identified. The most significant reported to date (1975) are the oceans and the degradation of chlorophyll. None of these processes, however, appears to yield anything like the amount of the gas produced by oxidation of methane. It thus appears that the hydroxyl radical largely controls the balance of carbon monoxide in the troposphere. It has been suggested that the converse is the case — that methane and carbon monoxide may control the hydroxyl concentration (56). However, the suggestion carries the requirement that global carbon monoxide should already have increased significantly, which is not borne out by the available evidence.

Global concentrations of carbon monoxide have little bearing on urban atmospheres, where the concentration of the gas may exceed 60 mg/m^3. Although carbon monoxide plays a small role in the chemistry of urban atmospheres, it is removed chiefly by the physical processes diffusion and dispersion. To reduce the concentration of the gas in cities, its sources, principally motor vehicle exhaust, must be controlled directly. Through 1976, it appeared that controls on automotive exhaust in fact had resulted in a downward trend in the concentrations of carbon monoxide in the air of most cities (1, 57).

Photochemical smog

Photochemical smog was first recognized as a serious air pollution problem in Los Angeles County, Calif., in the late 1940's. The research that explained its origin — the action of sunlight on mixtures of hydrocarbons and nitrogen oxides — was a milestone in the study of the chemistry of polluted air. The large population of automobiles in Los Angeles is a significant source of hydrocarbons and nitrogen oxides in the air. The city enjoys regular sunshine, which provides the ultraviolet light that initiates the photochemical reaction. Diffusion and dispersion of pollutants are limited horizontally by low winds and surrounding mountains and vertically by the inversion that often lies over the area.

Since photochemical smog* was identified in Los Angeles, it has become a serious problem in most major urban areas in this country and in many abroad. Its impact, moreover, can extend hundreds of kilometers downwind. The manifestations of smog include haze and lachrymatory (tear-inducing) substances, and it is characterized by a relatively high level of oxidants**, chiefly ozone (O_3). Photochemical smog in Los Angeles has been considered serious at ozone levels above 0.5 ppm (1070 $\mu g/m^3$).

The chemical transformations involved in a smog system are extremely complex. The phenomenon is by no means fully understood, but important advances continue to be made (58). An ultimate goal of the study of photochemical smog is airshed models that can be used to predict reliably the degree to which smog will be eased by various emission control measures. Such models are necessary to evaluate pollution control strategies and options in transportation, land use, and urban growth.

An important photochemical cycle that is characteristic of smog chemistry involves nitric oxide, nitrogen dioxide, and ozone (Fig. 2). The cycle is initiated by the dissociation of nitrogen dioxide by sunlight into an oxygen atom and nitric oxide. The oxygen atom reacts rapidly with molecular oxygen (O_2) to form ozone, which in turn reacts rapidly with nitric oxide to form nitrogen dioxide again. This reaction cycle does not change the relative concentrations of the three reactants, nitric oxide, nitrogen dioxide, and ozone. When hydrocarbons are present, however, they enter the reaction scheme via an hydroxyl radical (HO) chain to effect a net conversion of nitric oxide to nitrogen dioxide with a concomitant buildup of ozone.

In urban atmospheres in the early morning, nitric oxide generally is found at much higher concentrations than nitrogen dioxide because it is emitted in much larger amounts by combustion processes, as in power plants and automobiles. As the day wears on, the concentration of nitric oxide declines as photochemical reactions convert it to nitrogen dioxide (Fig. 3). When the nitrogen dioxide reaches its peak concentration, ozone begins to appear for the first time in significant amount, beginning a second phase of smog generation.

Hydroxyl radicals. A longtime problem in smog chemistry involved the rate of consumption of hydrocarbons associated with the conversion of nitric oxide to nitrogen dioxide. Extensive work in the laboratory showed a rate of loss far higher than could be accounted for by the rates of reaction of hydrocarbons with oxygen atoms and ozone. The discrepancy was explained in 1972 in terms of a

*"Smog" is a word coined several decades ago from "smoke" and "fog" to describe the characteristic, highly polluted fogs of London. For no obvious reason it was appropriated, and is used widely, to describe the photochemically generated pollution haze first identified in Los Angeles. Ironically, if either smoke or fog is present in quantity, this variety of smog cannot occur.

**"Oxidant" is a general term meaning substances, such as oxygen and ozone, that oxidize other substances. In air pollution usage, "oxidant" usually means a compound that oxidizes a particular chemical reagent used to measure it (generally potassium iodide). The EPA analytical method is an instrumental technique based on the reaction between ozone and ethylene.

hydroxyl-radical chain reaction that previously had not been considered (59). In this chain, the hydroxyl reacts first with the hydrocarbon, initiating a series of complex reactions in which several molecules of nitric oxide are converted to nitrogen dioxide, with the hydroxyl being regenerated at the end of each cycle. The cycle is repeated many times; thus the hydroxyl radical can be regarded as a catalyst in that it promotes the conversion without itself being consumed. A simplified representation of this cycle is included in Fig. 2.

The dominant role of hydroxyl-radical chains in the conversion of nitric oxide to nitrogen dioxide has been confirmed indirectly by measurements of relative rate constants for reactions of a number of different hydrocarbons with hydroxyl radicals. These relative rate constants correlate very closely with the rates of conversion of nitric oxide to nitrogen dioxide observed for these hydrocarbons in the laboratory. Excepting a few unreactive hydrocarbons — methane, ethane, propane, acetylene, and benzene — all hydrocarbons studied appear to contribute significantly to the conversion of nitric oxide to nitrogen dioxide through the hydroxyl-radical chain mechanism. This is in contrast to reactions with oxygen atoms and ozone, where the unsaturated (hydrogen-deficient) straight-chain hydrocarbons display substantially larger rate constants than do other classes of hydrocarbons. Recently, measurements of hydroxyl radical in a laboratory photochemical reactor have shown concentrations that agree with the concentrations deduced from chemical mechanism considerations (60). These results provide direct confirmation of the hydroxyl-radical chain mechanism. In addition, further studies have been made of the relationship between hydrocarbon reactivity and hydroxyl rate constants (61, 62). The reactivities of the various hydrocarbons can serve as a guide to effective control measures.

Another discovery relating to hydroxyl-radical chains is that significant rates of conversion of nitric oxide to nitrogen dioxide can be obtained in laboratory studies without adding hydrocarbons. This occurs because the hydroxyl is an important intermediate in the conversion and, presumably, because minute amounts of hydroxyl-radical precursors are adsorbed on the walls of the reaction

FIGURE 2

Simplified reaction scheme for photochemical smog

NO_2	+	Light	$\longrightarrow$	$NO + O$
O	+	O_2	$\longrightarrow$	O_3
O_3	+	NO	$\longrightarrow$	$NO_2 + O_2$
HO	+	$RH(+O_2)$	$\longrightarrow$	RO_2
RO_2	+	NO	$\longrightarrow$	$RO + NO_2$
RO	+	O_2	$\longrightarrow$	HO_2 + aldehydes, ketones, etc.
HO_2	+	NO	$\longrightarrow$	$HO + NO_2$

Notes:

a. R represents a hydrocarbon radical.

b. Hydroxyl radicals (HO) come from a variety of sources, such as reactions of oxygen atoms and ozone with hydrocarbons, photolysis (light-induced breakdown) of aldehydes and nitrous acid (HONO), etc.

c. This reaction scheme is illustrative, not definitive. Research is still in progress on the detailed chemistry of the smog-forming process.

vessels. The presence of these precursors probably explains the apparent differences in smog reactions observed by different laboratories or with different smog-chamber walls (63).

The hydroxyl-radical chemistry involved in smog formation is related closely to the hydroxyl chemistry that determines the natural balance of methane and carbon monoxide in the global atmosphere. One difference between the cycles is their duration. The characteristic times are of the order of a day for smog chemistry, a month for the carbon monoxide chemistry, and a year for the methane chemistry. The newly developed ability to measure hydroxyl-radical concentrations directly in the atmosphere (55) should help markedly to upgrade our understanding of smog chemistry as well as of the more remote global chemistry.

Ozone formation. In the second phase of smog generation, marked by the appearance of ozone, nitrogen dioxide is consumed and the concentration of ozone rises to a maximum. The ultimate fate of the nitric oxide and nitrogen dioxide is not fully understood. One class of compounds formed from the nitrogen oxides (NO_x) are organic and inorganic nitrates (64). Of these, the peroxy nitrates, such as peroxy acetyl nitrate (PAN) and peroxy benzoyl nitrate (PBzN), have received much attention because they are powerful lachrymators. These compounds, however, account only partially for all of the nitrogen oxides lost in smog systems. More recent studies have shown that nitrogen oxides also are converted into inorganic nitrogen compounds, such as ammonium nitrate, and perhaps into nitrogen.

In the third and final phase of smog chemistry, the primary pollutants — hydrocarbons and nitric oxide — are essentially used up, and the system is

FIGURE 3

Variation of concentrations of nitric oxide, nitrogen dioxide, and ozone with time of day in Los Angeles, July 19, 1965

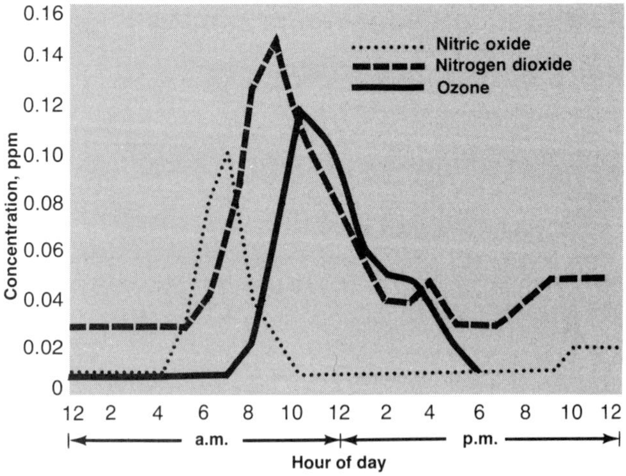

Source: "Air Quality Criteria for Photochemical Oxidants," National Air Pollution Control Administration, Publication Number AP-63, Washington, D.C., 1970.

characterized by a high concentration of ozone. The level of ozone that will result from differing reactant mixes in the atmosphere is one of the most critical questions to be answered; the relevant ozone chemistry is comparatively poorly understood. For ozone to form in the lowest layers of the atmosphere, nitrogen oxides must be present. The first three reactions of Figure 2 yield a low ozone level whenever nitrogen dioxide and sunlight are present, but for high oxidant levels to develop, hydrocarbons must also be present. A few alternative mechanisms have been suggested, but they have not been observed in the atmosphere.

Oxidant control. Urban atmospheres contain both hydrocarbons and nitrogen oxides, and we must learn what relative degree of control of these compounds would most effectively reduce oxidant/ozone levels. The addition of nitric oxide to a reacting air mass at one location, for example, clearly will reduce the concentration of ozone there because the two react rapidly to form nitrogen dioxide; overcontrol of nitric oxide emissions might be a mistake. It is possible, however, that the higher nitrogen dioxide concentration will result in higher ozone levels in the air mass as it moves downwind. Some evidence that this occurs has been found in the Los Angeles Basin. Vehicle emission control measures there have reduced ambient concentrations of hydrocarbons, but the levels of nitrogen oxides have remained nearly the same. In downtown Los Angeles, therefore, ozone concentrations have trended downward with the rise in the ratio of nitrogen oxides to hydrocarbons in the air (65). At points downwind, however, ozone concentrations have trended upward because of further photochemical aging of the air mass (66).

In any event, it is certain that present and anticipated emission control measures will reduce substantially the very high levels of ozone that occur occasionally in urban atmospheres, and attention then will shift to the minimums that we can expect realistically to achieve. The present goal is the National Ambient Air Quality Standard: 0.08 ppm (160 $\mu g/m^3$) for one hour, not to be exceeded more than once a year. Our growing understanding of atmospheric chemistry suggests, however, that this goal may not be attainable. The reason in part is that ozone concentrations in that range are insensitive to reductions in emissions of ozone-related hydrocarbons and nitrogen oxides. Field studies have shown, for example, that ozone levels in areas remote from emissions sources often exceed the air quality standard (67). These "background" levels of ozone may be due partly to intrusion of the gas into the troposphere from the stratosphere, where it is formed photochemically from oxygen. A more important source of high ozone levels in rural areas may be the transport of ozone formed in urban areas under meteorological conditions (a high pressure system) that prevent its dispersion into the global atmosphere.

To resolve these questions, we must improve our understanding of the ozone balance, and a promising development in this quarter is the Fourier transform infrared spectrometer. With this new instrumental technique, many constituents of smog can be measured simultaneously at minute concentrations (68), and new, important species such as pernitric acid (HO_2NO_2) can be identified (69). Also, in 1973, smog constituents were measured simultaneously at ground level and aloft for the first time in the Los Angeles Reactive Pollutant Program (LARPP), sponsored by the Coordinating Research Council and EPA (70). In this study, air masses labeled with radar-reflecting balloons were tracked across the Los Angeles Basin, and air-mass compositions were measured systematically at specific times and locations. The resulting extensive body of new data is providing an opportunity to model the generation of photochemical smog more precisely than ever in the past (71, 72).

Aerosol formation in smog

Smog chemistry yields not only gaseous products, but also particulate matter — the aerosols that are largely responsible for the visibility-reducing haze associated with smog. The chemical processes that transform gaseous molecules into solids are at least as complex as gas-to-gas transformations and are even less well understood. The problem is complicated by the presence of significant amounts of aerosols that are not formed photochemically. The atmosphere of Los Angeles, for example, contains soot and ash generated directly by combustion processes, sea salt and soil dust generated by natural processes, oil droplets and lead particles from autos, and numerous other kinds of primary particles.

A major reason for our imperfect knowledge of smog-related (and other types of) aerosols has been the lack of methods for determining their chemical compositions, but the 1970's have seen marked progress in this area. Analytical techniques that show promise in this application include high resolution mass spectrometry, photoelectron spectroscopy (ESCA), Fourier transform infrared spectroscopy (FTS), and x-ray fluorescence spectroscopy (73). Analyses of atmospheric aerosols by ESCA, for example, give results that differ significantly from those obtained previously by wet chemical analyses: the latter had shown particulate sulfate to be largely ammonium sulfate, but ESCA shows that only part of it is in that form; ESCA shows also that much less particulate nitrate is present than would be deduced from the earlier, wet-chemical methods.

An important indicator of the origin and age of aerosols is their size distribution. Aerosols produced by photochemical processes are originally much smaller than other particles because they are formed by chemical reactions at the molecular level. Their diameters lie in the submicrometer range; particles in this size range scatter light and reduce visibility much more efficiently than larger ones.

Aerosol mechanisms. Working out the mechanisms by which these submicrometer aerosols are produced from gases is a particularly challenging aspect of smog chemistry. For the formation of sulfate aerosols, for example, current knowledge suggests three candidate mechanisms: photochemical oxidation of gaseous sulfur dioxide; absorption of sulfur dioxide into water droplets followed by oxidation of the resulting sulfite ($SO_3^{=}$); and adsorption of sulfur dioxide on soot or other noncarbonate carbonaceous particles, followed by oxidation in the presence of oxygen and water adsorbed on the particles. (Sulfate aerosol also may form in nonsmoggy air, but the process is much slower.) For the formation of nitrate aerosol, the most likely mechanism appears to be linked with the absorption of gaseous nitrogen oxides into airborne particles containing moisture. As mentioned earlier, however, smog reactions yield organic and inorganic nitrates, such as PAN and nitric acid, and these substances also may play an important role in the formation of nitrate aerosol.

Important progress in our understanding of aerosols in urban atmospheres has come from the California Aerosol Characterization Experiment (ACHEX) (22, 32). These field studies in the Los Angeles Basin were sponsored by the state of California in 1972–73. Analysis of the ACHEX data is still in progress, but already it has further delineated the importance of sulfates, nitrates, ammonium compounds, and water to the composition of aerosols. In addition, new carbonaceous compounds and organic acids have been identified in the aerosol samples. The analyses to date support the theory that most of the sulfur and nitrogen compounds in the aerosols originate as gases and are transformed in the atmosphere to aerosols by reactions that are not exclusively photochemical. Also, particulate sulfate was found to be distributed more uniformly over the Los Angeles Basin than nitrates, which were more concentrated in the eastern part of the basin.

The 1969 Pasadena Smog Study detected two size regions in the aerosol mass distribution: a fine-particle region at 0.1 to 1.0 μm with a peak at 0.2 to 0.3 μm; and a coarse-particle region at 1.0 to 100 μm with a peak at about 20 μm (74–76). The ACHEX study, using improved and probably more accurate instruments, also detected two size ranges, but with a dividing line at 1.0 μm and peaks at 0.4 μm and 6 to 10 μm (22). The fine particles, as mentioned earlier, result from combustion and gas-to-particle conversions and are primarily man-made; the coarse particles are primarily from the earth's crust and oceans and include man-made material like dust stirred up by automobiles (77–79).

The ACHEX data strongly support the need eventually to change the instrumentation used to monitor urban aerosols. The current (1978) EPA method, the high-volume sampler, determines total weight of particles per unit of volume; it does not characterize the sample in terms of weights of particles in various size ranges. The ACHEX results, on the other hand, reinforce the view that submicrometer particles are the most important to the problems of pollution and visibility.

Stratospheric ozone

The 1970's have seen great concern over the possibility that some pollutants may deplete the ozone layer in the stratosphere. The total amount of ozone in the stratosphere is small; the maximum concentration is a few parts per million. If all of the ozone overhead were by some means sifted from the air that dilutes it and brought to sea level, it would make a layer no more than 3 or 4 mm thick. This ozone, however, plays a key role in determining the temperature of the stratosphere and in preventing harmful ultraviolet light from the sun from reaching the earth's surface (80). These effects are due to the optical properties of ozone, which absorbs light in key portions of both the ultraviolet and infrared spectra. Hence a change in the total amount of ozone could change the temperature of the stratosphere and also the light in the near ultraviolet that reaches the surface. This near ultraviolet light is the part of sunlight that causes sunburn and, very likely, skin cancer as well. It may also have a role in the development of lower life forms, both plant and animal.

The present ozone level in the stratosphere reflects a balance among the continuous generation of ozone by sunlight, its transport from the tropics (where most of it is formed) toward the poles, and its destruction by at least three processes. Some is broken down directly by sunlight in a slightly different spectral range from that in which it is formed; much is destroyed by interaction with nitric oxide; and a small fraction is removed by interaction with chlorine atoms, a process that appears to be quite closely analogous to that with nitric oxide.

In the nitric oxide chain, the nitric oxide reacts with ozone to form nitrogen dioxide (the same reaction discussed in the section on photochemical smog). The nitrogen dioxide, in turn, reacts with atomic oxygen to form oxygen molecules and to regenerate the initial nitric oxide. The latter may be considered a catalyst because it takes part in the process without itself being consumed. This catalytic chain reaction, according to present theories, destroys well over half of the ozone consumed in the natural ozone balancing process.

Global levels. The global concentration of ozone displays large natural variations (81). The total amount of ozone overhead at a particular location is affected on a daily basis by the meteorology of the stratosphere and on a seasonal basis by the larger motions of the stratospere that originate in the change in inclination of the sun. A third cycle, of about two years, is caused by a still unexplained reversal of wind direction in the stratosphere about every other year. Additional cycles, the shortest of which is 11 years, are caused by changes in the number of sunspots. When sunspots are at a minimum, the sun's magnetic field is also at a minimum, and the earth is less protected from cosmic rays from outer space. These cosmic rays, entering the stratosphere, cause the formation of significant amounts of nitrogen oxides, which tend to depress ozone levels. At a sunspot maximum, the solar magnetic field deflects cosmic rays, and the ozone concentration increases. Recent evidence (82) suggests that during historical periods of a good fraction of a century there have been no sunspots whatever. It is not clear what these periods have meant in terms of ozone levels; however, they must be considered in evaluating claims of catastrophic impacts from relatively minor changes in stratospheric ozone.

The present concern about ozone depletion began with the prospective advent of supersonic transport aircraft, which inject nitric oxide directly into the stratosphere. The concern spread further with the hypothesis that chlorofluorocarbon aerosol propellants and refrigerants might carry chlorine into the stratosphere.

Nitrogen oxides. The effect on the ozone shield of nitrogen oxides emitted by high-flying aircraft became an issue in the U.S. in 1971 (83). Early in 1975, the conclusion that nitrogen oxides from SST's would indeed deplete the ozone shield emerged independently from major studies of the problem by the Climatic Impact Assessment Program of the U.S. Department of Transportation and by a committee of the National Academy of Sciences (84, 85). The estimated degrees of depletion differed, however, because of uncertainties about atmospheric processes and differences in approach. The smallest loss projected — for a mixed fleet of Anglo-French Concordes and Soviet TU-144's — was less than the minimum detectable change (0.5%); the largest — for 300 to 400 SST's of the U.S. design — was 10%. The projected 10% loss, viewed as potentially serious, would be reached over a period of years, and the ozone shield then would stabilize at the depleted level. Later calculations gave estimates so much lower than these as to discount the SST as a serious threat to stratospheric ozone (86). The reductions came mainly from consideration of chlorine-containing substances newly found to have important effects on the pertinent reactions.

The American SST program was dropped in the spring of 1971 when federal financing was withdrawn after Congressional hearings. The record suggests strongly that the principal reasons for the withdrawal of federal support were the high first-cost and projected low profitability of the aircraft and not its stratospheric impact. The Concorde and the TU-144 were flying in small numbers by 1976. They seemed to be confirming the U.S. economic analysis, and it appeared unlikely that hundreds of these aircraft would ever become airborne. Efforts to block the aircraft from the U.S. again cited possible environmental harm, but probably were more persuasive in citing the high noise levels. In retrospect, the issue of the SST probably has been more important in raising the specter of global pollution from the unilateral actions of individual nations, and in turning the attention of the scientific community to the chemistry of the stratosphere, than in its ultimate impact on the affairs of the world.

Chlorofluoromethanes. The idea that chlorofluoromethanes may carry chlorine into the stratosphere and thereby deplete stratospheric ozone was hypothesized in 1974 (87, 88) and accepted qualitatively by 1976. These compounds of chlorine, fluorine, and carbon are extremely inert (unreactive), which accounts largely for their utility as spray can propellants and refrigerants (other propellants used in spray cans include nitrous oxide, carbon dioxide, and certain hydrocarbons). Because the chlorofluoromethanes are so inert, however, their global concentration has increased gradually ever since they were first produced, during the 1930's (89–92). The compounds diffuse from the troposphere into the stratosphere, where ultraviolet radiation breaks them into fragments that include chlorine atoms. Chlorine reacts with ozone to form chlorine oxide and oxygen, and the chlorine oxide reacts with atomic oxygen to form oxygen molecules and regenerate the initial chlorine atoms. (As noted earlier, the process is closely analogous to the chain process involving nitrogen oxides.) The degree of ozone depletion by this mechanism, however, remains problematical. As pointed out below, significant amounts of chlorine compounds from natural sources undergo analogous reactions, and the magnitude of their impact is not yet clear.

A committee of the National Academy of Sciences estimated that the indefinite continuation of production and release of the two most widely used chlorofluoromethanes (trichlorofluoromethane, F-11, and dichlorodifluoromethane, F-12) at 1973 levels would result in a 3 to 3.75% ozone depletion in 40 to 50 years and a steady-state depletion of 6 to 7.5% at some time thereafter (93, 94). These were mid-range estimates; the actual total depletion, the committee said, could be as little as 2% or as much as 20%. The uncertainty stemmed from gaps in the understanding of known processes. If as yet unknown processes were at work, the range could be still larger. [By most reckonings, it is difficult to credit figures approaching or beyond 20%, since they would demand present depletions large enough to be detected. Statistical arguments have been proposed that changes much larger than 1% would be detectable today, and no such changes have been found (95)]. In view of the present uncertainties, the amount of research now under way to reduce them, and the minor incremental impact of a short delay in instituting controls, the NAS study recommended that research on the problem be continued, that no action be taken immediately, and that a regulatory decision be made in no more than two years (i.e., in 1978).

Other chlorine compounds. The amount of chlorine distributed to the stratosphere by the two chlorofluoromethanes, the NAS study concluded, is roughly matched by the total contributed by three other compounds: hydrochloric acid, chloromethane (methyl chloride), and carbon tetrachloride. This conclusion would connote that the levels of ozone depletion by the two sets of compounds also are comparable. Hydrochloric acid was stated to be natural in origin and present in relatively unimportant amounts as a primary contaminant. (Far more is formed in the stratosphere as the end product of the reaction series involving chlorine and ozone, and this secondary hydrochloric acid is the form in which chlorine is believed ultimately to leave the stratosphere and return to the troposphere, thus terminating the chain of reaction with ozone.) Necessarily ignored for lack of data was the possible injection of large amounts of hydrochloric acid by occasional very large volcanic eruptions. Chloromethane is believed to come from natural processes in the ocean and from smoldering combustion, both natural and otherwise, on land. More questionable is the status of the highly inert carbon tetrachloride (96–98). Man-made sources have been emitting the compound since about 1907, and its background concentration in both troposphere and stratosphere appears to approximate that of the chlorofluoromethanes. A basic issue is whether man-made emissions account for all of the compound that the global atmosphere evidently contains.

A key question with all of these inert compounds is whether processes exist that destroy them before they reach the stratosphere and enter the chlorine cycle. The question is not easily answered, since processes very slow by ordinary criteria would have a large effect on processes as sluggish as the pertinent ones in the stratosphere. A tropospheric process that could remove the average molecule of carbon tetrachloride or chlorofluorocarbon in 25 years, for example, would decrease the stratospheric impact of these molecules by two thirds. There is in fact evidence that just such a process may exist (99, 100), but it is certainly not unequivocal. The problem is that answering the question about destructive processes requires the extremely accurate measurement of substances in air at concentrations of a fraction of a part per billion. At such concentrations, it is difficult even to perform duplicate analyses that agree; the production of accurately known standard concentrations to calibrate the analytical techniques is still more difficult.

When these analytical uncertainties are added to uncertainties in the rates of stratospheric reactions and still larger uncertainties in the mathematical models of the mixing of these compounds into and within the stratosphere, it is small wonder that the overall question of ozone depletion remains unresolved. Nevertheless, in mid-1977, the Food and Drug Administration, EPA, and the Consumer Product Safety Commission announced plans to phase out nearly all uses of chlorofluorocarbon propellants on a two-year timetable ending in April 1979; final regulations appeared in March 1978 (100a).

Carbon dioxide

Carbon dioxide is a natural constituent of the air, a part of the natural carbon cycle, and, with water, an end product of the combustion of carbon-containing materials. Man generates an enormous amount of the gas by burning fossil fuels, so that it is also a pollutant, although not commonly treated as such because its emissions are uncontrollable and its toxicity is trivial. Man is dwarfed by nature as a source of carbon dioxide, but even so his contribution is raising the global level by an amount that could cause mean global temperature to rise (101–103).

Solar energy reaching the atmosphere as short-wavelength radiation passes through carbon dioxide to the earth if it is not absorbed by other species. The gas absorbs the long-wavelength infrared radiation emitted by the earth, however, and radiates it back to earth. Thus if heat energy is arriving from the sun at a constant rate, an increase in the atmospheric level of carbon dioxide will reduce the amount of energy radiated by the earth into outer space, and global temperature will rise. The real consequence of this so-called greenhouse effect is obscure because temperature is subject to so many additional influences, both man-made and natural. We have seen earlier, for example, how particles may affect temperature. Carbon dioxide is, however, the substance that has been studied most thoroughly for possible geochemical effects.

Sources, behavior. The main natural sources of carbon dioxide are the oceans, bacterial oxidation of dead plant and animal material, and metabolic processes in plants and animals. Carbon dioxide takes part in no significant chemical reactions with other substances in air; its main removal mechanisms are absorption in the oceans and biological uptake, including photosynthesis. The movement of carbon dioxide through the natural cycle involves a variety of carbon-containing substances and chemical and physical processes. Some of the processes, such as plant and animal metabolism, may occur quickly; others, such as those that proceed in the oceans and ocean sediments, may require thousands of years. Thus man-made emissions of the gas, though small by natural standards, have been entering the air faster than the natural cycle can adjust to them. Combustion of fossil fuels, for example, releases carbon dioxide far more rapidly than the fuels can form from carbon-containing (organic) materials in sediments.

Man has been burning fossil fuels for centuries, but the rate at which he does so has been rising sharply since about 1900. Worldwide emissions of carbon dioxide from combustion are expected to quadruple between 1950 and 1980. About half of the man-made carbon dioxide emitted during the 20th century has been taken up by the natural sinks; the other half remains in the atmosphere. Measurements of the background concentration of the gas show that it rose from about 312 ppm (612 mg/m^3) in 1958 to about 332 ppm (651 mg/m^3) in 1976. The level is believed to have been about 296 ppm (580 mg/m^3) in 1900; it is predicted to reach at least 375 ppm (736 mg/m^3) by 2000.

The rising global concentration of carbon dioxide cannot alone explain the measured trends in global temperature. The temperature climbed about 0.8°C between 1850 and 1940, but then fell about 0.3°C by the early 1970's, although a warming trend may exist in the southern hemisphere (104). The increase up to 1940 could be laid to carbon dioxide; the decline thereafter, while the carbon dioxide level was still rising, could be laid to the larger, opposing effect of an uptrend in turbidity (particle concentration). This explanation, however, discounts changes in other important factors: the amount of water vapor in the air, the nature of the earth's surface, interactions of the oceans with the atmosphere, and the intensity of the solar radiation that reaches the atmosphere. Determination of the actual effects of carbon dioxide and other contaminants on temperature and climate requires mathematical models that account reliably for all of the relevant influences. The continuing effort to devise and upgrade such models calls for consistent progress in computers and in measurements of the pertinent parameters over adequate periods.

Pollutant Removal

The foregoing discussion illustrates the sorts of processes and time-spans generally involved in the removal of pollutants from the atmosphere. It will be seen that transformations as a rule yield two types of end products: normal constituents of the air; or particles that can be removed by gravity, precipitation, or atmospheric mixing to the ground. In addition, a number of pollutants are removed to varying degrees in their original forms.

The relative speeds of removal of gaseous contaminants are approximated roughly by the speeds at which they dissolve in water. Highly soluble gases like sulfur dioxide and ammonia may be removed very rapidly at rates determined almost entirely by how quickly they are mixed to the ground. On the other hand, a highly soluble gas that fills an air mass to a significant depth may travel some hundreds of kilometers before all of it diffuses to the ground. Removal may be very rapid if a good deal of surface is available for pollutant molecules to encounter. Thus forests quickly purify contaminated air that penetrates them; the walls of buildings may speed removal in the center of a city.

When pollutant gases are transformed into particles, their ultimate rates of removal decrease markedly, barring washout by precipitation. The particles formed initially are extremely small; they tend to settle from the air far less rapidly than individual gas molecules diffuse to the ground. It has been estimated, for example, that the transformation of sulfur dioxide into sulfuric acid (aerosol) prolongs the removal period to at least 10 times that for the untransformed gas. Eventually, however, small particles either coagulate to settleable size or are washed out by rain, which is rather less efficient in removing gases. Available evidence shows that, with no rain, sulfate particles can travel thousands of kilometers, but that with rain along the entire track of the plume they are removed essentially completely in the first 100 km (105).

It seems safe to assume that other very soluble gases, such as hydrogen fluoride, hydrogen chloride, and ammonia, will behave like sulfur dioxide. That is, in air, they will be transformed into fluoride, chloride, and ammonium aerosols. These gases have not been studied nearly so completely as sulfur dioxide, however, and the data even for sulfur dioxide are far from definitive. The problem is complex, because rates of both transformation and removal are affected by temperature, humidity, and other pollutants. Still less is known of the oxides of nitrogen, which pose a particularly difficult problem because they interconvert in a very complex fashion. Ozone apparently is destroyed by contact with surfaces, and carbon monoxide, besides its reaction with hydroxyl radical, is absorbed slowly by soils.

Major studies of the transport, transformation, and removal of air pollutants include EPA's Project MISTT (Midwest Interstate Sulfur Transformation and Transport) and the Electric Power Research Institute's SURE (Sulfate Regional Experiment). Project MISTT, conducted during 1973–76, involved extensive measurement of urban and power-plant plumes originating in the St. Louis area (105a). Urban plumes were sampled in three dimensions out to 250 km from their sources and power-plant plumes out to 60 km. A variety of parameters were measured, but the emphasis was on various forms of sulfur. The SURE program also involves extensive measurement, over the northeastern United States (105b). The program began in 1976 and will end in 1980; the emphasis, again, is on various forms of sulfur.

Despite the progress that has been made, we still have little really quantitative information on the rates of transformation and removal of any pollutant. Without such information, the mathematical models discussed earlier will not predict the dispersal of contaminants accurately. It is probable that these gaps in our knowledge are far more important to such predictions than are the deficiencies in the models themselves. The problems are complex, but the stakes are extremely high. An investment of as much as $100 million in the necessary research would much more than repay itself by reducing pollution-control costs, but current (1978) spending is far below that level.

Recommendation A3: *Laboratory studies of the fundamental chemical processes in the atmosphere should be continued with particular emphasis on the application of new means of measuring transient and low-concentration* *species. Processes of interest include formation of aerosols in urban air and the reactions of trace substances in the stratosphere.*

Recommendation A4: *Studies similar to those of Recommendation A3 should be conducted in real atmospheres at concentrations of constituents ranging from those found in urban air to those found in remote global atmospheres.*

Recommendation A5: *The coupling of chemical behavior with meteorological phenomena such as diffusion and dispersion should be given greater emphasis in terms of the effects of contaminants on the urban, regional, and global atmospheres.*

Recommendation A6: *Further support should be devoted to research in heterogeneous chemistry, such as aerosol chemistry, both in the laboratory and in real atmospheres.*

Recommendation A7: *Field investigations of atmospheric aerosols should be designed to integrate within the same time-frame measurements of size-dependent concentrations in the size range from 0.003 μm to 50 μm; inorganic and organic composition, including trace elements and gases; and meteorological factors.*

Recommendation A8: *Studies of persistent chemicals that potentially could affect the stratosphere should be continued and selectively intensified to provide answers in the shortest reasonable time.*

Recommendation A9: *Evaluations of the effects of contaminants on the urban and global atmospheres and of the changes to be expected from regulatory actions should be reviewed continuously on the basis of new scientific understanding of such problems.*

Recommendation A10: *Air-monitoring programs and field experiments must always include quality assurance of the resulting atmospheric data as an integral part of the effort.*

SOURCES AND CONTROL OF AIR POLLUTION

We have defined air pollutants as materials that enter the air directly or indirectly as a result of human activity. They become a problem generally not because they exist, but because they exist in high concentrations among high concentrations of humans and other receptors, animate and inanimate. Virtually every human activity — even physical exertion — causes some degree of air pollution, but the principal source is combustion: the rapid combination of oxygen with other substances, notably carbon, and the consequent release of energy.

Perfect combustion of a pure hydrocarbon fuel with the correct amount of pure oxygen yields only carbon dioxide and water. A combustion process is rarely perfect, however, and the reactants are rarely pure. Sulfur in the fuel is converted to sulfur oxides. The air that usually supplies the necessary oxygen also supplies nitrogen, a very small amount of which is converted to nitrogen oxides, mainly nitric oxide. The products of combustion with less than the correct amount of oxygen will include carbon monoxide and hydrogen. Some of the fuel may escape oxidation wholly or partly. Basically, then, emissions from a combustion process depend on its inputs and its efficiency, and this in general is true of any process.

Many technological and economic factors affect the selection of means of barring the emission of a pollutant. In the end, however, there are six alternatives.

- Select process inputs that do not contain the pollutant or its precursors.
- Remove the pollutant or its precursors from the process inputs.
- Operate the process so as to minimize generation of the pollutant.
- Remove the pollutant from the process effluent.
- Replace the process with one that does not generate the pollutant.
- Use less of the product whose manufacture generates the pollutant.

Agriculture

Agricultural activities, at least at certain times of the year, denude large areas of soil. If they do so during periods of drought and high wind, the resulting air pollution can be enormous and have great economic impact. Witness the "dust bowl" years of the 1930's in the American Middle West. The problem can be minimized by improving weather forecasts and using them intelligently in planning schedules of plowing. For the longer run, further study and wider use of "no-till" farming seems promising.

Animal husbandry on a large scale, as in feedlots, also can cause air pollution, primarily by objectionable odors. A number of pesticides and other agricultural chemicals become airborne under certain conditions; some of them are extremely active biologically and can present an acute air-pollution hazard. Such pesticides generally occur in relatively large particles, however, and do not travel far, so that the hazardous areas are small. The primary means of controlling pollution by both odors and pesticides are changes of practice. Many feedlots that cause severe odor problems are simply overcrowded; many problems with pesticides result from incorrect use. The solution in both cases is a combination of education and awareness.

Agriculture also contributes to air pollution because it is a major user of energy. Excepting the solar energy used in photosynthesis, each calorie of food for actual consumption in the U.S. in 1970 represented about 10 calories of energy input to the food system, which in that year accounted for something more than 13% of the nation's total consumption of energy (106). About a quarter of the energy used in the food system was consumed on the farm in the form of tractor fuel, fertilizers and pesticides, electricity for irrigation pumps, and other appurtenances of agriculture. Man obtains this energy almost entirely by combustion, the leading source of air contaminants. More efficient use of energy in agriculture, therefore, would decrease air pollution as well as conserve resources.

Recommendation A11: *Increased attention must be given to the energy cost of agriculture, and means must be sought to decrease it.*

Industry

Industry generates a variety of air pollutants specific to the processes involved, as well as a substantial fraction of five criteria pollutants: carbon monoxide, hydrocarbons, nitrogen oxides, sulfur oxides, and particles (Table 7, Table 2). Control of these contaminants is based on emission limits embodied in State Implementation Plans and federal New Source Performance Standards and National Emission Standards for Hazardous Air Pollutants. Emission limits place ceilings on the releases of specific pollutants from specific classes of sources: they do not in general specify the means of compliance (Table 8).

Establishing and meeting emission standards for industrial sources is a complex and time-consuming process. Common problems include simple noncompliance and faulty State Implementation Plans. A problem with particles is "fugitive dust" released through doors, windows, and similar vents not subject to emission limits, which apply as a rule only to intended exhaust systems such as stacks. New or improved control technology is a continuing need. Progress in these and related areas comes slowly at best; one indication is the failure to meet various primary standards of air quality in many Air Quality Control Regions, a failure not due, however, to industrial sources alone.

Emission factors

Pollutants from manufacturing fall into two major categories: process wastes and nonprocess wastes. Process wastes are specific to actual processes; nonprocess wastes are those such as office and shipping wastes that are common to all industries, and they will not be considered here. An example of process

Table 7

Some emissions associated with selected industries

Industry	Pollutants
Chlor-alkali	Chlorine, mercury (mercury cells only)
Copper smelters	Particles, sulfur oxides
Explosives (TNT)	Particles, sulfur oxides, nitrogen oxides, nitric acid mist, sulfuric acid mist
Ferroalloys	Particles
Integrated steel mills	Particles, carbon monoxide, gaseous and particulate fluorides
Kraft (sulfate) pulp	Particles, sulfur oxides, carbon monoxide, hydrogen sulfide, methyl mercaptan, dimethyl sulfide, dimethyl disulfide
Metallurgical coke	Particles, sulfur oxides, carbon monoxide, hydrocarbons, nitrogen oxides, ammonia
Petroleum refining	Sulfur oxides, carbon monoxide, hydrocarbons, nitrogen oxides, aldehydes, ammonia
Portland cement	Particles, sulfur oxides, nitrogen oxides
Primary aluminum (including bauxite)	Particles, gaseous and particulate fluorides

Source: "Compilation of Air Pollutant Emission Factors," 2nd ed., Environmental Protection Agency, Publication No. AP-42, Research Triangle Park, N.C., February 1976.

wastes can be seen in the production of metallic copper from copper sulfide, a naturally occurring ore. The refining process converts the undesired sulfur to gaseous sulfur dioxide, which is released into the air. With it go particles of copper and of other metals that are impurities in the ore. These materials are discharged into the air for one of two reasons: they are not perceived as valuable enough to repay the cost of recovering and using or selling them; or the manufacturer is sufficiently specialized that he considers such products extraneous to his business and therefore has never considered recovering them.

A tool used commonly to estimate typical process emissions in an industry is the "emission factor": the weight of a given pollutant produced per unit weight of the desired product. EPA has published extensive tabulations of emission factors for various processes along with some data on how emissions are modified by various air pollution control devices (Table 9) (107). The growing use of control equipment in a variety of industries, however, is creating a need for new sets of tables, and these to date have been appearing only slowly. The existing tabulations, moreover, are deficient in two respects: they contain only limited

Table 8

Examples of new source performance standards

Source & pollutant	Allowable emissions
Coal-burning steam generators (more than 63 x 10^6 kcal/hr of heat input); pulverized wet bottom	
Particles	0.18 g/10^6 cal heat input
Sulfur dioxide	2.2 g/10^6 cal heat input
Nitrogen oxides (as NO_2)	1.26 g/10^6 cal heat input
"Floating roof" storage tanks Hydrocarbons	If true vapor pressure under storage conditions exceeds 78 mm (1.52 psia) mercury but is no greater than 570 mm (11.1 psia) mercury, the vessel must be equipped with a floating roof or its equivalent
Municipal incinerators Particles	0.18 g/m^3, corrected to 12% CO_2
Nitric acid plants Nitrogen oxides (as NO_2)	1.5 kg/metric ton of 100% acid produced
Sulfuric acid plants Sulfur dioxide	2.0 kg/metric ton of 100% acid produced
Sulfuric acid mist (as H_2SO_4)	0.075 kg/metric ton of 100% acid produced

Source: "Compilation of Air Pollutant Emission Factors," 2nd ed., Environmental Protection Agency Publication No. AP-42, Research Triangle Park, N.C., February 1976.

Table 9

Emission factors and controls for sulfate pulping (unit weights of air-dried unbleached pulp)

Source	Type control	Particles		Sulfur dioxide		Carbon monoxide		Hydrogen sulfide		Organic sulfur compounds	
		lb/ton	kg/MT	lb/ton	kg/MT	lb/ton	kg/MT	lb/ton	kg/MT	lb/ton	kg/MT
Digester relief and blow tank	Untreated	—	—	—	—	—	—	0.1	0.05	1.5	0.75
Brown stock washers	Untreated	—	—	0.01	0.005	—	—	0.02	0.01	0.2	0.1
Multiple effect evaporators	Untreated	—	—	0.01	0.005	—	—	0.1	0.05	0.4	0.2
Recovery boiler and direct contact evaporator	Untreated	150	75	5	2.5	2–60	1–30	12	6	1	0.5
	Venturi scrubber	47	23.5	5	2.5	2–60	1–30	12	6	1	0.5
	Electrostatic precipitator	8	4	5	2.5	2–60	1–30	12	6	1	0.5
	Auxiliary scrubber	3–15	1.5–7.5	3	1.5	2–60	1–30	12	6	1	0.5
Smelt dissolving tank	Untreated	5	2.5	0.1	0.05	—	—	0.04	0.02	0.4	0.2
	Mesh pad	1	0.5	0.1	0.05	—	—	0.04	0.02	0.4	0.2
Lime kilns	Untreated	45	22.5	0.3	0.15	10	5	0.5	0.25	0.25	0.125
	Scrubber	3	1.5	0.2	0.1	10	5	0.5	0.25	0.25	0.125
Turpentine condenser	Untreated	—	—	—	—	—	—	0.01	0.005	0.5	0.25
Miscellaneous sources	Untreated	—	—	—	—	—	—	—	—	0.5	0.25

Note: Table has been simplified by the deletion of a number of explanatory footnotes.

Source: "Compilation of Air Pollutant Emission Factors," 2nd ed., Environmental Protection Agency Publication No. AP-42, Research Triangle Park, N.C., February 1976.

data on trace emissions (as opposed to major emissions like sulfur oxides) (108); and they do not in general consider the effects of altitude.

The situation with trace emissions can be seen, again, in the production of metallic copper from its ore. The amount of sulfur dioxide discharged per ton of smelted copper is known with reasonable accuracy, but far fewer data exist on the traces of selenium dioxide discharged simultaneously. Concern for the trace composition of both fuels and stack gases has emerged only recently, and much more information, and more systematic compilation, are needed to translate the concern into effective action. The need is doubly pressing because many trace substances may be far more active biologically than the primary contaminants.

The failure of present emission factors to take account of altitude is questionable because in some cases there is reason to expect that the chemistry of the discharge depends on altitude. In nearly all processes that generate high temperature, for example, a small amount of atmospheric nitrogen combines with oxygen to form the pollutant nitric oxide. The precise amount depends on the concentrations (partial pressures) of the nitrogen and oxygen, so that less nitric oxide is formed at high altitudes than at low. In combustion processes, more than enough air usually is provided for complete combustion, yet it is certainly possible that the lower concentration of oxygen at altitude could result in incomplete reaction. Thus some combustion processes may yield less carbon dioxide and more carbon monoxide at high altitudes than at low. These effects appear not to have been investigated in the context of emission factors.

The need is growing for more precise evaluation of emission factors for use in long-range planning, making land-use decisions, and modeling pollutant behavior when developing control strategies. As the need grows, the uncertainties become more important. Clearly a sustained effort will be needed soon in the determination of emission factors for trace contaminants and of the effects of control devices and altitude on emission factors in general.

Emission offset

In December 1976 EPA issued guidelines for a new "emission offset" policy designed to permit industrial growth in polluted areas (109); the approach was incorporated in the Clean Air Act Amendments of 1977. Under this policy, a new source of air pollutants may be added in an area that already is violating ambient air quality standards, provided that: emissions from the new source are controlled with the best technology available; the new emissions are more than offset by reductions of emissions from sources already in the area; the tradeoff yields a net gain in air quality. It is too early to predict how the offset policy will work out, but the concept raises interesting possibilities. Control of emissions from small industry, for example, has been one of the most intractable problems, not because the technology does not exist, but because its cost is high relative to the capitalization of small enterprises. However, a large corporation that wishes to expand in a polluted area may well find that it can achieve the necessary offset at least cost not by reducing its own emissions but by installing and maintaining controls on other, smaller emitters that could not afford them on their own. The offset policy also could spur breakthroughs in more speculative areas. One of these is "cogeneration" of electric power from waste process heat: since no additional emissions would result, and an equivalent amount of power would not be required from utilities, the outcome would be more efficient use of energy and less pollution.

The offset policy illustrates the potential value of learning to build reliable predictive models of air pollution situations. The policy is being implemented on a mass basis: each kilogram of emissions from a new source must be offset by eliminating more than one kilogram of emissions from existing sources. Ideally, the offset should be in terms of effects. For example, one can visualize situations in which controlling 10 kg would compensate for new emissions of a metric ton, and vice versa. Offsetting in terms of effects, however, would require extremely good knowledge of air movements, atmospheric chemistry, and the other elements of the predictive model, and in general such knowledge is not yet available.

Control alternatives

In popular opinion, air pollution usually is best controlled by adding at the end of the process a device that prevents contaminants from entering the atmosphere. In fact, add-on devices are undesirable except where a valuable by-product can be recovered or as a last resort. They are used so widely today largely because in many instances they seem to be the only way to avoid excessive pollution. Contaminants collected at the end of a process still must be disposed of, often by means that can convert the original air-pollution problem into a soil- or water-pollution problem. And almost without exception, add-on equipment consumes additional power. For such reasons, gas-cleaning devices should not be considered until at least two alternatives — demand modification and process modification — have been explored and found wanting.

Demand modification. Pollution control by demand modification tends not to be a straightforward procedure. The demand for, say, disposable plastic tableware would decline if the consumer used the ware more than once before discarding it. Pollution caused by the manufacturing process would decline accordingly. Reuse, on the other hand, would require more dishwashing and a consequent increase in detergent production and use and in the associated discharges. Similarly, keeping automobiles longer in service decreases pollution near the manufacturing plants, but older cars may consume more petroleum products and create more pollution locally than new ones. Changes in demand also would affect employment and profits in the enterprises involved. Demand modification and air pollution evidently would not interact as simply as one might wish, but virtually none of the kinds of tradeoffs illustrated here has ever been examined quantitatively. The potential for reducing pollution and conserving resources would appear nevertheless to warrant a demand for higher quality and more durability in practically everything that we use.

Process modification. Process modification, a second alternative to gas-cleaning devices, may take many forms. The enormous emissions of open-hearth steel furnaces are nearly uncontrollable. They are declining in general, however, as open-hearths are replaced by basic-oxygen and electric-arc furnaces, which together were accounting for some 85% of U.S. steel output by 1977. Emissions of highly reactive solvents often are controlled by substituting other, less reactive solvents. Such substitution can cause economic dislocation, but it should not be serious if adequate time is allowed for the change and if the regulations are reasonably uniform over adequately large market areas. Uniform regulations avoid unfair competitive advantages and simplify planning, especially for consumer products. If specifications for solvents for domestic paints vary widely over normal marketing areas, for example, manufacturers are compelled unnecessarily to compound quite different paints for contiguous geographical areas.

Gas cleaning. As noted earlier, air pollution in industry often can be controlled only by equipment that removes the contaminants at the end of the manufacturing process. Many such devices exist (110, 111). The basic designs may be adapted to many specific applications, but in each they display two characteristics: the size and cost of the equipment increase with the volume of gas to be cleaned per unit of time; the cost of removing the contaminant tends to rise exponentially with the degree of pollutant removal (efficiency). These principles apply, with varying force, to both of the two broad classes of pollutants — particles, and gases and vapors.

Particles may be removed from gas streams by four basic types of equipment: mechanical collectors, fabric filters, wet scrubbers, and electrostatic precipitators. Generally speaking, the technology exists to remove particulate matter with any degree of efficiency desired. Besides the increase in cost that accompanies rising efficiency, however, removal becomes more difficult and costly as particle size decreases (Table 10). The large particles that are responsible for esthetic effects, such as soiling of surfaces, are removed fairly easily. Less readily trapped are particles below about 5 μm in diameter, and these are the ones that are most easily taken into the lungs and are most potent biologically. The technology is poorest, perhaps, for particles of around 0.2 μm in diameter, and these exert the maximum effect on visibility. In considerable measure they have been ignored, because collection efficiency is calculated as a rule on a mass basis, and in many processes the submicrometer particles make up a relatively small fraction of the total mass of emissions.

There is clearly a need to learn more of the size distribution of particles from all processes, of the differential efficiencies of particle collectors as a function of particle size, and of the effects of differential efficiencies on effluent composition. A need exists also to evaluate separately the control of particles in the respirable and optically significant size ranges and to design pollution control equipment specifically to collect them more efficiently.

Pollutant gases and vapors are in general more difficult to control than particles. The basic methods are oxidation, adsorption, and absorption. Organic materials can be oxidized, by direct flame or catalysis, but care must be taken that the oxidation is complete. Incomplete oxidation leads to aldehydes and carbon monoxide, which in many cases are more objectionable than the original materials. Organic vapors may be adsorbed on solids, notably activated carbon; synthetic zeolites (molecular sieves) are used to adsorb sulfur dioxide and nitrogen oxides from acid-manufacturing plants and mercury vapor from chloralkali plants. These systems are used primarily where the material adsorbed is recovered at savings that exceed or substantially reduce the cost of treatment. Acidic gases such as hydrogen chloride, hydrogen fluoride, sulfur dioxide, hydrogen sulfide, and nitrogen oxides can be adsorbed, at least theoretically, by alkaline scrubbing solutions. The degree of success, however, varies widely. Nitrogen oxides in particular dissolve very slowly, except in a few rather complex organic solvents. Nitric oxide is not removed at all by alkaline solutions; it must be oxidized first to the dioxide. The nitrogen oxides are thermodynamically unstable — given sufficient time, they will revert to nitrogen and oxygen at ambient temperatures — so that the possibility of decomposing them catalytically has been investigated to some extent. The resulting reaction is very sensitive to the other gases present, however; it can yield nitrous oxide, whose global concentration is attracting interest, or ammonia, which can be a very undesirable pollutant.

One of the serious problems in gas cleaning is a surviving tradition that dilution of the stack gas with air is desirable. As a result of this belief, many contaminants enter gas-cleaning equipment at concentrations so low that they cannot easily be removed. A little-explored possibility would be to avoid dilution, so that pollutants would react more rapidly with scrubbing agents. Reducing the amount of gas to be treated also would reduce the size and cost of the scrubbing equipment.

Table 10

Efficiencies of particle collection equipment in relation to particle size[a]

	Efficiency, %[b]					
		Particle size range, μm				
Type of collector	Overall	0 to 5	5 to 10	10 to 20	20 to 44	>44
Baffled settling chamber	58.6%	7.5%	22%	43%	80%	90%
Simple cyclone	65.3	12	33	57	82	91
Long-cone cyclone	84.2	40	79	92	95	97
Multiple cyclone (12-in. diameter)	74.2	25	54	74	95	98
Multiple cyclone (6-in. diameter)	93.8	63	95	98	99.5	100
Irrigated long-cone cyclone	91.0	63	93	96	98.5	100
Electrostatic precipitator	97.0	72	94.5	97	99.5	100
Irrigated electrostatic precipitator	99.0	97	99	99.5	100	100
Spray tower	94.5	90	96	98	100	100
Self-induced spray scrubber	93.6	85	96	98	100	100
Disintegrator scrubber	98.5	93	98	99	100	100
Venturi scrubber	99.5	99	99.5	100	100	100
Wet-impingement scrubber	97.9	96	98.5	99	100	100
Baghouse	99.7	99.5	100	100	100	100

a. Based on standard silica dust with the following particle size and weight distribution:

Particle size range, μm	Percent by weight
0 to 5	20
5 to 10	10
10 to 20	15
20 to 44	20
>44	35

b. The collection efficiencies shown are useful for illustration, but are idealized. In practice, collection efficiencies depend very much on design and operational factors at specific sites and may also be influenced by nonideal effects such as rapping losses in electrostatic precipitators, entrainment in wet scrubbers, and seepage in baghouses.

Source: "Compilation of Air Pollutant Emission Factors," 2nd ed., Environmental Protection Agency, Publication No. AP-42, Research Triangle Park, N.C., February 1976.

Tall stacks. A last-ditch alternative to gas-cleaning devices in meeting air quality standards is dispersion by means such as tall stacks. Unlike gas-cleaning devices, dispersion techniques merely redistribute pollutants. Legal interpretations of the Clean Air Act, however, permit their use as an interim measure under State Implementation Plans. The general rule is that tall stacks and related technology may be used where emission-limiting techniques are available but their application would be economically unreasonable, providing the user also takes specified action toward the eventual adoption of adequate emission-control technology (112).

Recommendation A12: *Further data must be obtained on emission factors for air pollution sources, especially with regard to emissions from control devices, emissions of less common and trace elements, the composition of emitted particles as a function of particle size, and the effects of altitude on emissions. Studies are needed on particle-control devices to determine differential efficiencies as a function of particle size and the effects of these differential efficiencies on effluent composition.*

Electrical Power Generation

The generation of electrical power in this country is responsible for more than half of the sulfur dioxide emitted into the atmosphere and for major fractions of the nitrogen oxides and particulate matter. Almost half of the power is obtained by burning coal (Table 11), and the fraction is certain to rise. Nuclear power generation is growing much more slowly than expected only a few years ago owing to social resistance, questions about reactor safety, radioactive pollution, and economics, and because of other factors, such as the shortage of risk capital. Solar and other new forms of energy will contribute little to electric power generation for decades, although an indirect contribution by replacing electric heating with solar heating is feasible in many new residences today. Residual fuel oil will become more costly and less plentiful for generating electricity; natural gas already is almost totally unavailable for the purpose. The nation's

Table 11

Primary sources of energy in the United States, 1977

	Fraction supplied	
Source	Electric utilities	Total consumption
Coal	46.4%	18.8%
Oil	16.8	48.6
Natural gas	14.4	25.9
Nuclear	11.8	3.5
Hydroelectric	10.4	3.2
Other[a]	0.2	—
	100%	**100%**

a. Geothermal, wood, waste.

Source: *Monthly Energy Review*, April 1978, U.S. Department of Energy, Washington, D.C.

reserves of coal, in contrast, are adequate for some centuries, and that fuel, therefore, seems certain to become the predominant source of electrical power during the foreseeable future (113).

One important way to reduce air pollution in cities is to supply as much of their energy as possible in the form of electricity generated at significant distances from population centers. Emissions still must be controlled, however. Efforts are under way to develop more efficient, less-polluting combustion systems for coal and processes for converting it to more convenient, less-polluting liquid and gaseous fuels (114, 115), but their impact will be minor until at least 1985. It must be considered that such conversions and nearly all alternatives consume energy, so that valid assessment of alternatives requires honest energy accounting for each. Most electrical power plants, for example, operate at an efficiency of around 40% or less — a definite energy penalty is exacted in transforming the chemical energy of coal into electrical energy. On the other hand, although the inefficiency of electrical-resistance heating (about 30%, starting with coal at the generating plant) sounds appalling, many older coal-burning furnaces for domestic heating operate at efficiencies of less than 20%, not including the energy penalties of delivering the coal and removing the ash. Resistance heating could become a sound alternative, therefore, especially as more efficient, less-polluting generating plants are developed.

The emissions produced by burning coal depend on its composition, which varies with its source, and on the efficiency of combustion. Emissions typical of bituminous coal burned in a utility power plant appear in Table 12 (anthracite coal is an insignificant source of electrical power). In addition to the substances shown in Table 12, some fraction of the trace elements in coal is emitted in association with the particulate matter or fly ash (108). Furthermore, a number of toxic trace elements, such as lead, cadmium, and selenium, appear to concentrate preferentially in the smaller, more readily respirable particles, which are collected less efficiently than large ones by gas cleaning equipment (116). EPA's performance standards for utility power plants limit only their emissions of particulate matter, sulfur dioxide, and nitrogen oxides.

Fly ash

All coal contains mineral matter that, after combustion, remains behind as ash. A fraction of this ash determined by the coal and boiler design is sufficiently fine to be entrained by the combustion airstream and emitted as fly ash. The past decade has seen great improvement in the technology of fly ash removal, although it is still a problem with some coals. This is true particularly of low-sulfur coals, which are common in the western U.S., because they yield an ash whose composition gives it relatively high electrical resistivity. Fly ash traditionally has been removed by electrostatic precipitators located at the flue-gas exit or "cold side" of the air preheater (which transfers heat from outgoing combustion gases to incoming combustion air). Such installations lose efficiency with high-resistivity ash, however, and so must be modified in one of two ways (117). The first is to use a cold-side precipitator of European design with a very large collection area, which counterbalances the reduced collection efficiency; the second is to use a precipitator of moderate collection area on the hot side of the preheater, where the higher temperature (315° - 400 °C) reduces the resistivity of the ash, which thus is more readily collected. Collection methods less sensitive than precipitators to the composition of the ash include water scrubbing and bag

Table 12

Typical emissions from combustion of bituminous coal

Furnace size, 10^6 Btu/hr (2.52×10^5 kcal/hr) heat input	Particles[a]		Sulfur oxides[b]		Carbon monoxide		Hydro-carbons[c]		Nitrogen oxides		Aldehydes	
	lb/ton coal burned	kg/metric ton coal burned	lb/ton coal burned	kg/metric ton coal burned	lb/ton coal burned	kg/metric ton coal burned	lb/ton coal burned	kg/metric ton coal burned	lb/ton coal burned	kg/metric ton coal burned	lb/ton coal burned	kg/metric ton coal burned
Greater than 100 (Utility and large industrial boilers)												
Pulverized												
General	16A	8A	38S	19S	1	0.5	0.3	0.15	18	9	0.005	0.0025
Wet bottom	13A[d]	6.5A	38S	19S	1	0.5	0.3	0.15	30	15	0.005	0.0025
Dry bottom	17A	8.5A	38S	19S	1	0.5	0.3	0.15	18	9	0.005	0.0025
Cyclone	2A	1A	38S	19S	1	0.5	0.3	0.15	55	27.5	0.005	0.0025

a. The letter A indicates that the weight percentage of ash in the coal should be multiplied by the value given. Example: If the factor is 16 and the ash content is 10 percent, the particulate emissions before the control equipment would be 10 times 16, or 160 pounds of particles per ton of coal (10 times 8, or 80 kg of particles per metric ton of coal).

b. S equals the sulfur content (see footnote a).

c. Expressed as methane.

d. Without fly-ash reinjection.

Source: Environmental Protection Agency, "Compilation of Air Pollutant Emission Factors," 2nd ed., Environmental Protection Agency Publication No. AP-42, Research Triangle Park, N.C., February 1976.

filters. Bag filters that have been used at small installations (118–120) are being scaled up for large utility boilers.

Sulfur dioxide

A significant fraction of the sulfur in coal is converted in the combustion process to sulfur dioxide. The ash, depending on its composition, reacts with some of the gas. The balance enters the exhaust stream, and in many areas sulfur dioxide from electrical power plants is the major air pollutant. A number of approaches to control have been tried with varying success. Emission standards for sulfur dioxide generally can be met by burning coal that contains 1% sulfur or less (which creates the fly-ash-collection problem of the preceding paragraph). Most of the nation's low-sulfur coal is in the West, and moving it to the major markets in the Midwest and East is relatively costly, although a significant amount is moved nevertheless. Much of the low-sulfur coal in the East is earmarked for the metallurgical industries. In high-sulfur (3 to 6%) coals, a significant fraction of the sulfur may be present as mineral sulfides. These are much denser than coal and so can be removed from it by mechanical means. In most coals, though, at least half of the sulfur is in organic form and an integral part of the coal itself and cannot be removed by such simple methods. The chemical conversion processes mentioned earlier break down the molecules in coal so that organic sulfur can be removed from the resulting liquids or gases, but these processes are some years from commercial use. The remaining alternative, and apparently the cheapest, is to burn high-sulfur coal and remove the sulfur dioxide from the stack gases (121).

Flue-gas treatment. By about 1970, a number of processes for removing sulfur dioxide from stack gases were at about the same stage of development, and all seemed promising. It appeared that an electric utility soon would be able to select from a range of techniques the one best suited to its circumstances, but such is not yet the case. By 1976, only lime scrubbing had proved satisfactory with high-sulfur coals and lime or limestone scrubbing with low-sulfur coals, although other processes were actively in development (Tables 13, 14). Flue-gas desulfurization units were operating, being built, or planned at facilities whose generating capacity totaled some 42,000 megawatts; lime or limestone scrubbing was associated with more than 75% of that capacity. By early 1977, EPA listed some 30 systems as operational and 86 under construction or planned; lime and limestone predominated and were expected to do so until well into the 1980's (122). It can be said that in any area where a major reduction of sulfur dioxide emissions is necessary, it can be accomplished by lime or limestone scrubbing. This assumes that the consumers affected are willing to pay an associated increase in electrical costs estimated at from 5 to 30%.

Lime and limestone scrubbing, although they remove sulfur dioxide effectively, are poor long-range solutions to the problem environmentally because their end product is a solid (calcium sulfite/calcium sulfate) that must be disposed of in a suitable way. More advanced systems recover sulfur as the element or as sulfuric acid; those based on scrubbing regenerate the scrubbing compound, using much less of it and producing far less solid waste than lime and limestone scrubbing. Sulfur is best recovered as the element, which is easier and cheaper to store and ship than sulfuric acid. The present market would not absorb the abrupt injection of an amount of recovered sulfur essentially equal to total current consumption. The demand for sulfur is increasing rapidly, however, and could well accommodate the gradual increase in supply that would attend an orderly transition from disposal to recovery of power-plant sulfur over the course of 10 to 15 years. Still, utilities hesitate to adopt abatement methods whose costs depend in part on success in the novel — to them — business of marketing sulfur and its compounds. The relative long-term costs of the various sulfur-oxides control techniques for utilities remain in fact difficult to assess (123, 124). Large-scale operating experience is scanty, even for lime and limestone scrubbing, and the technology is changing comparatively rapidly.

Flue-gas desulfurization at utilities remained relatively uncommon in 1976. Equipment was operational on slightly more than 5% of the nation's coal-fired generating capacity, and the fraction was expected to reach about 10% in 1980. This approximate doubling, however, will require an approximate tripling of desulfurization capacity because of the expected expansion of coal-fired generating facilities. As a rule of thumb, roughly 70% of the electrical generating capacity on line in the U.S. at the end of 1976 can be retrofitted at costs that

Table 13

Utility flue-gas desulfurization (FGD) systems in the U.S. — January 1976

Status	Number of units	Megawatts (MW) capacity
Operational	21	3796
Under construction	20	7026
Planned		
Contract awarded	10	3761
Letter of intent	10	3911
Requesting or evaluating bids	7	3837
Considering FGD systems	40	19,797
Total	**109**	**42,128**

Process	MW[a] 1976	MW[a] 1980
Limestone	4077	11,221
Lime	3442	7716
Sodium carbonate	375	375
Magnesium oxide	365	1091
Sodium sulfite	115	1830
Catalytic oxidation	110	110
Lime/limestone	30	1470
Dilute acid	23	23
Activated carbon	20	20
Double alkali	20	20
Total	**8577**	**24,416**

a. FGD megawattage for which a process has not been selected is not included.

Note: In 1976, about 65% of FGD megawattage was retrofit, and about 35% was on new plants. By 1980, 30% and 70% will be retrofit and new installations, respectively.

Source: *Environ. Sci. Technol.*, **10,** 416 (1976).

Table 14

Chemistry of selected flue-gas desulfurization processes

FGD process	Reagent	End product	Principle of operation
Limestone scrubbing	Limestone (LS)	$CaSO_3/CaSO_4$ sludge	LS slurry reacts in scrubber absorbing SO_2 and producing insoluble sludge
Lime scrubbing	Lime	$CaSO_3/CaSO_4$ sludge	Lime slurry reacts in scrubber absorbing SO_2 and producing insoluble sludge
Sodium sulfite scrubbing	Sodium carbonate makeup (regenerated)	Sulfuric acid or sulfur	Soluble sodium sulfite absorbs SO_2 in scrubber. The sodium bisulfite produced is thermally regenerated yielding sodium sulfite and SO_2 for either acid or sulfur production
Double alkali scrubbing	Sodium carbonate makeup (regenerated) Lime	$CaSO_3/CaSO_4$ sludge	Soluble sodium sulfite absorbs SO_2 in scrubber. The sodium bisulfite produced is reacted with lime precipitating $CaSO_3/CaSO_4$
Magnesium oxide scrubbing	Magnesium oxide (regenerated)	Sulfuric acid	Magnesium oxide slurry absorbs SO_2 in scrubber. The magnesium sulfite produced is thermally treated yielding MgO and SO_2 for acid production
Citrate scrubbing	Sodium citrate-citric acid (regenerated)	Sulfur	Sodium citrate-citric acid solution absorbs sulfur dioxide. Reaction with hydrogen sulfide precipitates elemental sulfur. Hydrogen sulfide prepared from part of recovered sulfur
Catalytic oxidation	None	Sulfuric acid	Sulfur dioxide in flue gas is converted catalytically to sulfur trioxide, which reacts with moisture present to form 80% sulfuric acid

Source: Dunham, J.T., Rampacek, C., Henrie, T.A., "High-Sulfur Coal for Generating Electricity," *Science*, **184**, 346 (1974).

would raise consumer's electrical bills by the 5 to 30% mentioned earlier. Roughly 20% can be retrofitted at substantially higher costs. The remaining 10% for practical purposes cannot be retrofitted at all: the reasons are various, but include the need to demolish business-district buildings worth many times the cost of the generating station to make room for air-cleaning equipment. It is thus probable that some 30% of the nation's electrical generating capacity will have to complete its useful life before it can be replaced by controlled generating plants or alternative power sources.

Nitrogen oxides

Nitrogen oxides emitted by coal-burning power plants are formed primarily from nitrogen in the combustion air. Significant amounts of the oxides can be formed as well from nitrogen in the coal, which may contain up to 2% of the element in the form of organic compounds. The only practical means of limiting emissions of nitrogen oxides at present is to minimize their formation. This can be done by modifying the combustion process to reduce peak flame and furnace temperatures, the time that combustion gases remain at peak temperature, and the amount of excess air in the combustion zone (125). Such changes, on the other hand, tend to reduce the efficiency of combustion and increase the emission of other pollutants, such as carbon monoxide and hydrocarbons. The conflict can be resolved by changing the design and operation of conventional combustion equipment, but only up to a point. The best that can be achieved is a reduction of approximately 50% in emissions of nitrogen oxides. The measures described can meet current emission standards, and U.S. utilities have been applying them in slowly increasing numbers since as long ago as the late 1950's (126).

Combustion modification is unlikely to be able to meet the more stringent standards that may be necessary in the future, however, and work is under way on processes for removing nitrogen oxides from flue gases (127). One approach is to inject ammonia to reduce nitrogen oxides to nitrogen; a second method is to use an oxidant such as ozone to convert nitric oxide to nitrogen dioxide, which is then absorbed in an alkaline medium. An ammonia injection process had been demonstrated in Japan by the end of 1977; evaluations of flue-gas-cleaning processes were under way in the U.S. (128).

The Automobile

With the spread of air pollution control devices in industry and of clean-burning natural gas for heating homes, many cities have reached the point where the automobile, including trucks and buses, accounts for at least 90% of the total weight of air pollutants emitted. A city that attributes less than 90% of its emissions to cars can be assumed almost invariably to be attributing more than the typical percentage to emissions from nonautomotive sources and thus to have a worse-than-typical pollution problem overall. In any event, vehicular emissions pose a massive problem. Americans rely almost totally on the automobile for transportation. More than 135 million cars, trucks, and buses were registered in the United States in 1976, and about 109 million of that number were passenger cars. The classical gasoline-fueled engine used in more than 98% of these vehicles is a relatively inefficient converter of chemical energy. It rarely extracts more than 15% of that available in its fuel, and the measures taken thus far to clean up auto exhaust have, on the whole, reduced the engine's efficiency.

The automobile is the source of a number of air contaminants. The amounts of pollutants in the exhaust depend on factors like the design and condition of the engine, the mode in which it is operating (idle, acceleration, etc.), and the composition of the fuel. The main offenders are carbon monoxide and hydrocarbons resulting from incomplete combustion, nitrogen oxides formed during combustion, and lead from antiknock compounds in the gasoline. The exhaust also contains compounds of phosphorus and other elements present in fuel additives and lubricants and in the fuel itself, as well as the carbon dioxide and water that result from complete combustion. In addition, cars emit hydrocarbons that evaporate from the fuel system and that escape around the piston rings ("blowby") and exit from the crankcase. The pattern of emissions from an uncontrolled automobile appears in Figure 4.

Emission standards

California set the nation's first limits on automotive pollutants, for crankcase hydrocarbons, beginning with the 1961 model year. Federal standards first took effect with the 1968 model year and have grown increasingly stringent. The Clean Air Act of 1970 established ultimate exhaust standards to be imposed on hydrocarbons and carbon monoxide by model year 1975 and on nitrogen oxides by model year 1976. The schedule was not met, however. Vehicle manufacturers have been granted various suspensions of standards, a procedure provided for by the law; the Energy Supply and Environmental Coordination Act of 1974 and the Clean Air Act Amendments of 1977 specified further delays. The act of 1970 permitted California, in the face of "compelling and extraordinary conditions," to set standards more stringent than the federal standards with the approval of EPA, and the state consistently has done so; the amendments of 1977 extended this right to the other states.

The status of exhaust emission standards for passenger cars appears in Table 15. In addition to automobiles, the law imposes exhaust limits on light-,

FIGURE 4

Pattern of emissions from an uncontrolled automobile

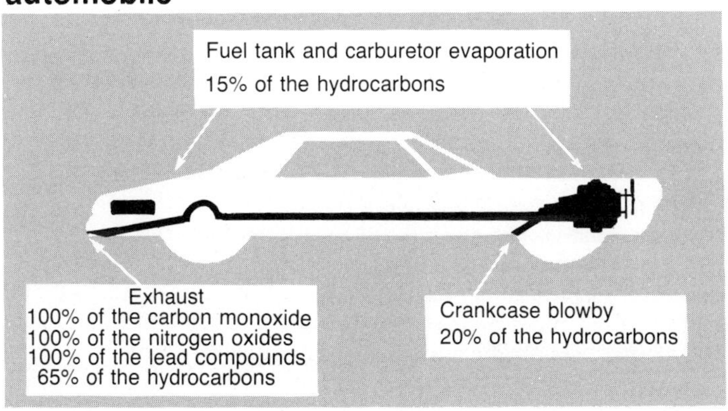

medium-, and heavy-duty vehicles with both diesel and gasoline engines. Federal standards have barred crankcase emissions of hydrocarbons since 1968 for light-duty passenger vehicles and trucks and since 1976 for heavier vehicles. Evaporative emission standards have been in effect since 1972 for light-duty passenger vehicles and trucks.

Emission controls

Vehicle manufacturers may meet emission standards by any means they choose. They have controlled crankcase emissions with "positive crankcase ventilation" systems, which recycle crankcase ventilation air and blowby gases to the engine intake. They have controlled evaporative emissions with tighter seals on gas caps and other fixtures and with canisters of activated carbon that adsorb and store hydrocarbon vapors for recycle to the engine. The major problem is the control of exhaust emissions, which presents a series of dilemmas: leaner carburetor settings — higher ratios of air to fuel — decrease exhaust carbon monoxide and hydrocarbons, but increase combustion temperature and thus the formation of nitrogen oxides; higher compression improves the efficiency of combustion, thereby decreasing carbon monoxide and hydrocarbons, but requires more lead in the gasoline and, again, increases combustion temperature and nitrogen oxides.

The exhaust control systems on the vehicles of the early 1970's were a monument to careful engineering. Manufacturers optimized air-fuel ratios, spark timing, and other variables so as to minimize exhaust carbon monoxide and hydrocarbons; with these adjustments they integrated systems that recycle part of the exhaust gases to the engine to decrease combustion temperature and nitrogen oxides. Such measures were effective, and they avoided major redesign of the engine. They increased fuel consumption significantly, however, and made the engine much more sensitive to maladjustment and environmental

Table 15

Exhaust emissions limits for passenger cars[a] (grams per mile)

	Federal			California[b, c]		
Year	HC	CO	NOx	HC	CO	NOx
1976	1.5	15.0	3.1	0.9	9.0	2.0
1977–79[b]	1.5	15.0	2.0	0.41	9.0	1.5
1980[c]	0.41	7.0	2.0	0.41	9.0	1.0/1.5[d]
1981	0.41	3.4[e]	1.0[f]	0.41	3.4	1.0
1982 and later	0.41	3.4[e]	1.0[f]	0.41	7.0	0.4

a. Emissions from a typical, light-duty uncontrolled car, in grams per vehicle mile, are 8.7 for hydrocarbons, 87.0 for carbon monoxide, and 3.5 for nitrogen oxides.

b. In 1979 and later, California must obtain EPA waiver for standards more stringent than federal standards.

c. In 1980 and later, any state may impose California standards upon obtaining EPA's approval.

d. 1.5 allowable if certified by the California Air Resources Board optional 100,000-mile durability procedure.

e. Waiverable to 7.0 through 1982.

f. Waiverable to 1.5 for diesel engines or innovative technology; waiverable only for innovative technology after 1984.

Source: Environmental Protection Agency.

change. In Denver, for example, it was nearly impossible to locate an automobile that did not violate federal emission standards purely because of the lower density of the air a mile above sea level.

Catalysis. By about 1970, the automobile industry knew that first-generation techniques would be unable to meet the stringent exhaust standards expected by mid-decade. A new approach was needed, and the one selected in the U.S. was catalysis. The catalytic converter was used on some 70% of the cars of the 1975 model year, the first ones to be equipped with it, and on 85% of those of the 1976 model year. The devices are part of the exhaust system; on 1975 and 1976 cars they contain pellets or blocks of catalyst based on the noble metals platinum and palladium. The catalyst promotes the oxidation of carbon monoxide and hydrocarbons that pass through the converter in the exhaust gases. It does not affect nitrogen oxides, but control of those compounds by different catalysts was being tested by 1976.

By 1978, catalytic control of carbon monoxide and hydrocarbons seemed to be working, although few data were in hand on the long-term performance of the converters in consumer hands. Catalysis, like first-generation methods, avoided major redesign of the engine. Also, the approach led to better fuel economy, because the engine could be tuned more to that end than to control pollutants, leaving the converter to deal with the consequent higher emissions in the exhaust. It was not certain, however, that the gain could survive the stricter emission standards scheduled for later years, especially for nitrogen oxides, which pose new problems in catalytic control (6, 129). The converters used in 1975–76, moreover, were found to generate sulfuric acid aerosol, and they call for gasoline free of lead, which inactivates the catalyst.

Catalytic control of nitrogen oxides in auto exhaust has progressed relatively slowly. Chemically it involves reduction of the compounds to nitrogen, as opposed to the oxidation desired with carbon monoxide and hydrocarbons. One approach to the problem is the dual-bed converter, in which a reduction catalyst and an oxidation catalyst function in sequence. A second approach is the single, three-way catalyst, which promotes an oxidation-reduction reaction between the carbon monoxide and hydrocarbons on the one hand and the nitrogen oxides on the other (130). U.S. auto makers have favored the three-way catalyst (131).

The most effective three-way catalysts are based on a combination of platinum and rhodium. Ruthenium is being studied, however, as a replacement for the more costly and less plentiful rhodium. For a three-way system to perform adequately, the carbon monoxide, hydrocarbons, and nitrogen oxides must enter the converter in relatively fixed proportions (the stoichiometric ratio). The standard carburetor cannot control the air-fuel ratio precisely enough to meet this requirement, which demands an advanced metering technique such as electronic fuel injection. One difficulty here is that domestic industry needs an estimated three to five years' lead time to produce enough advanced metering devices to equip the number of cars made in this country annually. The 1977 amendments extended through 1980 the nitrogen oxides emission standard in effect in 1977, which U.S. auto makers had been meeting with engine adjustment and exhaust gas recirculation. It thus appears that catalytic control of nitrogen oxides emissions will not be used widely on American production cars until at least model year 1981, when a tighter standard is scheduled to take effect. However, some Ford and General Motors cars of model year 1978 for sale in California were equipped with three-way catalysts, and more are expected to be introduced in 1979 and 1980.

Sulfuric acid mist. In late 1972, when it was really too late to change the course of development of the catalytic muffler, the devices were found to yield an unwelcome by-product, sulfuric acid aerosol (132). Gasolines sold in the U.S. are uniformly very low in sulfur. The catalytic converter with an oxidizing catalyst, however, transforms part of the sulfur into sulfuric acid mist (133). The amount emitted is relatively small, particularly in comparison to secondary sulfate pollutants originating with emitted sulfur dioxide. Nevertheless, concern arose that widespread use of catalyst-equipped vehicles might produce hazardous, localized levels of sulfuric acid along roadways. Later studies indicated, however, that the problem is not as serious as was thought at first (134–136).

Lead in fuel. EPA acted early in 1973 to forestall inactivation of catalytic converters by lead. The agency issued regulations designed to make unleaded gasoline generally available by July 1, 1974; to prevent inadvertent use of leaded fuel, it ruled that the gas filler inlet on catalyst-equipped cars must be compatible only with a special nozzle to be used only on pumps carrying unleaded gasoline. Under the regulations, some 120,000 gas stations — more than half of those in the nation — should be offering at least one grade of unleaded fuel.

These measures to protect catalysts complemented a separate, much broader EPA move against lead in gasoline, also initiated in 1973 and intended to protect the public health. The agency held that airborne lead from human activities was reaching hazardous levels and that 90% of it came from the antiknock compounds, lead tetraethyl and lead tetramethyl, in gasoline. At the time, the lead content of gasoline averaged 2 to 3 g per gallon nationwide; the new EPA regulations called for a graduated reduction to 0.5 g per gallon after Jan. 1, 1979. Implicit in the action were two problems: the existence of a hazard to public health was not scientifically clear-cut, so that EPA's authority to rule as it had was legally questionable; and a large fraction of the cars on the road required premium gasoline, so that refiners would have to maintain octane ratings by replacing lead with high-octane petroleum fractions, which would dictate marked change in their operations.

Opponents of lead reduction quickly challenged EPA's regulations. The issue was not resolved until June 1976, when the Supreme Court refused to review an appeals court decision favoring the agency. The decision held in part that the Clean Air Act empowered EPA to act to protect the public health without necessarily establishing "rigorous step-by-step proof of cause and effect." With this, EPA resumed enforcement of the phase-down schedule, which had been suspended during most of its life, but extended the deadlines. The ultimate limit on lead in gasoline, 0.5 g per gallon, is to take effect after Oct. 1, 1979, nine months later than originally scheduled (EPA data put the national average at 1.72 g per gallon in the second quarter of 1976). The agency extended the schedule to give refiners time to increase their production capacity for the high-octane aromatic hydrocarbons needed to replace lead. The compounds are used in the lead-free gas required for catalyst-equipped cars and also as feedstocks for the petrochemicals industry. EPA determined that existing capacity for aromatics was insufficient to supply these markets as well as the new one created by the phase-down program.

Nontraditional engines

The apparent succcess of catalytic control of exhaust emissions may have postponed but has not obviated the production of nontraditional types of automotive engines in the U.S. (137, 138). Rotary and diesel engines are made abroad and sold in cars in this country in limited numbers, and American makers were marketing diesel-driven models in small numbers in 1978. In terms of emissions, the rotary engine offers no special advantage, and the diesel, although inherently a low emitter of carbon monoxide and hydrocarbons, has not yet proved capable of meeting stringent standards for nitrogen oxides. A more realistic candidate for mass production in the U.S. is a spark-ignition engine that burns very lean air-fuel mixtures. One such engine, using the stratified-charge concept, was introduced in the 1975 model year by Japan's Honda Motor Co. Basically, the engine uses a very rich mixture in the vicinity of the spark plug, to maintain easy ignition, and a far leaner mixture in the rest of the cylinder. It demands more complicated carburetion and injection than the conventional engine, as well as modification of the cylinder head and engine proper. On the other hand, a small engine of this type has displayed very low emissions over 50,000 miles without use of a catalytic muffler, air injection, or exhaust gas recirculation. The engine may, however, require exhaust gas recirculation to meet the limit on nitrogen oxides scheduled for 1981.

Longer-range candidates for powering personal vehicles are the gas turbine and external combustion engines. The latter are of two types: the Rankine engine, best known in the form of the steam engine, transfers the heat from combustion to a condensible working fluid in a separate, closed system; the Stirling engine is similar in concept, but uses a noncondensible working fluid such as a gas at high pressure. All three engines are inherently low-emitting and could meet the 1981 standards readily without external controls. Each could use a variety of liquid hydrocarbon fuels. None of the three has been proved capable yet of matching the blend of performance, durability, weight, and cost displayed by current automotive engines, but each has sufficient potential to warrant continued study and development.

For the much longer run, it can legitimately be questioned whether the United States or, in fact, the other western nations can continue to dedicate such a large fraction of their resources to personal transportation. It has been estimated, for example, that fully 20% of this country's labor force is involved directly or indirectly in producing and servicing the automobile. The number of revenue passengers on public transit nationwide declined some 70% between 1945 and 1975. The energy crisis brought a very slight upturn, and a few cities have achieved gains of 30 to 40% since 1975. Nevertheless, persistence and real innovation will be needed to persuade a significant fraction of the population to break away from its present dependence on the automobile.

Recommendation A13: *Efforts to develop more efficient, less inherently polluting vehicular engines should be increased, along with selective development of new concepts in public transportation.*

Space Heating

Domestic and commercial space heating accounts for a relatively small fraction of total national emissions of criteria pollutants. The effects of space heating on the quality of urban air, however, are much greater than the fraction would indicate, because the emissions enter the air in crowded areas at relatively low heights above the ground. Space heating additionally is a difficult abatement problem because of the large numbers of relatively small sources involved.

The only economic means of reducing sulfur oxides emissions from space heating is the use of low-sulfur fuels. To reduce particle emissions the only economic means are to avoid the use of fuels, such as bituminous coal, that produce relatively large emissions, and to upgrade the efficiency of combustion equipment. A long-range alternative is "district heating," in which domestic and commercial spaces are heated with hot air, electricity, steam, or high-temperature water (204° C, 18.6 atm) from central production units large enough to use advanced pollution abatement technology economically. District heating with steam has been used to a very limited degree in this country, and "all-electric" homes and buildings embody the concept in restricted form. District heating of both space and water is much more common in Europe (139). The systems there typically are based on hot water and rely extensively on waste heat from steam-electric and nuclear power plants, refuse incinerators, and other sources.

Air Pollution and the Energy Crisis

The abatement of air pollution cannot logically be considered apart from the now-familiar energy crisis. The scrubbers used to remove sulfur dioxide from power-plant stack gases, for example, consume from 1.5 to 5% of the power generated, and consumption of power by these and other air-cleaning devices has been used in many areas as an argument against pollution control. The argument is valid only in the short run. In the long run, the energy cost of such devices will often be far offset by gains in efficiency that will reduce the consumption of energy resources, the generation of pollutants, and the amount of gas to be cleaned per unit of energy extracted.

Historically, energy has cost too little to create any incentive to conserve it. Mounting costs, however, make it increasingly attractive to use hitherto wasted heat. Many manufacturing processes, for example, discharge water or steam to the environment at temperatures that would permit its energy content to be used to heat homes and buildings, defrost access roads, generate power, and perform a number of other tasks now handled by separate systems. There seems no theoretical reason why even rather low-level process heat from industries could not be put to work in some way (140). There may well be practical lower limits in terms of both temperature and process scale. Waste heat recovery of this type has been done sporadically in the past; today, the need to obtain maximum energy from fuel that must be burned anyway puts a premium on developing the necessary concepts systematically and applying them as rapidly as possible.

We noted earlier that electric utilities are reluctant to enter a new business — chemicals — by becoming sources of sulfur. Similarly, many manufacturers are reluctant to become utilities by purveying their waste process heat or electricity they generate with it. A leveling of these institutional barriers, by whatever means, would certainly increase employment: few utilities nowadays employ chemical engineers, and not many more manufacturers employ electrical engineers specializing in power generation (105).

Probably the most overlooked aspect of the energy crisis is that it stems not from short supply but from technological inflexibility. The nation's reserves of fossil fuels include enough coal to last for centuries and substantial deposits of oil shale. Abroad are other reserves of fossil fuels, such as the Athabasca tar sands of Canada. Enough sunlight strikes 1% of the land surface of the U.S. to supply our present energy needs. It would even be feasible to grow enough trees in this country to meet our total demand for energy by burning wood. Given the solutions to technological problems, power will be available from nuclear fission and fusion, geothermal sources, winds and waves, and the temperature and salinity gradients and tides in the oceans. For fossil fuels we have the promise of more efficient, less-polluting combustion systems such as the fluidized bed and magnetohydrodynamics. The frequent problem with the available sources of energy — and the basis of the energy crisis — is that we are not technologically flexible enough to harness them efficiently in the forms in which they occur (141).

A power plant designed to burn gas or oil does not work well on coal. A coal furnace will not burn wood efficiently. The automobile must have a liquid fuel to function effectively. Even an immediate shift to electrical heating for all new homes would leave millions of existing homes requiring gaseous fuel for heating space and water and a smaller number for cooking as well. Thus for several decades at least, fuels must be available in liquid and gaseous forms. For this reason, research and development on the liquefaction and gasification of coal is necessary despite the energy penalty incurred by such processes. Some critics suggest, for example, that oil from shale may supply less energy than is required to extract it. Even should this turn out to be the case, the process may well become advantageous nevertheless if it yields a liquid fuel at less net energy loss than is incurred by liquefying coal.

Sporadic energy sources

Energy is available sporadically from numerous natural sources. They include solar energy and manifestations of it such as winds, waves, and, probably, the salinity and temperature gradients in the oceans, as well as tides, which reflect forces exerted by both sun and moon. To harness such sources effectively, we need efficient means of accumulating surplus energy when it is available and storing it for use during periods of darkness, calm, or other interruption of supply. Current methods of accumulating energy are inefficient and unwieldly. It has apparently proven impossible, for example, to design a battery-powered automobile that can travel more than about 100 miles without recharging. Existing lead-acid batteries are nearly self-defeating because of their great weight; nearly all of the remaining electrode combinations for storage batteries require metals that are in short supply or pose safety problems or both.

One way that energy might be accumulated is in the form of hydrogen liberated from water. Electrolysis of water is a relatively inefficient way to release the hydrogen, but a number of chemical or thermal cycles that could be driven electrically are potentially more efficient. Storage of energy in the form of hydrogen is very promising and merits far more intensive investigation than it is receiving today. The combustion of hydrogen in air produces some nitrogen oxides, but little else in the way of by-product air pollutants. A serious question, however, is whether hydrogen could be distributed safely through the existing pipeline system. The element permeates many metals, including some of those used in pipelines, making them extraordinarily brittle and easily broken. A second serious question concerns the safety of fueling automobiles with high-pressure tank hydrogen, which would be necessary to approximate the range provided by a tankful of gasoline. The explosion hazard is often overrated, but the hazard of puncturing a tank containing any gas at very high pressure is not overrated in the slightest.

It is evident that more efficient means of storage would greatly increase the usefulness of many inconstant sources of energy. The level of research on the problem is clearly inadequate and should be increased. This does not mean that more intensive research and development on solar energy should be postponed until the energy-storage problem is solved. Even present solar-energy technology, if used to heat homes, for example, would decrease domestic gas consumption significantly. The major problem is that economical ways have not been found to retrofit existing houses.

Recommendation A14: *Cost-risk figures for nuclear and other forms of power generation should be recomputed to be certain that true costs and hazards have been explicitly included.*

Recommendation A15: *Studies of coal purification, liquefaction, and gasification should be given high priority by the agencies and industries involved, with proper stress on the cleanliness of both processes and products.*

Recommendation A16: *Additional federal funds should be channelled through the Department of Energy to studies on intermittent energy sources (wind, waves, sun, tides, etc.), coordinated with research and development on energy accumulation systems, including hydrogen generation.*

Recommendation A17: *Economic incentives should be developed at the federal level to encourage systematic utilization of waste heat by its producers and by other prospective consumers, as well as resource conservation in general.*

EFFECTS OF AIR POLLUTANTS

The effects of air pollutants result mainly from their encounters with and removal by receptors or targets, including man and other animals, vegetation, and materials or objects of economic value. The sole exceptions are the optical effects exerted by contaminants while airborne; among these are reduction of visibility by particulate matter and the potential impact of particles and carbon dioxide on the global flow of radiant energy and thus on weather and climate.

Effects on Man and Animals

Information about the effects of air contaminants on man and other animals has come from both epidemiological and laboratory studies. The epidemiologist attempts to associate an effect, in a large human population, with its cause; the laboratory worker starts with a cause and attempts to determine its effects. The two types of studies should complement each other. Ideally, they should provide dose-response relationships for each air contaminant and should delineate the influence of other components of the atmosphere on individual dose-response relationships. At the current state of knowledge, however, the data are more indicative than definitive (4, 142, 143).

The acute toxicological effects of many air contaminants are understood reasonably well, but the effects of exposure to heterogeneous mixtures of gases and particles at very low concentrations are only just beginning to be comprehended. The limits of our understanding for the criteria pollutants were documented in the studies made to support the national air quality standards established in 1971 (20, 144–147). Indications of our current understanding of the effects of air contaminants on health and physiological response appear in Table 16 and in the following selected observations.

Sulfates. Sulfur dioxide has been implicated in a number of serious incidents of air pollution. The results of numerous epidemiological studies, moreover, definitely associate the combination of sulfur dioxide, sulfates, and particles with health effects of varying severity. Two theories have been proposed for the action of sulfur dioxide. The first is that airborne particles catalyze rapid oxidation of the gas to sulfur trioxide, which combines with water in the air to form droplets of sulfuric acid which are inhaled. The second theory is that sulfur dioxide adsorbed on airborne particles is inhaled and later desorbed deep in the lungs in amounts significantly larger than would be the case if the gas alone were inhaled.

Pertinent to these theories is an epidemiological study of 1970–71 by EPA's Community Health and Environmental Surveillance System (CHESS). The work included analyses of the chemical makeup of particulate matter as well as measurements of its total mass. The results suggested that adverse effects on health were associated more consistently with exposure to suspended sulfate than with exposure to sulfur dioxide or to total suspended particles (148). Suspended sulfate can include or originate with sulfuric acid, so that the CHESS finding would appear to favor the sulfuric acid theory for the action of sulfur dioxide. It does not, however, preclude the alternative theory, that health is harmed by inhalation of sulfur dioxide adsorbed on particles. The true significance of the CHESS observation, in fact, is uncertain for two reasons: questions about the atmospheric chemistry involved — specifically, how and to what extent and in what compounds sulfate is formed in the atmosphere; and doubts about the validity of the data gathered by the air monitoring instruments used. In addition, the CHESS investigation suffered from computer-based modeling that did not account for all of the relevant parameters.

Particles. Observations of animals exposed to gas-aerosol systems (149–151) show that particulate matter may serve as a respiratory irritant and thus enhance the respiratory action of other contaminants. Particles smaller than 1μm in diameter are considered the most potent irritants. Particulate matter also may provide an adsorbent surface for other pollutants, and it may catalyze their conversion to other forms.

Concern is growing over the synergistic — greater than additive — effects of iron oxides with sulfur dioxide and with carcinogens (152). The various oxidation states of manganese (153) suggest that it might catalyze the oxidation of other air pollutants to less desirable forms (154, 155). Many specific substances encountered as respirable particles are considered potential hazards to health. Vanadium (156), for example, inhibits a number of enzyme systems (157–160); inhalation of vanadium has been correlated with low cholesterol levels (161) and upset of ascorbic acid (vitamin C) metabolism (162) in man. Chromium, cadmium, nickel, arsenic, asbestos, and particulate polycyclic organic matter have been indicted as environmental carcinogens (163–174). The accumulation of lead by the body and its possible effects on health have been the subject of vigorous debate (175–177).

The foregoing discussion of individual chemical species in particulate matter highlights a major weakness of air quality standards, such as the current federal standard, based on generalized particulate matter. In Los Angeles, for example, 100$\mu g/m^3$ of smog is highly unpleasant because 95% of the particles are in the respirable size range. On a typical day in Denver, on the other hand, a similar

Table 16

Health effects of major air pollutants

Pollutant	Effects	Estimated effects threshold ($\mu g/m^3$)	National ambient standards ($\mu g/m^3$)
Sulfur dioxide	Aggravation of respiratory diseases including asthma, chronic bronchitis, and emphysema; reduced lung function; irritation of eyes and respiratory tract; increased mortality	300 to 400 (short-term) 91 (long-term)	365 80
Suspended sulfate	Aggravation of respiratory diseases including asthma, chronic bronchitis, and emphysema; reduced lung function; irritation of eyes and respiratory tract; increased mortality	8 (short-term) 8 (long-term)	none none
Total suspended particles	Directly toxic effects or aggravation of the effects of gaseous pollutants; aggravation of asthma or other respiratory or cardiorespiratory symptoms; increased cough and chest discomfort; increased mortality	70 to 250 (short-term) 100 (long-term)	260 75
Nitrogen dioxide	Aggravation of respiratory and cardiovascular illnesses and chronic nephritis	141 (long-term)	100
Carbon monoxide	Reduced tolerance for exercise, impairment of mental function, impairment of fetal development, aggravation of cardiovascular diseases	23 (8 hour) 73 (1 hour)	10 40
Photochemical oxidants	Aggravation of respiratory and cardiovascular illnesses, irritation of eyes and respiratory tract, impairment of cardiopulmonary function	200 (short-term)	160

Source: "Proceedings of the Conference on Health Effects of Air Pollutants," National Academy of Sciences — National Research Council, prepared for the Committee on Public Works, U.S. Senate, U.S. Government Printing Office, Washington, D.C., November 1973.

"Sixth Annual Report of the Council on Environmental Quality, U.S. Government Printing Office, Washington, D.C., December 1975.

concentration is usually innocuous because 95% of the particles are larger than the respirable size range. In the Rocky Mountain area, total suspended sulfate tends to be high, because of the gypsum in the soils, but physiologically insignificant: the gypsum itself is largely innocuous, and most of the sulfate is in the larger, nonrespirable particles. But in New York City, most of the suspended sulfate is ammonium sulfate, a potent bronchoconstrictor, and it is concentrated in the smaller, respirable particles. For such reasons, an air quality standard based on gross particulate matter is clearly inadequate in terms of health (177a). At the same time, establishing standards for particles that consider both size range and individual chemical species, and monitoring ambient air in support of such standards, present technical difficulties that are not readily resolved.

Carbon monoxide. Carbon monoxide is an important contaminant in urban air principally because it can interfere with the normal transport of oxygen to the tissues by the hemoglobin in red blood cells. The compound interferes by combining with hemoglobin to form carboxyhemoglobin, which cannot take up oxygen. The reaction is reversible: if the level of exposure to carbon monoxide declines, a proportionate amount of carboxyhemoglobin reverts — with a half-life of about five hours — to hemoglobin that again functions normally. The background level of carboxyhemoglobin in humans is equivalent to about a 0.5% reduction in the oxygen-carrying capacity of the blood. Exposure for one hour at the concentration of the one-hour national air quality standard for carbon monoxide produces a level of about 1.5%. A study of carboxyhemoglobin levels in 29,000 blood donors in 18 metropolitan areas in the U.S. was conducted during 1969–72 (178). In nonsmoking adults, the levels ranged from 0.4 to 1.5% in suburban areas and from 0.8 to 3.2% in urban areas; in pack-a-day smokers, the levels were two to three times higher.

Dogs exposed to high concentrations of carbon monoxide over long periods — carboxyhemoglobin levels reached 20% — have displayed changes in the brain and central nervous system (179). The changes in the brain were regarded as secondary results of myocardial (heart-muscle) changes (180); changes in the electrocardiograms of rats and dogs exposed to carbon monoxide in fact have been reported (180, 181).

In both man and animals, exposure to carbon monoxide causes changes in electroencephalographic patterns, which portray the electrical activity of the cerebral cortex (180, 182). The data from such studies thus far have contributed little to our understanding of brain function, but refinements in encephalographic techniques have opened new avenues for research (183). Investigations of this kind may be significant in view of reports that exposure to low levels of carbon monoxide — carboxyhemoglobin levels 2 to 3% above background — might impair human performance, including the ability to judge short intervals of time (184, 185).

Epidemiological studies of motor vehicle accidents have not established a causal role for carbon monoxide (186, 187). A statistical study of auto accidents in California found that carbon monoxide was not a significant factor, but did indict photochemical oxidants (188). Laboratory studies, on the other hand, have indicated that exposure to carbon monoxide at the levels experienced by urban motorists impairs the ability to remain highly alert in a monotonous environment (189, 190).

Oxidants. The apparent statistical correlation between photochemical oxidants and auto accidents (188) may reflect many variables, among them the level of eye irritation. Human test panels have named as eye irritants acrolein, formaldehyde, and peroxyacyl nitrate; each is found in photochemical smog. Eye irritation, and the compounds and their precursors in smog that cause it, however, are not fully understood. The nature of the precursors may be involved (191).

In photochemical smog the predominant oxidant is ozone, and the concentration of oxidants is expressed in ozone equivalents. The peroxyacyl nitrates (PAN's), although present in lesser amounts, are included as oxidants.

In prolonged exposures to ozone under occupational conditions, ozone levels of about 590 $\mu g/m^3$ (0.3 ppm) produce nasal and throat irritation. In experimental appraisal of short-term exposures of man, concentrations of 980 to 1960 $\mu g/m^3$ (0.5 to 1.0 ppm) of ozone produce changes in pulmonary function: increased airway resistance, decreased vital capacity, decreased carbon monoxide diffusing capacity, and decreased forced expiratory volume.

Information on the effects of the PAN's is very limited. PAN appears to be about as lethal to mice as nitrogen dioxide, more lethal than sulfur dioxide, and less lethal than ozone. Limited observations indicate that exposure of man to PAN increases oxygen uptake during exercise.

In an effort to evaluate the effects of exposure to oxidants, numerous experiments have been performed using photochemical smog obtained from the ambient atmosphere or from the irradiation of automobile exhaust. Laboratory animals exposed to ambient air high in oxidants showed an increase in lung-flow resistance. Similarly, guinea pigs exposed to irradiated automobile exhaust showed a marked increase in total expiratory flow resistance during the exposure. The change in flow resistance was not observed when the animals were exposed to unirradiated exhaust. For the irradiated automobile exhaust the increase in flow resistance occurred at low oxidant levels compared to the concentration of ozone (1960 $\mu g/m^3$ or 1 ppm) required to produce a similar change in flow resistance. Thus the data suggest that other components in the exhaust played a role (192).

Test animals exposed to ozone have decreased resistance to infection as seen in shortened survival time and increased mortality (193). Similarly, synthesized photochemical smog increases the mortality of mice from streptococcal pneumonia (194). Since the ozone alone produced the effects at concentrations below that of the oxidants from the photochemical smog, it would appear that the decreased resistance to infection could be attributed to ozone. This conclusion is uncertain, however, because increased mortality from respiratory infections also is observed after exposure to nitrogen dioxide (195), which is present in smog.

The combination of substances, including residual hydrocarbons and nitrogen oxides, in photochemical smog creates experimental difficulties in establishing cause-and-effect relationships between particular biological responses observed and specific components of the mixture. The problem is complicated further by the fact that the effects of each component of the mixture may not be assumed to be independent of the effects of the others.

Nitrogen dioxide. Nitrogen dioxide has been associated with adverse effects on health more closely than any of the other nitrogen oxides. In animals the compound exerts its primary toxic effects on the lungs and can be associated with increased susceptibility to respiratory infection and emphysematous changes. The effects of community exposure to nitrogen dioxide were examined in a study in Chattanooga, Tenn., during November 1968–April 1969 (196–198). The study covered 4043 individuals in 871 families and produced the principal support for the federal primary standard for ambient nitrogen dioxide. A precise dose-response relationship for nitrogen dioxide could not be established from the Chattanooga data, but the study did yield significant observations. The ventilatory performance (lung function) of second-grade children in areas exposed to high levels of nitrogen dioxide was significantly lower than that of children in control (lower exposure) areas. The illness-incidence rates in the high nitrogen dioxide areas were consistently higher than the rates in the control areas for each family segment (all family members, second graders, siblings, mothers, fathers).

The incidence of acute respiratory disease in the Chattanooga study was observed when the average 24-hour nitrogen dioxide concentration in the high nitrogen dioxide areas was between 117 and 205 $\mu g/m^3$ (0.062 to 0.109 ppm) and the average suspended nitrate level was 3.8 $\mu g/m^3$ or greater. Subsequently, the monitoring method for nitrogen dioxide was found to be faulty, and the estimated nitrogen dioxide exposures were revised upward. Most affected were the data from the neighborhood where a significant excess of acute respiratory illness was recorded in the least-exposed of the three high-nitrogen-dioxide-exposure areas. The revised data, however, were not considered sufficient reason to revise the federal air quality standard for nitrogen dioxide (7).

Nitrosamines. Besides nitrogen dioxide's effects on the lungs, the compound conceivably could play a role in the formation of nitrosamines in ambient air. Many of the nitrosamines are carcinogenic in laboratory animals. They have not been believed to be present in air, but recently the potent carcinogen dimethylnitrosamine has been detected in two locations (199, 200). At some measurement sites the source of the compound evidently was nearby chemical plants that handle dimethylnitrosamine or a closely related compound, but at other sites such emissions were considered unlikely to be a significant factor. The source at these latter sites was termed "unknown," but the reaction of nitrogen dioxide with other compounds in ambient air to produce dimethylnitrosamine has not been excluded.

Mixed pollutants. Considerable work has been done and is in progress on the effects of mixed pollutants and mixed gas-aerosol systems on animals (149–151). Such work has shown in general that mixed pollutants may act in several different ways. They may produce an effect that is additive, amounting to the sum of the effects of each pollutant acting alone; they may produce an effect that is greater than additive (synergism) or less than the simple additive (antagonism); or they may produce an effect that differs in some other way from the simple additive.

Work with whole animals is complicated further by protective effects (201, 202) and cross tolerances (203). The problem of tolerance is complicated still further by the possibility that it is species-related. Thus the effects of contaminants on the whole animal are not independent of each other, and the animal's total response involves its past history and its overall reponse to its environment. Physiological stress, for example, could be a factor in the changes in alkaline phosphatase activity found in the livers of rats that had inhaled acrolein, ozone, nitrogen dioxide, formaldehyde, or sulfur dioxide (204). The same is true of changes in blood serum protein and alkaline phosphatase of the lung found in whole animals exposed to ozone and nitrogen dioxide (205). Similarly, increased susceptibility to bacterial and viral infections follows exposure to auto exhaust (194), ozone (193), and nitrogen dioxide (195). The possibility that this increased susceptibility to infection is a secondary result of the degranulation of the lung mast cells, as has been observed after inhalation of nitrogen dioxide (206), has not been examined.

Cell response. The complexities of research on whole animals have stimulated work on simpler biochemical systems designed to explain the basic mechanisms of the responses of cells to toxic or irritating substances. Such work appears to be necessary to evaluate the effects of long-term, low-level exposure to individual components of mixtures of gases. It has included research on various bacteria, individual tissues, and isolated enzyme systems, using both individual gases and mixtures of gases. Ozone, nitrogen dioxide, peroxyacyl nitrate, and formaldehyde, for example, have been found to inhibit the activity of a number of enzymes in isolated systems, but with differences in rate, extent, and mode of action (207–209). Evidence has been found of correlations between the initial concentrations of reactants in gaseous mixtures and the rates at which the mixtures kill or inhibit the growth of bacteria (210). No correlation was observed with the chemical analysis of the mixtures after they had been caused by irradiation to react photochemically (211).

Despite the progress made in these and other experiments, the study of mechanisms of action is barely out of its infancy. Research on the biochemistry of the lung, which is crucial to true understanding of many mechanisms, is at about the same point (212). The evaluation of biological response, moreover, has been impeded by the limitations of the analytical chemical methods currently available. Such evaluations require methods more precise, more accurate, and more specific than ever required in the past. Subclinical intoxication with consequent predisposition to accident or disease is an uncharted area. Laboratory research and epidemiology have done much to explain the effects of air contaminants on man and animals, but what is known is plainly miniscule in the light of what must be learned.

Esthetics. At the other end of the spectrum, little work has been done on those effects of air pollution on humans generally described as "esthetic" or psychological. Such effects include depression caused by perpetual grayness, irritation at loss of scenic views, and nausea caused by offensive odors. The few opinion surveys made to date on such effects have generally been poorly documented in terms of pollutant concentrations. Often they have lacked adequate controls for the effect of the questioner in bringing to people's attention matters that they might normally ignore.

Recommendation A18: *Better understanding of the reactions of sulfur oxides and nitrogen oxides in the atmosphere must be developed to make it possible to appraise accurately the potential physiological impact of those contaminants.*

Recommendation A19: *The effects of the inhalation of metals, which have been examined only at high concentrations, should be examined at low and, where known, ambient concentrations and the consequent body burdens determined.*

Recommendation A20: *Further research should be done on the effects of combinations of air pollutants, both in vitro and in vivo, in carefully defined systems at known humidities and temperatures.*

Recommendation A21: *The effects of air contaminants on tissue oxidation must be probed more deeply, along with the uptake and catabolism (destructive metabolism) of contaminants by the tissues.*

Recommendation A22: *Epidemiological programs, such as EPA's now-terminated CHESS Program, ought to be supported by stronger overall analytical capabilities, firmer computer-based models, the development of measurement techniques for subclinical responses, and the determination of additional pollutants, some of which may actually account for effects that are being correlated with other, known contaminants.*

Recommendation A23: *Further research should be pursued on the responses of air contaminants to photoirradiation and the importance of these responses in the development of adverse effects on health.*

Recommendation A24: *Studies of the biochemistry and biophysics of the lung should be pressed to provide the knowledge needed to examine more specifically the in vivo effects of exposure to ambient concentrations of air pollutants.*

Recommendation A25: *More precise and specific analytical methods should be developed for determining specific biological responses to individual air contaminants, such as the conversion of hemoglobin to carboxyhemoglobin by carbon monoxide.*

Effects on Vegetation

For more than a century, scientists have been studying the effects of air pollution on vegetation (213–222). During most of this period they were concerned primarily with visible effects, such as dead areas (necrosis) in foliage, and the associated declines in growth and yield. The past decade or so, however, has seen clear recognition of the long-suspected reductions in growth and yield that occur unaccompanied by visible symptoms (214, 215, 223, 224). The distinction is pertinent to the development of air quality standards, which in part has involved assessments of economic loss from plant damage. Estimates of the loss to agriculture and forestry have ranged up to $1 billion nationwide, but they have been based mainly on visible damage. We have as yet no means of appraising reductions in growth and yield that occur when visible symptoms are not present or are not recognized.

Major air pollutants that produce visible damage to vegetation include sulfur dioxide and sulfuric acid mist (225–228), fluorides (229–232), photochemical oxidants (146, 219, 223, 224, 233, 234), and nitrogen oxides (221, 235). Visible damage also can be produced by ethylene (236), hydrogen chloride (237), and chlorine (238). Damage by roadside salt (for snow and ice removal), converted to aerosol by passing traffic, has become an international problem (239–241). The human health controversies on the ingestion of heavy metals with vegetables and fruits are far from settled, but questions about plant uptake of heavy metals from air and soils are becoming more tractable (242–244).

Growth responses. The reality and significance of the effects of air pollutants on vegetative growth are no longer in question. Sulfur dioxide has been reported to reduce the growth of pine (245), rye grass (246), and a variety of other species (247). Fluoride also reduces the growth of a number of species (230). Ethylene alters the aging process in some types of vegetation, with serious effects on growth (236). Ambient nitrogen dioxide has been shown to reduce yield in oranges (248) and ambient lime dust to reduce the growth of trees (249).

The growth effects of the photochemical-smog complex are especially clear-cut. Many of the relevant investigations have turned around the possible synergism of ozone and sulfur dioxide (250–254). Regardless of the question of synergism, simple addition, or antagonism, however, photochemical oxidants are associated with definite and significant generalized reductions of growth and yield in the field (255, 256). If reductions in yield reported for potatoes (256) apply only in part to other crops, and if the reductions in soybeans reported for controlled fumigations (253) apply only in part to field conditions, the potential impact of more smog is rather frightening. An estimate published in 1976 put the cost to the consumer of agricultural losses to "oxidant" pollutants at as much as $600 million annually in the two or three preceding years (234), and this estimate made little allowance for the generalized losses in productivity noted above.

Recommendation A26: *Studies of the effects of air pollutants, especially in combination, on plants must be continued, and studies of generalized losses in productivity should be expanded. The species studied should include those important in wild ecosystems as well as in agriculture and forestry.*

Effects on Wild Ecosystems

A major shortcoming of our knowledge of the biological effects of air pollution is the sparseness of information on the sensitivities of organisms that do not clearly have economic value. Funds and manpower have been inadequate to study everything, and priority has been given to effects that can be related directly to costs. In the long run, however, it is important to know the susceptibilities of many organisms of no obvious practical value. All forms of life are interrelated; the ultimate survival of man may depend quite as much on the welfare of lichens and foxes as on that of wheat and hogs.

Recommendation A27: *Studies on members of wild ecosystems should be integrated carefully with other research on the biological effects of air pollutants.*

Effects on Materials

A number of air contaminants damage materials. Ozone cracks rubber, weakens fabrics, and causes dyes to fade; hydrogen sulfide tarnishes silver; smoke dirties laundry; sulfur dioxide is a source of sulfuric acid, which corrodes metals. These and similar types of damage may lead to shortened service life, substitution of more corrosion-resistant materials or formulations, unusual protective measures such as humidity control, and other costly consequences.

The economic impact of air pollution on materials undoubtedly is extensive, but assigning an accurate dollar value to it is difficult and perhaps impossible except in specific instances. Still, estimates must be made to justify the costs of pollution control. One such estimate put the total cost of materials damage at $1.7 billion in 1970 (257). A second estimate, for the same year, attributed $1.22 billion in materials damage to ozone alone (234).

MEASUREMENT OF AIR POLLUTANTS

The costs and effectiveness of air pollution control programs depend to a marked degree on our ability to detect and measure contaminants. Although Chapter 2 deals in some depth with air analysis, at least a few of the problems it poses warrant brief treatment here as well.

The earliest work on air pollutant analysis was done by scientists and engineers experienced in measurements related to industrial hygiene. Without derogating their pioneering work in the least, it can be said that they were accustomed to working with higher concentrations and less complex mixtures than normally are found in ambient air. The result was that many of the early measurements of air pollutants were made with techniques of marginal sensitivity and poor specificity for the substances of interest.

Manual methods. These early methods were largely manual, and some of them survive today. Probably the outstanding example is the high-volume sampler with glass-fiber filter for collecting atmospheric particulate matter, together with some of the subsequent analyses for the particles collected. Nearly everyone in the field continues to rely on these techniques, whose imprecision can create serious problems. Some extraordinarily well designed epidemiological studies of the toxicity of airborne sulfates, for example, have suffered from analyses of sulfate whose results may range from half to twice the true values. In justice to the scientists involved, their continued dependence on old methods is largely by default: better methods have not been devised. Clearly it is time to make a major investment in identifying defective manual techniques and in developing better ones employing newer technology and chemical knowledge.

Air monitors. The situation is somewhat similar with automatic air monitors. Many of the first generation of these instruments also were adaptations of devices developed for use in industrial hygiene. The second generation of air monitors were essentially robot chemists, generally based on measurement of the intensity of colors formed in appropriate reactions then emerging gradually from analytical research. These colorimetric reactions in themselves were more sensitive and specific than earlier methods. Automating them, however, required for the most part that the chemistry involved be simplified, often by sacrificing those steps that resulted in specificity. The loss was balanced to some extent by prefilters that removed some of the interfering substances, but the resulting devices tended to be cantankerous and to need nearly constant attention to function reliably.

The automatic monitors of today belong largely to a third generation. They employ more specific sensors, such as ion specific electrodes and spectrally unique chemiluminescence detection. As a rule they no longer depend on the virtually impossible feat of maintaining a constant ratio between the flow of sampled air and the flow of liquid reagent; liquid reagents, in fact, are rarely used. Completeness of specificity for some sensors remains in question, however. And a new problem has arisen: concern for the long-range transport and consequent diffusion of pollutants, and in some cases the setting of very low ambient air quality standards, have led to demands for sensitivities that are not yet available. The real alternative is to return to manual sampling and collect enough material to permit analyses by less sensitive methods. It is thus unfortunate that the shift of manpower to the development of monitoring instruments has reduced significantly the numbers of analysts concerned with manual laboratory techniques. For reasons outside the scope of this report, one of the few programs remaining in the field, at the National Center for Atmospheric Research, was terminated in 1973 (258–261). This problem emphasizes the importance of Recommendation A2.

Particles. The development of second and third generation monitoring instruments for particles has scarcely begun; the only system used routinely is the very questionable high-volume sampler. It has become clear in recent years that the effects of particles are very much contingent on their size spectrum and on the distribution of chemical species within that spectrum. A few collectors exist that separate particles by size, but all are subject to substantial uncertainties. The subject needs a great deal more emphasis. Particularly useful starts have been made in the development of a low-pressure impactor that permits fractionation of smaller particles than has previously been possible (262) and of the virtual impactor, which does not suffer problems of "bounce-off" or reentrainment of particles that should have been collected but failed to adhere to the collecting surface (263). In the absence of suitable instruments, special interest has been shown in optical techniques that, while they do not exclude the large irrespirable particles, nevertheless give primary weight to those particles in the size range having the greatest physiological effect. The past few years have seen particular use of one such device, the integrating nephelometer.

Nephelometry

Reduction in visibility is a vivid manifestation of air pollution, and the understanding and control of visibility-reducing aerosols has been the goal of many air pollution control programs. Only in the past few years, however, has instrumentation become available for studying visibility in a manner compatible with other aerosol measurements (264). One such device is the angular integrating nephelometer (Fig. 5). Basically, the instrument measures the scattering of light by suspended particles (265–269); the results can be correlated with visibility (visual range) (270, 271), particle size (272, 273), particle concentration (270, 274, 275), and even the color of the aerosol and objects viewed through it (276–278).

The integrating nephelometer is a good means of measuring fine particles, including visibility-reducing aerosol (0.1 to 1.0 μm) and much of the respirable aerosol (under 5 μm). The device also is useful for studying the chemistry of visibility-reducing aerosol. The light-scattering values (coefficients) measured by the instrument can be plotted against relative humidity to produce curves or "humidograms" whose shape indicates whether the particles tend to be hygroscopic (absorb water) or deliquescent (form a solution when exposed to water vapor) (279). These properties in turn correlate strongly with the chemical composition of the particles (280). Composition also can be inferred by mixing a

FIGURE 5
Schematic of integrating nephelometer

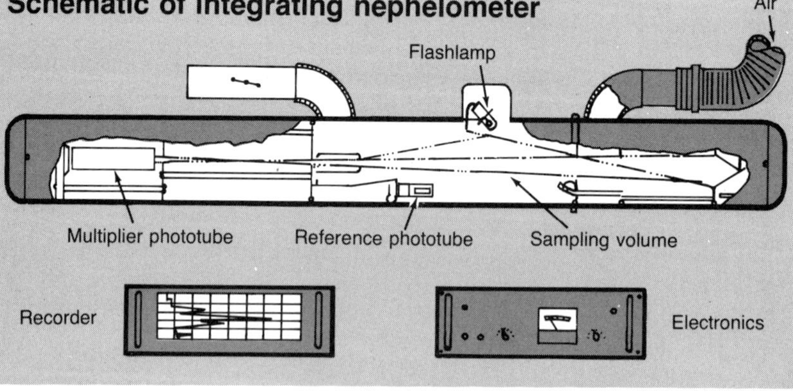

reactive gas with the aerosol and detecting chemical reaction by changes in the shape of the humidogram. This technique has been used, with ammonia as the reactive gas, to detect sulfate aerosol in the air and to qualitatively detect ammonium bisulfate and sulfate from the gas phase reactions in the air (281). Work is under way to refine this technique into a quantitative method for measuring sulfate aerosol. Still another approach is to heat the sample entering the nephelometer (282). The volatility of the material can be determined by the reduction in light-scattering coefficient as the temperature rises; the amount of liquid water, organics, and water of hydration can then be deduced from the reduction in light-scattering coefficient in each temperature range.

Recommendation A28: *The development of inexpensive methods for measuring atmospheric aerosol in both fine and coarse regions in real time should be stimulated.*

Recommendation A29: *Methods for measuring sulfuric acid aerosol in the atmosphere should be developed.*

Recommendation A30: *A national data base should be established systematically for visibility measured instrumentally, so that trends in visibility (and air pollution) can be established more precisely and reproducibly than is possible with the visual estimates of visibility used largely today.*

Recommendation A31: *The relationships among visibility and the sources of air pollution in an urban area should be defined. A by-product would be regional visibility models analogous to, but probably quite different from, the rollback and other types of models used to relate ambient concentrations of pollutants to emissions from their sources.*

Recommendation A32: *National regulations should be established for ambient visibility, which is an obvious effect of air pollution noticed by most people, but priority should remain on health-oriented regulations.*

INSTITUTIONAL PROBLEMS

The scientific and technological problems of air pollution and its abatement must be solved, but that alone will not control air pollution in the real world. It is also necessary that the cost of control be less than the benefits of cleaner air and that this cost-benefit ratio be perceived as such by the public, including especially the decision-makers: corporate management and government officials. Frequently, legal barriers must be overcome. We noted earlier, for example, that, other things being equal, avoiding an emission by process change is preferable to controlling it with some kind of device. Control equipment, however, is subject to a subsidy in the form of rapid tax write-offs, while process changes are not. Thus a well-intentioned law actually favors the less desirable decision.

Probably the major problems of recent years in the orderly achievement of cleaner air arose from certain unforeseen consequences of the federal Clean Air Act of 1970 and of the various state laws passed in response to it. In virtually every case the legislators acted on the basis of the best information available to them at the time. In the light of new information and of other developments, substantial changes have been made in the act of 1970 by the Clean Air Act

Amendments of 1977. Some of the earlier problems will be eased by the amendments, but others will persist and new ones no doubt will come with time.

One of the major strengths and weaknesses of the Clean Air Act of 1970 was the setting of firm and optimistic deadlines — ranging from 1975 to 1977 — for achieving a variety of air quality goals. This approach required in effect that in many cases new inventions and technology had to be devised by specified dates. In principle the requirement was unreasonable because invention cannot be scheduled. In practice the approach stimulated considerable scientific and technological effort. The result, notwithstanding missed deadlines, has been real progress toward the goal of clean air. Nevertheless, various institutional or semiinstitutional problems require attention if the momentum is to be maintained.

The deadlines and the legislative arrangements for meeting them comprised an awesome array of machinery, and two major breakdowns occurred, along with minor problems. First, virtually no segment of industry appears to have taken the deadlines very seriously at the outset; the necessary large research outlays lagged the passage of the act by two or three years. Secondly, the act made EPA responsible for writing State Implementation Plans for states that seemed unable to do so for themselves and for providing much critical information to states that wished seriously to undertake their own planning. The combination of timing, personnel needs, and funding simply overstressed the agency's capabilities. Target dates came and went with even the most cooperative states unable to act because of lack of firm planning data promised them by EPA. States that acted without optimum knowledge were attacked bitterly by industry and, occasionally, by EPA itself for doing so, even though the action was necessary to meet deadlines. Industry by now perceives the seriousness of its problem, and the amendments of 1977 extended the various deadlines affecting industry and the states. EPA remains in charge of an awesome array of machinery, however, and must be given the resources needed to operate it effectively.

Enforcement orders

Whenever a new air pollution regulation is passed, time is required for the affected portion of industry to comply with it. The lag may range from a few months to engineer and implement process changes to several years to develop specific solutions to unique control problems. An example of the latter is the alfalfa dehydration industry. By 1971, most states required all manufacturing processes to achieve a maximum visual opacity of emissions below 20%. Most of industry met the standard with little difficulty. Alfalfa dehydrators, however, are still unable to do so continuously because of the variable composition of the input alfalfa and the small profit margin of their operations.

The normal means of permitting necessary delays in such cases is the issuance of variances, permits for specific sources to continue in violation while they are solving their problems. This procedure of making each source justify its continued violation, if equitably administered, strikes a near optimum between total inflexibility and total permissiveness, although there are other routes to this goal. EPA and the courts, however, construed the act of 1970 to prohibit state-issued variances after the various attainment dates. After those dates, only EPA could issue variances, which, however, were called enforcement orders. The situation was demoralizing to states in direct proportion to their own aggressiveness in enforcing air pollution regulations. A few states that were unenthusiastic from the beginning probably were happy to pass the postdeadline onus to the Federal Government. But states that tried hardest to maintain local control were faced with a regulation that said, in effect, that if they failed to meet deadlines they lost control of their destinies, while the Federal Government was entitled to fail indefinitely to achieve the same goals.

This situation appears to have been eased by the 1977 amendments, which authorize both the states and, after 30 days' notice, EPA to issue enforcement orders. Such orders must require compliance as soon as possible, but with a few exceptions not later than July 1, 1979. After that date no enforcement order may be issued unless it imposes a delayed compliance penalty in the form of monthly payments equal to at least the cost of compliance. These (and certain other) provisions of the amendments amount to a ceding of power from EPA to the states. EPA remains ultimately in charge, however, so that the effect of the amendments on EPA-state relationships will depend very much on how they are administered.

Compliance problems

Another problem concerns automobiles. Under the 1977 amendments, the most stringent emission requirements authorized by law will be imposed in model year 1981. The amendments also specify that federal primary air quality standards must be attained by the end of 1982. If an area has "especially severe" problems with oxidants and carbon monoxide, the deadline may be as late as the end of 1987. The average life of an automobile in the U.S. is about 10 years. With this turnover rate and the rate of growth in the number of automobiles, the maximum impact of emission standards comes about eight years after they take effect. Short of seriously disruptive regulations, therefore, most major cities will be unable to meet the federal air quality standards for automotive pollutants by the 1982 deadline. For such cities, the states may then have to enforce alternative strategies, such as traffic limitations, that may become unnecessary by the 1987 deadline or soon thereafter.

In similar vein, the National Ambient Air Quality Standards must be reviewed by the end of 1980 and revised as indicated. It would be highly desirable to set the standard for particulate matter on the basis of a large and a small size range, with the dividing line at perhaps 3 μm, and EPA is moving in that direction. The effect of such a standard would differ with the locale. In Los Angeles, for example, the problem is predominantly small particles, while in Denver it is often large particles; control measures would differ correspondingly. Until a new standard takes effect, however, EPA must require compliance with the existing standard. If the ultimate costs of control are not to be unduly high, therefore, the interim level of enforcement should be balanced carefully against the likely requirements of a new standard, which in addition should be developed as rapidly as possible.

It can be argued that whether the ambient air quality standards will be met in a given area is not so important a question as the level of improvement in air quality to be expected from incremental regulatory requirements. In many areas, for example, it is unlikely that the existing primary oxidant standard will be met even with extreme and costly regulation of hydrocarbons and nitrogen oxides. It would seem more reasonable to define the standard in such a way as to permit assessments to be made of the degree of oxidant reduction that will result from a variety of control schemes. Reliable means of implementing this approach for photochemical oxidant are not yet available, but the important point is to seek to avoid regulations that are unlikely to improve air quality to any significant degree.

Future needs

Many of these institutional problems could not have been foreseen in 1970 or arose from data unobtainable at that time. The deadlines set, while questionable in detail, have served at least to mobilize a great deal of effort toward a cleaner environment, however belatedly. EPA has moved to some of its more intractable stances not of its own volition, but under orders from the courts, following legal interpretations of points in the Clean Air Act that were not necessarily clear when the act was written. Finally, the energy crisis has been seized upon, not always in complete honesty, as an excuse further to delay source compliance beyond reasonable deadlines. The upshot is that even those most sincerely desiring the earliest practicable achievement of clean air have found themselves in untenable positions. It must be emphasized that in most cases these circumstances resulted not from bad faith, but from physical realities insufficiently explored when the legislation was passed.

It is certainly clear that the amendments of 1977 were necessary. The more reasonable deadlines that they embody were needed, but it is to be hoped that they are administered in such a manner as to avoid removing pressure to comply from sources that reasonably can do so. EPA should not be given additional tasks without being given the additional funds and personnel required to discharge them. In the recent past, funds have been allocated without parallel authorization to add personnel, with the result that the agency has had to assign studies well within its own capabilities to less qualified outside contractors. Federal enforcement must be structured so as to reward, not penalize, the states that take initiative. Intermediate deadlines must be synchronized with the availability of information adequate to support informed action. Too often, dates have been mandated by law or administratively with no firm grasp of the time and resources needeed to obtain the background information necessary to make the required decisions.

It is plain that the "environmentalists" are not unilaterally to blame for the energy crisis, however convenient it may be to accuse them. A far greater crisis is in the offing if we do not begin to react promptly and sensitively to unfavorable environmental change caused by our airborne effluents. No matter what the technological possibilities may be, they will pass into use only as the legal structure permits. We also need a serious, tenacious, and well-researched input to the legislative and administrative processes from the competent portion of the scientific community.

LITERATURE CITED

1. "National Air Quality and Emissions Report, 1976," Environmental Protection Agency, Research Triangle Park, N.C., 1977.
2. "Progress in the Prevention and Control of Air Pollution in 1976," Annual Report of the Administrator of the Environmental Protection Agency to the Congress of the United States, Washington, D.C., 1977.
3. *Fed. Regist.*, **42**, 63076, Dec. 14, 1977.

4. "Proceedings of the Conference on Health Effects of Air Pollutants," National Academy of Sciences—National Research Council, Committee Print, Committee on Public Works, U.S. Senate, 93rd Congress, U.S. Government Printing Office, Washington, D.C., November 1973.

5. Chaput, L.S., "Federal Standards of Performance for New Stationary Sources of Air Pollution," *J. Air Pollut. Control Assoc.*, **26,** 1055 (1976).

6. "Progress in the Prevention and Control of Air Pollution in 1975," Annual Report of the Administrator of the Environmental Protection Agency to the Congress of the United States, Washington, D.C., 1976.

7. Hauser, T.R., Shy, C.M., "Position Paper: NO_x Measurement," *Environ. Sci. Technol.*, **6,** 890 (1972).

8. Lovelock, J.E., Lodge, J.P., Jr., "Oxygen in the Contemporary Atmosphere," *Atmos. Environ.*, **6,** 575 (1972).

9. Lovelock, J.E., "Gaia As Seen Through the Atmosphere," *Atmos. Environ.*, **6,** 579 (1972).

10. Lovelock, J.E., Margulis, L., "Atmospheric Homeostasis By and For the Biosphere: The Gaia Hypothesis," *Tellus*, **26,** 2 (1974).

11. Margulis, L., Lovelock, J.E., "Biological Modulation of the Earth's Atmosphere," *Icarus*, **21,** 471 (1974).

12. Otterman, J., "Anthropogenic Impact on the Albedo of the Earth," *Clim. Change*, **1,** in press (1977).

13. Schneider, S.H., "The Genesis Strategy," Plenum, New York, N.Y., 1976, p. 81 ff.

14. Friedman, I., "The Amazon Basin, another Sahel," *Science*, **197,** 7 (1977).

15. Lodge, J.P., Jr., et al., "Atmospheric Trace Chemistry in the American Humid Tropics," *Tellus*, **26,** 250 (1974).

16. Breeding, R.J., et al., "Background Trace Gas Concentrations in the Central United States," *J. Geophys. Res.*, **78,** 7057 (1973).

17. de Nevers, N., Morris, J.R., "Rollback Modeling — Basic and Modified," *J. Air Pollut. Control Assoc.*, **25,** 999 (1975).

18. "Patterns and Perspectives in Environmental Science," National Science Board, U.S. Government Printing Office, Washington, D.C., 1972, pp. 335–336.

19. Schiermeier, F.A., "RAPS' Field Measurements Are In," *Environ. Sci. Technol.*, **12,** 644 (1978).

20. "Air Quality Criteria for Particulate Matter," National Air Pollution Control Administration, Publication No. AP-49, Washington, D.C., 1969.

21. Cadle, R.D., "Particulate Matter in the Lower Atmosphere," in "Chemistry of the Lower Atmosphere," S.I. Rasool, ed., Plenum Press, New York, N.Y., 1973.

22. Hidy, G.M., et al., "Summary of the California Aerosol Characterization Experiment," *J. Air Pollut. Control Assoc.*, **25,** 1106 (1975).

23. Cadle, R.D., Charlson, R.J., "Worldwide Air Pollution and Its Effects," in "Atmospheric Chemistry: Problems and Scope," National Academy of Sciences, Washington, D.C., 1975.

24. Chylek, P., Coakley, J.A., "Aerosols and Climate," *Science*, **183,** 75 (1974).

25. Ensor, D.S., et al., "Influence of Atmospheric Aerosol on Albedo," *J. Applied Meteor.*, **6,** 1303 (1971).

26. Idso, S.B., Brazel, A.J., "Planetary Radiation Balance As a Function of Atmospheric Dust: Climatological Consequences," *Science*, **198,** 731 (1977).

27. Mitchell, J.M., "Summary of the Problem of Air Pollution Effects on Climate," in "Man's Impact on the Climate," Matthews, Kellogg, and Robinson, eds., MIT Press, Cambridge, Mass., 1971.

28. Reck, R., "Aerosols in the Atmosphere, Calculation of the Critical Ratio," *Science*, **186,** 1034 (1974).

29. Weiss, R.E., et al., "Application of Directly Measured Aerosol Radiative Properties to Climate Models," Optical Society of America Aerosol Topical Meeting, Williamsburg, Va., Dec. 13, 1976.

30. Yamamoto, G., Tanaka, M., "Increase of Global Albedo Due to Air Pollution," *J. Atmos. Sci.*, **29**, 1405 (1972).

31. Kellogg, W.W., et al., "The Sulfur Cycle," *Science*, **175**, 587 (1972).

32. Altshuller, A.P., "Atmospheric Sulfur Dioxide and Sulfate: Distribution of Concentrations at Urban and Nonurban Sites in the United States," *Environ. Sci. Technol.*, **7**, 709 (1973).

33. Charlson, R.J., et al., "Sulfate Aerosol: Its Geographical Extent in the Midwestern and Southern United States," *Science*, **195**, 979 (1977).

34. Gartrell, F.E., Thomas, F.W., Carpenter, S.B., "Atmospheric Oxidation of SO_2 in Coal-Burning Power Plant Plumes," *Am. Ind. Hyg. Assoc. J.*, **24**, 113 (1963).

35. Gerhard, E.R., Johnstone, H.F., "Air Pollution Studies — Photochemical Oxidation of Sulfur Dioxide in Air," *Ind. Eng. Chem.*, **47**, 972 (1955).

36. Johnstone, H.F., Coughanowr, D.R., "Absorption of Sulfur Dioxide from Air. Oxidation in Drops Containing Dissolved Catalysts," *Ind. Eng. Chem.*, **50**, 1169 (1958).

37. Johnstone, H.F., Moll, A.J., "Air Pollution — Formation of Sulfuric Acid in Fogs," *Ind. Eng. Chem.*, **52**, 861 (1960).

38. Boone, R.E., Brice, R.M., "Continuous Measurements of Acid Aerosol in the Atmosphere," paper 65-119 presented at the 58th Annual Meeting of the Air Pollution Control Association, Toronto, Canada, June 1965.

39. Thomas, M.D., "Sulfur Dioxide, Sulfuric Acid Aerosol and Visibility in Los Angeles," *Int. J. Air Water Pollut.*, **6**, 443 (1962).

40. Likens, G.E., "Acid Precipitation," *Chem. Eng. News*, **54**, 48, 29 (1976).

41. Dochinger, L.S., Seliga, T.A., "Acid Precipitation and the Forest Ecosystem," *BioScience*, **26**, 564 (1976).

42. Scriven, R.A., Fisher, B.E.A., "The Long Range Transport of Airborne Material and Its Removal by Deposition and Washout — I. General Considerations," *Atmos. Environ.*, **9**, 49 (1975).

43. Pierotti, D., Rasmussen, R.A., "The Atmospheric Distribution of Nitrous Oxide," *J. Geophys. Res.*, **82**, 5823 (1977).

44. Noxon, J.F., "Nitrogen Dioxide in the Stratosphere and Troposphere Measured by Ground-Based Absorption Spectroscopy," *Science*, **189**, 547 (1975).

45. Weiss, R.F., Craig, H., "Production of Atmospheric Nitrous Oxide by Combustion," *Geophys. Res. Lett.*, **3**, 751 (1976).

46. Brice, K.A., Eggleton, A.E.J., Penkett, S.A., "An Important Ground Surface Sink for Atmospheric Nitrous Oxide," *Nature*, **268**, 127 (1977).

47. Spicer, C.W., Schumacher, P.M., "Interferences in Sampling Atmospheric Particulate Nitrate," *Atmos. Environ.*, **11**, 873 (1977).

48. Rasmussen, R., *J. Geophys. Res.* (in press); personal communication.

49. "Cleaning Our Environment: The Chemical Basis for Action," a report by the Subcommittee on Environmental Improvement, Committee on Chemistry and Public Affairs, American Chemical Society, Washington, D.C., 1969.

50. An extensive review of this subject appears in *Tellus*, **26**, 1–297 (1974), based on papers presented at the "CACGP (Commission on Atmospheric Chemistry and Global Pollution) Symposium on Trace Gases," sponsored by International Association of Meteorology and Atmospheric Physics, Mainz, Germany, April 2–5, 1973.

51. Weinstock, B., "The Residence Time of Carbon Monoxide in the Atmosphere," *Science*, **166**, 224 (1969).

52. McConnell, J.C., McElroy, M.B., Wofsy, S.C., "Natural Sources of Atmospheric CO," *Nature*, **233**, 187 (1971).

53. Weinstock, B., Chang, T.Y., "The Steady-State Concentration of Carbon Monoxide in the Troposphere," *Environ. Biogeochem.*, **1**, 39 (1976).

54. Chan, W., et al., *Chem. Phys. Lett.*, **45**, 240 (1977).

55. Wang, C.C., et al., "Hydroxyl Radical Concentrations Measured in Ambient Air," *Science*, **189**, 797 (1975).

56. Sze, N.D., "Chlorine Nitrate: Its Role in Atmospheric Ozone," presented at the 172nd National Meeting of the American Chemical Society, San Francisco, Calif., Aug. 29–Sept. 3, 1976.

57. "The National Air Monitoring Program: Air Quality and Emissions Trends," Annual Report, Vol. I., EPA-450/1-73-001-a, U.S. Environmental Protection Agency, August 1973.

58. Finlayson, B.J., Pitts, J.N., Jr., "Photochemistry of the Polluted Troposphere," *Science*, **192**, 111 (1976).

59. Niki, H., Daby, E.E., Weinstock, B., "Mechanisms of Smog Reactions," *Adv. Chem. Ser.*, **113**, 16 (1972).

60. Weinstock, B., et al., *Int. J. Chem. Kinetics* **VIII**, 765 (1976).

61. Wu, C.H., Japar, S.M., Niki, H., *J. Environ. Sci. Health-Environ. Sci. Eng.*, All (2), 191 (1976).

62. Pitts, J.N., Jr., et al., "Reactivity Scale for Atmospheric Hydrocarbons Based on Reaction with Hydroxyl Radical," *Environ. Sci. Technol.*, **10**, 692 (1976).

63. Bufalini, J.J., Kopczynski, S.L., Dodge, M.C., *Environ. Lett.*, **3**, 101 (1972).

64. Altshuller, A.P., Bufalini, J.J., *Environ. Sci. Technol.*, **5**, 39 (1971).

65. Altshuller, A.P., "Evaluation of Oxidant Results at CAMP Sites in the United States," *J. Air Pollut. Control Assoc.*, **25**, 19 (1975).

66. Tiao, G.C., Box, G.E.P., Hamming, W.J., "Analysis of Los Angeles Photochemical Smog Data: A Statistical Overview," APCA Preprint No. 73-79, 66th Annual Meeting of the Air Pollution Control Association, Chicago, Ill., June 1973.

67. Ripperton, L.A., et al., "High Ozone Concentrations in Non-Urban Air," Paper 37, Preprints, Division of Environmental Chemistry, 167th National Meeting of the American Chemical Society, Los Angeles, Calif., April 1974.

68. Hanst, P.L., et al., "A Spectroscopic Study of Pasadena Smog," Paper 54, Preprints, Division of Environmental Chemistry, 167th National Meeting of the American Chemical Society, Los Angeles, Calif., April 1974.

69. Niki, H., et al., *Chem. Phys. Lett.*, **45**, 564 (1977).

70. The Los Angeles Reactive Pollutant Program (LARPP) data can be obtained from the National Technical Information Service (NTIS): Modelers' Archive PB 255 506 and Guide to Modelers' Archive PB 255 507, Springfield, Va.

71. Calvert, J.G., "Test of the Theory of Ozone Generation in Los Angeles Atmosphere," *Environ. Sci. Technol.*, **10**, 248 (1976).

72. Calvert, J.G., "Hydrocarbon Involvement in Photochemical Smog Formation in Los Angeles Atmosphere," *Environ. Sci. Technol.*, **10**, 256 (1976).

73. Hidy, G.M., et al., "Overview of the California Aerosol Characterization Experiment," Symposium on Sources and Evolution of the Atmospheric Aerosol, Paper 6, Preprints, Division of Environmental Chemistry, 167th National Meeting of the American Chemical Society, Los Angeles, Calif., April 1974.

74. Whitby, K.T., Husar, R.B., Liu, B.Y.H., "The Aerosol Size Distribution of Los Angeles Smog," *J. Colloid. Interface Sci.*, **39**, 177 (1972).

75. "Abatement of Particulate Emissions from Stationary Sources,"COPAC-5, National Academy of Engineering — National Research Council, Washington, D.C., 1972.

76. Ensor, D.S., et al., "Multiwavelength Nephelometer Measurements in Los Angeles Smog Aerosol. I: Comparison of Calculated and Measured Light Scattering," *J. Colloid Interface Sci.*, **39**, 242 (1972).

77. Husar, R.B., Whitby, K.T., "Growth Mechanisms and Size Spectra of Photochemical Aerosols," *Environ. Sci. Technol.*, **7**, 241 (1973).

78. Heisler, S.L., Friedlander, S.K., Husar, R.B., "The Relationship of Smog Aerosol Size and Chemical Element Distributions to Source Characteristics," *Atmos. Environ.*, **7**, 633 (1973).

79. Dzubay, T.G., Stevens, R.K., "Applications of X-ray Fluorescence to Particulate Measurements," presented at the Second Joint Conference on the Sensing of Environmental Pollutants, Washington, D.C., December 1973.

80. Proceedings of the Fourth Conference on the Climatic Impact Assessment Program, Feb. 4–7, 1975. Sponsored by the U.S. Department of Transportation, DOT-TSC-OST-75-38. Available from the National Technical Information Service, Springfield, Va.

81. London, J., Kelley, J., "Global Trends in Total Atmospheric Ozone," *Science*, **184**, 987 (1974).

82. Eddy, J.A., "The Maunder Minimum," *Science*, **192**, 1189 (1976).

83. Johnson, H., *Science*, **173**, 517 (1971).

84. Grobecker, A.J., Coroniti, S.C., Cannon, R.H., Jr., "Report of Findings: The Effects of Stratospheric Pollution by Aircraft," DOT-TST-75-50, National Technical Information Service, Springfield, Va., 1974.

85. Climatic Impact Committee, "Environmental Impact of Stratospheric Flight: Biological and Climatic Effects of Aircraft Emissions in the Stratosphere," National Academy of Sciences, Washington, D.C., 1975.

86. Wuebbles, D.J., Chang, J.S., Duewer, W.H., "The Effect of the SST in a Chlorinated Stratosphere," paper presented at the 173rd National Meeting of the American Chemical Society, New Orleans, La., March 1977.

87. Molina, M.J., Rowland, F.S., "Stratospheric Sink for Chlorofluoromethanes: Chlorine Atom-Catalysed Destruction of Ozone," *Nature*, **249**, 810 (1974).

88. Cicerone, R.J., Stolarski, R.S., Walters, S., "Stratospheric Ozone Destruction by Man-Made Chlorofluoromethanes," *Science*, **185**, 1165 (1974).

89. Lovelock, J.E., "Atmospheric Fluorine Compounds as Indicators of Air Movements," *Nature*, **230**, 379 (1971).

90. Lovelock, J.E., Maggs, R.J., Wade, R.J., "Halogenated Hydrocarbons In and Over the Atlantic," *Nature*, **241**, 194 (1973).

91. Wilkniss, P.E., et al., "Atmospheric Trace Gases in the Southern Hemisphere," *Nature*, **245**, 45 (1973).

92. Su, C.-W., Goldberg, E.D., "Chlorofluorocarbons in the Atmosphere," *Nature*, **245**, 27 (1973).

93. Committee on Impacts of Stratospheric Change, "Halocarbons: Environmental Effects of Chlorofluoromethane Release," National Academy of Sciences — National Research Council, Washington, D.C. 1976.

94. "Halocarbons: Effects on Stratospheric Ozone," Panel on Atmospheric Chemistry, National Academy of Sciences — National Research Council, Washington, D.C., 1976.

95. Hill, W.J., Sheldon, P.N., Tiede, J.J., "Analyzing Worldwide Total Ozone for Trends," paper presented at the 172nd National Meeting of the American Chemical Society, San Francisco, Calif., Aug. 29–Sept. 3, 1976.

96. Altshuller, A.P., "Average Tropospheric Concentration of Carbon Tetrachloride Based on Industrial Production, Usage, and Emissions," *Environ. Sci. Technol.*, **10**, 596 (1976).

97. Singh, H.B., Fowler, D.P., Peyton, T.O., "Atmospheric Carbon Tetrachloride: Another Man-Made Pollutant," *Science*, **192**, 1231 (1976).

98. Galbally, I.E., "Man-Made Carbon Tetrachloride in the Atmosphere," *Science*, **193**, 573 (1976).

99. Pack, D.H., et al., "Halocarbon Behavior from A Long Time Series," *Atmos. Environ.*, **11,** 329 (1977).

100. Jesson, J.P., testimony before the Ad Hoc Subcommittee on the Upper Atmosphere, Committee on Aeronautical and Space Sciences, U.S. Senate, December 1976.

100a. *Fed. Regist.*, **43,** 11301, March 17, 1978.

101. Schneider, S.H., Kellogg, W.W., "The Chemical Basis for Climate Change," in "Chemistry of the Lower Atmosphere," S.I. Rasool, ed., Plenum Press, New York, N.Y., 1973.

102. "Man's Impact on the Global Environment," Report of the Study of Critical Environmental Problems, The MIT Press, Cambridge, Mass., 1970.

103. "Energy and Climate," National Academy of Sciences — National Research Council, Washington, D.C., 1977.

104. Damon, P.E., Kunen, S.M., "Global Cooling?", *Science,* **193,** 447 (1976).

105. "Air Quality and Stationary Source Emission Control," Commission on Natural Resources, National Academy of Sciences, National Academy of Engineering, National Research Council, prepared for the Committee on Public Works, U.S. Senate, Serial No. 94-4, March 1975, U.S. Government Printing Office, Washington, D.C.

105a. Wilson, W.W., et. al., "Sulfates in the Atmosphere — A Progress Report on Project MISTT," EPA-600/7-77-021, Environmental Protection Agency, Research Triangle Park, N.C., March 1977.

105b. Perhac, R.M., "Sulfate Regional Experiment in Northeastern United States: The 'SURE' Program," presented at the International Symposium on Sulfur in the Atmosphere, Dubrovnik, Yugoslavia, Sept. 7–14, 1977.

106. Steinhart, J.S., Steinhart, C.E., "Energy Use in the U.S. Food System," *Science,* **184,** 307 (1974).

107. "Compilation of Air Pollutant Emission Factors," 2nd ed., Environmental Protection Agency, Air Pollution Technical Information Center, Research Triangle Park, N.C., 1973.

108. "Emission Factors for Trace Substances," Environmental Protection Agency, EPA-450/2-73-001, Research Triangle Park, N.C., 1973.

109. *Fed. Regist.*, **41,** 55524, Dec. 21, 1976.

110. "Air Pollution," Stern, A.C., ed., 2nd ed., 3 vol. Academic Press, New York, N.Y., 1968.

111. "Air Pollution and Industry," Ross, R.D., ed., Van Nostrand Reinhold Co., New York, N.Y., 1972.

112. *Fed. Regist.*, **41,** 7450, Feb. 18, 1976.

113. Osborn, E.F., "Coal and the Present Energy Situation," *Science,* **183,** 477 (1974).

114. White, P.C., "ERDA's Fossil Energy Activities," *Environ. Sci. Technol.*, **10,** 746 (1976).

115. "Coal Conversion Activities Picking Up," *Chem. Eng. News,* **53,** 48, 24 (1975).

116. Natusch, D.F.S., Wallace, J.R., Evans, C.A., Jr., "Toxic Trace Elements: Preferential Concentration in Respirable Particles," *Science,* **183** (1974).

117. Sparks, L.E., "Electrostatic Precipitator Operations for Collection of High Resistivity Fly Ash," Paper 5, Conference on Particulate Collection Problems in Converting to Low Sulfur Coals, Environmental Protection Agency, EPA-600/7-76-016, Research Triangle Park, N.C., 1976.

118. "Fine Particle Electrostatic Precipitation: A Proceedings," *J. Air Pollut. Control Assoc.*, **25,** 98 (1975).

119. "Fine Particle Scrubbing: A Proceedings," *J. Air Pollut. Control Assoc.*, **24,** 926 (1974).

120. "The User and Fabric Filtration Equipment," *J. Air Pollut. Control. Assoc.*, **26,** 16 (1976).

121. Dunham, J.T., Rampacek, C., Henrie, T.A., "High-Sulfur Coal for Generating Electricity," *Science*, **184**, 346 (1974).

122. Herlihy, J., "Flue Gas Desulfurization in Power Plants — Status Report," Environmental Protection Agency, Washington, D.C., April 1977.

123. Yan, C.J., "Evaluating Environmental Impacts of Stack Gas Desulfurization Processes," *Environ. Sci. Technol.*, **10**, 54 (1976).

124. Devitt, T.W., et al., "Estimating Costs of Flue Gas Desulfurization Systems for Utility Boilers," *J. Air Pollut. Control Assoc.*, **26**, 204 (1976).

125. Blakeslee, C.E., Burbach, H.E., "Controlling NO_x Emissions from Steam Generators," *J. Air Pollut. Control Assoc.*, **23**, 37 (1973).

126. *NO_x Control Review*, Vol. 2, No. 2, Spring 1977, Environmental Protection Agency, Research Triangle Park, N.C.

127. *Chem. Eng.* (NY), Feb. 14, 1977, page 33; April 11, 1977, page 84; April 25, 1977, page 70.

128. *NO_x Control Review*, Vol. 3, No. 1, Winter 1977–78, Environmental Protection Agency, Research Triangle Park, N.C.

129. Starkman, E.S., "Emission Control and Fuel Economy," *Environ. Sci. Technol.*, **9**, 820 (1975).

130. Gandhi, H.S., et al., "Laboratory Evaluation of Three-Way Catalysts," Paper 760201, SAE Automotive Engineering Congress and Exposition, Detroit, Mich., 1976.

131. Guiles, R., "Three-Way Catalysts Prep for Auto-Exhaust Cleanup," *Chem. Eng.* (NY), Sept. 13, 1976, page 110.

132. Misch, H.L., letter to R.L. Sansom, Feb. 5, 1973; Pierson, W.R., Hammerle, R.H., Kummer, J.T., SAE Paper 740287 (1974). SAE *Trans.*, **83**, 1233 (1975).

133. Pierson, W.R., "Sulfuric Acid Generation by Automotive Catalysts," *Chem. Tech.*, **6**, 332 (1976).

134. Cadle, S.H., et al., "General Motors Sulfate Dispersion Experiment: Experimental Procedures and Results," *J Air Pollut. Control Assoc.*, **27**, 33 (1977); Chock, D.P., "General Motors Sulfate Dispersion Experiment: Assessment of the EPA HIWAY Model," *ibid.*, **27**, 39 (1977); Wilson, W.E., et al., "General Motors Sulfate Dispersion Experiment: Summary of EPA Measurements," *ibid.*, **27**, 46 (1977).

135. Tanner, R.L., Newman, L., "Chemical Speciation of Sulfate Emission from Catalyst Equipped Automobiles under Ambient Conditions," Division of Environmental Chemistry, Preprints, Vol. 16, No. 2, Paper 22, 172nd National Meeting of the American Chemical Society, San Francisco, Calif., Aug. 29–Sept. 3, 1976.

136. Groblicki, P.J., General Motors Research Publication GMR-2127, April 12, 1976.

137. "Report by the Committee on Motor Vehicle Emissions," National Academy of Sciences — National Research Council, Washington, D.C., November 1974.

138. "Should We Have a New Engine?" Summary (Vol. 1) and Technical Reports (Vol. II), Publication JPL SP 43-17, Jet Propulsion Laboratory, California Institute of Technology, Pasadena, Calif., August 1975.

139. Karkheck, J., Powell, J., Beardsworth, E., "Prospects for District Heating in the United States," *Science*, **195**, 948 (1977).

140. "Waste Heat Management Guidebook," National Bureau of Standards Handbook 121, U.S. Government Printing Office, Washington, D.C. 1977.

141. "Proceedings, 12th Intersociety Energy Conversion Engineering Conference," American Chemical Society, Washington, D.C., 1977.

142. "Effects of Chronic Exposure to Low-Level Pollutants in the Environment," Congressional Research Service, Library of Congress, Committee Print, Committee on Science and Technology, U.S. House of Representatives, 94th Congress, U.S. Government Printing Office, Washington, D.C., November 1975.

143. "The Costs and Effects of Chronic Exposure to Low-Level Pollutants in the Environment," Hearings before the Subcommittee on the Environment and the Atmosphere of the Committee on Science and Technology, U.S. House of Representatives, U.S. Government Printing Office, Washington, D.C., November 1975.

144. "Air Quality Criteria for Sulfur Oxides," National Air Pollution Control Administration, Publication No. AP-50, Washington, D.C., 1969.

145. "Air Quality Criteria for Carbon Monoxide,"National Air Pollution Control Administration, Publication No. AP-62, Washington, D.C., 1970.

146. "Air Quality Criteria for Photochemical Oxidants," National Air Pollution Control Administration, Publication No. AP-63, Washington, D.C., 1970.

147. "Air Quality Criteria for Nitrogen Oxides," National Air Pollution Control Administration, Publication No. AP-84, Washington, D.C., 1971.

148. "Health Consequences of Sulfur Oxides: A Report from the CHESS Program, 1970–1971," Report No. EPA-650/1-74-004, Environmental Protection Agency, Research Triangle Park, N.C., 1974.

149. Frank, N.R., Amdur, M.O., Whittenberger, J.L., "A Comparison of the Acute Effects of SO_2 Administered Alone or in Combination with NaCl Particles in the Respiratory Mechanics of Healthy Adults," *Int. J. Air Water Pollut.*, **8,** 125 (1964).

150. Fairchild, E.J., Murphy, S.D., Stokinger, H.E., "Protection by Sulfur Compounds Against the Air Pollutants Ozone and Nitrogen Dioxide," *Science,* **130,** 861 (1959).

151. "Health and Air Pollution Subject of New Studies," *Environ. Sci. Technol.,* **2,** 246 (1968).

152. Stokinger, H.E., Coffin, O.L., "Biological Effects of Air Pollutants," in "Air Pollution," Vol. I, Academic Press, New York, N.Y., 1968.

153. "Manganese," Committee on the Medical and Biological Effects of Environmental Pollutants, National Academy of Sciences — National Research Council, Washington, D.C., 1974.

154. Bracewell, J.M., et al., "The Catalytic Oxidation of Sulfur Dioxide in Solution at Concentrations Occurring in Fog Droplets," in "Air Pollution, Proceedings of the Symposium on the Physico-Chemical Transformation of Sulfur Compounds in the Atmosphere and the Formation of Acid Smogs," Mainz, Germany, June 8–9, 1967. Organization for Economic Cooperation and Development Directorate for Scientific Affairs (1967).

155. Stein, K.C., et al., "Catalytic Oxidation of Hydrocarbons," *Ind. Eng. Chem.,* **52,** 671 (1960).

156. "Vanadium," Committee on Medical and Biologic Effects of Environmental Pollutants, National Academy of Sciences — National Research Council, Washington, D.C., 1974.

157. Aiyer, A.S., et al., "Effect of Vanadium Administration on Coenzyme Q Metabolism in Rats," *Proc. Soc. Exp. Biol. Med.,* **107,** 914 (1961).

158. Anbar, M., et al., "Effect of Pyridoxal 5-Phosphate in the Presence of Vanadyl Ions on the Deiodination of Thyroxine," *Nature,* **196,** 1213 (1962).

159. Azarnoff, D.L., et al., "A Specific Site of Vanadium Inhibition of Cholesterol Biosynthesis," *Biochem. Biophys. Acta* **51,** 397 (1961).

160. Rifkin, R.J., "In Vitro Inhibition of Na^+-K^+ and Mg^{2+} ATPases by Mono-, Di-, and Trivalent Cations," *Proc. Soc. Exptl. Biol. Med.,* **120,** 802 (1965).

161. Lewis, C.E., "The Biological Actions of Vanadium. I. Effects upon Serum Cholesterol Levels in Man," *A.M.A. Arch. Ind. Health,* **19,** 419 (1959).

162. Watanabe, H., et al., "Some Clinical Findings on Vanadium Workers," *Japan J. Ind. Health* (Kawasaki), **8,** 23 (1966).

163. Hueper, W.C., Payne, W.W., "Experimental Cancers in Rats Produced by Chromium Compounds and their Significance to Industry and Public Health," *Am. Ind. Hyg. J.,* **20,** 274 (1959).

164. Hueper, W.C., Payne, W.W., "Experimental Studies in Metal Carcinogenesis: Chromium, Nickel, Iron, Arsenic," *Arch. Environ. Health*, **5**, 445 (1962).

165. Gunn, S.A., et al., "Specific Response of Mesenchymal Tissue to Carcinogenesis by Cadmium," *Arch. Pathol.*, **83**, 493 (1967).

166. Heath, J.C., et al., "Cadmium as a Carcinogen," *Nature*, **193**, 592 (1962).

167. Kazantzis, G., et al., "The Induction of Sarcoma in the Rat by Cadmium Sulphide and by Cadmium Oxide," *Brit. J. Cancer*, **20**, 190 (1966).

168. Sunderman, F.W., Donnelly, A.J., "Studies of Nickel Carcinogenesis Metastasizing Pulmonary Tumors in Rats Induced by the Inhalation of Nickel Carbonyl," *Am. J. Pathol.*, **46**, 1027 (1965).

169. Hueper, W.C., "Occupational and Non-Occupational Exposure to Asbestos. IV. Human Exposure to Asbestos: Community Studies," *Ann. N.Y. Acad. Sci.*, **132**, 184 (1965).

170. "Chromium," Committee on Medical and Biological Effects of Environmental Pollutants, National Academy of Sciences — National Research Council, Washington, D.C., 1974.

171. "Particulate Polycyclic Organic Matter," Committee on Biological Effects of Atmospheric Pollutants, National Academy of Sciences — National Research Council, Washington, D.C., 1972.

172. "Asbestos: The Need for and Feasibility of Air Pollution Controls," Committee on Biological Effects of Atmospheric Pollutants, National Academy of Sciences — National Research Council, Washington, D.C., 1971.

173. "Vapor-Phase Organic Pollutants: Volatile Hydrocarbons and Oxidation Products," Committee on Medical and Biologic Effects of Environmental Pollutants, National Academy of Sciences — National Research Council, Washington, D.C., 1976.

174. "Nickel," Committee on Medical and Biologic Effects of Environmental Pollutants, National Academy of Sciences — National Research Council, Washington, D.C., 1975.

175. Goldsmith, J.R., Hexter, A.C., "Respiratory Exposure to Lead: Epidemiological and Experimental Dose-Response Relationships," *Science*, **158**, 132 (1967).

176. Kehoe, R.A., "Lead Intake from Food and from the Atmosphere," *Science*, **159**, 1000 (1968).

177. "Airborne Lead in Perspective," Committee on Biologic Effects of Atmospheric Pollutants, National Academy of Sciences — National Research Council, Washington, D.C., 1971.

177a. "Airborne Particles (Medical and Biologic Effects of Environmental Pollutants)," Committee on Medical and Biologic Effects of Environmental Pollutants, National Academy of Sciences — National Research Council, Washington, D.C. 1977.

178. Stewart, R.D., et al., "Carboxyhemoglobin Levels in American Blood Donors," *J. Am. Med. Assoc.*, **229**, 1187 (1974).

179. Lewey, F.H., Drabkin, D.L., "Experimental Chronic Carbon Monoxide Poisoning of Dogs," *Am. J. Med. Sci.*, **208**, 502 (1944).

180. Lindenberg, R., et al., "An Experimental Investigation in Animals of the Functional and Morphological Changes from Single and Repeated Exposure to Carbon Monoxide," presented at the American Industrial Hygiene Conference, Washington, D.C., May 13–17, 1962.

181. Roussel, A., Stupfel, M., Bouley, G., "Trends of Experimental Research on Air Pollution in France," *Arch. Environ. Health*, **18**, 613 (1969).

182. Grudzinska, B., "Electroencephalographic Patterns in Cases of Chronic Exposure to Carbon Monoxide in Air," (Polish) *Folia Med. Cracov.*, **5**, 493 (1963). (Eng. summary, pp. 514–515).

183. "Effects of Chronic Exposure to Low Levels of Carbon Monoxide on Human Health, Behavior, and Performance," Committee on Effects of Atmospheric Contaminants on Human Health and Welfare of the National Academy of Sciences and National Academy of Engineering, Washington, D.C., 1969.

184. Beard, R.R., Wertheim, G.A., "Behavioral Impairment Associated with Small Doses of Carbon Monoxide," *Am. J. Public Health*, **57**, 2012 (1967).

185. Beard, R.R., Grandstaff, N., Paper Presented at the New York Academy of Sciences Conference on Biological Effects of Carbon Monoxide, New York, N.Y., January 12–14, 1970.

186. Clayton, G.D., Cook, W.A., Fredrick, W.C., "A Study of the Relationship of Street Level Carbon Monoxide Concentration to Traffic Accidents," *Am. Ind. Hyg. Assoc. J.*, **21**, 46 (1960).

187. Chovin, P., "Carbon Monoxide: Analysis of Exhaust Gas Investigations in Paris," *Environ. Res.*, **1**, 185 (1967).

188. Ury, H., "Photochemical Air Pollution and Automobile Accidents in Los Angeles," *Arch. Environ. Health*, **17**, 334 (1968).

189. Horvath, S.M., Dahms, T.E., O'Hanlon, J.F., "Carbon Monoxide and Human Vigilance: Deleterious Effect of Present Urban Concentrations," *Arch. Environ. Health*, **23**, 343 (1971).

190. Beard, R.R., Wertheim, G.A., paper at Fifth Rochester International Conference on Environmental Toxicity, 1972.

191. Glasson, W.A., Heuss, J.M., "Synthesis and Evaluation of Potential Atmospheric Eye Irritants," *Environ. Sci. Technol.*, **11**, 395 (1977).

192. Murphy, S.D., et al., "Effects on Animals of Exposure to Auto Exhaust," *Arch. Environ. Health*, **7**, 60 (1963).

193. Stokinger, H.E., "Ozone Toxicology. A Review of Research and Industrial Experience: 1954–1964," *Arch. Environ. Health*, **10**, 719 (1965).

194. Coffin, D.L., Blomer, E.J., "Acute Toxicity of Irradiated Auto Exhaust. Its Indication by Enhancement of Mortality from Streptococcal Pneumonia," *Arch. Environ. Health*, **15**, 36 (1967).

195. Purvis, M.R., Ehrlich, R., "Effect of Atmospheric Pollutants on Susceptibility to Respiratory Infection: II Effects of Nitrogen Dioxide," *J. Infec. Dis.*, **113**, 72 (1963).

196. Shy, C.M., et al., "The Chattanooga School Children Study: Effects of Community Exposure to Nitrogen Dioxide. Incidence of Acute Respiratory Illness," *J. Air Pollut. Control Assoc.*, **20**, 582 (1970).

197. Shy, C.M., et al., "The Chattanooga School Study: Effects of Community Exposure to Nitrogen Dioxide. Methods, Description of Pollutant Exposure and Results of Ventilatory Exposure Testing," *J. Air Pollut. Control Assoc.*, **20**, 539 (1970).

198. Pearlman, M.E., et al., "Nitrogen Dioxide and Lower Respiratory Illness," *Pediatrics*, **47**, 391 (1971).

199. Epstein, S.S., et al., "N-Nitroso Compounds: Detection in Ambient Air," *Science*, **192**, 1328 (1976).

200. Fine, D.H., et al., "N-Nitroso Compounds in the Ambient Community Air of Baltimore, Maryland," *Anal. Lett.*, **9**, 595 (1976).

201. Matzen, R.N., "Effect of Vitamin C and Hydrocortisone on the Pulmonary Edema Produced by Ozone in Mice," *J. Appl. Physiol.*, **11**, 105 (1957).

202. Stokinger, H.E., "Evaluation of the Hazards of Ozone and Oxides of Nitrogen — Factors Modifying Acute Toxicity," *A.M.A. Arch. Ind. Health*, **15**, 181 (1957).

203. Mountain, J.T., et al., "Biochemical Effects of Ozone and Nitrogen Dioxide on Laboratory Animals," paper presented at the 138th National Meeting of the American Chemical Society, New York, N.Y., September 1960.

204. Murphy, S.D., Davis, H.V., Zaratzian, V.L., "Biochemical Effects in Rats from Irritating Air Contaminants," *Toxicol. Appl. Pharmacol.*, **6**, 520 (1964).

205. Scheel, L.D., et al., "Physiologic, Biochemical, Immunologic, and Pathologic Changes Following Ozone Exposure," *J. Appl. Physiol.*, **14,** 67 (1959).

206. Thomas, H.V., Mueller, P.K., Wright, G., "Response of Rat Lung Mast Cells to Nitrogen Dioxide Inhalation," *J. Air Pollut. Control Assoc.*, **17,** 33 (1967).

207. Estes, F.L., Pan, C., "Response of Enzyme Systems to Photochemical Reaction Products," *Arch. Environ. Health*, **10,** 207 (1965).

208. Mudd, J.B., "Enzyme Inactivation by Peroxyacetyl Nitrate," *Arch. Biochem. Biophys.*, **102,** 59 (1963).

209. Mudd, J.B., McManus, T.T., "Products of the Reaction of Peroxyacetyl Nitrate with Sulfhydryl Compounds," *Arch. Biochem. Biophys.*, **132,** 237 (1969).

210. Estes, F.L., "The Effect of Initial Concentration of Reactants on the Biological Effectiveness of Photochemical Reaction Products," *Atmos. Environ.*, **1,** 159 (1967).

211. Estes, F.L., "Analysis of Air Pollution Mixtures: A Study of Biologically Effective Components," *Anal. Chem.*, **34,** 998 (1962).

212. "Air Pollution and the Lung," E.F. Aharonson, et al., eds., Halsted Press, John Wiley & Sons, New York, N.Y., 1976.

213. Darley, E.F., Middleton, J.T., "Problems of Air Pollution in Plant Pathology," *Ann. Rev. Phytopathol.*, **4,** 103 (1966).

214. Brandt, C.S., Heck, W.W., "Effects of Air Pollutants on Vegetation," in "Air Pollution," Vol. 1., 2nd ed., A.C. Stern, ed., pp. 401–443, Academic Press, New York, N.Y., 1968.

215. Heggestad, H.E., Heck, W.W., "Nature, Extent and Variation of Plant Response to Air Pollutants," *Adv. Agron.*, **23,** 111 (1971).

216. Garber, K., "Luftverunreinigungen und ihre Wirkungen," Gebruder Borntrager, Berlin, 1967.

217. Berg, H. "Immissionsschaden (Gas-, Rauch-, und Staubschaden)," in "Handbuch der Pflanzenkrankheiten," Bd. I, Die Nichtparasitaren Krankheiten, 4 Lief. Paul Porey, Berlin, pp. 1–169, 1970.

218. Treshow, M., "Environment and Plant Response," McGraw-Hill, New York, N.Y., 1970.

219. Garber, K., "Luftverunreinigungen, eine Literaturubersicht." Berichte Nr. 102, der Eidg. Anstalt, forst. Versuchswesen, Birmensdorff ZH. 1972.

220. van Haut, H., Stratmann, H., "Farbtafelatlas uber Schwefeldioxid — Wirkungen an Pflanzen," W. Girardet, Essen. 1970.

221. Jacobson, J.S., Hill, A.C., "Recognition of Air Pollution Injury to Vegetation: A Pictorial Atlas," Air Pollution Control Association, Pittsburgh, Pa., 1970.

222. "Air Pollution Damage to Vegetation," Naegele, J.A., ed., *Adv. Chem. Ser.* No. 122, American Chemical Society, Washington, D.C., 1973.

223. Heck, W.W., "Factors Influencing Impression of Oxidant Damage to Plants," *Ann. Rev. Phytopathol.*, **6.** 165 (1968).

224. Dugger, W.M., Ting, I.P., "Air Pollution Oxidants — Their Effects on Metabolic Processes in Plants," *Ann. Rev. Plant Physiol.*, **21,** 215 (1970).

225. "Effects of Sulfur Oxides in the Atmosphere on Vegetation," Environmental Protection Agency, EPA-R3-73-030, Research Triangle Park, N.C. 1973.

226. Daines, R.H., "Sulfur Dioxide and Plant Response," *J. Occup. Med.*, **10,** 516 (1968).

227. Spedding, D.J., "Uptake of Sulfur Dioxide by Barley Leaves at Low Sulfur Dioxide Concentrations," *Nature*, **224,** 1229 (1969).

228. Guderian, R., "Untersuchungen uber quantitativ Beziehungen zwischen dem Schwefelgehalt von Pflanzen und dem Schwefeldioxidgehalt der Luft. *Z. Pflanzenkh. Pflanzenschutz.*, **77,** 200–220; 289–308; 387–399 (1970).

229. Brandt, C.S., "Report on Fluorides," prepared for National Air Pollution Control Administration: U.S. Department of Agriculture, Washington, D.C. (mimeo) 1970.

230. "Fluorides," Committee on Biologic Effects of Atmospheric Pollutants, National Academy of Sciences — National Research Council, Washington, D.C., 1971.

231. Treshow, M., "Fluorides as Air Pollutants Affecting Plants," *Ann. Rev. Phytopathol.*, **9,** 21 (1971).

232. Brandt, C.S., "Ambient Air Quality Criteria for Hydrogen Fluoride and Fluorides," VDl — Berichte, Nr. 164, 23–31 (1971).

233. Linzon, S.N., Costonis, A.C., "Symptoms Caused by Photochemical Air Pollution Injuries to Forest Trees," *Mit. Forstl. Bundes – Versuchsanst., Wien.*, **92,** 71 (1971).

234. "Ozone and Other Photochemical Oxidants," Committee on Medical and Biologic Effects of Environmental Pollutants, National Academy of Sciences — National Research Council, Washington, D.C., 1976.

235. "Air Quality Criteria for Nitrogen Oxides," Environmental Protection Agency, Publication No. AP-84, Washington, D.C., 1971.

236. Ables, F.B., Heggestad, H.E., "Ethylene: An Urban Air Pollutant," *J. Air Pollut. Control Assoc.*, **23,** 517 (1973).

237. "Hydrochloric Acid and Air Pollution: Bibliography," Environmental Protection Agency, AP-110, Washington, D.C., 1971.

238. Brennan, E., Leone, I.A., Holmes, C., "Accidental Chlorine Gas Damage to Vegetation," *Plant Dis. Rep.* **53,** 873 (1969).

239. Hofstra, G., Hall, R., "Injury on Roadside Trees: Leaf Injury on Pine and White Cedar in Relation to Foliar Levels of Sodium Chloride," *Canad. J. Bot.*, **49,** 613 (1971).

240. Horace, V.W., Cohen, E., "Salt Damage to Vegetation in the Washington, D.C., Area During the 1966–67 Winter," *Plant Dis. Rep.*, **52,** 350 (1968).

241. Evers, F., "Schaden an Bestanden durch Auftausalze," *Allg. Forstg*, **50,** 90 (1970).

242. Daines, R., Motto, H., Chelko, D., "Atmospheric Lead, Its Relationship to Traffic Volume and Proximity to Highways," *Environ. Sci. Technol.*, **4,** 318 (1970).

243. Broyer, T., Johnson, C., Paull, R., "Some Aspects of Lead in Plant Nutrition," *Plant Soil*, **36,** 301 (1972).

244. Lagerwerff, J., Specht, A., "Contamination of Roadside Soil and Vegetation with Cd, Ni, Pb and Zn," *Environ. Sci. Technol.*, **4,** 583 (1970).

245. Linzon, S.N., "Economic Effects of Sulfur Dioxide on Forest Growth," *J. Air Pollut. Control Assoc.*, **21,** 81 (1971).

246. Bell, J.N.B., Clough, W.S., "Depression of Yield in Rye Grass Exposed to Sulfur Dioxide," *Nature*, **241,** 47 (1973).

247. Guderian, R., "Reaktionen von Pflanzengemeinschaften des Feldfutterbaues auf Schwefeldioxidwirkungen," *Schriftenr. der Landesanst. Immissions-Bodennutzungsschutz. LNRW*, Nr. 4, 80–100 (1967).

248. Thompson, C.R., Kats, G., Hensel, E.G., "Effects of Ambient Levels of Nitrogen Dioxide on Navel Oranges," *Environ. Sci. Technol.*, **5,** 1017 (1971).

249. Brandt, C., Rhoades, R.W., "Effects of Limestone Dust Accumulation on Lateral Growth of Forest Trees," *Environ. Pollut.*, **4,** 207 (1973).

250. Menser, H.A., Heggestad, H.E., "Ozone and Sulfur Dioxide Synergism: Injury to Tobacco," *Science*, **153,** 424 (1966).

251. Tingey, D.T., Heck, W.W., Reinert, R.A., "Effects of Low Concentrations of Ozone and Sulfur Dioxide on Foliage, Growth and Yield of Radish," *J. Am. Soc. Hort. Sci.*, **96,** 369 (1971).

252. Tingey, D.T., et al., "Foliar Injury Responses on Eleven Plant Species to Ozone/Sulfur Dioxide Mixtures," *Atmos. Environ.*, **7,** 201 (1973).

253. Tingey, D.T., et al., "Chronic Ozone or Sulfur Dioxide Exposures, or Both, Affect the Early Vegetative Growth of Soybean," *Can. J. Plant Sci.*, **53,** 875 (1973).

254. Heck, W.W., Personal Communication (1974).

255. Heggestad, H.E., "Variations in Response of Potato Cultivars to Air Pollution," *Phytopathology*, **60,** 1015 (1970).

256. Heggestad, H.E., "Photochemical Air Pollution Injury to Potatoes in the Atlantic Coastal States," *Amer. Potato J.*, **50,** 315 (1973).

257. Waddell, T.E., "The Economic Damages of Air Pollution," Environmental Protection Agency, EPA-600/5-74-012, U.S. Government Printing Office, Washington, D.C. 1974.

258. Lodge, J.P., Jr., "When NOT to Use On-Line Air Pollution Instruments," *Instrum. Technol.*, **18,** (12), 28 (1971).

259. Kehoe, T.J., "On-Line Experiences with Analyzers," *Instrum. Technol.*, **19,** (6), 33 (1972).

260. Lodge, J.P., Jr., "Analysis Dilemma," *Instrum. Technol.*, **19** (6), 4 (1972).

261. Wade, N., "Center for Atmospheric Research: Monument to Science's Old Life-Style," *Science*, **182,** 36 (1973).

262. Hering, S.V., Flagan, R.C., Friedlander, S.K., "A New Low-Pressure Impactor: Determination of the Size Distribution of Aerosol Sulfur Compounds," paper presented before Division of Environmental Chemistry, American Chemical Society, New Orleans, La., March 20–25, 1977.

263. Dzubay, T.G., Stevens, R.K., "Ambient Air Analysis with Dichotomous Sampler and X-Ray Fluorescence Spectrometer," *Environ. Sci. Technol.*, **9,** 663 (1975).

264. George, D.H., Zeller, K.F., "Visibility Sensors in Your Air Quality Program," presented at the Second Joint Conference on the Sensing of Environmental Pollutants, Washington, D.C., December 1973.

265. Lin, C.-I, Baker, M., Charlson, R.J., "Absorption Coefficient of Atmospheric Aerosol," *Appl. Opt.*, **12,** 1356 (1973).

266. Butcher, S.S., Charlson, R.J., "An Introduction to Air Chemistry," Academic Press, New York, N.Y., 1972, pp. 195–198.

267. Ensor, D.S., Waggoner, A.P., "Angular Truncation Error in the Integrating Nephelometer," *Atmos. Environ.*, **4,** 481 (1970).

268. Rabinoff, R.A., Herman, B.M., "Effect of Aerosol Size Distribution on the Accuracy of the Integrating Nephelometer," *J. Appl. Meteor.*, **12,** 184 (1973).

269. Charlson, R.J., et al., "Monitoring of Atmospheric Aerosol Parameters with the Integrating Nephelometer," *J. Air Pollut. Control Assoc.*, **19,** 943 (1969).

270. Samuels, H.J., Twiss, S., Wong, E.W., "Visibility, Light Scattering and Mass Concentration of Particulate Matter," State of California Air Resources Board, 1973.

271. Horvath, H., Noll, K.E., 'The Relationship between Atmospheric Light Scattering Coefficient and Visibility," *Atmos. Environ.*, **3,** 543 (1959).

272. Mie, G., "Optics of Turbid Media Especially for Metal Colloidal Sols," *Ann. Phys.* (Leipzig), **25,** 377 (1908).

273. Bullrich, K., "Scattered Radiation in the Atmosphere," *Advan. Geophys.*, **10,** 99 (1964).

274. Charlson, R.J., Ahlquist, N.C., Horvath, H., "On the Generality of Correlation of Atmospheric Aerosol Mass Concentration and Light Scatter," *Atmos. Environ.*, **2,** 455 (1968).

275. Noll, K.E., Mueller, P.K., Imada, M., "Visibility and Aerosol Concentration in Urban Air," *Atmos. Environ.*, **2,** 465 (1968).

276. Ahlquist, N.C., Charlson, R.J., "Measurement of the Wavelength Dependence of Atmospheric Extinction Due to Scatter," *Atmos. Environ.*, **3,** 551 (1969).

277. Charlson, R.J., "Multiwavelength Nephelometer Measurements in Los Angeles," *J. Colloid Interface Sci.*, **39,** 240 (1972).

278. Porch, W.M., et al., "Blue Moon: Is This a Property of Background Aerosol," *Appl. Opt.*, **12,** 34 (1973).

279. Covert, D.S., Charlson, R.J., Ahlquist, N.C., "A Study of the Relationship of Chemical Composition and Humidity to Light Scattering by Aerosols," *J. Appl. Meteor.*, **11**, 968 (1972).

280. Pueschel, R.F., Charlson, R.J., Ahlquist, N.C., "On the Anomalous Deliquescence of Sea-Spray Aerosols," *J. Appl. Meteor.*, **8**, 995 (1969).

281. Charlson, R.J., et al., "Sulfuric Acid–Ammonium Sulfate Aerosol: Optical Detection in the St. Louis Region," *Science*, **184**, 156 (1974).

282. Pueschel, R.F., Bodhaine, B.A., Mendonca, B.G., "The Proportion of Volatile Aerosols on the Island of Hawaii," *J. Appl. Meteor.*, **11**, 968 (1973).

Chapter 5

The Water Environment

5 / The Water Environment

INTRODUCTION

The water environment is confined and unevenly distributed over the earth, so that man has had to learn to manage it to a much greater degree than has been the case with air. That the need for water management in the U.S. has outrun the application of the available technology is due less to ignorance than to negligence. Ignorance exists, certainly. The biological processes used for decades to treat wastewaters are still in many ways highly empirical. The complex chemistry of natural waters (1) is far from being well understood. But despite the deficiencies in fundamental knowledge, it has long been technologically feasible to enhance markedly the quality of the nation's waters.

Many organizations have worked for years at water quality management, the complex of activities — scientific, engineering, legal, administrative — required to control the quality of water. These groups include more than 3000 soil and water conservation districts in the 50 states, Puerto Rico, and the Virgin Islands; interstate compacts like the Ohio River Valley Water Sanitation Commission; the seven interstate river basin commissions set up since 1976 under the federal Water Resources Council; and the International Joint Commission established by the U.S. and Canada under the Boundary Waters Treaty of 1909. Only in the past two decades, however, have federal legislative and administrative measures evolved that require comprehensive use of the available knowledge in upgrading and maintaining the quality of the water environment. To expand and refine the science and technology that these measures can bring into play, steady progress must be made in our fundamental understanding in four broad areas:

- The origin and dispersion of water pollutants, their degradation or conversion to other chemical and physical forms, and their ultimate fate.
- Treatment processes and other means of abating water pollution where generation of the pollutants cannot be avoided.
- The effects of water pollutants on plant and animal life and on inanimate objects.
- The means of detecting and measuring water pollutants and their effects (see also Chapter 2).

Background

A body of water, like an air mass, is a dynamic system, regularly assimilating a range of solids, liquids, and gases, both natural and man-made. Natural surface waters, moreover, normally teem with living organisms, which can powerfully affect the course of events in a given water system. All of these substances, living and nonliving, may flow, disperse, and interact chemically and physically before they reach a sink such as the ocean or a receptor such as a fish. En route from source to sink or receptor they may assume a variety of chemical and physical forms.

Water as H_2O is a chemical compound of unvarying composition, and in this sense natural waters are never pure (Table 1). Water is "polluted" when man's activities have made it unsuitable for a particular use. In other words, the nature and extent of pollution are defined by the intended use of the water. Natural waters, unaffected by man, are also unsuitable sometimes for certain uses, such as public water supply or propagation of fish and wildlife. In the public mind

Table 1

Range in values of water quality constituents for natural surface waters in the United States

	Tenth percentile[a] value	Median value[b]	Ninetieth percentile[c] value
Dissolved calcium (Ca), mg/l	1.8	11	71
Bicarbonate (HCO_3), mg/l	5	40	270
Alkalinity as $CaCO_3$, mg/l	4	28	220
Dissolved sulfate (SO_4), mg/l	1.8	5.8	57
Dissolved chloride (Cl), mg/l	0.6	2.3	9.3
Dissolved orthophosphate as P, mg/l	0.00	0.02	0.05
Dissolved solids, residue at 180° C, mg/l	24	63	354
Hardness as $CaCO_3$ (Ca, Mg), mg/l	6	37	360
Noncarbonate hardness, mg/l	0	4	41
pH	6.0	7.4	8.4
Water temperature, ° C	0.0	10.5	21.5
Turbidity, Jackson Turbidity Units	1	3	15
Dissolved oxygen, mg/l	7.8	10.1	12.4
Biochemical oxygen demand (BOD), mg/l	0.2	0.7	1.8
Total coliform bacteria, colonies/100 ml	2	67	1420
Fecal coliform bacteria, colonies/100 ml	0	9	183
Fecal streptococci, colonies/100 ml	1	22	422

a. Ninety percent of values are equal to or greater than this value.
b. Fifty percent of values are equal to or greater than this value.
c. Ten percent of values are equal to or greater than this value.

Source: Data come from samples collected during water years 1974 and 1975 from 57 stream basins relatively unaffected by man and expected to remain in their present or natural condition. The 57 stream basins comprise the U.S. Geological Survey's nationwide Hydrological Bench-Mark Network.

such waters are not "polluted" or even "naturally polluted." The important point, however, is that the quality of a body of water may be naturally good or bad, depending on the intended use. When man does use water, on the other hand, he almost surely changes its quality, although the effect may be only a minor one such as a slight rise in temperature.

This country uses an immense and growing amount of water. The total in 1975 averaged 3720 billion gallons per day (bgd) or more than 17,000 gal per person per day (2) (Table 2). This was about 11% more than in 1970 and 98% more than in 1950. Of the 3720 bgd used daily in 1975, 3300 bgd or 89% was used to produce hydroelectric power. Water used in this way is not degraded and is available for reuse or dilution downstream; if it is excluded from the total, usage per person becomes about 1900 gal daily, about 8% more than in 1970 and 46% more than in 1950. The remaining 420 bgd was divided among four categories: public supply, 7%; rural, 1%; irrigation, 35%; and self-supplied industrial, 57%. The sources of the 420 bgd were fresh surface water, 63%; fresh groundwater, 20%; saline surface water, 16%; saline groundwater 0.24%; and reclaimed sewage, 0.12%. The degree of reuse of water in 1970 is evident in a comparison of the total used (3720 bgd) with the total available or average annual streamflow (1200 bgd). About 95 bgd of the fresh water used was "consumed" or made totally unavailable for reuse. This was about 8% of the average annual streamflow in the 48 coterminous states. Irrigation of crops in the West accounted for about 78% of the total water consumed.

Determining the relative significance of man-made sources of water pollution is complicated by the fact that contaminants often enter water in complex mixtures of many substances whose specific chemical identities are largely unknown. For practical purposes, this difficulty can be bypassed in part by describing waste streams in terms of certain collective characteristics. One of these characteristics is biochemical oxygen demand (BOD), a measure of the weight of dissolved oxygen consumed in the biological processes that degrade organic matter in surface waters. Another collective characteristic is the weight of suspended solids, only part of which is settleable, in the water. Comparisons of manufacturing and domestic wastes in terms of these collective characteristics can be deceptive. One reason is that industrial wastes are much more likely than domestic wastes to contain "refractory" chemicals, those that resist biological degradation. Such chemicals include toxic organic compounds that must be converted to other, less harmful compounds before they are discharged, or trace elements that must be removed in the treatment process.

Water-quality information is collected by numerous federal, state, and local government agencies as well as by private organizations. The two largest water-quality data banks are the storage and retrieval (STORET) system of the Environmental Protection Agency and the National Water Data Storage and Retrieval System (WATSTORE) of the U.S. Geological Survey (USGS) in the Department of the Interior. The federal water-data collection effort is coordinated by the Geological Survey's Office of Water Data Coordination (OWDC). Annually OWDC prepares a federal plan that describes the total data-collection effort and each federal agency's data collection plans through the budget year. With the federal Water Resources Council, OWDC has developed a nationally consistent set of state maps showing hydrologic units suited to water resources planning and development.

In 1974, USGS began operating the National Stream Quality Accounting Network (NASQAN), which assesses the quality of the nation's rivers on a

Table 2

Water withdrawals and uses in the United States (billion gallons per day)

	1950	1960	1970	1975	Percent increase or decrease 1970–75[a]
Total population (millions)	150.7	179.3	205.9[b]	217.5[c]	5.6
Total withdrawals	200	270	370	420	11.7
Public supplies	14	21	27	29	7.9
Rural domestic and livestock	3.6	3.6	4.5	4.9	10.3
Irrigation	110	110	130	140	10.9
Self-supplied thermoelectric power use	40	100	170	190	18.0
Other self-supplied industrial use	37	38	47	44	−5.6
Sources from which water was withdrawn					
Fresh groundwater	34	50	68	82	21.7
Saline groundwater	[d]	0.38	1.0	1.0	−6.0
Fresh surface water	160	190	250	260	5.1
Saline surface water	10	31	53	66	30.9
Reclaimed sewage	[d]	0.1	0.5	0.5	2.2
Water consumed by nonhydroelectric power uses	[d]	61	87[e]	95[e]	9.9
Water used for hydroelectric power	1100[f]	2000[f]	2800[f]	3300[f]	20.7

a. Calculated from original unrounded computer printout figures for the two years.
b. Including Puerto Rico.
c. Including Puerto Rico and Virgin Islands.
d. Data not available.
e. Freshwater only.
f. This water is not degraded and is available for reuse or dilution downstream.

Note: Partial figures may not add to totals because of independent rounding.

Source: Murray, C.R., Reeves, E.B., "Estimated Use of Water in The United States in 1975," U.S. Geol. Survey Circ. 765, U.S. Geological Survey, Washington, D.C., 1977.

uniform basis. By 1977, NASQAN had 345 monitoring stations, each measuring more than 50 water-quality variables; some 525 stations were expected to be operating by 1979. In 1976, USGS established the National Water Data Exchange (NAWDEX), a nationwide service to help users identify and acquire water data. NAWDEX maintains computer files that in 1977 identified more than 300 organizations that collect water data and described water data available from more than 200,000 sites nationwide. The computer files are accessible to NAWDEX users through terminals at 52 local assistance centers.

Legislative framework

The primary legislative basis for managing the water environment is the Federal Water Pollution Control Act of 1956, as amended by the Water Quality Act of 1965 (Public Law 89-234), the Federal Water Pollution Control Act Amendments of 1972 (Public Law 92-500), and the Clean Water Act of 1977 (Public Law 95-95). In addition, ocean dumping is controlled by the Marine Protection, Research and Sanctuaries Act of 1972 (Public Law 92-532). Together these measures constitute the most sweeping legislative attack yet on water pollution in this country. Furthermore, the Safe Drinking Water Act of 1974 (Public Law 93-523) for the first time provided for the setting of national drinking water quality standards to replace the Public Health Service Standards of 1962. The latter applied legally only to water supplies for interstate carriers, although they served as the recognized standards throughout the nation.

The 1972 amendments to the act of 1956 grew from the realization that the quality of the nation's waters was barely holding its own, if that, under then-current laws (3). The amendments envisioned two national goals:

- To eliminate, by 1985, the discharge of pollutants into the nation's waters.
- To achieve wherever possible, by July 1, 1983, water that is clean enough for recreation, including swimming, and for the protection and propagation of fish and wildlife.

The 1972 amendments extended the federal pollution control effort to all U.S. waters; before, federal law had applied only to interstate waters. The amendments also established stringent controls and deadlines for industrial and municipal pollution; shifted the emphasis from ambient water quality standards toward effluent standards; and created a system of permits for discharges into the nation's waters.

Under the 1972 amendments, industries were to be treating their effluents with the "best practicable" water pollution control technology by July 1, 1977; they were to be using the "best available" technology by July 1, 1983. The meanings of "best practicable" and "best available" depend on guidelines issued by the Environmental Protection Agency (EPA). The guidelines are flexible: they may reflect the nature of the industry involved; the age of the specific facility; the cost of control; and similar factors. The discharge of toxic materials is controlled by effluent standards devised by EPA. The agency can, where it sees fit, ban absolutely the discharge of toxic substances in any amount. The 1972 law itself applied an absolute ban to discharges of high-level radioactive wastes, or radiological, chemical, or biological warfare materials. The 1972 amendments required municipal sewage treatment plants to be employing "best practicable" treatment to qualify for a federal construction grant (after June 30, 1974). With a few exceptions, all sewage treatment plants were to be providing a minimum of secondary (biological) treatment after July 1, 1977; they must also meet whatever more stringent effluent standards may be imposed by EPA or a state to meet water quality or treatment standards or compliance schedules. All publicly-owned wastewater treatment plants were to be using "best practicable" treatment by July 1, 1983. The law provided also for areawide waste treatment management plans in urban industrial areas with serious water pollution problems. Once such a plan is established, publicly-owned treatment plants in the area must be part of the plan to qualify for a federal construction grant.

The Water Quality Act of 1965 had amended the act of 1956 to require states and other jurisdictions to establish quality standards for interstate waters. These ambient standards, in turn, were the basis of implementation and enforcement plans devised by the states to control water pollution. The 1972 amendments retained the ambient water quality standards program and extended it to intrastate as well as interstate waters (4). At the same time, however, the amendments created a national water permit system. The effect was to shift the emphasis in water pollution control from maintenance of ambient standards to limitations on the nature and amounts of contaminants in all point-discharges to the nation's waters.

Under the permit scheme, the National Pollutant Discharge Elimination System (NPDES), no pollutant may be discharged to water from any point source without a permit. The program is designed to be administered by the states, subject to approval by EPA. (The Army Corps of Engineers issues permits for disposal of dredge-and-fill wastes at specified sites, subject to EPA veto in some circumstances.) Among other requirements, applicants for discharge permits must specify the physical and chemical nature of the effluents involved.

The 1972 amendments to the Federal Water Pollution Control Act were superimposed on the Marine Protection, Research and Sanctuaries Act of 1972. The latter, in addition, requires permits for transporting and dumping materials in the ocean. Dumping sites are designated by EPA and, where possible, are beyond the Continental Shelf. EPA and the Corps of Engineers administer the permit system; the Coast Guard is responsible for surveillance and enforcement. The research objective of the law is to minimize or end, by 1983, all dumping of wastes into the oceans, coastal waters, and the Great Lakes or their connecting waters. The Secretary of Commerce is responsible for research under the law, in cooperation with EPA and the Coast Guard. The law also authorizes the Secretary of Commerce to designate marine sanctuaries — areas whose conservation, recreational, ecological, or esthetic values warrant their restoration and preservation.

The 1977 amendments. The Clean Water Act of 1977 made significant changes in the Federal Water Pollution Control Act, but did not alter its basic structure. Among other revisions, the act of 1977 relaxed the deadlines set by the 1972 amendments, but strengthened the regulation of toxic water pollutants. Also, for regulatory purposes it classified water pollutants as conventional, nonconventional, and toxic. Conventional pollutants include, but are not limited to, suspended solids, fecal coliform bacteria, and substances that exert biochemical oxygen demand or affect pH (acidity/alkalinity). Nonconventional pollutants are those not classified as conventional or toxic. The act specified as toxic a list of 65 compounds or families of compounds (5); EPA may revise the list, by adding or removing compounds, subject to legal challenge.

The act gives EPA three options for dealing with industrial dischargers who missed the July 1, 1977, deadline for best practicable treatment. Depending on the specific circumstances, the agency may require compliance in a "reasonable'' time; it may require compliance no later than Jan. 1, 1979; or it may require that the facility discharge into a publicly owned treatment works at the earliest practicable date, but no later than July 1, 1983.

For industrial discharges of conventional pollutants, the original deadline of July 1, 1983, for best available treatment was extended by the act to July 1, 1984. Also, the required treatment was revised to "best conventional pollutant control technology," whose stringency is to reflect a reasonable balance between the costs and benefits of treatment. The level probably will be somewhere between best practicable and best available treatment for other than conventional pollutants. For nonconventional pollutants, the deadline for best available treatment was revised to three years after effluent limitations are established or July 1, 1984, whichever is later, but in no case later than July 1, 1987. For toxic pollutants, industrial dischargers must apply best available technology by July 1, 1984; EPA may require zero discharge of a toxic pollutant where appropriate.

For municipal wastewater treatment, the act of 1977 allows EPA to extend the July 1, 1977, deadline for a minimum of secondary treatment to as late as July 1, 1983, on a case-by-case basis. Also, the act allows the requirement for secondary treatment to be modified for conventional pollutants where specified conditions are met and the facility discharges to marine waters. Furthermore, the act requires that 4% of construction grant funds allotted to rural states for wastewater treatment facilities be set aside for "alternative or unconventional systems" for municipalities of 3500 or less or for highly dispersed areas of larger communities. The provision recognizes the difficulties of establishing conventional secondary treatment facilities in dispersed communities.

Drinking water. The Safe Drinking Water Act of 1974 made EPA responsible for establishing national standards of quality for drinking water. It made the states responsible for enforcing the standards and for supervising public water supply systems, both publicly and privately operated, and sources of drinking water. The law defines a "public water system" as one that provides piped water for human consumption and has at least 15 service connections or regularly serves an average of at least 25 persons daily at least 60 days out of the year. Some 240,000 such systems were operating in the U.S. in 1977.

The act provides for primary regulations to protect public health and secondary regulations related to the taste, odor, and appearance of drinking water. It provides for the protection of groundwater sources of drinking water. It provides also for research on drinking-water problems including, specifically, research on viruses and carcinogens.

Progress in Water Management

Pollution control

The nation's effort in water pollution control was producing halting progress by the late 1970's. Water quality had clearly improved in a number of areas, but remained poor in many others (6, 7); it was apparent that the goals of the federal legislative program would not be achieved for some years.

By September 1976, EPA had identified almost 65,000 point-source dischargers subject to the National Pollutant Discharge Elimination System and had issued some 53,000 discharge permits (7). Of the permits issued, about two thirds were for industrial and agricultural sources and one third for municipal sources. The July 1, 1977, deadline for best practicable treatment was missed by an estimated 15% of the 4400 major industrial dischargers and by many smaller industrial dischargers. Only about one third of the nation's municipal wastewater facilities met the July 1, 1977, deadline for secondary treatment as the minimum; the National Commission on Water Quality estimated that municipal wastewaters would not uniformly be receiving a minimum of secondary treatment until at least a decade after the deadline (8).

The gradual resolution of these difficulties with point sources would be a major step toward cleaner water. The estimated reductions in BOD and total suspended solids that would be achieved for selected sources appear in Table 3. Nonpoint or areal sources, on the other hand, comprise an immense problem that remains to be dealt with satisfactorily. Such sources account for an estimated one third of the BOD and one half of the total suspended solids generated nationwide; their relative contributions become even larger as point-source pollution is abated (Table 4).

Drinking water standards

Under the Safe Drinking Water Act of 1974, EPA established the National Interim Primary Drinking Water Standards (Table 5) that took effect in June 1977. The standards are designed to protect human health. They are enforceable by the Federal Government, but the intent of the law is that the states assume primary responsibility for enforcement. In March 1977, EPA proposed secondary drinking water regulations (Table 6), which also are required by the act. The regulations are not health-related, but cover contaminants that may affect the odor, palatability, or other characteristics of drinking water. They are not enforceable by the Government, and their use by the states is optional.

EPA requires public water supply systems to monitor their water routinely for adherence to the primary standards. For coliform bacteria and turbidity, this requirement took effect with the standards, in June 1977. For the other contaminants, the requirement was to be phased in gradually. The agency believed that most water supplies were already meeting the primary standards when they took effect. Systems that do not meet the standards must take corrective measures that can include the installation of additional treatment equipment or even a switch to a new source of raw or treated water. Individual systems may be granted variances or exemptions to cope with economic and other problems, so long as an unreasonable risk to public health is not involved.

The interim primary standards for drinking water were modeled on the Public Health Service Standards of 1962. The latter had protected the public health well, but were based on the premise that public water supplies would be drawn from natural waters. This premise has grown steadily less tenable. The growing number of wastewater discharges and the shrinking distances between them and the water supply intakes have made sources of public water supply increasingly more liable to pollution not only by untreated wastes but by the normal residual contaminants in treated wastewaters. Water-supply treatment plants have had to cope with a growing load of contaminants, including some that they are not designed to remove. When the interim primary standards were established, the available data and technology were insufficient to support workable standards significantly more comprehensive than those of 1962. However, it was anticipated that the standards would be expanded. To provide guidance, the National Academy of Sciences assessed the existing data on the health effects of more than 150 contaminants (9).

Table 3

Estimated national generation and discharge of biochemical oxygen demand and total suspended solids by selected sources, 1973 (millions of pounds)

	Generated		After in-place technologies		After best practicable technology[a]		After best available technology[a]	
	BOD	TSS	BOD	TSS	BOD	TSS	BOD	TSS
Pulp and paper mills	3872.83	5082.06	2023.63	2391.79	387.29	406.57	193.64	152.45
Organic chemicals	1980.17	0.0	1326.72	0.0	186.14	0.0	39.60	0.0
Petroleum refining	218.30	64.77	115.88	29.77	10.92	1.49	4.36	0.52
Iron and steel	78.62	12167.86	39.31	2567.50	5.90	820.29	1.18	68.09
Inorganic chemicals	11.74	2395.90	7.86	1150.03	5.63	693.87	4.04	347.14
Plastics and synthetics	335.05	241.51	224.48	115.92	33.50	26.56	6.70	2.41
Textile mills	693.75	529.73	382.69	387.18	69.39	58.30	22.78	12.25

a. About 85% of major industrial dischargers were using best practicable technology by July 1, 1977. For BOD and TSS, the Clean Water Act of 1977 replaced best available technology with the somewhat less stringent "best conventional pollutant control technology," to be achieved by July 1, 1984.

Source: "The National Residuals Discharge Inventory," National Commission on Water Quality, National Research Council, Washington, D.C., January 1976.

Table 4

Estimated national generation and discharge of biochemical oxygen demand and total suspended solids by point and nonpoint sources, 1973 (billions of pounds)

	BOD				TSS			
	1973 Generation	1973 Discharge	BPT/ST[a] Discharge	BAT/BPWTT[a] Discharge	1973 Generation	1973 Discharge	BPT/ST[a] Discharge	BAT/BPWTT[a] Discharge
Point Sources (PS)	21.2	10.1	2.7	1.3	1160	729	14.3	13.4
% of National total	68	51	22	12	45	34	1	1
Industrial	8.0	4.3	0.9	0.4	1150	723	12.4	12.3
% of PS	38	43	33	31	99	99	88	93
Municipal	13.2	5.8	1.8	0.9	14.1	6.0	1.7	0.8
% of PS	62	57	67	69	1	1	12	7
Nonpoint Sources (NPS)	9.8	9.8	9.8	9.8	1430	1430	1430	1430
% of National total	32	49	78	88	55	66	99	99
Urban runoff	3.3	3.3	3.3	3.3	59.9	59.9	59.9	59.9
% of NPS	34	34	34	34	4	4	4	4
Nonirrigated agriculture	6.5	6.5	6.5	6.5	1370	1370	1370	1370
% of NPS	66	66	66	66	96	96	96	96
National totals	**31.0**	**19.9**	**12.5**	**11.1**	**2600**	**2160**	**1450**	**1450**

a. BPT/ST = best practicable technology for industrial sources and secondary treatment for municipal sources. BAT/BPWTT = best available technology for industrial sources and best practicable waste treatment technology for municipal sources. About 85% of major industrial dischargers were using best practicable technology by July 1, 1977, and about 33% of municipal wastewater treatment facilities were using secondary treatment or better. For BOD and TSS, the Clean Water Act of 1977 replaced best available technology with "best conventional pollutant control technology," to be achieved by July 1, 1984.

Source: "The National Residuals Discharge Inventory," National Commission on Water Quality, National Research Council, Washington, D.C., January 1976.

In February 1978, EPA proposed two additions to the primary drinking water standards (10). Chloroform and related organic compounds of the trihalomethane class would be limited to a maximum combined concentration of 0.10 milligrams per liter (mg/l) in drinking water; the concentrations of other organic compounds would be controlled by requiring that water-supply treatment include filtration by granular activated carbon. The trihalomethane standard would apply at first only to community water systems serving more than 75,000

Table 5

National interim primary drinking water standards

Contaminant	Maximum level
Inorganic	mg/l
Arsenic	0.05
Barium	1.
Cadmium	0.010
Chromium	0.05
Fluoride[a]	1.4 (90.5°F) – 2.4 (53.7°F and below)[a]
Lead	0.05
Mercury	0.002
Nitrate (as nitrogen)	10.
Selenium	0.01
Silver	0.05
Organic[b]	
Endrin	0.0002
Lindane	0.004
Methoxychlor	0.1
Toxaphene	0.005
2,4-D	0.1
2,4,5-TP (Silvex)	0.01
Radioactivity	pCi/l[c]
Naturally-occurring:	
Radium-226, -228	5
Gross alpha particle activity	15
Man-made: The average annual concentration of beta-particle and photon radioactivity shall not produce an annual dose-equivalent to the total body or to any internal organ greater than 4 millirem per year.[c]	
Turbidity	1 Turbidity Unit
Microbiological contaminants	
No more than four coliform bacteria per 100 ml. The arithmetic mean of all samples cannot exceed one per 100 ml.	

a. Fluoride level varies with annual average of maximum daily air temperature at location of water system. The rationale is that people consume more water in hotter climates and that, with fluorides, a relatively small margin exists between the levels that produce beneficial effects and those that produce adverse effects.

b. Endrin, lindane, methoxychlor, and toxaphene are insecticides; 2,4-D and silvex are herbicides.

c. A picocurie (pCi) is a standard unit of measurement of radioactivity. A millirem is one thousandth of a rem, which is a unit measure of dose of radioactivity, weighted to reflect potential for producing biological damage to man.

Source: Environmental Protection Agency (effective June 24, 1977).

people and would have to be met within 15 months of the time the proposed regulation became final; smaller systems would be phased into the program as feasible. The carbon-filtration requirement would apply only to systems serving more than 75,000 people and would have to be met within five years of the time the regulation became final. Systems that draw their water supply from unpolluted sources would not have to meet the requirement if they could show that it was unnecessary to protect public health.

The trihalomethanes are suspected of being carcinogenic in humans. Those detected in drinking water are chloroform, bromoform, dibromochloromethane, and bromodichloromethane. The compounds are formed by the reaction of naturally occurring organic substances with the chlorine used by most water supply systems as a disinfectant. They have been found in nearly all water supplies tested. The concentrations are generally low, however, and can be controlled by modifying the treatment process, changing disinfectants, or removing the compounds with an adsorbent. EPA's carbon-filtration requirement

Table 6

Proposed secondary drinking water regulations

Contaminant	Proposed maximum level	Principal effects
Chloride	250 mg/l	Taste
Color	15 Color units	Appearance
Copper	1 mg/l	Taste, fixture staining
Corrosivity	(Noncorrosive)	Deterioration of pipes, unwanted metals in drinking water
Foaming agents	0.5 mg/l	Foaming, adverse appearance
Hydrogen sulfide	0.05 mg/l	Taste, odor
Iron	0.3 mg/l	Taste, brown stains on laundry and fixtures
Manganese	0.05 mg/l	Taste, brown stains, black precipitates
Odor	3 Threshold odor number	Odor
pH	6.5–8.5	Corrosion below 6.5; incrustations, bitter taste, lowered germicidal activity of chlorine over 8.5
Sulfate	250 mg/l	Taste, laxative effects
Total dissolved solids	500 mg/l	Taste, reduction in life of hot water heaters, precipitations in cooking utensils
Zinc	5 mg/l	Taste

Source: Environmental Protection Agency (proposed March 31, 1977).

is aimed at the many synthetic organic compounds that have been detected in drinking water supplies in recent years, at levels lower than those of the trihalomethanes. A number of these compounds are suspected of being human carcinogens, but knowledge of their health effects at the concentrations found is too limited to support numerical standards of concentration. EPA, therefore, proposed filtration by activated carbon, a good adsorbent for organic compounds. The agency planned to establish numerical standards when sufficient data became available.

FLOW, DISPERSION, DEGRADATION

A substance moves through a water system by flow or convection and disperses by diffusion or mixing. As the substance travels from source to sink or receptor it may be degraded or converted to other chemical and physical forms by chemical, biological, and physical processes. This overall transport process determines in large measure the effects of waterborne substances on quality throughout a water system; in surface streams it is fundamental to the ecology of the system. Clear understanding of organic degradation phenomena, in particular, is needed to decide whether specific pollutants should be allowed to enter surface-water or groundwater systems or even sinks, such as the oceans.

It is possible, though often extremely difficult, to describe the effects of the transport process mathematically. Computer simulation and other high-speed computational techniques can often be brought to bear on such problems. Encouraging progress has been made in developing mathematical descriptions of the physical processes of flow, convection, and dispersion in groundwater, streams, estuaries, and coastal waters. The major difficulty in describing pollutant transport mathematically appears to be our incomplete understanding of the factors that control the physical and chemical interactions of solids, gases, and liquids. The necessary insights demand knowledge not only of the controlling factors, but also of the nature and amounts of the substances that are interacting.

Amounts and Composition of Wastes

Rational planning for water quality management in regions or drainage basins requires detailed inventories of the composition and volume of all pollutants from all significant sources. Such inventories have been compiled for short stretches of a few rivers and, partially, for a few water systems, such as the Ohio River, the Delaware River, and Lake Erie (11, 12). But for almost no pollutants have we had adequate balances of the amounts that enter particular water systems and the amounts that leave. Gross inventories are available of the sources and amounts of municipal and some industrial wastes and of nonpoint discharges, but the data are far from definitive (13). Some better data may emerge in time from the National Pollutant Discharge Elimination System, which requires that applicants for permits specify the physical and chemical character of their discharges. For point sources, at least, this provision potentially could yield a much more comprehensive inventory of the compositions and amounts of contaminants that are entering U.S. waters. A second potential source of improved data is the National Eutrophication Survey of 1972–75. This study involved analyses of about 100,000 samples from more than 800 lakes, 4200 tributaries and outlets, and about 1000 sewage treatment plants. Data from the survey began to appear during 1976 (14, 15).

The main inorganic constituents of most wastewaters include ions such as sodium, potassium, ammonium, calcium, magnesium, chloride, nitrate, nitrite, bicarbonate, sulfate, and phosphate. Inorganic constituents also include heavy metals such as copper, zinc, cadmium, lead, nickel, and cobalt (16). Until recently, the specific organic compounds in waterborne wastes were largely unidentified. The exceptions included pesticides, synthetic detergents, and, to a lesser extent, phenolic substances and carboxylic acids. Even the main classes of organic compounds in untreated and treated domestic wastes were known only partially (17–21). The years since about 1970, however, have seen an accelerating effort to detect and measure organic pollutants using sophisticated analytical technology (22, 23). Notable among the methods used has been the combination of gas chromatography and mass spectroscopy, with computer-processing of the data generated.

These analyses have detected large numbers of individual organic compounds in the effluents from municipal and industrial waste treatment plants and also in natural waters and public water supplies (24–26). At a municipal sewage treatment plant, for example, more than 150 organic compounds were separated from the primary effluent and about 50 from the secondary effluent (27). In the drinking water supply of New Orleans, 94 organics were identified specifically (28). The concentrations being measured are in general quite low, at the level of micrograms per liter. The evidence indicates that perhaps 80% of these organic compounds are man-made; the remainder occur naturally or result from reactions of man-made compounds in water.

Still to be obtained are definitive data on the compositions of municipal storm water runoff and discharges from combined sewer systems (29). Combined systems collect storm water as well as raw sewage, and the stream bypasses the treatment plant when the system is overloaded. The studies that have been made show that urban storm water runoff at times is at least equal to domestic sewage in concentrations of suspended solids, coliform bacteria, chemically oxidizable organics, organic and inorganic nitrogen, and inorganic phosphorus. Particles, particularly those of colloidal and smaller size, make up a large part of the pollutants discharged from many sources, but important characteristics like particle size distribution often are not known. Even for municipal wastewater and secondary treatment plant effluent the data on particles are too scarce to provide the insights that might help to upgrade sewage treatment processes. A particle size classification scheme for wastewater has been devised using the electron microscope, centrifugation, and filtration (30). The scheme classifies particulate matter in four size ranges:

Group	Size Range (diameter)
Settleable solids	Larger than 100 micrometers (μm)
Supracolloidal solids	1 to 100 μm
Colloidal solids	1 nanometer (nm) to 1 μm
Soluble solids	Less than 1 nm

This classification scheme has served as a basis for studying the changes induced by secondary treatment in the colloidal material in wastewater (31). Colloidal organic material in particular has been associated with certain problems in advanced treatment, such as fouling of ion exchange resins and electrodialysis membranes. The study at hand showed that secondary treatment decreases the concentration of colloidal materials, both organic and inorganic, in wastewater, although much of the colloidal matter in secondary effluent is formed during biological treatment. Thus advanced treatment problems associated with colloids in secondary effluent probably reflect the nature of the colloids — not an increase in their concentration — resulting from biological treatment.

Biological Aspects

The fate of matter dissolved or suspended in water is influenced strongly by living organisms, both physically and, through metabolic processes, chemically. Plants and masses of clams can markedly impede the flow through irrigation canals. Less dense masses of organisms may dampen waves in lakes, create eddies in moving water, and trap suspended substances. Flow velocities, in turn, profoundly affect the kinds and quantities of organisms in aquatic systems. Organisms remove dissolved and suspended substances from water to sustain their metabolic processes and return substances to the water by respiration and decomposition. A community of organisms that is growing and multiplying — as during the warmer, sunnier days of the year — removes more matter from the water than it returns. Some such material is simply adsorbed passively by organisms, but most of it they metabolize as cell building blocks and sources of energy. Microorganisms, mainly algae and bacteria, account for most of the biological uptake of dissolved materials. Microorganisms, in fact, may largely control the concentrations of certain elements dissolved in water. Notable examples are carbon, nitrogen, and phosphorus, which are major elements in cells.

Particularly important to water quality are the bacteria that oxidize organic compounds to obtain energy and carbon for synthesizing cellular material. On the one hand they degrade organic contaminants in water; on the other they deplete the oxygen dissolved in water. Dissolved oxygen is especially critical to the aquatic environment: few forms of life can survive without it; and the element is not highly soluble in water (whereas dry air contains about 23% oxygen by weight, water at 0°C and atmospheric pressure can accommodate only about 0.001% dissolved oxygen).

The potential for oxygen depletion affects the amount of organic waste that should be discharged to a water system, such as a stream, and the points at which it should be discharged. Waste treatment must be handled so that the degradation process does not reduce the level of oxygen dissolved in the water to the point where other functions of the water will be impaired. The classical pattern of oxygen depletion downstream from a source of organic waste is the "oxygen sag curve" (32, 33). These patterns vary both diurnally and seasonally, and proper management of organic wastes requires that the nature of the variations be determined. This is especially true of the timing and duration of the least desirable conditions that might be encountered.

Organic compounds degrade naturally at rates that depend on the nature and amounts of the compounds, the kinds and numbers of bacteria involved, and variables like temperature and dissolved oxygen concentration. In natural systems all of these factors are dynamic. At the present state of knowledge their complex interactions can be modeled mathematically at a generalized level only. Thus an empirical method must be used to assess the effect of biodegradable organics on dissolved-oxygen concentration in flowing waters. The potential oxygen demand of the waste is estimated by laboratory testing. Also determined is the reaeration capacity of the water, the rate at which it can obtain oxygen from the air. The dissolved-oxygen levels are then estimated by difference.

The most widely accepted measure of the potential oxygen demand of organic substances is biochemical oxygen demand.* BOD is determined in the laboratory by measuring the amount of oxygen consumed in a fixed time and under prescribed conditions as microorganisms oxidize organic materials in a sample of waste. The standard test (34) is conducted for five days (hence BOD_5) at 20°C. Where domestic sewage only is involved, empirical equations have been developed that permit data from the standard BOD test to be extrapolated to other times and temperatures (32, 33). Similar conversions can be made for other organic wastes of fixed composition. They require, however, that relationships be established through a series of laboratory determinations for different times and under a sequence of conditions. Of two substances having identical standard BOD's, for example, one may have exerted its total oxygen demand in a matter of hours, while the other, such as domestic sewage, may have exerted only 70% to 80% of its total demand during the full five days of the test. Such differences clearly are important in estimating the effects of organic wastes on the oxygen balance of a particular water system. It should be noted also that BOD tests tell little or nothing of the fate of specific organic compounds that enter a stream or other body of water, including those that resist biodegradation in varying degree.

Because of the time required to determine BOD, methods have been worked out for estimating the oxygen-consuming potential of organic materials more quickly. The most widely used of these are COD (chemical oxygen demand) and TOC (total organic carbon). The COD test, which uses strong oxidizing agents, measures the equivalent oxygen demand of organic compounds that are biologically degradable (and of many that are not). The TOC test involves rapid combustion of carbon and measurement of the resulting carbon dioxide. Both the COD and TOC tests oxidize organics more completely than do microorganisms, and thus give higher oxygen demands than does the BOD test. For wastes of constant

*A more detailed explanation of biochemical oxygen demand is as follows (34): "The oxygen demand of sewage, sewage plant effluents, polluted waters, or industrial wastes is exerted by three classes of materials: (a) carbonaceous organic material usable as a source of food by aerobic organisms; (b) oxidizable nitrogen derived fron nitrite, ammonia, and organic nitrogen compounds which serve as food for specific bacteria (e.g., *Nitrosomonas* and *Nitrobacter*); and (c) certain chemical reducing compounds (ferrous iron, sulfite, and sulfide) which will react with molecularly dissolved oxygen. In raw and settled domestic sewage, most — and, for practical purposes, all — of the oxygen demand is due to the first class of materials and is determined by the biochemical oxygen demand (BOD) test. . . . In biologically treated effluents, a considerable proportion of the oxygen demand may be due to oxidation of Class (b) compounds and will also be included in the BOD test. Class (c) materials present may not be included in the BOD test unless the test is based on a calculated initial dissolved oxygen. It should be understood that all three of these classes will have a direct bearing on the oxygen balance of the receiving water and must be considered in the discharge of a waste to such a water.

"If wastes consisted only of raw or treated domestic sewage, measurement of the oxygen load on a receiving water would be simple. Unfortunately, this is not always the case, because most wastes are complex in nature and may contain organic compounds not easily amenable to biologic oxidation. When such compounds are present, the usual methods of seeding and the standard incubation period of 5 days will fail to assess the effect these wastes may have at some point below their point of discharge.

"Complete stabilization of a given waste may require a period of incubation too long for practical purposes. For this reason, the 5-day period has been accepted as standard . . . ."

chemical composition, however, quantitative conversion factors can be set up to relate BOD to oxygen demand determined by other means.

Reaeration of an aquatic system, the second key to dissolved-oxygen level, depends on three parameters: the rate of mixing of air and water; the deficit of oxygen (below equilibrium) in the water at the air-water interface; and the amount of oxygen transferred to solution as a result of photosynthesis. Reaeration capacity must be determined empirically; it can be estimated for a particular aquatic system by comparison with hydraulically similar systems where it has been approximated already.

Organics are not the only substances whose fate in water may be affected by biological activity. Living cells are composed of nitrogen, phosphorus, and other elements in addition to carbon. The extent to which these other building blocks are taken up from solution depends on both their relative proportions in organisms and their relative abundances in solution. The uptake, release, and transformation of materials from solution are governed also by biochemical reactions related to the flow of energy through living communities. Some elements, moreover, such as vanadium, are sequestered preferentially by certain aquatic organisms for reasons that are not yet understood.

At our present state of knowledge, then, the fates of a few materials in certain natural environments can be predicted with reasonable accuracy. More generally, the myriad biological interactions possible, and the complexity of aquatic systems, dictate an empirical approach to such questions. The difficulties are illustrated by the unexpected discovery in 1970 that bacteria in bottom environments can convert metallic mercury to soluble methylmercury — the latter can then be taken up by living organisms including fish. A second illustration is the extensive studies that have been needed to trace the movement of radionuclides through aquatic communities. These examples display both the incompleteness of our understanding and the complexity of defining quantitative pathways and rates of movement of materials as they pass through natural systems.

Particulate Matter

Particles are ubiquitous in natural waters, where they vary widely in concentration, in chemical and biological properties, and in size, shape, density, and other physical properties (35). Particles can adsorb and otherwise bind various substances, can serve as sites for bacterial growth, and can act in other ways that may strongly affect the transport of pollutants. Detailed data are not extensive on size distribution and properties of particulate matter in water. A promising start toward deeper understanding has been made, however, using centrifugal methods of separation (36). Such methods have been used to separate particles in natural waters into organic and inorganic fractions and to characterize them in terms of size, density, sedimentation rate, and other properties. The technique also can determine the distribution of a contaminant among the various fractions. (EPA and the U.S. Geological Survey, by agreement, treat particulate matter of less than 0.45 μm in diameter as dissolved, except for research purposes.)

The transport of contaminants by waterborne particles has been studied to a limited degree. Study of the transport of pesticides, for example, has been stimulated by the fact that low concentrations of some of them, such as DDT, are present in many surface waters of the U.S. Many pesticides are not very soluble in water and often are adsorbed (or absorbed) quickly by suspended or sedimented substances. A significant fraction of the pesticide residues in streams may be in continuous transport on suspended particles or in sediments (37). Sediments in lakes evidently can act as a reservoir from which a pesticide enters the water to an extent that depends on its solubility, its concentration in the sediment, the type of sediment, and the degree of adsorption (38). Notwithstanding the research on pesticides, however, the transport of pollutants by particulate matter is not a well-explored area.

Transport in Soil and Groundwater

Knowledge of the transport of substances in soil and groundwater is growing steadily more important for two reasons: the trend toward spreading dilute wastewater on land as a final step in treatment; and the increased use of underground space for emplacing especially troublesome wastes. A related and growing problem is pollution of groundwater, which normally is found in aquifers — highly permeable, subsurface strata of rock, sand, or gravel.

Intensive effort during the past several decades has provided enough knowledge to permit water to be applied experimentally on land to recharge fresh groundwater lying not too far beneath the surface. Where the applied water is treated wastewater, the physical, chemical, and biological processes that occur in the upper layer of soil constitute a final or "polishing" step in the treatment process (39–42). Much of what is known of groundwater recharge is empirical, however, and large gaps exist in our understanding of the processes involved. Under the proper conditions, soils are a reasonably good medium for the biological degradation of organic wastes and nitrates; soil particles remove other materials by adsorbing them or straining them from the water. But dissolved materials that are not readily removed by soil or rock particles may travel long distances in groundwater and are potential sources of pollution. The movement of the applied water into the ground, moreover, can be slowed by slimes that may form or by solids that plug the passages in the soil and rock. The solids may be introduced in the water or formed by its interactions with substances in the soil.

Natural underground space — the openings in soil and rock — has been termed a natural resource that requires much more thorough appraisal than has yet been made if it is to be used intelligently (43). One use for underground space is storage of useful fluids for later recovery. A second use, which in some locations may compete with the first for the available space, is permanent disposal of fluid wastes that are difficult to treat. The practice of injecting wastes into underground space, such as deep rock formations, is increasing rapidly; by the mid-1970's, this country had an estimated 300 documented waste injection wells. It is thus vital that we learn to monitor and predict the fate of the injected wastes, whose exact disposition and composition underground are often unknown under present practice. To monitor these wastes we must understand the movement of the groundwater that contains them. Such knowledge often is not available, is costly to acquire (44), and is inherently limited in the information it provides. Predicting the movement of the wastes requires even more detailed knowledge. Monitoring wells, furthermore, are expensive to install. Many states did not require in the past that underground wastes be monitored; they merely regulated the construction and operation of injection wells to avoid pollution of surface waters. This is changing rapidly, however, as the need for information for groundwater management becomes more clear.

Virtually all underground space is occupied by a fluid, such as groundwater, brine, petroleum, natural gas, or air. A gas or liquid injected from the surface displaces the original fluid, changing the pressure distribution around the injection well. Continued injection forces this pressure effect outward from the well at a speed determined by the rate of injection and the nature of the underground formation. Under certain conditions the pressure can displace rock formations. This may be done by design, as in the use of hydraulic fracturing to create additional underground space; or it may result accidentally from the injection of wastes into a seismically active zone. An example of the latter condition occurred in Denver, Colo., in the late 1960's. A series of tremors afflicted the area as a consequence of the injection of 150 million gallons of wastes into a 12,000-ft. well over a period of several years.

Groundwater is polluted commonly when liquid or solid wastes are disposed of or stored for a time on land. A notable offender is the unlined holding basin for liquid waste. The types of pollution caused by solid wastes, however, are not well defined as yet (45). Groundwater polluted from surface sources like these may reach the surface and contaminate surface waters.

The U.S. Geological Survey has compiled in the past few years more than 800 examples of groundwater pollution caused by human activities. They include instances of pollution from surface sources, from wastes emplaced underground, and by salt water drawn into groundwater aquifers when fresh water was withdrawn from them. As the nation turns more and more to groundwater as a primary or supplementary source of water, its protection grows ever more significant; in 1977, it was estimated that more than half the population depended on groundwater for its drinking water supply. And because groundwater moves as a rule very slowly, it is much more difficult than surface water to clean up once it is polluted.

Sinks for Waterborne Substances

Research on the sources, transport, and effects of water pollutants generally has taken precedence in the past over research on the sinks: oceans, lakes, their sediments, and deep underground formations. In the past decade or two the sinks, particularly the oceans, have begun to receive more attention. The oceans cover some 70% of the earth's surface. They constitute an enormous sink that contains at least traces of all of the naturally occurring elements (46).

The rivers of the world deliver an estimated 4 billion tons of dissolved matter annually to the seas (47), together with 20 billion tons of suspended sediment (48). These river waters contain an average of 120 mg/l of the major dissolved solids. Such solids include ions like bicarbonate, calcium, sulfate, chloride, sodium, magnesium, potassium, and nitrate. The oceans themselves average about 35,000 mg/l of dissolved solids. They contain in addition large amounts of dissolved gases and dissolved organic matter that ranges from 0.1 mg/l in isolated water masses at intermediate to great depth up to 10 mg/l near the shores.

The effects of contaminants on the seas and other sinks depend on an intricate web of factors such as currents, mixing processes, and geochemical and biological events. Biological activity, for example, can cause differences of up to 100-fold between one water mass and another in the concentrations of some elements, including oxygen, carbon, phosphorus, iron, silicon, nitrogen, and mercury. Chemical reactions occur unceasingly in the oceans at the interfaces between water and sediment and between water and air, and the compositions of all three phases are affected accordingly. Various phosphorus compounds in lakes may move from water to sediment and back again in accordance with equilibria that affect the supply of phosphorus for nourishing algae and other life forms. Clays and hydrous oxides of manganese and iron in river sediment may act to remove heavy metals like zinc from the water. The sinks, in brief, are part of the cycle followed by waterborne substances and, at the current level of knowledge, perhaps the least understood part of that cycle.

Marine disposal and pollution

The sheer size of the marine environment makes it potentially the biggest water pollution problem of all. The seas are a natural sink not only for billions of tons annually of dissolved matter and sediment, but for almost all the materials of industrial civilization as well. These inputs may change oceanic processes in ways that at times or in some places impair the utility of the water to man. Some such changes could be irreversible, moreover, and have consequences unpredictable today.

Wastes discharged directly to U.S. marine waters are estimated conservatively to exceed 50 million tons per year. About 80% of this amount is dredge spoils, 10% is industrial wastes, 9% is sewage sludge, and the remaining 1% is miscellaneous. The total is but a small fraction of waste production nationwide. The volume of wastes dischargeable to the seas is growing sharply, however, as a result of rising population, particularly on the coasts, and per-capita consumption rates. Generally tougher air and water standards, meanwhile, are creating pressure to divert additional wastes to the oceans.

Many of the marine pollution problems in the world today occur off the coasts near large urban areas. The problems include concentrations of toxic substances; rapid uptake of contaminants by marine organisms; heavy deposits of materials on the near-shore bottom (benthic) environment, if not on the beaches; and excessive growth of undesirable organisms. Such problems result in part from the relatively poor natural flushing in near-shore waters, but they stem primarily from high rates of input of wastes. Most of them can be remedied — or could have been avoided — by discharging the wastes farther at sea. Pollutants that enter the water at one point can always be carried elsewhere, but there is still good reason to introduce them at sites where the resulting damage, if any, will be minimized.

This approach to intentional discharge of wastes to the oceans can be summed up in terms of "allowable sites" and "acceptable materials." Less clear-cut is the scientific basis for implementing the pertinent regulations (49). Thousands of measurements and extensive research on oceanic processes have been undertaken in recent years to relate man's activities to the condition of the seas, but our understanding, especially of the open ocean, remains imperfect at best (50).

Allowable sites. The marine environment can be imagined to consist of four zones: the coastal zone, the surface zone, the open ocean, and the seabed. The selection of allowable discharge sites must be conditioned by the fact that events in these zones are mutually exclusive only in a gross sense; the zones interact unceasingly among themselves, both physically and chemically, just as the marine ecosystem overall interacts with the terrestrial and atmospheric ecosystems. Thus, although this maneuvering of waste inputs beyond the coastal zone is generally required under current regulations, the practice will not necessarily solve all marine pollution problems.

The surface zone may be an allowable discharge point for certain kinds of wastes. The primary consideration is that surface waters are the scene of important marine processes; they should not be contaminated with materials that interfere physically or chemically with the productivity and perpetuation of fish and other marine organisms. A second factor is incidental inputs of pollutants, as by atmospheric fallout and washout, spills and other discharges from vessels, and mixing with coastal waters. Materials from the deep ocean and seabed, transported by a variety of processes, also may enter the surface zone and congregate there for an indefinite period. And the need to avoid visual pollution of the surface zone further affects its suitability as a discharge site.

The seabed beyond the coastal zone might well be an allowable discharge site in some circumstances. It is a natural point of accumulation for many materials regardless of man's wishes. Substances deposited on the seabed, furthermore, may remain in one place for comparatively long periods, thus providing a degree of waste management unavailable when contaminants are distributed in the surface zone of the open ocean. It may be possible, in addition, to constrain barge and pipeline discharges to the point where they approximate a direct input to the seabed. For the most part, however, materials in the coastal zone, surface waters, or deep open ocean interact to a degree before they accumulate on the seabed. The time spans of these phenomena vary greatly, ranging from minutes to possibly years, depending on the extent of chemical and biological transformation and reutilization. The seabed also is the scene of chemical and biological processes that might be impaired by premature accumulation of inordinate amounts of wastes. The biological processes are thought to be more concentrated toward shallower regions, but relatively little is known of such processes in the bottom waters of the deep ocean. An effort to inventory materials discharged to the seabed might provide the options of recovering or covering them should surveillance show the practice to be undesirable. A delicate balance is involved, on the whole, in choosing allowable discharge sites on the seabed. The technological limits imposed by great depths tip the balance one way; the need to avoid the active and sensitive coastal zone tip it the other.

The open ocean — locations other than the coastal zone, surface zone, and seabed — may be an allowable discharge point for a number of waste materials. Its advantages include opportunities to disperse wastes in a large mass of water and long travel and reaction times before the wastes infringe on the coastal margins. Long reaction times may be important to successful use of the open ocean because biodegradation and other waste-stabilization processes may be slowed noticeably by the low temperatures, high pressures, and physical separation from the chemically and biologically active boundaries of the sea.

Acceptable materials. The acceptability of materials for discharge at allowable sites depends on their behavior in the marine environment. Their behavior, in turn, depends in greater or lesser degree on which of the four zones of the ocean is chosen for disposal. This assessment is costly and difficult, where it can be made at all, so that acceptability often is determined by laboratory simulations of one or more facets of a given material's behavior. A frequently used test of this sort is the median lethal level as found for particular organisms. Such tests are imprecise, however, and tell nothing of the material's other effects on the health of the marine community at hand. The results are used primarily, therefore, as a screen to preclude the discharge of acutely toxic materials. Those that do not display median toxicity during the period of the test, usually 96 hours, or that can be diluted to below the toxic level at a "reasonable" distance from the point of discharge, may be judged to have passed the screen. That judgment does not guarantee, however, that such substances will not harm the marine environment. Two of the most important factors that must be assessed in addition to a material's acute toxicity are its persistence and its tendency to build up in the food chain by the process called bioaccumulation.

Persistence is the degree to which a material resists biological or other forms of stabilization. A substance that is stabilized readily might be acceptable for discharge in any coastal location; one that persists for a long time not only requires more thorough predischarge evaluation, but also may be restricted to a single discharge site. The persistence of materials can be categorized conveniently — if quite arbitrarily — at three levels of half-life: less than 48 hours; 48 hours to six months; and more than six months. Materials with long half-lives clearly must be tested more elaborately than those with short half-lives; if they are found to be marginally acceptable, or dangerous, they must be dealt with differently. The products of degradation or stabilization also must be evaluated to determine the acceptability of the parent waste (although present marine pollution regulations do not all make the point specifically). Plant and animal matter and wastes are examples of short half-life materials. The fraction of these primarily organic substances that is refractory — and survives municipal waste treatment or periods of residence in reservoirs — might fit the medium half-life category. The synthetic organic compounds made in petrochemical plants include many that might fit the long half-life category. "Persistence" in the same sense does not apply to inorganic compounds, but the concern similarly is to determine the lifetimes of the compounds (or their ions) in the marine environment.

The importance of bioaccumulation to acceptability is evident in the familiar (and persistent) chlorinated hydrocarbon pesticides. These compounds, discharged from an ocean outfall, for example, can be diluted to below the lethal concentration in the short time required to travel several thousand feet. But over long periods certain of the compounds or their breakdown products can accumulate in the food chain to the point where, apparently, they harm marine organisms. The same food chain extends to marine organisms for human consumption. The regulations on the acceptability of materials for ocean dumping recognize bioaccumulation as a screening mechanism in addition to tests of lethal toxicity. But at the time the pertinent regulation was promulgated, late in 1973, no standard technique was available for estimating bioaccumulation in marine organisms.

Although current federal regulations strictly prohibit the discharge to marine waters of a very few materials, such as mercury, cadmium, and chlorinated hydrocarbons, they recognize that small amounts of such substances may be discharged. They could be dumped inadvertently, for example, in wastes collected from municipal and industrial operations. It was necessary, therefore, to establish "acceptable amounts" of these materials. Thus the amount or concentration of a material, either in raw waste or marine receiving waters, however arbitrarily determined, is now integral to the determination of the material's acceptability for discharge.

It was noted earlier that the 1977 amendments to the Federal Water Pollution Control Act modified the requirement for secondary treatment of municipal wastewaters for facilities discharging to marine waters. Specifically, under the amendments it may be possible for coastal communities, in certain cases, to postpone or bypass conventional biological secondary treatment if they can demonstrate achievement of specified water quality criteria and maintenance of desired water uses, including propagation of indigenous fish, shellfish, and wildlife. With the approval of EPA and the state, the discharger may obtain a permit modification for up to five years. During that period, the discharger must monitor to demonstrate continuously that the partially treated wastes are not harming the marine environment. This situation provides an excellent opportunity to collect the research information called for elsewhere in this report in order to improve the scientific base for regulating waste discharges to the marine environment.

Marine research goals. Continuing research will be required to support and improve marine pollution control activities and strengthen the relevant legislation. Essential to this effort is a global marine-water-quality monitoring program. Only through such a program can we assess the merit of existing pollution controls and formulate new ones, if indeed they are needed. The research tasks of a global monitoring scheme should be distinguished from those of coastal monitoring related to day-to-day implementation of regulations. A global program requires improved analytical chemical techniques, certainly, but the ultimate goal is to learn to assess and describe ecological change, if it occurs, in quantitative terms. Ecosystem studies, in both prototypal and simulated environments, must be pursued to relate detrimental or beneficial changes to the forms and concentrations of contaminants in the system. These cause-and-effect studies will lead to improved and more stringent effluent requirements, which can be based only on scientific understanding of the water quality necessary to protect marine life. This understanding in hand, research can be delineated to improve today's waste treatment technologies, which were developed with objectives in mind other than protecting the marine environment.

Recommendation W1: *Regional inventories should be made of pollutants from all sources that are known or expected to be important as quickly as data from the National Pollutant Discharge Elimination System and other relevant programs can be made available. The current effort to identify the specific chemical compounds, particularly organic compounds, present in both wastes and natural waters should be pressed vigorously. Chemical and biological research on natural waters, polluted or not, should be emphasized.*

Recommendation W2: *In implementing Recommendation W1, a particular effort should be made to identify and compile data specifically relevant to sources and potential sources of public water supplies. This effort should be organized administratively so that individuals responsible primarily for research on sources of public water supplies are directly involved.*

Recommendation W3: *Fundamental research should be expanded on the action of natural mixed populations of bacteria and other organisms on specific compounds. Such research will require the development of analytical methods for identifying and quantifying the specific compounds produced by biological degradation. More and more, the collective chemical parameters now in common use will prove inadequate for understanding the behavior of complex natural systems.*

Recommendation W4: *Comprehensive investigations of naturally occurring and pollutant particles in water should be undertaken to determine such* *characteristics as size, charge, composition, and adsorptive properties. Expanded knowledge of particles would be important in studies of sedimentation, erosion, and certain waste treatment processes, as well as in work on transport.*

Recommendation W5: *Research on improved mathematical descriptions of natural water systems subject to pollution should be strongly supported. The chemical-biological complexity of those systems requires that such research be highly interdisciplinary, involving scientists from disciplines such as chemistry, chemical engineering, civil engineering, biology, and ecology. Sheer complexity may ultimately limit the generality of the mathematical descriptions that can be developed, but it is essential to find a proper balance between the difficulty of understanding natural water systems and the need to describe them mathematically.*

Recommendation W6: *Systematic studies should be encouraged on the flow and reactions of forms of phosphorus and nitrogen, and various organic substances, in soil and groundwater.*

Recommendation W7: *More emphasis should be placed on investigation of the transport and long-term deposition of pollutants in the oceans, lakes, and deep underground formations. The initial requirement is improved analytical methods for identifying and measuring specific chemical compounds.*

MUNICIPAL WASTEWATER TREATMENT

Historically the main goal of treating municipal wastewater has been to reduce its content of suspended solids, oxygen-demanding materials, and bacteria before discharging the water to the environment. More recently this goal has expanded to include reductions in dissolved inorganic compounds, particularly compounds of phosphorus and nitrogen, and in potential toxicants. Recent years also have seen more stress on improving the means of ultimate disposal of the solid residues from municipal treatment processes.

The basic principles and technology of the unit operations and unit processes used in treating municipal wastewater have been well established through years of practice. Refinements in design and improved operation are required, however, to meet the more stringent standards of today. Besides domestic sewage, municipal wastewater usually contains storm-water drainage, commercial wastes, and frequently industrial wastes. The domestic component generally predominates, but the quantities of commercial and industrial wastes may be large enough to warrant special consideration in the design of treatment processes. The removal of contaminants like dissolved inorganic compounds, for example, will call increasingly for the use of processes described in the section on Physicochemical Processes and Advanced Waste Treatment. Such methods in general will reinforce, not replace, the basic methods used now.

Basic Primary and Secondary Treatment

The basic methods of treating municipal wastewater fall into two classes:

- Primary treatment: grit removal, screening, grinding, flocculation, sedimentation.
- Secondary treatment: biological oxidation, using processes such as activated sludge and the trickling filter.

The conventional wastewater treatment scheme of Figure 1 is subject to a number of variations, but normally the process starts with removal or comminution (pulverization) of large solids and removal of grit. Gravity sedimentation in the primary clarifier removes settleable solids and grease and scum as primary sludge. Clarified wastewater from primary treatment goes to secondary treatment. There, microorganisms carry out in a controlled manner the assimilation and degradation process that breaks down organic matter (just as it is broken down in nature).

In the activated sludge process, a widely used secondary treatment, the wastewater is aerated to supply oxygen for the microorganisms. The solids or activated sludge formed in the process are removed by sedimentation, and the clarified effluent is discharged to the receiving waters. Some activated sludge, consisting principally of active microorganisms, is returned to the aeration tank and mixed with incoming wastewater to serve as an inoculum or starter culture. The excess activated sludge must be disposed of. As a final step, the treated water may be chlorinated to destroy pathogenic organisms.

In the trickling filter process, another means of secondary treatment, wastewater from the primary sedimentation basin is distributed over a bed of coarse material such as broken stones, slag, or synthetic media. A zoogleal slime — a gelatinous mass created by growing bacteria — forms on the bed material, and organic compounds dissolved in the wastewater are oxidized as they trickle down through the bed. The effluent is collected in an underdrain system and flows to the final clarification tank, where particulate solids that slough from the filter are removed, usually by gravity separation. The thickness of the layer of bacterial slime is also controlled by the grazing of protozoans and insect larvae, especially in trickling filters that are operated at low hydraulic loads.

Various modifications of the conventional activated sludge and trickling filter processes also are used to treat municipal wastewaters. However, the basic principle of aerobic biological oxidation applies to all such modifications.

Other methods used to treat wastewaters are aerated lagoons, with diffused air or mechanical aeration systems, and waste stabilization ponds in which bacteria and algae in symbiotic relationship oxidize organic material. Wastewater must be retained for relatively long periods in aerated lagoons and stabilization ponds, which thus require more land than the activated sludge or trickling filter processes; performance also is affected significantly by climatic conditions, particularly sunlight, temperature, and wind. Ponds and lagoons commonly are used by small communities and account for some 25% of all publicly owned wastewater treatment works. Although they are generally classed as secondary treatment, few can meet the corresponding effluent limitations on suspended solids. As a result, EPA in late 1977 adjusted these limitations for ponds and lagoons so as not to compel small communities unreasonably to switch to more complex and expensive treatment technology.

The inventory of municipal sewage treatment works in the U.S. for 1968 showed some 140 million people living in communities served by sewer systems; about 6.8% of the wastewater from sewered communities was untreated. Estimates by EPA show that wastewater treatment kept pace with the rise in the population served by sewers during 1968–72 (Table 7). More rapid progress is required under the effluent limitations imposed on publicly owned treatment plants by the Federal Water Pollution Control Act Amendments of 1972. In general, the effluent limitations are based on conventional secondary treatment. EPA studies indicate that as of Jan. 1, 1976, about 36% of the population of the

FIGURE 1

Basic primary and secondary treatment scheme[a]

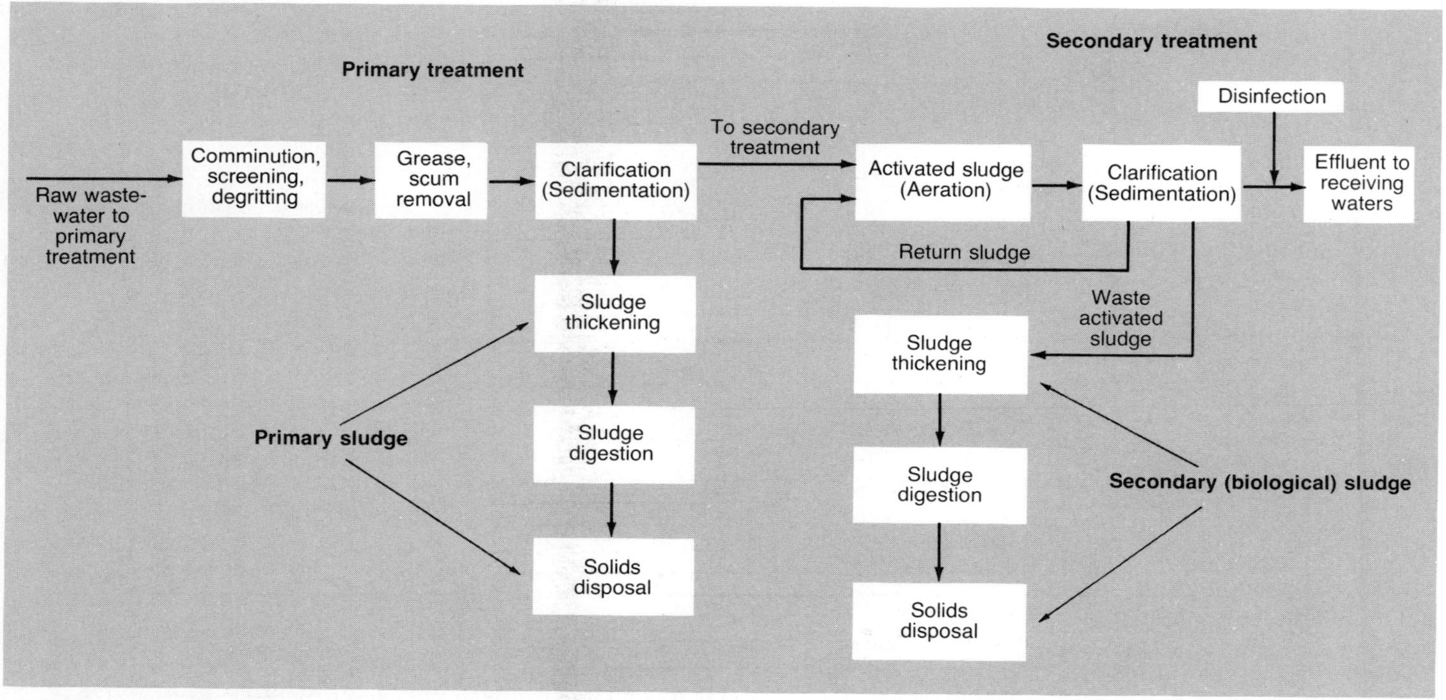

a. Treatment systems for municipal wastewaters are subject to modifications, such as the use of the trickling filter instead of the activated sludge process in the secondary treatment section of the scheme shown above. All secondary systems, however, rely on aerobic biological oxidation of organic matter.

Table 7

Wastewater treatment for population served by sewers

	Fiscal year				
	1968	1969	1970	1971	1972
Population served by sewers (millions)	140	144	148	152	156
BOD wastes treated by municipal plants (billions of pounds)	14.1	14.8	15.4	16.1	16.9
Level of treatment (Percent of population served by sewers):					
Untreated	7	7	6	6	5
Primary	31	30	28	25	24
Secondary	62	63	66	68	70
Advanced	<1	<1	<1	<1	<2

Source: "Research and Demonstration Programs to Achieve Water Quality Goals: What the Federal Government Needs to Do," Comptroller General of the United States, B-166506, Washington, D.C., 1974.

states and territories was served by secondary treatment or better (51). A special analysis of the EPA data showed that 83% of the population of 25 municipal areas was receiving sewage treatment services, but only 58% of the total wastewater flow was receiving secondary treatment or better (7). The 25 areas were scattered across the nation and had a total population of 22.4 million (1970 census).

An important aspect of the effluent limitations is their potential effect on the costs of treatment. The approximate performance of conventional treatment of municipal wastewaters appears in Table 8; the cost of treatment in relation to the percent of BOD and COD removed is shown in Figure 2. The figure shows in part that the cost of removing the approximately 10% of the COD that remains after secondary treatment is about twice the cost of secondary treatment itself. At locations where secondary treatment may prove unable to meet water quality standards, advanced biological or physicochemical processes will be required. EPA has estimated from 1976 data that to meet overall requirements for the population expected in 1990 would require capital expenditures of $34.3 billion on treatment plants alone (51). An additional $61.6 billion would be required for related construction, including sewage collection facilities, and another $54.1 billion for storm-water control facilities.

The treatment and related requirements of current federal law should stimulate not only the construction, expansion, and upgrading of wastewater treatment works; they should also stimulate research and development directed toward treatment processes that perform better and at lower cost than those used now. Municipalities in addition must find practical controls for pollutants from the final disposal of sewage sludge and from combined sewers and storm waters.

Sludge handling and disposal

Handling and disposing of sludges are the most troublesome aspects of wastewater treatment. Often they account for 25 to 50% of the capital and operating costs of a treatment plant (52). Production of municipal treatment sludges exceeds 100 million tons annually and is rising with the move toward a

Table 8

Performance of conventional municipal wastewater treatment processes

	Efficiency of treatment percent removal[a]	
	Primary	Primary plus secondary
Biochemical oxygen demand	32%	90%
Chemical oxygen demand	36	86
Suspended solids	50	90
Total nitrogen	20	40
Total phosphorus	10	30

a. The values shown may vary widely with the compositions of the wastes in specific localities. A medium-strength municipal wastewater would have a five-day biochemical oxygen demand (BOD_5) of about 200 mg/l; treatment by the activated sludge process would yield an effluent containing about 20 mg/l of BOD_5.

minimum of secondary treatment nationwide. Restrictions on ocean disposal and the increasing costs of labor and land required for land disposal are making the situation even more difficult.

The magnitude of the problem can be seen in the operations (1976) of the Metropolitan Sanitary District of Greater Chicago, which serves 5.4 million people plus a nondomestic equivalent of 4.5 million. Four activated sludge treatment works process some 1.3 billion gallons of wastewater per day and produce more than 670 tons of solids daily on a dry basis. Solids disposal costs about $21 million per year or 38% of total operating and maintenance costs.

FIGURE 2

Percent of pollutants removed and related costs

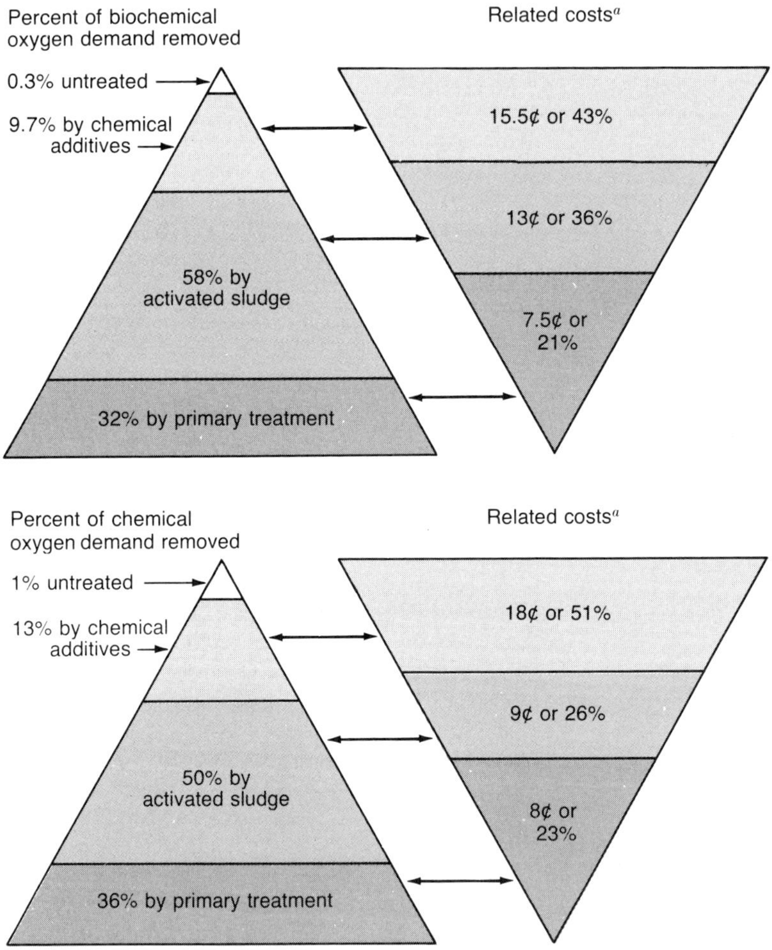

a. Cost per 1000 gal of sewage treated.

Source: "Research and Demonstration Programs to Achieve Water Quality Goals: What the Federal Government Needs to Do," Comptroller General of the United States, B-166506, Washington, D.C., 1974.

Sludge is of two general types: primary sludge, containing the raw wastewater solids removed in the primary sedimentation basins; and secondary or biological sludge, containing mainly the excess microorganisms produced in the activated sludge and trickling filter processes. Both kinds of sludges contain pathogens, numerous organic compounds, and many of the chemical elements, including toxic elements (53). The volumes and solids contents of sludges produced normally by municipal treatment plants are shown in Table 9. The excess microorganisms in biological sludges contain water internally and flocculate into a structure that has a strong affinity for water. It is thus difficult and costly to dewater biological sludges for ultimate disposal. Costs are particularly high for the activated sludge process because of the large volume of sludge produced. Even more sludge is produced when treatment levels are upgraded beyond secondary treatment or where phosphorus must be removed. Advanced treatment processes involve addition of lime or salts of aluminum or iron. The yields of sludge from primary, secondary, and tertiary phosphorus removal systems appear in Table 10.

Sludge handling and disposal have four general objectives:

- To convert organic matter to a relatively stable form.
- To reduce the volume of the sludge by removing liquids.
- To destroy or control harmful organisms.
- To obtain by-products whose use or sale reduces the overall cost of processing the sludge.

To achieve these goals, treatment works typically use some or all of a series of processes: concentration or thickening, stabilization, conditioning, dewatering, heat drying, combustion, and final disposal.

Thickening. Liquid sludge is concentrated or thickened to reduce the volume to be handled in subsequent processes. If a given amount of activated sludge is thickened from 1% solids to 2% solids, its volume is cut in half, with consequent savings in several areas. Pumping and transport cost less; tanks can be smaller; lesser amounts of chemicals are required for conditioning; and less heat is required for digestion, drying, and combustion (54). Sludges can be thickened by gravity, by flotation, or by centrifugation.

Gravity thickeners are circular tanks to which raw primary or activated sludge is fed continuously. The solids settle slowly, and the concentrated sludge toward the bottom of the tank is agitated gently by a rotating mechanism to dislodge gas bubbles and prevent bridging of solids. Liquor from which the solids have settled is removed over a peripheral weir at the top of the tank, and concentrated sludge is withdrawn from the bottom of the tank. Gravity thickeners are designed to process 500 to 800 gallons per day (gpd) of raw sludge per square foot of horizontal cross section; solids loadings range from 5 to 15 $lb/day/ft^2$ for mixtures of primary and activated sludges to 20 to 30 $lb/day/ft^2$ for primary sludge only (55).

Flotation thickening may be acomplished by dispersed air, dissolved-air vacuum flotation, or dissolved-air pressure flotation (DAF). DAF is the most widely used flotation method for thickening activated sludge. The process consists of mixing air and a liquid stream — such as flotation tank, primary, or plant effluent — under pressures of 40 to 80 psi, providing a brief time for the air to go into solution, and releasing the pressurized stream into a tank with the activated sludge at atmospheric pressure. The release of pressure causes air to leave solution and form minute bubbles that attach themselves to and are enmeshed in the sludge flocs. The bubbles carry the solids upward to form a sludge blanket at the surface of the liquid in the flotation tank. The thickened sludge is skimmed off mechanically, and the effluent water is discharged from below the sludge blanket. Dissolved-air flotation of activated sludge usually yields sludge concentrations of 4 to 6%. Certain organic compounds (cationic polyelectrolytes) are used sometimes in air flotation systems to encourage flocculation and thus to aid in thickening and solids removal. Important variables in air flotation systems include the type and quality of sludge, pressure, recycle ratio, feed solids concentration, detention period, air-to-solids ratio, solids and liquid loading rates, and the use of chemical aids. To establish design criteria and predict performance for such systems these variables should be studied when possible both in the laboratory and in pilot plant.

Centrifugation for thickening sludge has seen only limited use compared with gravity and dissolved-air flotation. It has been used primarily where space limitations or sludge characteristics make other methods unsuitable. Solid base, disc, and basket centrifuges have been used, depending on the application (55).

Stabilization. Sludge is stabilized, by anaerobic or aerobic digestion, to reduce its volume, to decompose organic solids, to make the treated sludge less odorous, and to reduce the number of pathogenic organisms present. Anaerobic digestion, followed by partial dewatering of the digested sludge and disposal as landfill or soil conditioner, is a relatively low-cost means of handling sludge and continues to be used widely. Aerobic digestion, however, is replacing the anaerobic method more often than in the past for stabilizing municipal wastewater treatment sludges.

In anaerobic digestion, methane-producing bacteria play the dominant role, but also involved are a large group of anaerobic acid-forming bacteria and facultative bacteria (which can adapt to both aerobic and anaerobic conditions). Digestion is carried out mainly by two groups of organisms: the mesophilic group, which grows best at relatively low temperature; and the thermophilic group, which grows best at relatively high temperature. Optimum temperature ranges from 85° to 100°F for mesophilic digestion and 115° to 140°F for thermophilic digestion, so that digesters usually must be heated to operate efficiently. The optimum pH in both cases is 6.5 to 7.5, and the methane-producing bacteria essentially cease to function below pH 6.0. Heavy metals, such as lead, copper, chromium, and zinc, interfere with the digestion process.

The anaerobic gasification and liquefaction of the sludge yields about 15 ft^3 of gas per pound of volatile solids destroyed. About two thirds of the gas is methane; the rest is mainly carbon dioxide, with small amounts of hydrogen,

Table 9

Volumes and solids contents of sludges from municipal treatment

Wastewater treatment process	Percent moisture	Pounds dry solids per million gallons of sewage	Gallons of sludge produced per million gallons wastewater treated
Primary sedimentation	95	1250	2950
Trickling filter	92.5	476	745
Activated sludge	98.5	2250	19400

Source: "Wastewater Engineering," Metcalf and Eddy, Inc., McGraw-Hill Book Co., New York, N.Y., 1972. Copyright 1972 by McGraw-Hill Book Co.

Table 10

Additional sludge to be handled with chemical treatment systems

Treatment process	Percent moisture (mean values)	Pounds dry solids per million gallons (mean values)	Gallons of sludge produced per million gallons (mean values)
Lime addition to primary influent 350–500 mg/l	88.9	5630	8924
Aluminum addition to primary influent 13–22.7 mg/l	98.8	1323	23000
Iron addition to primary influent	97.75	2775	21922
Al^{+++} addition to aerator 9.4–23 mg/l	98.88	1180	13477
Fe^{+++} addition to aerator 10–30 mg/l	98.7	1705	18650
Lime addition to secondary effluent 268–450 mg/l	98.9	4650	53400
Aluminum addition to secondary effluent 16 mg/l	98	2000	12000
Iron addition to secondary effluent 10–30 mg/l	99.7	507	22066

Source: "Process Design Manual for Sludge Treatment and Disposal," U.S. Environmental Protection Agency, EPA 625/1-74-006, Washington, D.C., 1974.

nitrogen, and hydrogen sulfide. The gas has a heating value of 600 to 650 British thermal units (Btu) per cubic foot (60 to 65% of that of natural gas, which is mostly methane); it can serve as a fuel for boilers and for internal combustion engines that drive pumps, blowers, and electric generators. Although most treatment plants use sludge gas to heat the digesters and for similar purposes, the nation's energy problem has revived broader interest in the gas as a source of energy. Municipal plants using anaerobic digestion produce 0.6 to 1.0 ft^3 of gas per capita per day.

Despite the long history of anaerobic digestion, operational problems still arise, and better analytical tools are needed for controlling digester operation. The design criteria for anaerobic digesters have remained essentially unchanged for more than three decades with the exception of the introduction of controlled mixing.

Aerobic digestion, the second common stabilization method, resembles the activated sludge process with extended (in time) aeration. Organic primary sludges and waste activated sludge, in combination or separately, are aerated for 15 to 20 days in unheated reaction tanks. Microorganisms oxidize (metabolize) the biodegradable organics to produce cellular material that itself is oxidized in an "endogenous respiration" phase. The end products are carbon dioxide, water, ammonia, nitrate nitrogen, and a residue of inert compounds and organic solids that are not biodegradable under the conditions in the reaction tank. Aerobic digestion has certain advantages over the anaerobic method. The process is relatively simple to operate, has lower capital cost, produces a supernatant of lower BOD, does not produce objectionable odors, and yields a residue sludge that dewaters readily. Operating costs are higher than for anaerobic digestion because of the need for air or oxygen. Another consideration is that aerobic digestion does not produce by-product methane.

Additional research and plant operational data are needed to clarify the process design criteria for aerobic digestion. These criteria include the amount of air or oxygen required, residence time in the reaction tank, sludge loading values, the overall kinetics of the process, and costs.

Municipal wastewater sludges also may be stabilized by composting, either alone or mixed with municipal solid waste. The process has not been used widely in this country, however, principally because of the problem of marketing the resulting compost (55).

Sludge conditioning. Sludge is conditioned to improve its dewatering characteristics. It may be conditioned by elutriation and chemical addition, by heat treatment, or by both. Elutriation — mixing or washing the sludge with water — decreases the alkalinity of anaerobically digested sludge; this in turn decreases the amount of chemicals needed to destabilize the colloidal suspension of particles in the sludge so that flocculation can occur. Chemicals used in conditioning sludge include ferric chloride and ferrous sulfate, which act primarily as coagulants, and alum, lime, and organic polyelectrolytes. The polyelectrolytes cause flocculation by neutralizing surface charges on the dispersed particles. Heat treatment at 300° to 500°F and at pressures of 150 to 300 psi breaks down the gel structure of the sludge to yield water and residual solids. The biological cells are disrupted to release protoplasm. Proteins and zoogleal slimes are broken down and in part solubilized, leaving a residue of mineral matter and cell-wall debris that dewaters rapidly without chemical addition.

Dewatering. Of the methods used to dewater sludge, to reduce it from a fluid to a nonfluid form, drying on sand beds has been the most common. The increasing costs of land and labor, however, are leading to wider use of mechanical dewatering devices such as vacuum filters, filter presses, and centrifuges.

Vacuum filtration has been used for some years to dewater sludges, although it fell from favor between World War II and about 1960 because of operating problems and costs. Since 1960, vacuum filtration has come back strongly because of improved filter media, higher costs for competing methods, and the growing popularity of sludge incineration. Operation of vacuum filters depends on the nature of the sludge and the conditioning given it before it is applied to the filter. Many types of filter cloth are available, and the proper type should be selected by conducting filter leaf tests.

Plate and frame filter presses operate on a batch basis and have been used more widely in Europe than in the United States. Improvements in filter media and degree of automation, however, have led to more frequent consideration of filter presses for use with municipal sludges in this country. Other types of devices in use are moving screen concentrators, felt pressure filters, capillary dewatering systems, and rotating gravity concentrators.

The centrifuge also has been seeing wider use in the past decade for dewatering sludges, primarily because of improved design based on the solid bowl centrifuge. The scroll centrifuge is used most widely, but for special applications both the basket and disc centrifuges are proving successful. Although vacuum filtration remains the predominant mechanical dewatering device, the centrifuge has certain inherent advantages, including low capital cost, moderate operating cost, and low space requirement. The chief problem with the centrifuge is that the centrate (the liquid that it returns to the wastewater treatment system) often contains undesirable amounts of fine solids that can build up in the system. The problem is greater with the biological sludges from secondary treatment. Chemical flocculants can improve the quality of the centrate, but their use can increase operating costs significantly. Centrifuged sludge cake usually ranges from 6 to 35% solids, depending on the nature of the feed sludge, feed rate, bowl design, bowl speed, and chemical additions. Whenever possible, pilot tests should be conducted to determine optimum feed rate, solids capture, cake concentration, and polymer addition.

A number of unconventional methods have been studied for improving the dewatering characteristics of sludges. They include freezing and thawing, gamma irradiation, solvent extraction, electrical treatment, ultrasonic treatment, and treatment by bacteria. The aims of such work generally have been to increase the rate of production of dewatered sludge and to eliminate the need for conditioning chemicals. Thus far, however, none of these methods has shown sufficient promise to warrant application. Freezing and thawing improves the dewatering characteristics of sludge, evidently by breaking down the cell walls in the organisms that retain internal moisture. A small plant was built in England to condition water-treatment plant sludge by freezing, but the operating costs appeared to be impractically high for wastewater sludges.

Drying, combustion. Thermal means of sludge treatment include drying, incineration, and wet oxidation. Several wastewater treatment plants have used heat to dry activated sludge for use as a low-grade fertilizer. After mechanical dewatering, usually by vacuum filtration, the filter cake is blended with previously dried sludge and moves through a drum or kiln dryer along with hot gases to produce a sludge (fertilizer) that contains 8 to 10% moisture. The process is more expensive than incineration and because of limited demand for the product has not found widespread application. Interest is growing, however, in using sludge as fertilizer, and if the costs of heat drying can be reduced, sale of the dried product may make this process more attractive.

Incineration serves two main purposes: to reduce the volume of solids for final disposal, and to sterilize the final product. Incineration usually is the final step in a system that includes thickening and dewatering. The heating values of sludges vary from about 5300 Btu per pound of dry solids for anaerobically digested sludge and 10,000 Btu per pound of raw solids to 16,700 Btu per pound of grease and scum. Sludge with a moisture content of about 75% is fed to the incinerator. With a sludge that contains 3 lb of water for each pound of dry solids having a volatiles content of 75%, the heat required to evaporate the water nearly balances the heat from combustion of the dry solids. Where the sludge is so composed that evaporation requires more heat than is available from combustion, auxiliary fuel can be provided. The multiple hearth furnace and the fluidized bed incinerator are the major systems used (55). Air pollution control devices and ash handling facilities are necessary parts of an incineration system.

The wet oxidation process involves oxidation of wet sludges at elevated pressure and temperature. The wet sludge is ground to about ¼-in pieces and pumped under pressure with air from a compressor through a series of heat exchangers to a reaction vessel. As oxidation occurs in the reactor, the temperature rises. The mixture of gases, liquid, ash, and oxidized products is cooled in the heat exchangers and enters separators where the gases and solid residue are separated from the liquids. The solids are readily dewatered, and the liquid is returned to the treatment process. The wet oxidation process can operate at pressures of 150 to 3000 psi and at temperatures depending on the degree of oxidation to be achieved.

Final disposal. Disposal of liquid sludges on land, in lagoons, or in the ocean are the cheapest methods of disposing ultimately of sewage sludge, providing the treatment plant is within economical range of suitable land or the ocean. (Ocean disposal will be decreasing or, more likely, banned under current legislation and regulations.) Disposal by any of these means requires normally that the sludge be treated by one of the methods described previously to avoid problems with odors, insects, and water pollution. Liquid sludges commonly are spread on land, particularly by smaller treatment plants. The sludge contains nutrient values and can be used to fertilize or condition soil and to reclaim wasteland for uses such as parks or agriculture. Landfill is used to dispose of dewatered sludges of all types. Sludge can be transported to the point of disposal by truck, train, pipeline, or barge.

Despite the utility of these methods of sludge disposal, they are feeling the effects of changing economics, improved technology, and increasing urbanization. The majority of waste treatment plants being designed today in the U.S. are based on one of three general processing schemes:

- Dewater digested sludge mechanically and use it for landfill.
- Dewater digested sludge mechanically and dispose of it by thermal means, such as incineration.
- Dewater raw sludge mechanically and dispose of it by thermal means.

Sludge-handling options incorporating the foregoing schemes appear in Figure 3. Thermal disposal or combustion has good potential, because it seems likely to be able to cope with the sludge disposal problems of the future. The most effective means of sludge disposal, however, are determined by local conditions. The ability to evaluate alternative disposal schemes soundly requires thorough investigation of such questions as the value of liquid, dried, or

FIGURE 3

Options in handling municipal wastewater treatment sludges[a]

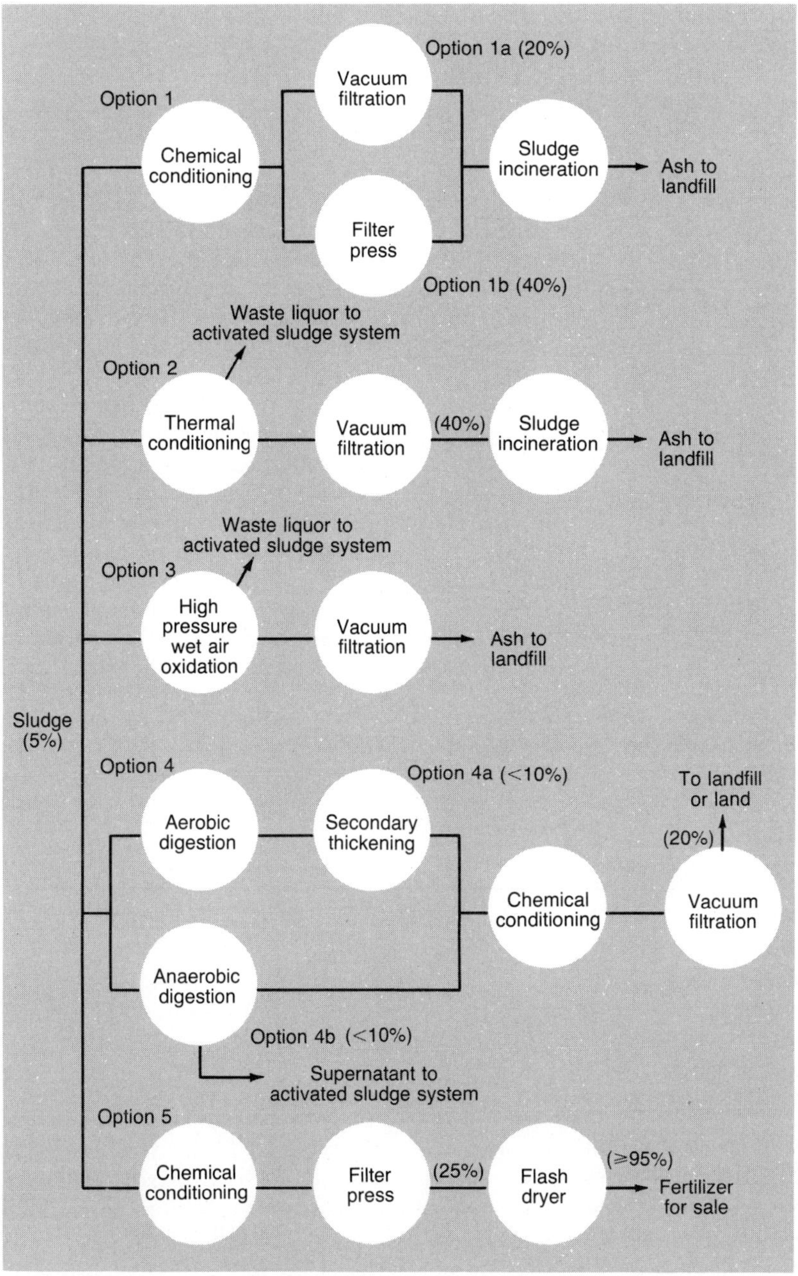

a. Percentages indicate weight-percent solids in sludge.

Source: Jones, J.L., et al, "Municipal Sludge Disposal Economics," *Environ. Sci. Technol.*, **11**, 968 (1977).

composted sludge as a fertilizer or soil conditioner; underground disposal as in abandoned mines; and land reclamation. Disposal on land, including use as a fertilizer or soil amendment, further requires consideration of the potentially adverse effects of leachate or runoff on ground and surface waters, uptake of cadmium and other toxic elements by crops, and similar issues (56, 57).

Urban Storm Water

Early sewerage systems often used sewers that carried storm-water runoff and sanitary sewage in the same conduit. These combined sewers were less costly than separate sewers for sanitary sewage and storm water. As it became necessary to treat sanitary sewage to protect receiving waters, the separate system became more common. In urban areas where combined systems existed and where treatment was provided for the sanitary sewage, combined flows during storms could exceed the capacity of the treatment plant, and overflow structures were used to divert the excess to the receiving stream. By 1967, more than 1300 municipalities with a population of 54 million were served by some 55,000 miles of combined sewers (58) containing some 20,000 overflow points (59).

The discharge of untreated sewage by combined systems is estimated generally to be about 3% of the amount that enters the system; it can be markedly higher than that during heavy precipitation. In addition to the potential pollution problems associated with untreated combined sewer overflows, storm water discharged directly from separate storm drains also contains significant amounts of BOD, COD, settleable solids, nutrients (nitrogen and phosphorus), toxic pollutants, and microorganisms (Tables 11, 12). The results of a study of several municipalities indicated that as much as 40 to 80% of the total annual BOD and COD entering receiving waters came from sources other than the treatment plants. One such source was storm runoff from urban land areas,

Table 11

Characteristics of combined sewer overflows[a]

Characteristic	Range of values
BOD_5 (mg/l)	30–600
Total suspended solids (mg/l)	20–1700
Total solids (mg/l)	150–2300
Volatile total solids (mg/l)	15–820
pH	4.9–8.7
Settleable solids (ml/l)	2–1550
Organic nitrogen (mg/l)	1.5–33.1
Ammonia nitrogen (mg/l)	0.1–12.5
Soluble phosphate (mg/l)	0.1–6.2
Total coliforms (no./100 ml)	20,000–90 x 10^6
Fecal coliforms (no./100 ml)	20,000–17 x 10^6
Fecal streptococci (no./100 ml)	20,000– 2 x 10^6

a. Selected data.

Source: Field, R., Struzeski, E.J., Jr., "Management and Control of Combined Sewer Overflows," *J. Water Pollut. Control Fed.*, **44**, 1393 (1972).

which carries accumulated debris, animal droppings, eroded soil, tire and vehicular exhaust residue, air pollution fallout, deicing compounds, pesticides and polychlorinated biphenyls, fertilizers and other chemical additives, decayed vegetation, heavy metals, and other pollutants (58). The other sources were storm sewer discharges, sewer overflows, sewer leaks, and treatment plant bypasses. So severe are these impacts that water quality planning and pollution abatement programs may fail unless storm-water loads are accounted for properly (60). Municipalities differ widely in extent of combined sewers and in rainfall-runoff relationships, which depend on geographical location. Thus the pollution load from storm-related flows must be considered case by case to assess the importance of such flows relative to other sources of pollution.

In some locations it may be necessary to separate combined sewer systems, but in many areas the cost of separation is impractically high. Alternative solutions include control of combined sewer overflows, reduction of infiltration, and treatment of overflows. A number of communities use excess storage capacity within the existing sewerage system to reduce the frequency and volumes of overflows. Computer techniques for remote monitoring and regulation of overflow structures also are being used to minimize overflows (61–63). Some communities collect at least part of the overflow from combined systems and send it to the wastewater treatment plant in periods of low flow. Concrete and steel holding tanks are used commonly for this purpose, but the cost and availability of land deter the installation of surface storage facilities.

Table 12

Characteristics of urban storm water[a]

Characteristic	Range of values
BOD_5 (mg/l)	1–>700
COD (mg/l)	5–3100
Total suspended solids (mg/l)	2–11,300
Total solids (mg/l)	450–14,600
Volatile total solids (mg/l)	12–1600
Settleable solids (ml/l)	0.5–5400
Organic nitrogen (mg/l)	0.1–16
Ammonia nitrogen (mg/l)	0.1–2.5
Soluble phosphate (mg/l)	0.1–10
Total phosphate (mg/l)	0.1–125
Chlorides (mg/l)	2–25,000[b]
Oils (mg/l)	0–110
Phenols (mg/l)	0–0.2
Lead (mg/l)	0–1.9
Total coliforms (no./100 ml)	200–146 x 10^6
Fecal coliforms (no./100 ml)	55–112 x 10^6
Fecal streptococci (no./100 ml)	200–1.2 x 10^6

a. Selected data.
b. With highway deicing.

Source: Field, R., Struzeski, E.J., Jr., "Management and Control of Combined Sewer Overflows," *J. Water Pollut. Control Fed.*, **44,** 1393 (1972).

On-site storage of storm water in small ponds or lakes in newly developed areas, on rooftops, in parking lots, and in other areas where ponding can be tolerated should also be considered. In planning the 80-acre Skyline Urban Renewal Project in Denver, the Denver Urban Renewal Authority required private developers to temporarily detain rain falling directly on their properties, which incorporate rooftop ponding and ponding in plazas (63). Another means of controlling pollution by storm water is subsurface storage, as in the Chicago Underflow Plan. This partly completed 10-year project calls for a system of tunnels to be built 150 to 300 ft below ground. When the plan was adopted, in fall 1972, its cost was put at $1.223 billion, about a quarter of the cost of separating the city's sewers.

A number of demonstration projects on storm-water treatment have been sponsored by EPA. These include evaluation of microstraining, high-rate filtration, dissolved-air flotation, physical-chemical treatment, biological processes, and detention-chlorination.

At Boston, Mass., a detention and chlorination system consisting of six parallel basins with a total retention capacity of 1.3 million gallons was activated 81 times during about two years (64). The system receives overflow from an intercepting sewer and, for the area served, provides a chlorine contact time in excess of 30 minutes for an estimated 80% of the overflows. For 15% of the storm events the system contains the overflows totally. The basins are drained into a downstream interceptor after each storm. During the first year of operation, about 0.55 yd^3 of coarse screenings (3½-in clear openings) and 9.8 yd^3 of fine screenings (½-in clear openings) were removed per 100 million gallons of combined sewage treated. Screenings are flushed directly to the downstream interceptor. Reductions in settleable solids averaged 85% and in suspended solids, 40%. Coliform reductions were reported to be usually 100%, and the chlorine demand of the influent ranged from 1.3 to 4.5 ppm. BOD results were erratic and showed little or no reduction. The facility cost $6.3 million to build; operating and maintenance cost was estimated at $65,000 annually, about one third of it for power.

At Racine, Wis., two modular, screening/dissolved-air flotation tanks costing $1.73 million were placed in operation in 1973 (64). The facility consists of a wet well to receive combined sewer overflows, and a mechanical screen followed by 50-mesh rotating-drum screens ahead of the flotation units, which use split flow with about 20% of the inflow pressurized. Provisions have been made for addition of chemical coagulants and chlorination.

At New Providence, N.J., although sanitary and storm sewers are separate, infiltration and inflow during wet weather raise flow rates to as much as 10 times the average during dry weather (64). To cope with this extreme variation, the city equipped the two trickling filters in its treatment plant to operate normally in series, but to shift automatically into parallel when precipitation raises flow to 2.8 million gallons per day (mgd). The wet-weather treatment capacity of this "dual-use" plant reinforces the dry-weather capacity at flows below 2.8 mgd, and both trickling filters are biologically active at all times.

At Mt. Clemens, Mich., combined sewer overflows from a 212-acre test area are treated in three lagoons in series (64). The lagoons average 8 to 10 ft deep and have retention times of four, eight, and seven days. The first lagoon serves as a combination storage basin and aeration lagoon, the second as an oxidation pond, and the third as an aeration lagoon. The effluent from the final lagoon is filtered and chlorinated before it is discharged to a river. Results from 28 stormflow operations showed BOD_5 and suspended solids removals higher than 90% and average effluent concentrations of 5.2 mg/l for BOD and 14.5 mg/l for suspended solids. The facilities are set within a park site and have been favorably received by the public. The lagoons or "lakelets" are free-form in shape, and the city planned to use the second and third for recreation. The cost of the facilities, exclusive of park improvements and land, was $1.08 million. Mt. Clemens planned to expand the operation to accommodate combined flows from four fifths of the city.

Water Treatment Plant Wastes

Plants that treat water for public water supply produce only a fraction of the amount of sludge produced by wastewater treatment. The sludge is largely inorganic and normally does not exert a troublesome oxygen demand in surface waters. Water treatment plants nevertheless face some of the same sludge disposal problems as wastewater treatment plants, including tighter restrictions on water pollution and a shrinking supply of cheap land that can be used for disposal. Sludge from water treatment has often been discharged to surface waters, but it can create turbidity and unsightly sludge banks and EPA has banned the practice. Disposal to lagoons is used, sometimes with subsequent removal of clarified liquid and use of the dried sludge as landfill. Some cities discharge the sludge to the sanitary sewer system. Others have used sand drying beds, vacuum filtration, centrifugation, or pressure filtration followed by disposal as landfill.

Modern pollution control measures at water treatment plants are illustrated by the Erie County Water Authority's Sturgeon Point Plant near Buffalo, N.Y. (65). In the mid-1970's, the plant's water treatment capacity was expanded from 60 mgd to 90 mgd; in the same period the plant was required to cease its long-time practice of discharging wastes to Lake Erie. The water treatment process at Sturgeon Point consists of prechlorination, coagulation with alum, postchlorination, pH adjustment with sodium hydroxide, and fluoridation. To achieve "zero discharge" of wastes, the water authority proceeded in two steps.

The first step consists of primary concentration of sludges in holding basins and clarifier-thickeners. The clarified water from this step is returned to the plant's raw water intake. For the maximum day, the clarifier-thickeners produce about 40,000 gal of coagulation-basin sludge containing 8000 lb of solids and 16,700 gal of filter-backwash sludge containing 2800 lb of solids. The combined sludge, 56,700 gal containing 10,800 lb of solids, has a concentration of about 2.3% solids. This sludge goes to the dewatering and subsequent facilities that constitute step two of the Sturgeon Point process.

The Erie County Water Authority studied 10 systems for dewatering, treating, and disposing of sludge at Sturgeon Point. Those considered most feasible were pressure filtration, precoat rotary-vacuum filtration, scroll centrifugation, freeze-thaw followed by vacuum filtration, and alum recovery followed by horizontal vacuum filtration. These studies and local economic considerations led to the selection of pressure filtration with ultimate disposal of dewatered sludge as landfill. (Alum can be recovered by treating coagulation-basin sludge with sulfuric acid to convert the hydroxide in the sludge to soluble aluminum sulfate. The process is not practiced generally in this country, but some large plants in Japan use it, before pressure filtration, to recover 50 to 70% of the alum added to the raw water.)

Sludge handling and disposal in the operation of lime–soda ash water-softening plants is also a problem that is becoming increasingly difficult as water quality requirements grow more stringent. Several cities calcine the sludge at about 1500°F, which converts the calcium carbonate to calcium oxide (lime) and carbon dioxide. To obtain lime of good quality, the sludge should be low in magnesium. Carbon dioxide produced in calcining can be used to recarbonate wet sludge, thereby improving the efficiency of centrifugation in separating magnesium hydroxide, which must be prevented from building up in the lime recycle system. Part of the carbon dioxide can be used to stabilize the water before filtration.

Recommendation W8: *Research should be expanded on new methods of handling wastewater treatment and water-treatment plant sludges with major emphasis on reducing costs. Comprehensive data should be developed on the economics and technology of complementary and/or alternative means of sludge disposal, such as combustion, use as fertilizer or soil conditioner, underground disposal, and land reclamation.*

Recommendation W9: *Systematic studies should be expanded on the quantity and quality of combined sewage, wastewater from nonpoint sources, and urban storm waters to provide a sound base for assessing alternative means of treatment. A major part of such studies should be consideration of the cost-effectiveness of the alternative methods.*

Recommendation W10: *Cost-effectiveness studies of physical-chemical treatment alternatives as applicable to municipal wastewater should be undertaken.*

Recommendation W11: *Research programs should be expanded on sewage treatment, primarily to seek innovations based on fundamental understanding of microbiological processes. Emphasis should be placed on the use of adequate chemical tools to develop data that will allow the biochemical and biological aspects of treatment processes to be interpreted more meaningfully and that can be used in process optimization.*

PHYSICOCHEMICAL PROCESSES AND ADVANCED WASTE TREATMENT

Advanced waste treatment includes, by common definition, any process or combination of processes that will remove more contaminants from municipal and industrial wastewaters than will conventional biological treatments. Advanced treatment, therefore, generally involves chemical or physicochemical processes. These treatments may be used to upgrade the effluent from conventional treatment, which contains on the order of 50 mg/l of organic contaminants and 700 mg/l of dissolved inorganic chemicals. Or, often more logically, they may be used to modify or replace conventional treatment.

The development of advanced waste treatment technology began seriously in the early 1960's in the Advanced Waste Treatment Program of the Division of Water Supply and Pollution Control, U.S. Public Health Service. This program has carried forward into the Environmental Protection Agency at the federal level and throughout the pollution control effort, both public and private. The main purpose of the work has been to upgrade wastewater treatment to achieve two goals: ultimate pollution control; and expanded water supply through water renovation and reuse (66).

A number of physicochemical processes, many of them used long and successfully in industry, have been screened and evaluated for advanced waste treatment potential. Several have proved particularly feasible for the purpose, both technologically and economically. These include coagulation, precipitation, filtration, adsorption, ion exchange, and certain membrane separation processes and chemical oxidations.

The selection of specific advanced treatment processes is complicated by the diversity and variability in concentration of the contaminants in wastewaters. The problem is quite unlike that of industrial processes, which may be designed to separate from water a relatively few dissolved components of relatively constant concentration. Instead, advanced waste treatment must contend with a variety of dissolved components, each of which may vary considerably in concentration over relatively short periods. Indeed, the variety of contaminants is so great, and their concentrations generally so low, that a reliable, comprehensive analytical methodology for determining the specific components of wastewaters has yet to be developed. Consequently, contaminants are grouped generally in classes that exert similar environmental impacts. The six major classes are suspended solids, dissolved organic compounds, phosphorus compounds, nitrogen compounds, microorganisms, and inorganic salts.

Apart from complete distillation, with its economic drawbacks, no single process will separate all pollutants from a "typical" wastewater. Most processes that have been considered for advanced treatment, in fact, have proven to be particularly well suited to separating certain classes or types of pollutants but not to separating others. It is difficult, therefore, to envision any single process that will renovate wastewater completely; it is more reasonable to envision renovation systems that combine a number of advanced processes or unit operations. The selection of processes for a specific treatment plant must be based on wastewater composition, effluent quality objectives, reliability requirements, resource availability, process economics, and restrictions on disposal of secondary wastes such as sludge, air pollutants, and ash.

Advanced Treatment

The development of physicochemical processes for advanced treatment centered at first on "tertiary" systems. These were designed to follow "primary" sedimentation and "secondary" biological treatment. This approach has several drawbacks, however. First, to install a tertiary system one must already have a primary and a secondary system. Second, adding tertiary to primary and secondary processes incurs capital and operating costs large enough to discourage such efforts. And third, the effectiveness of a tertiary process depends to a large extent on that of the biological secondary process whose effluent it treats; the problem here is that the biological processes are subject to upsets arising from the occasional presence of toxic contaminants and from variations in the composition and flow of incoming wastewater that often require it to be at least partly diverted.

The concept of applying independent physicochemical processes directly to raw wastewater rather than to secondary effluent, therefore, stems partly from the effectiveness and reliability of the processes and partly from the relative costs of direct and tertiary treatment. Direct independent physicochemical treatment (IPCT) has been demonstrated to be a technically and economically attractive alternative to conventional biological treatment of municipal wastewater (67–73). It has proved to be even more so for industrial waste treatment (74). Of the unit operations that make up IPCT systems, coagulation and adsorption on activated carbon are particularly noteworthy because of their key roles in advanced treatment schemes. Details of the principles and applications of these and the other physicochemical processes are available (75, 76).

Pretreatment

Raw wastewater entering a treatment plant contains coarse solids and abrasive materials that must be removed to avoid excessive operating and maintenance expense. Coarse solids are removed commonly by mechanically cleaned bar screens. They can then be incinerated, disposed of by landfill or in the municipal solid waste disposal system, or ground and returned to the wastewater to be removed in later processes. An alternative to screening is the comminutor, essentially an in-line sewage grinder. Coarse bar racks would still be required to protect the comminution equipment. Abrasive materials are removed normally in a grit chamber. The velocity of the sewage through the chamber is slowed to the point where heavy solids drop out, leaving sewage solids in suspension. The separated solids are removed by mechanical conveyors and may or may not be incinerated before disposal by landfill. Air is often diffused into the grit chamber, both to control turbulence and to reaerate septic (putrefying) sewage.

Coagulation, precipitation

Primary sedimentation of screened, degritted municipal wastewater rarely removes more than 50% of the suspended solids. By using chemical coagulants and controlled mixing, however, it is possible to remove 95 to 99% of the solids. Chemical coagulants, coupled with precipitation, can also remove most of the soluble phosphorus and the metal ions found normally in wastewater. Pilot plant studies have shown that a well designed chemical coagulation-precipitation system can achieve at least the equivalent of conventional secondary treatment (67–71).

Chemical clarification consists of chemical coagulation, flocculation, and sedimentation or flotation or both. Coagulation is the reaction of certain chemicals with constituents of wastewater to form insoluble precipitates and to destabilize small, colloidal particles suspended in the water. Flocculation, the agglomeration of coagulated particles, is promoted by agitating the wastewater gently. The particle size and density achieved by coagulation and flocculation greatly influence the results of sedimentation and flotation, the final steps in chemical clarification.

Common coagulants for wastewater treatment include lime, iron and aluminum salts, and synthetic polymers. The choice is based on the suitability of the coagulant for the particular waste, its availability and cost, and its effect on sludge treatment and disposal. Iron (ferric) chloride, for example, is available sometimes at low cost from the steel industry in the form of waste pickling liquor, and it presents no particular problems in anaerobic digestion of the sludge. Lime as a rule provides good clarification and a sludge that settles rapidly. Sludges from lime coagulation, moreover, can be thickened and incinerated to destroy most of the organic solids remaining from the sewage and to recover the lime for reuse. (The lime or calcium oxide appears in the sludge as calcium carbonate; incineration drives off carbon dioxide, leaving the regenerated lime.) Lime, therefore, is often the preferred coagulant for raw sewage; it may be used in combination with an organic polymer, a coagulant aid, or small amounts of iron or aluminum salts to aid flocculation and sedimentation and to decrease the amount of lime required. In addition, lime does not contribute undesirable inorganic anions such as chloride to the water. The point is important in view of the prospective reuse of water on a wide scale.

Lime coagulation may be done in one step or two, depending on the characteristics of the wastewater and the ultimate water quality goals. A one-stage process is generally suitable for removing solids and precipitating phosphorus (as phosphate) efficiently from waters of high hardness and alkalinities of more than about 200 to 250 mg/l as calcium carbonate. Where alkalinity is below about 200 mg/l, a two-stage process often is better. Even at low alkalinity, however, a one-stage operation combining lime with an organic polymer or with iron or aluminum salts may be satisfactory. In the two-stage process, lime is added in the first stage with the wastewater at high pH, usually above 11.0 to 11.5. Carbon dioxide (produced when the lime sludge is incinerated) is then added to the water to reduce its pH to about 10 and to precipitate excess calcium ions added in the first stage. This precipitate is coagulated and settled in the second stage by small amounts of polymer or iron or aluminum salts.

In an IPCT plant, chemical clarification normally follows pretreatment, but this is not always the case. Depending on how the sludge will be dealt with, a primary sedimentation step may precede chemical clarification.

The effectiveness of chemical coagulation and precipitation is evident in the results of pilot plant studies of two-stage lime clarification of raw municipal wastewater at the EPA-DC pilot plant in Washington, D.C. (71). The pilot plant was relatively large, operating at a nominal design flow of 100,000 gal of wastewater per day. The diurnal variation in flow was 3.2 to 1 (ratio of maximum flow to minimum flow). Clarification removed more than 90% of the suspended solids and 95% of the phosphorus and reduced both BOD and COD by 80%. The residual concentrations after treatment were: suspended solids, 6.3 to 22.7 mg/l; phosphorus, 0.16 to 0.45 mg/l; BOD, 13.0 to 31.4 mg/l; COD, 40.4 to 66.2 mg/l.

Similar efficiencies can be expected with other coagulants or combinations of coagulants. Chemical clarification will remove not only suspended solids and phosphorus, but many trace metal ions as well. The mechanisms involved are coagulation-precipitation and adsorption on chemical flocs. In pilot-plant studies, small amounts of four metals were added to secondary effluent, which then was treated by lime clarification (77). The treatment removed 94.5% of the cadmium, 9.3% of the chromium, 16.2% of the selenium, and 97.0% of the silver. Other studies were made of the removal of zinc, copper, and chromium from wastewaters by coagulation with alum, lime, iron (ferric) chloride, and an organic polymer (Cat Floc, a cationic polyelectrolyte) used normally either as a primary coagulant or as a flocculant aid (78). Alum and ferric chloride in general showed little promise in removing high levels of zinc, although ferric chloride did show promise in removing copper. Lime removed both zinc and copper consistently; residual concentrations were in the range of 0.2 mg/l of zinc and 0.01 mg/l of copper. None of the coagulants studied did well at removing hexavalent chromium, which thus would have to be controlled at its source.

Filtration

Filtration is used primarily to remove suspended solids that elude preceding operations. It may be a final treatment, or it may be a pretreatment for a process sensitive to suspended solids. Filter designs include simple screens, microstrainers, deep beds of granules such as gravel, and filters precoated with a filter aid such as diatomaceous earth. The type of filter selected depends on the application and the effluent quality desired. The basic mechanisms of filtration are straining, inertial impaction, and interception. Many other mechanisms come into play, however, when wastewater is filtered. They include sedimentation, physical and chemical adsorptions, coagulation-flocculation, and biological action. Variables that affect filtration efficiency include particle size and distribution; the concentration of suspended solids; the degree of flocculation and the shear strength of the floc; filtration rate and pressure; and the size, shape, and size distribution of the filter medium (sand, coal, gravel).

A bed-type filter is most efficient when the granules of filter medium are arranged so that wastewater flowing through the bed encounters the coarsest particles — and largest voids — first and then progressively smaller, more closely packed particles. This scheme exploits the full depth of the bed rather than just the surface. It also reduces the required frequency of backwashing — pumping water through the bed in the reverse direction to remove the solids filtered from the wastewater. Backwashing can still cause problems, however. The reverse flow of water "expands" the bed, in the process inverting the size distribution of the filter medium. The "ideal" filter has been approached by using particles of varying density. These multimedia filters combine larger, low-density particles with smaller, progressively higher-density particles. Backwashing of such filters causes some mixing of media, but does not invert the bed.

Adsorption

Adsorption on activated carbon is used to remove dissolved organic materials from wastewater. The process depends on attractive forces that impel dissolved organic molecules to attach themselves to the surface of the adsorbent. A major advantage of activated carbon for the purpose is its high surface area per unit of volume; its major disadvantage, perhaps, is its high initial cost. The material effectively adsorbs most organic substances that are degraded by conventional biological processes as well as many that resist biodegradation. Adsorption on activated carbon may follow conventional biological treatment or, with coagulation, it may replace biological treatment.

The utility of activated carbon in treating wastewater has grown steadily more apparent in recent years. In long-term pilot studies, for example, adsorption consistently produced effluents containing less than 5 mg/l each of TOC and BOD (69, 70); the same wastewater treated conventionally averaged 30 to 35 mg/l of both TOC and BOD. In six months of large-scale pilot work at the EPA-DC facility in Washington, D.C., carbon adsorption removed 95% of the TOC, BOD, and COD from wastewaters; average residual concentrations were 6 mg/l of TOC, 5 mg/l of BOD, and 13 mg/l of COD (71).

Carbon adsorption has been used widely for years to remove taste- and odor-causing organic compounds from public water supplies [and now, as noted earlier, is proposed for the removal of additional organics (10)]. The concentrations of these compounds are usually low, however, and treatment often is needed only intermittently or occasionally. Wastewaters, on the other hand, contain relatively large amounts of organic impurities and must be treated continuously. The initial costs of the large amounts of carbon involved, moreover, require that the material be regenerated and reused. These needs dictate the use of continuous adsorption systems based on granular activated carbon. The granular material has less adsorptive capacity than powdered activated carbon and costs more initially, but it is much the easier of the two to handle and regenerate, which gives it a decided economic and technological edge.

In the most common type of continuous system, wastewater flows by gravity or is pumped through fixed beds of carbon granules. The beds tend to compact in use and thus are termed "packed beds." The wastewater traverses the bed generally at rates of 2 to 8 gal per minute per square foot (of bed cross-section perpendicular to the direction of flow); contact time, the time a given parcel of water requires to traverse the bed, ranges from 30 to 60 minutes (calculated as if the bed contained no granules). These conditions usually give the optimum combination of adsorption efficiency, organic removal, and pumping cost. Carbon beds operating at the lower end of the flow range are designed as a rule for gravity flow of wastewater. In systems designed for higher flow rates the carbon must be contained in pressure vessels to withstand pumping pressure. A pressure system costs more than a gravity-flow system, but commonly occupies less area and adapts better to fluctuations in flow rate.

Packed-bed carbon adsorption systems filter suspended solids from wastewater and tend also to develop biological growths attached to the carbon. The beds must be backwashed regularly to relieve the consequent clogging. Complete removal of gelatinous biological growths in particular is best assured by a surface wash and air scour at the time of backwashing. Attached biological growth in packed beds can breed anaerobic conditions — no free oxygen and proliferation of microorganisms that subsist on oxygen combined in chemical compounds. One result is foul-smelling hydrogen sulfide in the product water. Addition of sodium nitrate to the carbon beds has been found to inhibit formation of hydrogen sulfide effectively (79, 80). Aeration of the feed helps to prevent anaerobic conditions, but it also accelerates biological growth, to the point where excessive backwashing is required. Biological growth can be controlled in most instances by chlorinating either the incoming wastewater or the water used to backwash the carbon beds.

Expanded beds. Packed beds of granular carbon normally operate effectively for long periods without clogging, so long as the water being treated contains few suspended solids. With municipal and industrial wastewaters, however, suspended solids and biological growths will usually clog the beds bit by bit, reducing efficiency and raising pumping costs. For this reason, interest has been growing in expanded-bed adsorbers. In these devices the water is pumped upward through the carbon particles fast enough to expand (or unpack) the bed. The technique minimizes fouling, plugging, and the associated rise in pumping costs. Because the expanded bed resists packing, it can accommodate smaller carbon particles than can a packed bed and thus take advantage of the somewhat higher adsorption rates of smaller particles (75). Comparative studies in both laboratory and pilot plant have shown that expanded carbon adsorption beds operate effectively over longer periods than do packed beds in both tertiary and IPCT applications (69, 70, 81–83).

The most significant potential benefit of expanded bed systems, perhaps, is the biological extension of the adsorptive capacity of the activated carbon. Because expanded beds require little maintenance, bacteria can develop and grow undisturbed on the carbon over extended periods of operation. These bacteria degrade some of the organic matter adsorbed on the carbon. In effect, they regenerate the carbon in place by clearing part of the adsorptive surface, which then can adsorb fresh organic material from solution. In expanded beds where microorganisms were allowed to develop fully, activated carbon has adsorbed more than its own weight of organic matter and 150% of its own weight as chemical oxygen demand (70). These "apparent" capacities are two or three times the true adsorptive capacity of the carbon. (Adsorptive capacity is expressed normally in weight-percent; activated carbon that can adsorb organics amounting to 50% of its own weight, for example, has an adsorptive capacity of 50 weight-percent.)

Raw sewage, settled but not chemically coagulated, has been fed to expanded-bed systems in which biological activity was encouraged (83). The results indicated no real advantage in the sequence and considerable disadvantage owing to fouling of the carbon by solids in the uncoagulated feed.

Countercurrent flow. The adsorptive capacity of the carbon usually is exploited most effectively by using an approach to countercurrent flow: the entering water, which is highest in dissolved organics, first contacts the carbon that has adsorbed the most organics; at the effluent end of the system, the departing — and cleanest — water is last in contact with the freshest carbon. In powdered-carbon systems, countercurrent flow is achieved by passing wastewater through a number of stages in series while the carbon moves in the opposite direction. In granular-carbon systems, the procedure usually can be simplified to avoid unnecessary handling. In most such systems, the first of a series of adsorption columns is removed from service when the carbon it contains is exhausted — loaded with organics — or nearly so. After being refilled with fresh carbon the column is placed at the effluent end of the series. The columns are not moved physically; the changes are made by appropriate piping and valving arrangements. Adsorption efficiency rises with the number of stages, but so do the cost and complexity of the piping and valving. The trick is to balance one against the other to achieve the best compromise.

Design factors. In designing a carbon adsorption system, then, a number of factors must be considered:

- Type of carbon: granular or powdered.
- Contact time.
- Flow rate.
- Configuration: series or parallel.
- Number of stages.
- Mode of operation: packed bed or expanded bed; pumped or gravity flow.
- Adsorption capacity.

For general planning purposes, activated carbon can reasonably be expected to have an adsorptive capacity of 50 weight-percent, in terms of COD, with no biological extension of capacity (Table 13). This is about equivalent to 500 lb of fresh activated carbon per million gallons of sewage treated. Results obtained with biologically-extended systems, however, suggest that it should be possible to reduce the carbon requirement by more than half (70).

Activated carbon can be effective also in removing many heavy metals from wastes. In pilot-plant work, carbon adsorption removed more than 96% of the silver, cadmium, and chromium from the effluent from secondary treatment (77). Significant amounts of selenium were removed in the same study. In other studies, adsorption has achieved residual concentrations of less than 0.1 mg/l, over a wide range of pH, for cadmium, chromium, cobalt, copper, iron, lead, manganese, mercury, nickel, silver, and zinc (84).

Regeneration. The initial cost of granular (and powdered) activated carbon is almost prohibitive for wastewater treatment unless the carbon is regenerated. It is feasible to regenerate granular carbon through at least 15 cycles. Exhausted carbon is removed from the adsorption system; the adsorbed organics are driven from it by controlled heating in a multiple-hearth or rotary-kiln furnace (at 1600° to 1800°F) in the presence of steam. During each regeneration cycle, some carbon is lost by burning and attrition; an additional amount is lost by alteration of the surface properties that makes the material ineffective as an adsorbent. The overall loss ranges from 5 to 10% of the weight of virgin carbon that would be required to restore the original capacity of the batch. For planning purposes, carbon makeup requirements can be put at 25 to 50 lb per million gallons of wastewater treated, with no allowance for biological extension in the adsorption system.

Regeneration processes for powdered activated carbon — with its lower initial cost and higher adsorptive capacity — are not yet out of the pilot-plant stage. The key factor will be holding the carbon loss at economical levels during regeneration. A successful process would be a significant step toward making a powdered-carbon adsorption system a technical and economical reality.

Activated carbon may be regenerated at the sewage treatment plant or at a central facility. Centralized regeneration, whether handled by an industrial firm or a regional public authority, would have several advantages: lower regeneration costs; better quality control; lower carbon loss; and elimination of a potential air-pollutant emission source at the site of the wastewater treatment plant.

Nitrogen removal

More than 90% of the nitrogen in raw domestic sewage ordinarily is in the form of dissolved ammonia (NH_3) or compounds from which ammonia or the ammonium ion (NH_4^+) are formed readily. It is cheaper and easier as a rule to keep nitrogen in these forms for removal, rather than to use nitrifying bacteria, for example, to convert it to nitrate. Ammonia and ammonium ion exist in equilibrium in water: at pH 7 (neutral), only the ammonium ion is present; at pH 11 (alkaline), only dissolved ammonia gas is present; at intermediate pH, both are present. Ammonia or its ion can be removed from wastewater physicochemically by one of three processes: ammonia stripping, ion exchange, and breakpoint chlorination.

In ammonia stripping, the wastewater is first raised to pH 11.0 or greater; the use of lime as a coagulant in the clarification step produces the proper pH automatically. The water is then contacted with air, which strips the gaseous ammonia from solution. Contact can be achieved with a standard industrial cooling tower or a specially designed stripping unit. One of several problems of the method is disposing of the stripped ammonia. The ammonia-laden air can be discharged directly to the atmosphere, as is done at Lake Tahoe (76), but this would be unacceptable in most areas since rainfall would carry much of the ammonia into surface waters. Incineration of the effluent airstream, which has been suggested, would release to the atmosphere more than 100 lb of nitrogen oxides per million gallons of wastewater treated. Ammonia stripping suffers also from formation of calcium carbonate scale on process equipment and, to operate efficiently, requires an ambient temperature well above 32°F.

Ion exchange can remove ammonium ion efficiently from wastewater (85). In pilot-plant studies, the ammonium-specific zeolite clinoptilolite has produced residual levels of less than 1.0 mg/l nitrogen as ammonium ion, although it did not consistently yield levels of 2.0 mg/l or below (71). With ion exchange, however, ultimate disposal remains a problem. If the ammonium-laden zeolite is regenerated with a lime slurry, for example, the end result is a water solution of ammonia gas, which presents the same disposal situation as ammonia stripping.

Breakpoint chlorination, unlike ammonia stripping and ion exchange, can reduce ammonia-nitrogen to any desired level down to about 0.1 mg/l. The breakpoint is the point where essentially all of the substances in the water that will react with the chlorine have done so, and free chlorine begins to appear. In the course of these reactions, ammonia in the water is converted to chloramines, and these compounds are converted in turn to nitrogen and chloride ion. The ammonia-nitrogen is thus released to the air as harmless nitrogen gas. The free chlorine remaining in the water is toxic, however, and must be converted to chloride. In a pilot study (72), breakpoint chlorination followed by adsorption of chloramines on activated carbon was used to promote the formation of free nitrogen and the conversion of free chlorine to chloride. The results indicated that the degree of ammonia removal could be controlled accurately by controlling the ratio of chlorine to ammonia in the feed water. A weight ratio of nine parts chlorine to one part ammonia-nitrogen yielded a residual ammonia-nitrogen concentration of less than 0.2 mg/l. The carbon adsorption step required 15 minutes of contact time after chlorination. One drawback of breakpoint chlorination is that, in milligrams per liter, it raises the chloride level of the effluent water by up to 10 times the amount of ammonia-nitrogen removed.

Table 13

Carbon adsorption capacities in IPCT[a] plants

Plant	Capacities[b] (weight-percent)[c] Total organic carbon	Chemical oxygen demand
Blue Plains, Washington, D.C.	15%	41%
Ewing-Lawrence, N.J.[d]	50	150
New Rochelle, N.Y.	20–24	60
Lebanon, Ohio	22	50
Owosso, Mich.	—	65
Salt Lake City, Utah	—	36

a. Independent physiochemical treatment.
b. Variation in capacities results from plant-to-plant differences in wastes, effluent criteria, number of stages, etc.
c. Weight of adsorbed organics as percentage of weight of carbon.
d. Biologically-extended, expanded-bed operation.

Source: Weber, W.J., Jr., Friedman, L.D., Bloom, R., Jr., "Biologically-Extended Physicochemical Treatment," Proceedings, Sixth Conference on Water Pollution Research, Jerusalem, June 18–24, 1973.

Dissolved inorganic solids removal

Renovation of wastewater requires that the concentrations of dissolved inorganic solids — minerals or salts — be reduced to the point where they will not build up in repeated cycles of reuse. The effluent from secondary (biological) treatment, for example, usually contains 300 to 400 mg/l more dissolved inorganics than the municipal water supply; the upper limit for palatable water is about 500 mg/l. The processes that have been studied for removing dissolved inorganics include primarily ion exchange, reverse osmosis, and electrodialysis. (In some cases these methods remove dissolved organics as well.) None of the processes has been applied yet, however, to large-volume flows of wastewater. And each produces a waste brine or a highly soluble residue that ultimately must be disposed of.

Ion exchange is used widely to soften or demineralize water and to recover useful by-products from industrial wastes. It involves a chemical reaction in which an ion from solution displaces an ion attached to the insoluble ion exchange material. This material may be natural, such as a number of zeolites, or one of various synthetic resins. The reaction is reversible, so that when the exchanger is exhausted it can be regenerated with a solution highly concentrated in the exchangeable ion attached to it initially. Cation (positive ion) exchangers, for example, might exchange their hydrogen or sodium ions with cations, such as metal ions, in solution; anion (negative ion) exchangers can exchange their hydroxyl ions for anions, such as chloride, in solution. The cation exchanger can be regenerated with an acid, such as sulfuric, and the anion exchanger with a base, such as sodium hydroxide. Ion exchange can remove most metal ions (and ammonia in the form of ammonium ion) from solution, but the concentrated waste from regeneration must be disposed of. (The process can remove negative ions from solution as well, but at impractically high cost.) Also, since this is an "exchange" process, some ions will remain dissolved in the product water. When metals are being removed, the ion remaining normally is sodium, either from direct exchange or from neutralization (adjustment of pH) of the product water with a sodium compound.

Reverse osmosis is essentially a high pressure (400 to 1500 psi) membrane filtration process. The membrane is semipermeable — it will pass water, but will not pass many dissolved substances. (The ideal membrane would pass only water.) If two solutions of different concentrations are separated by a semipermeable membrane, water will move through the membrane in the direction of higher concentration. Thus the difference in concentrations — the osmotic pressure — tends to equalize the two. The process is reversed by applying pressure to the more concentrated solution. The concentration of dissolved substances then tends to increase on that side of the membrane and decrease on the other side, where the product water accumulates. The pressure required to reverse the osmotic tendency increases with the difference in concentrations across the membrane. The purer the water on the product side of the membrane, therefore, the higher the pressure — and the energy expenditure — required on the waste-brine side to drive the purification process further.

Reverse osmosis can reduce dissolved solids to less than 60 mg/l. It has several drawbacks, however. Under optimum conditions, only 90% of the entering wastewater can be reclaimed; the remaining 10% must be disposed of as waste brine. Also, as hardness builds up in the waste brine, compounds of calcium can deposit on that side of the membrane, which reduces efficiency.

Electrodialysis also is a membrane process that can produce high-quality water for reuse. The method encourages the movement of contaminant ions through selective membranes by applying an electric potential across the electrodes of the dialysis unit. Negative ions move through anion-selective membranes toward the positive electrode; positive ions move through cation-selective membranes toward the negative electrode. (Nonpolar molecules are not affected.) The anion- and cation-selective membranes alternate in the electrodialysis cell; brine and purified water collect in alternating spaces between the membranes. A single pass through an electrodialysis unit can remove up to 50% of the total dissolved solids from the entering wastewater. Multiple-stage units can attain increasingly higher removals. Electrodialysis lacks the overall efficiency of reverse osmosis because it removes only ions from solution, while reverse osmosis removes water molecules from the wastewater. Operational problems are similar to those of reverse osmosis, including scale formation and waste brine disposal. In addition, electrodialysis requires a large source of electric power (direct current).

Each of the three processes — ion exchange, reverse osmosis, and electrodialysis — is less than optimum today for removing dissolved inorganic solids from wastewater. Each is capable of improvement, but reverse osmosis has the greatest potential. The basic need is improved membranes. The material used most widely in reverse osmosis is cellulose acetate prepared by special techniques. Efforts at improvement are concentrated on developing membranes of cellulose acetate or other materials that have longer operating life and that pass water more readily and various dissolved solids less readily.

Ultimate disposal of solids

Advanced waste treatment, like conventional treatment, produces wastes that ultimately must be disposed of. Processes like microscreening and filtration on diatomaceous earth, applied to secondary (biological) effluent, should produce only small volumes of wastes. The solids removed by these processes can be returned to the treatment plant, where they would slightly increase the volume of sludge. Chemical treatment for coagulating solids produces much larger amounts of waste. Often the volume can be reduced, however, by recovering processing chemicals, such as lime, for reuse. The activated carbon adsorption process yields little or no solid waste. The adsorbed contaminants are broken down and vaporized in the regeneration furnace (which becomes, therefore, a potential source of air pollutants).

The liquid wastes from processes like ion exchange, reverse osmosis, and electrodialysis present a more difficult problem. They may amount to 5 to 10% of the wastewater feed and contain from 5000 to 15,000 mg/l of total dissolved solids (compared with seawater at 35,000 mg/l). Concentration of such wastes to smaller volumes during the treatment process appears to be impossible because scale precipitates on membranes and heat transfer surfaces. Disposal in the oceans or salt lakes is one possibility. Another would be concentration by solar evaporation and subsequent disposal of the salts in the oceans or in sealed underground cavities. Many arid inland areas where demineralization — removal of dissolved solids — would be required also have low-cost land where solar evaporation ponds could be used.

Disposal of sludge and solids from advanced waste treatment, including physicochemical treatment, has been described in greater detail than the scope of this discussion permits (75, 76).

Bacteria, viruses

Reliable removal or inactivation of disease-causing bacteria or viruses in treated wastewater is growing ever more important, for two reasons: the waters that receive treatment plant effluent are being used increasingly for recreation, particularly in large urban areas; and direct reuse of treated water by humans would require means of preventing the numbers of organisms from building up in recycle systems.

Chemical clarification of wastewater can achieve high removals of bacteria, and subsequent filtration can further increase the level of removal. Chlorination is the standard disinfection process in the U.S.; in a properly operated plant it will destroy more than 99% of the bacteria. The effects of chlorination and other conventional processes on viruses are less well delineated. The indications are that chlorination, as usually practiced, does not produce a virus-free effluent, but that breakpoint chlorination very effectively inactivates viruses. Activated sludge removes viruses, but to what extent is uncertain. Activated carbon adsorbs certain viruses, but the process is reversible; the infectious properties of the desorbed viruses apparently are unimpaired. Coagulation and precipitation, as with lime or alum, is perhaps the best means of removing viruses short of chlorination. There is evidence also that viruses in treated wastewater applied to soil do not move through the soil into groundwater (86).

Viruses could be a serious problem in water reuse (87), and it appears that improved means of removing or inactivating them will be required to achieve virus-free effluents from advanced waste treatment. Further research is required on disinfection by chlorine or other oxidants, such as ozone (88, 89), iodine, and bromine. Since relatively high levels of such disinfectants might be required, more should be known of their reactions with the components of effluent streams to yield chlorinated organics or other undesirable compounds. Further work is needed also on the efficiencies of virus removal by processes such as adsorption, ion exchange, and filtration.

The Basic IPCT Process

The combinations of physicochemical processes that may be assembled to provide virtually any desired level of advanced treatment include one sequence that might be considered the basic independent physicochemical treatment process. The sequence consists of coagulation and precipitation to remove insoluble impurities, followed by adsorption on activated carbon to remove soluble organic impurities. This basic IPCT process has emerged as the sequence of greatest promise, both technologically and economically (68–71). Chemical clarification of raw sewage takes out significant amounts of both total and soluble organic matter, phosphates, and suspended solids; activated carbon adsorption commonly yields a clear effluent of low organic content.

Results at typical pilot-plant installations of the basic IPCT sequence — chemical clarification and activated carbon adsorption — appear in Table 14. In the late 1970's, on the order of 30 municipal treatment plants of the basic IPCT type were operating or in the design-construction stage in the U.S.; they ranged in capacity from about 5 million to 60 million gallons per day of wastewater. A substantially larger number of industrial waste treatment facilities were using at least the adsorption phase of IPCT.

A typical flow sheet for a basic IPCT plant appears in Figure 4. Coagulant is added to the raw sewage, and flocculation takes place in a chamber where moderate agitation is supplied during an average detention time of 15 minutes. Clarification occurs in a sedimentation basin with an average detention time of two hours. The clarified effluent passes through activated carbon in contacting basins for adsorption of dissolved organics. Expanded-bed units are preferred. They permit the use of simple, open-top concrete adsorption basins and relatively trouble-free operation. In addition, open tanks with overflow weirs or dams at the water level provide a supplemental means of aerating the wastewater during treatment, which helps to minimize the onset of anaerobic conditions in subsequent reactors. A typical plant, with a design capacity of 10 million gallons per day of wastewater, might have five parallel adsorption units of two stages each. When the granular carbon in the first stage of one unit is spent, the unit is taken offstream while the first-stage carbon is removed and regenerated. While the unit is offstream, each of the other four can run at 25% higher feed rate. The regenerated carbon is returned to its basin, which then becomes the second stage of that unit; the former second stage, whose carbon is now partially spent, becomes the first stage of the unit. Entering wastewater is then divided evenly among the five two-stage units until another carbon bed is spent.

The water resulting from such a treatment is lower in phosphates and organics and generally of higher quality than that produced by biological processes. IPCT effluent will meet most discharge requirements and will enhance the quality of most surface waters; with disinfection it will be suitable for many types of reuse. For some uses, a final filtration may be desirable to insure a crystal-clear effluent. This postfiltration removes any suspended matter that has been passed through or has been generated biologically in the carbon adsorption units.

Besides producing high-quality effluent consistently, the basic IPCT system is unusually stable and reliable. It withstands toxic wastes and shock loads far better than do biological systems, which are notoriously sensitive to environmental conditions. A biological process exposed even temporarily to a toxic material not only will lose efficiency but may not recover for several days or even

Table 14

Results of typical IPCT[a] treatment by coagulation and adsorption

Plant	Organic removal	Effluent concentration
Ewing—Lawrence, N.J.	95–98%	TOC[b] = 3–5 mg/l
Blue Plains, Washington, D.C.	95–98	TOC = 6
Lebanon, Ohio		
Granular carbon	97	TOC = 6
Powdered carbon	95	TOC = 11
New Rochelle, N.Y.	95	COD[c] = 8
Rocky River, Ohio	93	BOD[d] = 8
Salt Lake City, Utah		
Powdered carbon	91	BOD = 13
Owosso, Mich.	94	BOD = 8

a. Independent physicochemical treatment.
b. TOC — total organic carbon.
c. COD — chemical oxygen demand.
d. BOD — biochemical oxygen demand.

FIGURE 4

Typical flow sheet for treatment of wastewater by chemical clarification and adsorption[a]

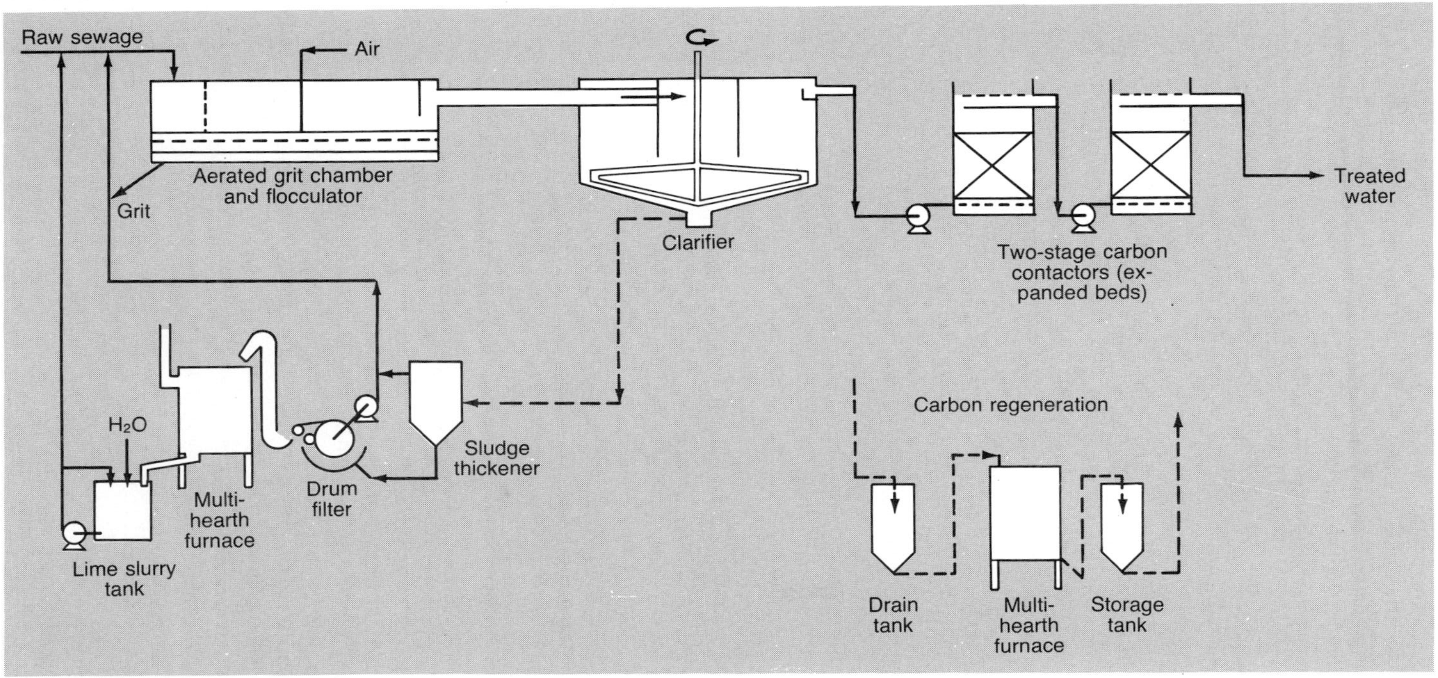

a. Single-stage coagulation and two-stage adsorption are shown, but more stages can be used in each case if desired.

Source: Weber, W.J., Jr., "Physicochemical Systems for Direct Wastewater Treatment," in "Applications of New Concepts of Physical-Chemical Treatment," Pergamon Press, Inc., New York, N.Y., 1972.

several weeks; hydraulic peaks can produce similar effects. Serious upsets are unlikely, however, in a physicochemical plant. When one does occur, the IPCT plant can be expected to recover immediately, once the cause of the upset is eliminated. This inherent stability is reflected also in greater flexibility in design and operation. Entire sections of a physicochemical plant can be cut in or out of the process stream as required, and temporary overloads, as may result from diurnal variations, can be accommodated with little effect. In addition to these advantages over biological systems, the basic IPCT process requires one quarter to one half the land area and has the potential for removing significant amounts of heavy metals from wastewaters.

Capital and operating costs

The basic IPCT system usually can produce higher effluent quality at lower capital cost and comparable operating cost than can conventional secondary (biological) treatment. A community of 85,000 people requires a treatment plant with a nominal design capacity of about 10 million gallons per day (mgd) of wastewater. A plant of this size embodying the basic IPCT sequence of Figure 4 would have a capital cost of about $3.5 million (1972 data); annual operating cost would be about 20 cents per thousand gallons, including amortization of invested capital at 6% for 24 years (90). These costs are based on a carbon exhaustion rate of 500 lb per million gallons of sewage. If the carbon exhaustion rate were halved, however, which seems feasible with extended-bed adsorption (70), the annual cost of carbon treatment would drop from 8.2 cents per thousand gallons to 6.8 cents. Total annual operating cost would drop correspondingly, from about 20 cents per thousand gallons to 18.6 cents. Not reflected here are the lower capital and operating costs of the regeneration furnace that would result from halving the rate of regeneration. Large-scale, long-term operating experience with the basic IPCT process remained sparse by the late 1970's. However, for a flow rate of 10 mgd, operating costs of 26 cents and 30 cents per thousand gallons had been estimated, respectively, from data for a large pilot plant (79) and a 10 mgd unit (80). These and the preceding values differ at least in part because of inflation.

Impact of Advanced Treatment

Although advanced treatment is being used successfully to renovate wastewaters, the total impact of such processes on the environment remains difficult to assess. The problem lies in balancing the benefits of the renovated water against factors such as the transfer of contaminants from water to air and soil and the consumption of energy and materials by the processes used. An example is provided by an evaluation of the treatment plant at South Lake Tahoe, in California, which combines primary, secondary, and tertiary processes to reclaim wastewater for transfer to a local reservoir (Fig. 5) (91).

Recommendation W12: *The basic IPCT system should be fully considered as an alternative to conventional biological treatment systems in every case where new wastewater treatment facilities are being designed and constructed.*

Recommendation W13: *Where existing conventional systems must be upgraded to meet higher effluent standards, design considerations should include conver-*

sion of existing physical plants to an IPCT system as an alternative to the addition of "tertiary" processes.

Recommendation W14: *Chemical and biological characterizations should be made of advanced treatment systems operating at the large pilot-plant or demonstration-plant level, including the identification of specific chemical compounds and studies of the effects of chlorine or other oxidants on organic residues. Such studies should be comprehensive research and development efforts, not simply demonstrations that the systems operate properly in terms of the traditional parameters.*

Recommendation W15: *Research should be intensified on the development of less costly physicochemical processes for ammonia removal.*

Recommendation W16: *Research should be intensified on the development of comprehensive design models to facilitate optimum design and selection of combinations of physicochemical processes for complex IPCT systems.*

Recommendation W17: *Research should be maintained at a high level on removal or inactivation of viruses by wastewater treatment processes and on other means of preventing virus buildup in recycle systems.*

INDUSTRIAL FACILITIES

It is difficult to generalize on industrial wastewaters because their characteristics can differ markedly both within and among industries. For the same reason industry treats its wastewaters by many methods (Table 15). They include all of those applied to municipal wastewaters and often are used in combination.

In the period 1954–68, the estimated volume of industrial wastewater treated in some way increased from about 13% of that discharged to about 30% (92). The total discharged increased by about one third in that period, however, so that the total discharged untreated actually rose slightly. Since then, considerable progress has been made. In 1973, industry discharged to receiving waters an estimated 54% of the BOD that it generated and 63% of the total suspended solids (Table 4) (13). These fractions should have declined markedly by 1977 because by then, as noted earlier, much of industry was applying best practicable treatment to its wastewaters. As we have seen, however, the impact of industrial (or other) discharges depends not only on their collective characteristics, such as BOD and suspended solids, but also on their content of specific inorganic and organic substances. Data on the discharge of such substances by industry were far from adequate by 1978, but they were beginning to accumulate, in part as the result of special EPA-sponsored studies of industrial wastewaters and of surface waters near heavily industrialized areas (28, 93).

Of the industrial wastewater treated before discharge in 1968, about 20% was treated in municipal plants (3); in 1977 it was estimated that industrial wastewaters may be discharged to as many as half of the nation's publicly owned treatment works (94). Industrial discharges may contain substances like trace metals and chemical compounds that upset or pass through conventional municipal treatment processes. That is why the 1972 amendments require that standards be established for pretreatment, where necessary, of wastes to be

FIGURE 5

Flow sheet of South Lake Tahoe wastewater treatment plant

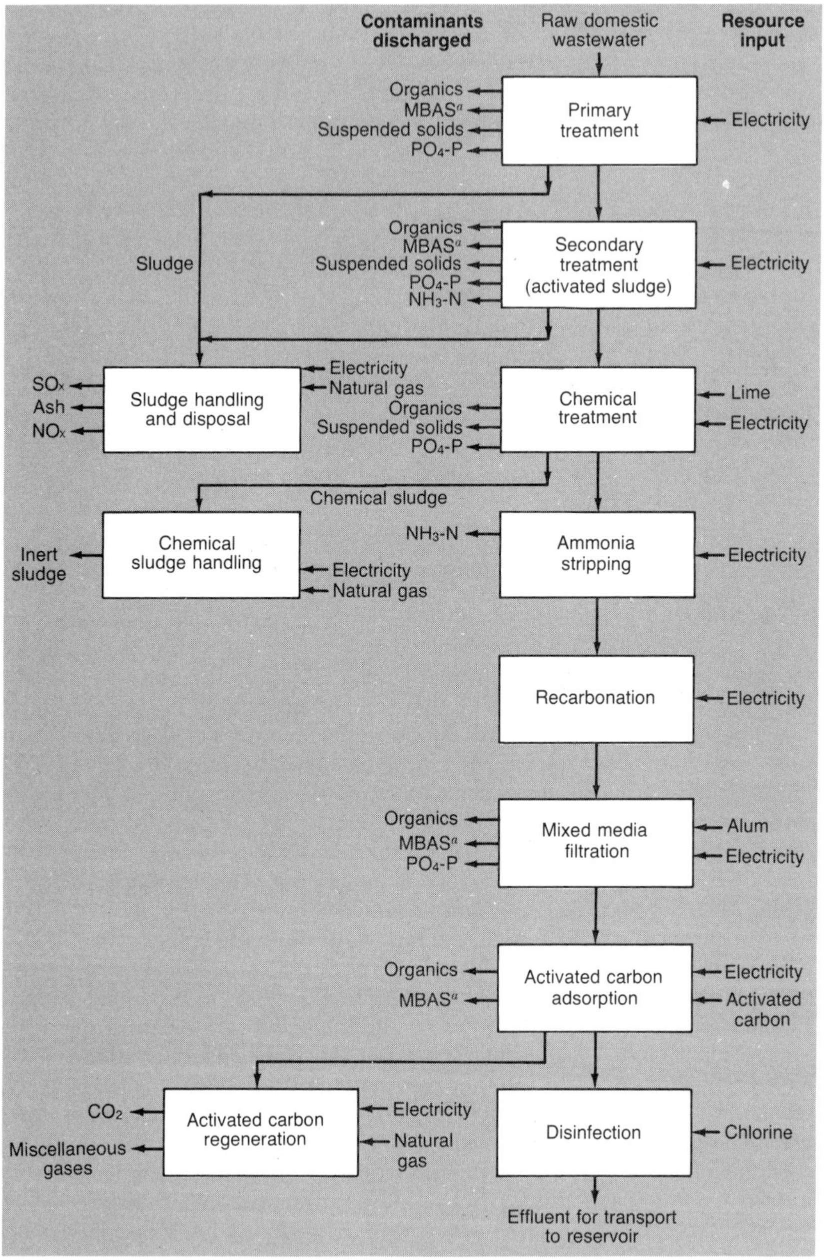

a. MBAS = methylene blue active substances — calculated and reported as alkyl benzene sulfonate, which originates in synthetic detergents.

Source: Antonucci, D.C., Schaumburg, F.D., "Environmental Effects of Advanced Wastewater Treatment at South Lake Tahoe," *J. Water Pollut. Control Fed.*, **47**, 2694 (1975).

discharged to public sewers. The amendments also include ground rules that can affect what companies must pay to use municipal treatment plants.

Pretreatment standards, and effluent limitation guidelines for point sources of pollutants, are intended in general to reflect the capabilities of best practicable or other required treatment technology. Dischargers must stay within the effluent limitation guidelines to qualify for the wastewater discharge permits issued by EPA or the states under the National Pollutant Discharge Elimination System. Applications for discharge permits require detailed information, including data on the amounts and compositions of the applicant's point discharges (95).

By mid-1976, EPA had promulgated some 43 effluent guidelines and proposed another 61 covering about 75 industry groups with numerous subcategories (14). The agency had promulgated or proposed pretreatment regulations for 30 industries with many subcategories, and in June 1978 promulgated new, general pretreatment regulations (94). By March 1976, some 27,000 discharge permits had been issued under the National Pollutant Discharge Elimination System; the permits covered about 67% of all industrial dischargers.

Industrial discharges of toxic water contaminants did not begin to come under specific control until 1977, when EPA issued regulations for six such substances: aldrin/dieldrin, DDT and related compounds, endrin, toxaphene, benzidine, and polychlorinated biphenyls (PCB's). In 1976 the agency began gathering data on health effects and control technology for the 65 substances specified as toxic by the Clean Water Act of 1977 (5). The results of these and related economic studies will be used to establish maximum discharge standards for the substances in at least 21 categories of industry. The standards are to be met by mid-1984, as noted earlier.

Water Recycle Ratio

Industry manipulates very large amounts of water, but returns almost all of it to receiving waters. Of the 15.5 trillion gallons that manufacturers withdrew in 1968, some 92% was discharged and only 8% was evaporated, incorporated into products, or otherwise consumed (made unavailable for reuse) (92). Cooling accounted for 65% of the water taken in; the rest was involved directly in processes in one way or another, and this process water was the vehicle for almost all of the contaminants in industrial wastewaters (96). As industry and population expand, the supply and quality of industrial water decline and the cost of using it rises. Superimposed on the pressures of growth have been those of pollution abatement, which intensified sharply during the 1970's. Industry has reacted to these economic incentives basically by raising the recycle ratio — the number of times each gallon of water is "used" between intake and discharge.

In 1954, manufacturers used about 1.8 gal of water per gallon taken in. By 1968 the ratio had reached 2.3 (92) and in 1975 was still at about that level (97), but further change was under way. Early in 1978, the U.S. Department of Commerce estimated on the basis of extensive data that the recycle ratio for all manufacturing would reach 8.4 by 1985 and 16.1 by the year 2000 (97). The department estimated also that, in the period 1975–2000, industry's water-intake rate would decline by some 67% and its discharge rate by 90%, despite a 140% increase in the water-usage rate. These improvements were expected to result from progressively more intensive reclamation and reuse of water, abetted by process change and recovery of salable or reusable materials.

Industry has long recirculated cooling water. Commonly it does so by using cooling towers to reduce the temperature of exhausted cooling water, which then is returned to the process-cooling equipment. In the 1970's, cooling accounted for some 70% of the "use" of water, and more efficient operation of cooling towers and the associated equipment was expected to yield much of the predicted increase in recycle ratio. Gains were foreseen also, however, from greater effort to reclaim process water for reuse. Zero discharge of water and pollutants has been approached, for example, in processes for making aluminum sulfate (alum), hydrofluoric acid, and wet-process phosphoric acid (98). In each case the method involved reclaiming and reusing process water previously discharged to receiving waters.

Process Change

New plants and processes tend to reduce the ratio of wastes to products, although there are exceptions. A study of integrated pulp and paper mills showed that, before treatment, "new" technology generated 70% less wastewater than "old" technology and less than half the amount of BOD and dissolved and suspended solids (99). In the iron and steel industry, on the other hand, "advanced" technology generated, before treatment, some 40% more wastewater than "old" technology and generally higher levels of contaminants. The reason was that steel products are becoming lighter and production rates higher. Wastes are generated in proportion to the surface exposed per unit of weight in

Table 15

Processes used to treat industrial wastewaters

Pollutant	Processes
Biodegradable organics	Aerobic biological (activated sludge), aerated lagoons, trickling filters, stabilization basins, anaerobic biological (lagoons, anaerobic contact), deep-well disposal
Suspended solids	Sedimentation, flotation, screening
Refractory organics	Carbon adsorption, deep-well disposal
Nitrogen	Maturation ponds, ammonia stripping, nitrification and denitrification, ion exchange
Phosphorus	Lime precipitation, aluminum or iron precipitation, biological coprecipitation, ion exchange
Heavy metals	Ion exchange, chemical precipitation
Dissolved inorganic solids	Ion exchange, reverse osmosis, electrodialysis

Source: "Water Quality Engineering for Practicing Engineers," Eckenfelder, W.W., Jr., Barnes & Noble, Inc., New York, N.Y., 1970, page 110.

rolling and finishing and in proportion to gas-liquid interfacial areas in iron-making and steelmaking; the newer technology tends to maximize these factors.

New processes have been developed traditionally to make better products at lower costs. A reduction in the amount of waste generated was often coincidental, excepting insofar as the process was designed to use materials more efficiently. As antipollution measures have grown more stringent, however, waste generation and disposal have been integrated more fully into process development. A case in point is the two types of electrolytic cell — diaphragm and mercury — used to make sodium hydroxide and chlorine from salt. (Diaphragm cells accounted for some 75% of U.S. chlor-alkali production capacity in 1976.) Each type of cell has its advantages, but owing in part to problems with water pollution by mercury, the diaphragm cell appeared to be gaining the edge by 1974 (100). Later, because the asbestos normally used in the diaphragm may pose a pollution problem, the environmental balance tilted toward the membrane cell, which in 1978 was nearing commercial use (101).

Another example of process change stimulated by pollution problems is the processing of potatoes, by weight the leading vegetable grown in this country. In the late 1960's many plants were processing 1 million pounds of potatoes daily; such a plant would generate about the same BOD as a city of 300,000 people. Some 75% of the BOD stemmed from the peeling operation — dipping in hot lye solution and peeling in a rotating reel with high-pressure water jets. Then a process was developed (102) that used a weaker lye solution, infrared heating, and mechanical peeling with rubber studs on rotating rolls. The overall effluent is a 15% solid waste that can be combined with other wastes, fermented to reduce pH, and fed to steers. The industry rapidly adopted the process, which uses less lye, produces no effluent, and reduces the loss of potato during peeling.

Materials Recovery

Recovery of salable or reusable materials has long been practiced in industry when economically justified. In making wood pulp, for example, recycle or sale of processing chemicals, tall oil, and turpentine are self-supporting operations as a rule. Spent sulfite pulping liquor, on the other hand, has not generally been an economical source of by-products in this country, although technically it can serve as a source of vanillin, ethyl alcohol, and other chemicals and as a nutrient in making torula yeast. Materials recovery is unlikely to become a panacea. It grows more attractive, however, as a cost-defraying or even profitable measure as the costs of industrial water creep upward. One instance is the use of activated carbon to recover the organic compound *p*-cresol, at a profit, from a liquid effluent for return to the process (103). The move grew from the need to minimize emission of the compound from wastewater to the air, where humans can detect it at concentrations above 1 ppb. Both reverse osmosis and electrodialysis are used commercially to recover salable food products from whey (104, 105), a residue of cheese manufacture and normally a pollutant of high BOD that is very difficult to treat. Similar opportunities are being pursued, experimentally or commercially, at a rising rate in a number of industries.

It was noted earlier that perhaps 15% of the major industrial dischargers and many smaller ones did not meet the mid-1977 deadline for best practicable treatment. In addition, it appeared that in some industrial categories the technologies on which EPA based the effluent limitation guidelines and discharge permits do not uniformly provide the anticipated level of control (8). But despite such difficulties it is apparent that the 1972 amendments led to diverse technological efforts to control water pollution by industry (106–109).

Recommendation W18: *Inventories should be compiled as quickly as possible of the specific substances in industrial wastewaters that are important or potentially important pollutants.*

Recommendation W19: *Information should be gathered and made widely available on the technology of joint municipal-industrial treatment of wastewaters, particularly physical-chemical treatment, for the guidance of companies and municipalities who wish to consider such an approach.*

AGRICULTURE

Agriculture, including commercial livestock and poultry operations, is the source of many organic and inorganic pollutants in surface waters and groundwater. These contaminants include sediment from the erosion of crop land and compounds of phosphorus and nitrogen that originate in part in animal wastes and commercial fertilizers. (Agricultural pesticides, which also are common contaminants of water, are covered specifically in Chapter 7.)

Sediment

Sediment transported from land through erosion and runoff is the nation's largest single water pollutant by weight. While sediment has long been associated mainly with agriculture, it is stemming increasingly from construction and other land-disturbing activities in urban areas (110). The most obvious effects of erosion and sediment are physical — degradation of sloping land, silting of lakes and reservoirs — but important chemical effects are involved as well. Soil particles washed into surface waters by runoff carry with them associated organic matter and adsorbed substances like fertilizers and pesticides. Once in the water the particles transport adsorbed substances and can affect their chemical and physical interactions with other substances present. And the particles themselves are made up of inorganic compounds that can exert chemical and physical effects.

Erosion and sediment are controlled, if imperfectly, primarily by physical means such as cultural and soil-stabilization practices. Control programs are pursued by several federal agencies, notably the U.S. Department of Agriculture (111). These programs include the assistance given by USDA's Soil Conservation Service to the more than 3000 soil conservation districts established under state laws.

Animal Wastes

A steadily growing water pollution problem is runoff from commercial livestock and poultry facilities during periods of rain or snow. The problem arises largely from the trend toward operations with many animals — beef and dairy cattle, swine, chickens, turkeys — confined in small areas. The leading example is beef cattle feedlots. In this country about 2500 feedlots carry more than 1000 cattle each and market some 70% of the nation's fed cattle. The total number of fed cattle marketed rose from 13 million in 1960 to 25 million in 1970. Most of the growth, moreover, involved facilities holding at least 10,000 cattle, and a

10,000-head feedlot generates annually about 100,000 tons of urine and feces (18,000 tons of dry manure).

Animal wastes are high in BOD, nitrogen, and phosphorus (Table 16) and often harbor pathogenic organisms like *Salmonella*. Commercial feeders as a rule contain their wastes and dispose of them on land, so that the main threat to natural waters is runoff and leaching. In the mid-1960's, for example, rainfall runoff from cattle feedlots reportedly accounted for more than 80% of the fish killed in one central state; the cause of death was depletion of dissolved oxygen by the high BOD in the runoff.

State laws generally prohibit wastewater discharges from feedlots except when precipitation exceeds the 24-hour high for a specified period. The period ranges from five to 25 years, depending on the state. The 1972 amendments to the Federal Water Pollution Control Act, however, classify feedlots as point sources, subject to EPA guidelines under the National Pollutant Discharge Elimination System. To make the regulatory task manageable, EPA established cutoff points for application of the NPDES permit system (112). The regulations apply, for example, to feeding facilities with more than 1000 beef cattle, more than 700 dairy cattle, more than 2500 swine, and so on. In general, the guidelines exempt feedlots that discharge only during 25-year, 24-hour storm events.

Animal wastes traditionally have been disposed of on land. Manure from cattle feedlots, for example, is stored and later spread on agricultural land as a low-grade fertilizer (mainly for its nitrogen content) and conditioner; liquid or slurried wastes and precipitation are held for two or three weeks and applied as fertilizer or irrigation water. This basic approach may involve various physical handling methods, including settling basins for liquids, as well as limited biological treatment in aerobic or anaerobic lagoons. Even the best land-disposal schemes, however, face the difficulties created by stricter regulation and large feeding operations. Feedlots tend to be near population centers, where the land required to confine wastes is relatively costly and problems with odors and insects are intensified. Uncertainties exist on the effects of high waste-loading rates on crops, on soils, and, via leaching, on groundwater. Better methods are needed to reduce costs, including those of land and labor, minimize nuisance, and improve sanitation.

The diverse efforts directed at these needs include the National Animal Feedlot Wastes Research Program (113). An early priority of the program was to characterize animal wastes (and runoff) in terms of factors like amounts produced, chemical composition, BOD, and decomposition products. By the mid-1970's the emphasis had shifted to measures such as closed-loop waste control, including resource recovery. The concept stems from the fact that land disposal already is unsatisfactory for many large feeding operations, a problem compounded by the spread of urban populations into rural areas. A possible solution is to process animal wastes for reuse. Cattle manure, for example, contains a significant fraction of nutritive material, mainly cellulose, that passes through the animal's digestive system untouched; in principle it could be converted to a feed supplement and, in effect, recycled through the cattle. Manure also can be converted to fuel (oil or gas) or even to synthesis gas (carbon monoxide and hydrogen) for making ammonia, the leading source of fertilizer nitrogen. The basic physical and chemical principles that would be involved in resource recovery from animal wastes are in hand; practical processes, and their economics in an antipollution context, remain to be developed and demonstrated.

Table 16

Pollutional characteristics of untreated animal wastes, summary of values

Animal	Animal weight (pounds)	Solids (pounds/day)	BOD (pounds/day)	Nitrogen (pounds/day)	Phosphorus as P_2O_5 (pounds/day)
Beef cow	1000	10.0	1.0	0.3	0.1
Dairy cow	1000	10.0	1.2	0.4	0.1
Swine	100	0.9	0.25	0.06	0.02
Poultry	5	0.06	0.015	0.003	0.003

Source: "Agricultural Practices and Water Quality," Proceedings of a Conference Concerning the Role of Agriculture in Clean Water, Iowa State University, Ames, Iowa, 1970, National Technical Information Service, U.S. Department of Commerce, Washington, D.C.

Phosphorus and Nitrogen

Runoff and leaching from barnyards, feedlots, and land treated with chemical fertilizers carry varying amounts of compounds of phosphorus and nitrogen into natural waters. The significance of these agricultural inputs, however, is not clear relative to those from other sources, such as raw or treated sewage and nitrate present naturally in soils. The concern over phosphorus is due to its role as a nutrient for algae and other plants whose excessive growth degrades surface waters. Nitrogen, too, is an important nutrient, but concern exists also over the nitrate in groundwater in many parts of the country (114). Both infants and ruminant livestock have grown ill and sometimes died from drinking water containing nitrate. The stomachs of both reduce nitrate to nitrite, which causes methemoglobinemia (blue-baby disease). Nitrate in drinking water was the cause of 328 cases of infant methemoglobinemia, including 39 deaths, reported in the U.S. in 1945–69. Only one of these cases involved a public water supply; all of the others were attributed to well water. (The federal primary drinking-water standards specify a maximum concentration of 10 mg/l of nitrate-nitrogen.)

There is evidence that much of the phosphorus in surface waters comes from sewage treatment plants and that phosphate in chemical fertilizers is a secondary source. Phosphate binds tightly to soil particles and enters surface waters from farm land mainly as a result of erosion and runoff. The amount of phosphate contributed by agricultural runoff and drainage water depends on factors like the type of soil and fertilizer and cultural practices, but definitive data on these relationships have not been developed.

Nitrate, unlike phosphate, leaches through soil into tile drainage and groundwater. There is evidence that chemical fertilizers are a significant source of nitrate in groundwater, and there is also evidence to the contrary (115). Further data are required to resolve the question. Domestic wastes, especially in septic tanks, are an important if not fully defined source of nitrate in groundwater in rural communities. Leachate from cattle pens and feedlots is a large potential source, particularly in restricted areas, but little is known of its overall significance.

Analyses of phosphate and nitrate in surface waters in Illinois (116, 117) suggest that nutrient problems caused by nitrogen and phosphorus might become common throughout the state. The sources of the nutrients vary with stream discharge and are difficult to identify, although significant amounts of nitrogen have been found in precipitation. Analyses by the U.S. Geological Survey of nitrate in six major rivers in 1950–70 showed an uptrend in three, a downtrend in two, and no clear trend in one. Many groundwater supplies, particularly in rural areas, exceed the 10 mg/l of nitrate-nitrogen specified as the maximum for drinking water. Analyses of 8844 wells in Illinois showed that 28% of those up to 8 m deep exceeded the limit; in one county, 81% of the dug wells and 34% of the drilled wells exceeded it (118).

Recommendation W20: *Strong support should be given to existing programs of research and development on economic means of treating and disposing of animal wastes in agriculture, particularly in the context of closed-cycle systems.*

Recommendation W21: *Basic data should be gathered systematically to delineate the relative importance of the various agricultural and rural sources of compounds of nitrogen and phosphorus in surface waters and groundwaters.*

ACID MINE DRAINAGE

Acid mine drainage pollutes an estimated minimum of 12,000 miles of streams in the U.S. The majority of the drainage originates in coal mines in eight states of the Appalachian region and three states of the Illinois coal basin, but some results also from the mining of copper, lead, zinc, iron, uranium, and other commodities (119). Mine drainage can be alkaline as well as acid, but the latter is by far the more widespread problem. The acidity harms or kills living organisms and corrodes piping and other structures; metals in mine drainage may harm or kill aquatic life; compounds of iron, the principal metal present, can coat the bottoms of streams, making them uninhabitable. Mine drainage contaminates groundwater, too, but its impact is not well defined. On the whole, the drainage problem can be expected to grow with the nation's increasing reliance on coal as a source of energy.

Control of pollution by mine drainage has been studied intensively only in the past decade — largely by industry, the state of Pennsylvania, and the Environmental Protection Agency. The resulting technology (120) remains crude, but it should remove active mines as serious sources of water pollution when federal and state discharge requirements are met. More than two thirds of the acid drainage, however, comes from abandoned mines, most of them underground, and the fraction will rise as discharges from active mines are brought under control. Drainage from abandoned mines, though subject to the technology used widely at active mines, is being combated as yet at relatively few sites, mainly by the state of Pennsylvania and EPA.

Many minerals, including coal, occur in association with iron and other metallic sulfides. The formation of acid mine water involves initially the oxidation of the sulfides to form, in a series of reactions, sulfates, sulfuric acid, iron oxides, and probably other compounds (121); the process evidently involves three kinds of oxidation — chemical, electrochemical, and bacterially catalyzed. The key to the overall rate of oxidation is the reaction of exposed sulfides with the oxygen in moist air. Water enters the mine and dissolves the oxidation products, and the resulting acidic solution, which may contain compounds of several metals, runs eventually into surface waters.

One means of dealing with acid mine water is to prevent or minimize its formation. Measures that can be used include flooding or sealing, to prevent air from entering the mine and oxidizing the sulfides, and excluding water from acid-producing strata or, when it does enter, minimizing its time of contact with reactive or oxidized surfaces. Such techniques are being practiced by industry, but often do not work well at their current state of development. Therefore, although prevention at the source holds the best long-term hope, it must be supplemented by methods of treating the acid mine water whose formation cannot be blocked.

Most of the 300-odd facilities (1977) that treat drainage from active mines do so by neutralizing it with lime (calcium oxide) or hydrated lime (calcium hydroxide). The two compounds are generally available and cost less than any other neutralizing agent but limestone. In the conventional process, the mine water is neutralized and then aerated for 15 to 30 minutes to oxidize soluble ferrous iron to the insoluble ferric form, which precipitates. The water and solids, mainly hydrated iron oxide, are separated in mechanical clarifiers or settling basins. The water is discharged; the voluminous sludge that remains is disposed of on land. Disposal sites include huge storage lagoons and, less often, abandoned deep mines.

The cost of lime has stimulated research and development on the use of the cheaper limestone (calcium carbonate) as a neutralizing agent. The effectiveness of limestone depends on the concentration and chemical form of the iron in the mine water. At first, the iron is in the ferrous form and cannot be precipitated completely (as ferrous hydroxide) below a pH of about 9.5. Treatment with limestone neutralizes the mine water, but forms carbonic acid in the process, thus limiting the pH to no more than about 6.5. At that pH, ferrous iron is soluble and does not oxidize rapidly, and little of it can be removed. Aeration will expel carbon dioxide and increase the rate of oxidation. Further oxidation, however, forms more acid, and more limestone is required to neutralize it. Limestone has certain advantages — it yields a relatively dense sludge, for example — but does not work well on mine waters that contain more than about 100 mg/l of ferrous iron. As a result, very few limestone treatment systems have been installed.

Research has been done on microbial and electrobiochemical means of abating pollution by acid mine water, but only on a laboratory scale (122, 123). Electrodialysis, flash distillation, ion exchange, and reverse osmosis have been studied for reclaiming potable water from acid mine drainage, but none has proved economically feasible. Two ion exchange plants were built for the state of Pennsylvania. Each is designed to produce potable water from 0.5 mgd of acid mine water, but neither has accumulated significant operating experience beyond the startup stage.

WATERCRAFT

Boats and vessels discharge a variety of water pollutants, including oil, litter, ballast and bilge waters, and sewage (sanitary wastes). Many states have had laws of varying impact on pollution by watercraft, but the first federal legislation to deal squarely with the problem was the amendments of 1970 and 1972 to the Federal Water Pollution Control Act. Parts of these amendments extend federal regulatory authority to discharges by watercraft of sewage, oil, and hazardous substances other than oil.

U.S. watercraft include some 8.8 million pleasure boats and nearly 4000 vessels of more than 100 tons gross weight. About 8.2 million of the pleasure boats are rowboats, canoes, and other small craft that are exempt from the regulations because they do not have installed toilet facilities. Indeed, concern exists over unburned fuel and other contaminants discharged to water by the 7.2 million outboard motors in use on small craft (124).

EPA standards of performance cover marine sanitation devices (MSD's) installed on vessels plying the navigable waters of the U.S. and limit the discharge of sewage from vessels into U.S. waters. The standards apply to more than 600,000 U.S. boats and ships with installed toilets, including about 550,000 recreational boats. They apply also to all foreign vessels with installed toilets that enter U.S. waters (a provision that raises difficult legal problems). The Coast Guard is responsible for promulgating certification procedures and design and construction requirements for MSD's to implement the standards.

The EPA standards took effect Jan. 20, 1977, for vessels on which construction began Jan. 30, 1975, or later; they take effect Jan. 30, 1980, for vessels on which construction began before Jan. 30, 1975. Under the standards, states can apply for the establishment of "no-discharge" waters, in which no sewage, treated or untreated, may be discharged by vessels equipped with a marine toilet; on waters not subject to "no-discharge" regulations, certified flow-through devices may be used. In meeting these standards, vessels must be equipped with Coast Guard-certified equipment of two general types: no-discharge devices (holding tanks, incinerators, or recirculating systems), from which wastes can be removed and treated at shore facilities; and discharge devices, which provide a specified minimum primary treatment before discharge.

Treatment equipment for watercraft includes incinerators and biological treatment devices as well as holding tanks and macerator-disinfectors. Since 1968 the Coast Guard has been conducting and sponsoring research and development on shipboard sewage treatment systems and on other aspects of marine environmental protection, such as shipboard pollution prevention systems that can handle bilge and ballast waters and other shipboard wastes (125, 126). Tests have been made, for example, of a full-scale prototype of a wastewater treatment system for a 20-man crew. In this system, briefly, incoming wastewater is filtered and the trapped coarse solids incinerated; organics are oxidized in an electrocatalysis unit; and the effluent is disinfected with sodium hypochlorite generated electrolytically from the saline wastewater. In its initial tests, the system reduced BOD by 88% and suspended solids by 91%. Other processes under study include membrane separation and wet oxidation. Both may prove useful in closed-cycle or no-discharge wastewater treatment systems. The Coast Guard effort extends to the evaluation and development of systems for all types of vessels, including recreational types.

EFFECTS OF WATER POLLUTANTS

Water quality criteria (127) reflect a very large body of knowledge of the effects of water pollutants. Much remains to be learned, however, in establishing the scientific bases for such criteria, which supply in turn the bases for the necessary abatement measures. The effects of pollutants depend on many interrelated variables: the acidity or alkalinity of the water; its temperature; the degree of dilution and mixing; chemical and biochemical processes; the rate of flow of the water; and the reinforcing or counteracting actions of contaminants on one another. Further complexity is introduced by the economic need to relate water quality criteria to the six generally recognized classes of water use: public supplies; freshwater aquatic life and wildlife habitat; marine aquatic life and wildlife habitat; industrial; recreation and aesthetic enjoyment; and agricultural. In this context, important gaps exist in our knowledge of the consequences of water pollution; notable instances include the effects involved in human health, eutrophication, and tastes and odors.

Human Health

At typical levels of exposure in this country, inanimate water pollutants seem unlikely on the whole to exert significant acute effects on human health. Constant vigilance is warranted against the odd exception. Nitrate in drinking water, for example, is known to have caused methemoglobinemia, sometimes resulting in death in infants. Cadmium in sludge-derived fertilizer can be taken up by crops; if ingested in sufficient amount, the metal can cause acute gastroenteritis and liver and kidney damage. Still, the hazard of acute toxicity is less worrisome than that of chronic effects from long-term exposure to low levels of contaminants (9, 128).

A clear-cut example of chronic damage involves methylmercury. The compound is formed in water by the action of bacteria on inorganic mercury, which enters the environment from natural sources, combustion of coal, and industrial processes such as mercury-cell chlorine-caustic plants. Methylmercury is fat-soluble and so can accumulate in fish and other organisms in the aquatic food chain. In humans it has a biological half-life of 70 days. In Japan in 1953–64, two localized outbreaks of a strange neurological ailment killed 49 people and left others permanently paralyzed. The cause was determined eventually to be chronic exposure to methylmercury, whose presence had been unexpected. The compound originated with an industrial source and had accumulated in fish and shellfish eaten by the victims. Mercury-contaminated fish have been found since in the U.S., although no harmful effects have been reported. The Food and Drug Administration has banned commercial sales of foods containing more than 0.5 mg/kg of mercury.

While mercury and other inorganic substances, such as arsenic, cadmium, lead, and selenium, might still invade the body by unexpected routes, their hazardous nature has long been known or strongly suspected. Less well defined is the epidemiological evidence for a relationship between cardiovascular disease and chronic ingestion of trace metals in drinking water (129). It is not clear, for example, why metals in water should be especially significant in view of the large amounts of them in the normal diet. Uncertain also is the extent of the threat from chronic ingestion of the many chlorinated and nonchlorinated organic compounds found lately in drinking water (9, 24, 26, 28, 130, 131). A few such compounds, at high concentrations, are carcinogenic in animals and so may be carcinogenic in man, but the long-term, low-level effects of most are unknown.

As water reuse grows more common, it becomes more important to single out potentially harmful substances and remove them at the source or by water treatment. The water reuse need not be intentional; refractory chemicals discharged by one community need only enter the water supply treatment plant of a community downstream. Determination of those substances that might pose a threat in long-term, low-level exposure requires sharply improved knowledge of the specific compounds that are entering receiving waters. It also requires considerable research in areas like epidemiology and long-term, low-level exposure of laboratory animals to a variety of water pollutants (132).

For example, the evidence that supports the primary standard for nitrate in drinking water, 10 mg/l of nitrate-nitrogen, is not definitive. Public health data indicate that the standard is reasonable, but the concentration of nitrate in water correlates only crudely with the frequency of clinical cases of infant methemoglobinemia. There has been only one report of the ailment in infants who drank from public water supplies, which are monitored regularly and in some cases may routinely exceed the standard. Private wells, however, are not monitored, and little is known of how nitrate concentration varies in the same well. Since well water is sampled as a rule only after nitrate injury occurs, the nitrate concentration conceivably could change, so that the analysis might not show the concentration that caused the injury. Nitrate is being found in more and more waters and is a good example of the need for better epidemiological, toxicological, and other kinds of evidence in setting water quality standards.

Inadequate knowledge of health effects is illustrated also by asbestos fibers, which during the 1970's have been detected in surface and drinking waters and in beverages and other products filtered through asbestos (133). Occupational data show that asbestos is a carcinogen when inhaled, but its effects on health when ingested are unknown.

Bacteria, viruses

The prospect of increased reuse of water, controlled or uncontrolled, has sparked growing concern in the past decade over the hazard of bacterial and viral diseases such as salmonellosis and infectious hepatitis. The feces and urine of warmblooded animals are the major sources of bacteria and viruses that can infect man; current bacteriological standards for both potable and recreational waters rely on counts of total coliform bacteria, which are taken to indicate a proportional level of fecal coliforms and viruses. Waterborne outbreaks of salmonellosis have occurred, however, where public water supplies met the standards, and salmonellae have been isolated from waters relatively low in both total and fecal coliforms (134). *Klebsiella*, a pathogenic coliform, has been isolated from chlorinated water supplies that passed standard bacteriological tests, which do not distinguish between that organism and other, nonpathogenic coliforms (135). These and other findings have cast doubt on the ability of total and fecal coliform tests, as now practiced, to indicate the bacteriological quality of water.

A major force behind the drive to control water pollution has been the demand for clean water for recreation. Public health officials have worked for years to maintain at bathing beaches the accepted bacteriological standards based on coliform counts (136). Coliforms come from many sources, however, and the fraction of the total in water that is of fecal origin may range from less than 1% to more than 90%. There is in fact no significant epidemiological basis for the total coliform standards used to assess the quality of bathing waters. Only three epidemiological studies have ever been made in this country to relate the bacteriological quality of natural bathing waters to human health (137). None of the three was conclusive or even highly suggestive as to a suitable standard. Eye, ear, nose, and throat ailments were the dominant illnesses reported. Gastrointestinal disturbances, the type that would be caused by fecal bacteria, accounted for 20% or less of the complaints. Swimmers showed distinctly higher rates of illness than nonswimmers, regardless of the bacteriological quality of the water, a point noted in other studies of illness among swimmers. Large amounts of money are being spent to meet bacteriological quality standards for recreational waters; we can have no assurance that it is being spent wisely until a sounder basis is established for such standards. Research is badly needed on the correlation of currently used or new indicator organisms with waterborne disease. Key questions include the rates of dispersion and inactivation of viruses in recreational waters.

The concern of recent years over waterborne viral diseases (138) has intensified the need to know more of the enteric viruses, those that occur in the gastrointestinal tract and feces of man and many lower animals. The concern exists despite the fact that, of the enteric-virus diseases, only infectious hepatitis has been shown to be transmittable by the water route, and in the U.S. no large outbreaks of that disease have originated in water. During 1971–72 in this country, for example, 11 outbreaks of infectious hepatitis affecting 266 people were attributed to water supplies; of these, four outbreaks affecting 80 people were attributed to municipal water supply systems (139). (Outside of the U.S., only one serious outbreak, in New Delhi, India, is known to have involved treated public water supply.) It has been suggested (140) that evidence for waterborne viral disease may be more apparent in secondary than in primary contacts, where the search thus far has concentrated. The small amount of virus ingested in water, the argument goes, would replicate and multiply but produce only subclinical symptoms in the initial host; clinical symptoms would appear in the secondary host. Uncertain also is the ability of water treatment processes to remove or inactivate viruses, although free chlorine is effective under the proper conditions, and ozone is promising as well (141). Assessing the performance of these disinfectants is difficult in part because techniques for assaying water for viruses at the levels at which they occur naturally are still in the development stage (142). Indeed, in the laboratory the infectious hepatitis virus cannot even be cultured, a necessary step in common virus assay methods. Furthermore, relatively little is known of the levels of enteric viruses likely to be present in drinking water that meets current bacteriological (coliform) standards (141).

It might be possible to use groundwater basins for water treatment and dilution, but the opportunity cannot be assessed intelligently because remarkably few studies have been made on viruses in reuse systems for protracted periods (143). Data on the movement of viruses in groundwater are also extremely limited (144).

Eutrophication

"Eutrophication" is derived from the Greek and means "well nourished;" in a strict sense it refers to the process of enriching a body of water with nutrients. Through common usage, however, "eutrophication" now refers most commonly to the entire process of enrichment and corresponding increases in plant growth, which can occur naturally as well as under the influence of man. More aquatic plants mean changes in the kinds and numbers of animal life, including fish. The process also may produce aesthetic problems like taste, odor, unsightly green scums of algae, and dense growths of rooted plants; oxygen depletion in the deeper waters and bottom sediments of lakes and impoundments; and other chemical changes such as precipitation of calcium carbonate in hard waters. Bodies of water with copious plant life are called eutrophic; those with moderate levels of plants mesotrophic; and those with very low levels of plants oligotrophic or dystrophic (where the water is also colored by yellow organic acids).

The eutrophication process may occur in estuaries, marine bays, and rivers, but most often is a problem in freshwater lakes and impoundments. Accelerated eutrophication often has been associated with rising human populations within lake basins and hence has been given the prefix "cultural." Among many examples of cultural eutrophication are Lake Washington in Seattle, the Madison Lakes in Wisconsin, and Lake Constance in Switzerland. The onset of increased algal growth in such lakes has been linked by strong evidence to the effluents from sewage treatment plants, input of untreated sewage, or both.

Limiting nutrients

Around the turn of the century, the German chemist Justus Liebig postulated that, at any given time, the growth of a plant is limited by the availability of some single substance. Agronomists found that phosphorus, nitrogen, and potassium are the nutrients whose availability most often limits the growth of plants on land. Potassium does not appear to be critical to plants growing in water, but attention has long focused on nitrogen and phosphorus as primary controls on growth in that medium.

Field and laboratory studies in recent years have refined the concept of limiting nutrients in several ways. Scientists have realized increasingly that, in natural communities of plants, different nutrients may limit the growth of different types of plants at the same instant in time. They have realized also that the growth-limiting nutrient for a single type of plant may change in a relatively short time: it may be, say, phosphorus one week and nitrogen the next. Most aquatic scientists now agree that no one substance is likely continuously to be the growth-limiting factor over a long (seasonal) period. Iron, cobalt, manganese, other trace elements, vitamins, and sometimes other macronutrients, such as sulfur, may affect not only the instantaneous rate of growth of algae, but the composition of the algal community as well.

Despite these principles, evidence is growing that, in most lakes and impoundments, phosphorus is the nutrient likely to be controlling the seasonal production of algae. In the National Eutrophication Survey, for example, phosphorus was found to be the limiting element for algal growth in 57% of the lakes sampled in 1972 (14). Nitrogen was the limiting element in 37% of those sampled, but it does not usually become limiting until the water already is rich in nutrients. Even where phosphorus inputs are small, that element is the easiest of the likely limiting substances to control and should, therefore, provide a key to regulating algal production. In fact, however, progress toward reducing the concentration of phosphorus in surface waters on a national basis has been limited at best. It appears, moreover, that nonpoint as well as point sources of phosphorus and other nutrients must be controlled to exert an appreciable impact on eutrophication.

A study of 17 Wisconsin lakes more than three decades ago led to the suggestion that when certain levels of inorganic phosphorus and inorganic nitrogen were exceeded during the winter, the period of minimum biological activity, algal blooms of nuisance proportions were likely during the following summer (145). (The levels were 0.015 mg/l of inorganic phosphorus and 0.3 mg/l of inorganic nitrogen.) This is a perfectly logical approach to the estimation of critical levels of nutrients. Such guideline values might even be extrapolatable to other bodies of water in similar hydrologic and climatic circumstances. These values have been misused repeatedly, however, as absolute concentrations indicating the trophic conditions of lakes, regardless of the time of year the concentrations are determined. Such interpretations disregard the dynamic nature of nutrient-algae relationships, where the critical factor in nutrient limitation is rate of supply versus rate of demand. In a completely mixed system, in fact, an algal population can grow and accumulate with no residual inorganic phosphorus or nitrogen in solution, so long as supply and uptake are exactly balanced. Further evidence that total phosphorus in the water column during the winter is a useful index of standing crops of phytoplankton the following summer has been reported from surveys of numerous Japanese and other, largely North American, lakes (146, 147).

A substantial controversy arose in the late 1960's over the possibility that carbon could be a limiting nutrient. The question may still be germane for extremely eutrophic waters, such as sewage oxidation ponds. Later research has shown convincingly, however, that in natural lakes and impoundments carbon is almost certainly available in superabundant amounts (148–150).

These remarks on eutrophication apply mainly thus far to problems associated with algae, particularly the minute, free-floating phytoplankton. Other kinds of algae grow attached to rocks, and one, *Cladophora*, has become a major nuisance in the lower Great Lakes. Higher plants, rooted in the bottom, may hinder recreation in shallower waters (less than 4 m) over soft bottoms. Rooted aquatic plants, given suitable bottoms, grow in lakes of any trophic state. They seem to become more dense, however, where nutrient concentrations are high. The kinds of plants that occur in trophic conditions probably differ as well, but both the differences and the details of the plants' nutrition remain poorly defined. When concentrations of inorganic phosphorus and nitrogen are very high, the free-floating phytoplankton generally outcompete (for nutrients) plant life that is rooted or otherwise attached.

Eutrophication often is portrayed as a natural, unidirectional process. Evidence is mounting, however, that for long periods (on a geological scale) lakes not exposed to human impacts may remain in trophic equilibrium. In a few cases they may even become more oligotrophic. It is now known also that the eutrophication process may be slowed or reversed by decreasing the supply of nutrients available to plant life (151–153).

Nutrient inputs to receiving waters have been controlled in several ways. The effluent from municipal sewage treatment plants has been diverted from the body of water (Lake Washington, the Madison lakes); phosphorus has been removed from wastewater by tertiary treatment (Lake Tahoe, the Great Lakes); and the receiving water has been diluted and flushed with low-nutrient water (Green Lake). Phosphorus has been controlled also in some areas by prohibiting the use of phosphates as builders in household laundry detergents. The restriction reduces by up to 40% the amount of phosphorus that enters sewage treatment plants (154).

Three areas of controversy surround phosphorus control. These are the relative quantitative importance of various sources of the element; the chemical and physical forms of phosphorus that ultimately are available to support plant growth; and the relative importance of phosphorus recycled from deep waters and bottom sediments to near the surface where it might support plant growth.

Nutrient budgets

Sources of phosphorus (or other nutrients) are usually treated in terms of nutrient budgets for particular basins and their receiving waters. The necessary numbers are difficult to estimate adequately because of the very large numbers of samples needed and the great importance of hydrology and climate. It is even more difficult to assess the contribution of phosphorus or other nutrients to a body of water by particular human activities on land, such as application of fertilizers, or even by point discharges of wastes. One reason is that nutrients may be lost in transport to the lake owing to uptake by stream life or stream-bank vegetation or to chemical precipitation.

The National Eutrophication Survey was an attempt to upgrade our knowledge of the trophic condition of the nation's waters and of the relevant nutrient budgets and sources (15, 155). Potentially more significant than this widespread but not very intensive sampling program is the North American Lakes Project and two similar efforts in Europe. These were begun in 1974. They are an attempt to bring together governmental and university scientists who were conducting in-depth studies of particular lakes, or groups of lakes, for systematic exchanges of current ideas and data. "Models" relating various combinations of parameters such as nutrient loading, lake depth, water retention time, lake trophic state, summer standing crops of algae, water transparency, and winter phosphorus concentrations have been reported to the North American group (156).

Linked to the problem of calculating phosphorus budgets is the question of which forms of the element are biologically significant and thus should be measured. We know, for example, that aquatic plants, including algae, can readily use phosphorus in soluble, inorganic forms. We know also that some forms of algae can recapture phosphorus from soluble organic compounds. Forms of phosphorus adsorbed on suspended particles settle to the bottoms of lakes and impoundments, and, in shallow water, would be available to rooted plants. There is little evidence, however, that any of this phosphorus becomes available to algae. Clays adsorb inorganic phosphates particularly efficiently and may actually "scrub" phosphorus from solution in this way as they settle through water.

It has long been known that phosphorus precipitated in various forms may reenter solution in waters that are anoxic (deficient in free oxygen). In deep lakes this phosphorus remains precipitated in the deeper water during the principle growing season for phytoplankton. It reenters solution and circulates throughout the water column during the autumn mixing period and reprecipitates during the fall and winter. It is unlikely that much of this phosphorus supports the growth of phytoplankton. The situation is much less clear in very shallow lakes, where vertical mixing occurs through most of the year.

Nonnutrient controls

When undesirably high concentrations of plants do appear, they can be controlled with chemicals. In some circumstances, the long-employed copper sulfate is still used effectively to control algal blooms. There is a growing trend, however, toward specially formulated organic herbicides. One need here is deeper understanding of the overall ecological effects of the compounds, including their significance to public drinking water supplies, actual or potential.

Rooted aquatic plants may be controlled also by mechanical harvesting, and considerable research and development is under way to develop optimal methods. It is critical that harvested plants be removed from the water; otherwise they release nutrients to the water when they decompose. Whenever growths of higher plants have been controlled, but not removed, increases in phytoplankton populations have followed.

Plants and fish

All animals depend ultimately on plants as a source of food. This means, for example, that, within limits, the yield of fish from a body of water will increase as plant production increases. The limits are determined by a number of complex factors, and aquatic scientists are just beginning to study the quantitative aspects of these relationships. The problem of seeking a balance between plant and fish production is complicated further by the fact that the trophic state of a natural body of water very much influences the kinds of fish that appear in it. In more eutrophic situations, less desirable species predominate.

Tastes and Odors

A number of substances, both man-made and natural, can cause tastes and odors in water and impart off-flavors to fish and shellfish. Only traces of some organic compounds are objectionable. Chlorophenol, for example, has been found to produce an unpleasant taste in fish at a concentration of 0.0001 mg/l (157).

Tastes and odors are caused primarily by organic materials; for public water supply, the best all-purpose treatment is adsorption on activated carbon, although oxidation to innocuous form with, say, chlorine or ozone is also used. Research, meanwhile, is directed in part at isolating and identifying the specific chemical compounds that cause tastes and odors and determining the sources of those compounds. One instance is the musty, earthy odors, typical of freshly plowed soil, that often occur in water supplies. A specific organic compound believed to be largely responsible for such odors has been isolated from bacteria of the actinomycetes group (158) and from two species of blue-green algae (159, 160); it has been detected also in the effluent from secondary sewage treatment plants. The compound has been found to be a dimethyl decalol (161), but commonly is called geosmin. Other odoriferous compounds have been isolated from actinomycetes, too (162, 163). The evidence suggests that the earthy odors in nature may be caused by several compounds that share common structural features. Should this be true, the number of variables involved in preventing or removing such odors might be considerably reduced.

Recommendation W22: *Research should be strongly supported on the health effects of known water pollutants in long-term, low-level exposure. Both laboratory and epidemiological work will be required.*

Recommendation W23: *To establish perspective in setting standards and regulations, efforts should be directed toward defining the risk of exposure to toxic substances in numerical terms based on currently known statistics such as those on morbidity and mortality.*

Recommendation W24: *Studies should be maintained on enteric viruses and their movement in soil and groundwater, using improved analytical techniques.*

Recommendation W25: *Correlations of bacterial indicator organisms with waterborne disease, particularly in recreational waters, should receive high priority.*

Recommendation W26: *Public water supply treatment methods should be upgraded through research on removal and destruction of potentially harmful substances not removed by waste treatment practices or bypassed with insufficient dilution during plant outage periods or in times of disaster such as power failures.*

Recommendation W27: *Investigations should be pursued of the fundamental chemical and biological parameters of eutrophication and its effects. Development of effective and economic long-term controls will depend on considerably improved knowledge of factors such as mass balances for significant nutrients; the forms in which those nutrients exist in water; natural population dynamics; potentially limiting nutrients in specific situations; and algal, bacterial, and plant physiology in general. Investigations of a variety of lakes to document responses to changes in nutrient (especially phosphorus) inputs are especially needed if effective, economic, long-term strategies for controlling eutrophica-*

tion are to be developed. Also critical to this question is a better understanding of nutrient sources, the transport of nutrients to standing bodies of water, and the biological reactivity of various chemical and physical forms of nutrients.

ALTERNATIVE MANAGEMENT STRATEGIES

The scope of this country's effort to manage its waters makes institutional problems unavoidable. A major problem has been the traditional separation of the management of water supplies, the management of wastewater collection and treatment, and the management of storm and flood waters at local, state, and federal levels of government. Each self-serving unit has its own goals, constituencies, and procedures, which creates a situation marked by complexity, inertia, and the absence of a clear-cut, overall strategy for managing water resources. The development of a national water policy (164) that stresses central administration of water programs based on watershed boundaries promises benefits that are unlikely to result from existing policies and institutions.

The federal assistance policies and the uniform treatment standards of the Water Pollution Control Act Amendments of 1972 have worked against the cost-effective cleanup of the nation's waters. Few communities have built wastewater treatment plants without the federal assistance (75% of construction costs) mandated under the 1972 amendments. The uncertain availability of these funds has been a bottleneck, slowing the rate at which treatment facilities are constructed. The availability of such funds also has eliminated the incentive for communities to fully finance their own treatment systems, especially where their actions would only benefit downstream users. While the 1972 amendments provide for regional planning of municipal wastewater treatment facilities, the overall management of water resources, including water supply, is neglected.

The management for water supplies does not recognize basinwide needs, and the Safe Drinking Water Act of 1974 imposes federal requirements that are insensitive to the needs and resources of local communities. Each community must establish water rights, and the monopolies thereby created can work hardships on neighboring communities. Residents of a watershed have no existing incentives to cooperate in establishing their water supplies. The Safe Drinking Water Act requires water suppliers to monitor their products to insure compliance with the standards for the specified contaminants. Many communities are too small to finance proper treatment and monitoring of their water supplies.

At least 80% of the drinking water produced by communities is not ingested but used in industry and for urban irrigation, washing cars, and flushing toilets. Highly treated wastewaters may be an alternative supply for these needs. But because of the fragmented management of the various elements of the water resource in a basin, few voices are raised for change, conservation, or innovation. The use of highly treated wastewaters for nonpotable purposes cannot easily be implemented under the existing system. This fragmentation and limitation of the existing approach to water-resource management were probably inevitable, given the country's diversity in political organization, geography, and concentration of population and industry. However, as the competition for water becomes more intense, regionalization and centralization of water management offer inherent advantages as long as broad federal guidelines exist to prevent regional degradation of the nation's waters.

Regionalized Water Management

During the past few years, water management has been regionalized and centralized on a river-basin basis in England and Wales. While these moves were stimulated by a particular set of conditions (165), their success makes them worth close examination in terms of future directions for water management in the U.S.

The present, regionalized system of water management in England and Wales (166, 167) is a product of three major pieces of legislation: the Water Act 1945, the Water Resources Act 1963, and the Water Act 1973. The organizational impact of these laws can be seen in selected statistics. In 1945, some 1186 drinking water supply operations in England and Wales were serving 40 million people; in 1974, only 187 facilities were serving 49 million people, and only 0.5% of the population was without public water supply. By 1971, water management was being handled by 1393 sewerage and wastewater disposal agencies and 29 river authorities in addition to the 187 water supply operations. On April 1, 1974, the facilities, other assets, and debts of these 1609 separate agencies were taken over by 10 regional water authorities in England and Wales. The authorities were based on hydrologic (watershed) boundaries; initially they had 65,000 employees overall and an annual budget of about $1.5 billion. Each authority is responsible within its region for the ownership, planning, design, construction, operation, and financing of facilities for:

- Conservation, augmentation, distribution, and use of water resources, including drinking water supplies.
- Provision of sewerage and treatment of wastewaters and other effluents.
- Restoration and maintenance of the quality of rivers and other inland waters, as well as of estuaries and coastal waters.
- Use of inland waters for recreation and preservation of the aesthetic values of such waters.
- Land drainage and flood prevention.
- Fisheries and navigation in inland waters.

The regional water authorities by intent are subject to a minimum of centralized direction. The main central body is the National Water Council, made up of the (appointed) chairmen of the 10 authorities and 10 additional members appointed for their special interests and competencies. Under the law, the council is responsible principally for testing equipment and training personnel. In practice, however, it has evolved into an important forum for making decisions on the implementation of water-management policy.

Although the current approach to water management in England and Wales dates only from April 1974, some of its potential advantages have been realized already. Very early, qualified technologists were made available to supervise the operation of small wastewater treatment plants that previously had been ignored. Construction of a large reservoir for water supply, approved earlier, was delayed when it was found that the area involved could be served readily by excess resources nearby. Regionalization clearly demonstrated its value during the severe drought of 1976, when it facilitated transfers of water to water-short areas (168).

With management on a river-basin basis, the locations and requirements for specific waste treatment facilities can be adjusted according to the ability of the receiving waters to assimilate wastewater and the demands of water users at the point of waste discharge. On the Thames, for example, a much higher level of wastewater treatment is required in the headwaters than in the estuary. In addition, a watershed management authority can facilitate intersubbasin water transfers without encountering state or federal bureaucracies. A basin authority can also address the issues of additional demands for withdrawal, improved water quality, or additional waste discharge, since all water users are regulated and each can be regulated to permit new entrants to share in the water resources of the basin. If all users of water in a basin must be accountable for their actions to all other users, and all users must share in the costs as well as the benefits of the water resources, the management of the water resource would be improved.

Each of the water authorities in England and Wales is required to support itself financially; support by the national exchequer was eliminated by the legislation that established the authorities. With the loss of national funds, and subsequent improvements in services and facilities, costs have risen substantially. In consequence, consumers have become much more aware of the true costs of water, and much attention is being paid to economic incentives and means of distributing charges equitably. In the U.S., on the other hand, the real costs of water and its disposal are masked by distortions arising in major part from the means of financing the facilities involved.

Britain and the U.S. differ in many ways, and their means of managing water resources must differ correspondingly. Even so, the system now in place in England and Wales warrants careful and continuing examination. Not the least of the advantages offered by regionalization is technical flexibility — the capacity to provide technical services to facilities that otherwise could not afford them, to select the best of alternatives based on all the needs of all members of a river-basin community and not on the needs of a special group, to use a given water resource for the purposes to which it is most suited. Water-management regions no doubt would be difficult to establish in this country, in part because hydrologic and political boundaries rarely coincide. But purposeful movement toward more regionalized operations under federally established policy seems essential to sound, long-term conservation of our water resources.

Recommendation W28: *High priority should be given to the development of innovative alternative strategies for the management of our water resources. In particular, the strategy of regionalization, including the consolidation of management of river basins through authorities responsible for water supply, wastewater disposal, and overall water-resources management and planning, should be carefully explored with its advantages and disadvantages enumerated. Recent British experience with consolidation of management warrants careful examination to determine its applicability to U.S. water management programs.*

LITERATURE CITED

1. "Equilibrium Concepts in Natural Water Systems," Gould, R.F., ed., Adv. Chem. Ser., **67,** American Chemical Society, Washington, D.C., 1967.
2. Murray, C.R., Reeves, E.B., "Estimated Use of Water in the United States in 1975," U.S. Geol. Survey Circ. 765, U.S. Geological Survey, Washington, D.C., 1977.
3. "Economics of Clean Water," 6th Annual Report to the Congress, U.S. Environmental Protection Agency, Washington, D.C., December 1973.
4. Guidelines for Water Quality Standards, Environmental Protection Agency, Washington, D.C., February 1973.

5. *Fed. Regist.*, **43,** 4108, Jan. 31, 1978.

6. "National Water Quality Inventory, 1975 Annual Report to Congress," Environmental Protection Agency, U.S. Government Printing Office, Washington, D.C., May 1976.

7. "Environmental Quality — 1977," Eighth Annual Report of the Council on Environmental Quality, U.S. Government Printing Office, Washington, D.C., December 1977.

8. "Report to the Congress by the National Commission on Water Quality," U.S. Government Printing Office, Washington, D.C., March 1976.

9. "Drinking Water and Health," National Academy of Sciences — National Research Council, Washington, D.C., 1977.

10. *Fed. Regist.*, **43,** 5756, Feb. 9, 1978.

11. "Delaware Estuary Comprehensive Study, Preliminary Report and Findings," U.S. Department of the Interior, Federal Water Pollution Control Administration, Washington, D.C., July 1966.

12. "Lake Erie Report. A Plan for Water Pollution Control," U.S. Department of the Interior, Federal Water Pollution Control Administration, Washington, D.C., August 1968.

13. Luken, R.A., Basta, D.J., Pechan, E.H., "The National Residuals Discharge Inventory," National Research Council, Washington, D.C., January 1976, National Technical Information Service, PB 252 288, Springfield, Va.

14. "Environmental Quality — 1976," Seventh Annual Report of the Council on Environmental Quality, U.S. Government Printing Office, Washington, D.C., September 1976.

15. "Nonpoint Source–Stream Nutrient Level Relationships: A Nationwide Study," Publ. No. EPA-600/7-77-105, Environmental Protection Agency, Corvallis, Ore., 1977.

16. Menzies, J.D., Chaney, R.L., "Waste Characteristics," in "Factors Involved in Land Application of Agricultural and Municipal Wastes," U.S. Department of Agriculture, Agricultural Research Service, National Program Staff Publ., Beltsville, Md., 18 (1974).

17. Bunch, R.L., Barth, E.F., Ettinger, M.B., "Organic Materials in Secondary Effluents," *J. Water Pollut. Control Fed.*, **33,** 122 (1961).

18. Hunter, J.V., Heukelekian, H., "The Composition of Domestic Sewage Fractions," *J. Water Pollut. Control Fed.*, **37,** 1142 (1965).

19. Painter, H.A., Viney, M., "Composition of a Domestic Sewage," *J. Biochem. Microbiol. Technol. Eng.*, **1,** 143 (1959).

20. Kolattukudy, P.E., Purdy, R.E., "Identification of Cutin, a Lipid Biopolymer, as Significant Component of Sewage Sludge," *Environ. Sci. Technol.*, **7,** 619 (1973).

21. Sridhar, M.K.C., Pillai, S.C., "Proteins in Wastewater and Wastewater Sludges," *J. Water Pollut. Control Fed.*, **45,** 1595 (1973).

22. Christman, R.F., Hrutfiord, B.F., "Water Quality Standards for Organic Contaminants: Analytical Limitations and Possibilities," Proc., 15th Water Quality Conference, University of Illinois, Champaign, Ill., February 1973.

23. "Identification and Analysis of Organic Pollutants in Water," Keith, L.H., ed., Ann Arbor Science Publishers, Inc., Ann Arbor, Mich., 1976.

24. Symons, J.M., et al., "National Organics Reconnaissance Survey for Halogenated Organics," *J. Am. Water Works Assoc.*, **67,** 634 (1975).

25. Jolley, R.L., "Chlorine-Containing Organic Constituents in Chlorinated Effluents," *J. Water Pollut. Control Fed.*, **47,** 601 (1975).

26. Kleopfer, R.D., Fairless, B.J., "Characterization of Organic Components in a Municipal Water Supply," *Environ. Sci. Technol.*, **6,** 1036 (1972).

27. Pitt, W.W., Jr., Jolley, R.L., Scott, C.D., "Determination of Trace Organics in Municipal Sewage Effluents and Natural Waters by High-Resolution Ion-Exchange Chromatography," *Environ. Sci. Technol.*, **9**, 1068 (1975).

28. Keith, L.H., "GC-MS Environmental Analyses: Water and Air," *J. Chromatog. Sci.*, in press (as of August 1978).

29. Field, R., Struzeski, E.J., Jr., "Management and Control of Combined Sewer Overflows," *J. Water Pollut. Control Fed.*, **44**, 1393 (1972).

30. Rudolfs, W., Balmat, J.L., "Colloids in Sewage. I. Separation of Sewage Colloids with the Aid of the Electron Microscope," *Sewage Ind. Wastes*, **24**, 247 (1952).

31. Rickert, D.A., Hunter, J.V., "Colloidal Matter in Wastewaters and Secondary Effluents," *J. Water Pollut. Control Fed.*, **44**, 134 (1972).

32. Linsley, R.K., Franzini, J.B., "Water Resources Engineering," 2nd ed., McGraw-Hill Book Co., New York, N.Y., 1972.

33. Clark, J.W., Veissman, W., Jr., "Water Supply and Pollution Control," International Textbook Co., Scranton, Pa., 1965.

34. "Standard Methods for the Examination of Water and Wastewater," 14th ed., American Public Health Association, Washington, D.C., 1975.

35. "Symposium on Particulates in Water," Division of Environmental Chemistry Preprints, Vol. 18, No. 1, 175th National Meeting of the American Chemical Society, Anaheim, Calif., March 12–17, 1978.

36. Lammers, W.T., "Biophysical Limnology. Separation of Suspended and Colloidal Particles from Natural Water," *Environ. Sci. Technol.*, **1**, 52 (1967).

37. Zabik, M.J., "The Contribution of Urban and Agricultural Pesticide Use to the Contamination of the Red Cedar River," Project Completion Report, Project No. A-012-Mich., Office of Water Resources Research, U.S. Department of the Interior, Washington, D.C., 1969.

38. Hamelink, J.L., Waybrant, R.C., Ball, R.C., "A Proposal: Exchange Equilibria Control the Degree Chlorinated Hydrocarbons Are Biologically Magnified in Lentic Environments," *Trans. Am. Fish. Soc.*, **100**, 207 (1971).

39. Bailey, G.W., "Role of Soils and Sediment in Water Pollution Control. Part 1. Reactions of Nitrogenous and Phosphatic Compounds with Soils and Geologic Strata," U.S. Department of Interior, Federal Water Pollution Control Administration, Southeast Water Laboratory, Athens, Ga., March 1968.

40. McGauhey, P.H., Krone, R.B., "Soil Mantle as a Wastewater Treatment System," Sanitary Engineering Research Laboratory Report No. 67-11, University of California, Berkeley, Calif., December 1967.

41. Robeck, G.G., et al., "Factors Influencing the Design and Operation of Soil Systems for Waste Treatment," *J. Water Pollut. Control Fed.*, **36**, 971 (1964).

42. Thomas, R.E., Schwartz, W.A., Bendixen, T.W., "Soil Chemical Changes and Infiltration Rate Reduction Under Sewage Spreading," *Soil Sci. Soc. Amer. Proc.*, **30**, 641 (1966).

43. McKelvey, V.E., "Underground Space — An Unappraised Resource," in "Underground Waste Management and Environmental Implications," Cook, T.D., ed., Amer. Assoc. Petroleum Geologists, Tulsa, Okla., 1972, p. 1–5.

44. Stallman, R.W., "Subsurface Waste Storage — the Earth Scientist's Dilemma," in "Underground Waste Management and Environmental Implications," Cook, T.D., ed., Amer. Assoc. Petroleum Geologists, Tulsa, Okla., 1972, p. 6–10.

45. Schneider, W.J., "Hydrologic Implications of Solid-Waste Disposal," U.S. Geol. Survey Circ. 601-F, U.S. Geological Survey, Washington, D.C., 1970.

46. "Chemistry and the Oceans," *Chem. Eng. News*, **42** (22), 1A (1964).

47. Livingstone, D.A., "Chemical Composition of Rivers and Lakes," in "Data of Geochemistry," Fleischer, Michael, ed., U.S. Geol. Survey Prof. Paper 440-G, U.S. Geological Survey, Washington, D.C., 1963.

48. Holeman, J.N., "The Sediment Yield of Major Rivers of the World," *Water Resour. Res.*, **4,** 737 (1968).

49. Ocean Dumping: Final Regulations and Criteria. *Fed. Regist.*, **38,** 198, Part II, 28610, Oct. 15, 1973.

50. "Baseline Studies of Pollutants in the Marine Environment and Research Recommendations," the International Decade of Ocean Exploration Baseline Conference, May 24–26, 1972, New York, N.Y., 1972.

51. "Cost Estimates for Construction of Publicly-Owned Wastewater Treatment Facilities — 1976 Needs Survey: Report to Congress," Publ. No. EPA-430/9-76-010; "Summaries of Technical Data (Categories I–IV) — 1976 Needs Survey," Publ. No. EPA 430/9-76-011; Environmental Protection Agency, Washington, D.C., 1977.

52. Burd, R.S., "A Study of Sludge Handling and Disposal," Federal Water Pollution Control Administration, Publication WP 20-4, Washington, D.C., May 1968.

53. Furr, A.K., et al., "Multielement and Chlorinated Hydrocarbon Analysis of Municipal Sewage Sludges of American Cities," *Environ. Sci. Technol.*, **10,** 683 (1976).

54. "Wastewater Engineering," Metcalf and Eddy, Inc., McGraw-Hill Book Co., New York, N.Y., 1972.

55. Technology Transfer "Process Design Manual for Sludge Treatment and Disposal," U.S. Environmental Protection Agency, EPA 625/1-74-006, Washington, D.C., October 1974.

56. Zenz, D.R., et al., "Environmental Impacts of Land Application of Sludge," *J. Water Pollut. Control Fed.*, **48,** 2332 (1976).

57. Hinesly, T.D., et al., "Effects of Annual and Accumulative Applications of Sewage Sludge on Assimilation of Zinc and Cadmium by Corn (*Zea* mays L.)," *Environ. Sci. Technol.*, **11,** 182 (1977).

58. Field, R., Struzeski, E.S., Jr., "Management and Control of Combined Sewer Overflows," *J. Water Pollut. Control Fed.*, **44,** 1393, (1972).

59. Field, R., "Combined Sewer Overflows," *Civil Eng.*, **43,** 2, 57 (1973).

60. "Total Urban Water Pollution Loads — The Impact of Storm Water," Environmental Control, Inc., prepared for Council on Environmental Quality, National Technical Information Service, PB 231 730, Springfield, Va., 1974.

61. Field, R., Szecley, P.S., "Urban Runoff and Combined Sewer Overflow," Literature Review, *J. Water Pollut. Control Fed.*, **46,** 6, 1209 (1974).

62. Field, R., Weigel, P., "Urban Runoff and Combined Sewer Overflow," Literature Review, *J. Water Pollut. Control Fed.*, **45,** 6, 1108 (1973).

63. Poertner, H.G., "Better Storm Drainage Facilities at Low Cost," *Civil Eng.*, **43,** 10, 67 (1973).

64. Lager, J.A., "Stormwater Treatment — Four Case Histories," *Civil Eng.*, **44,** 12, 40 (1974).

65. Westerhoff, G.P., Daly, M.P., "Water-Treatment Plant Wastes Disposal." Part I., *J. Am. Water Works Assoc.*, **66,** 5, 319 (1974); Part II., *Ibid.*, **66,** 6, 378 (1974); Part III, *Ibid.*, **56,** 7, 441 (1974).

66. "Water Reuse Highlights — A Summary Volume of Wastewater Reclamation and Reuse Information," American Water Works Association Research Foundation, Denver, Colo., 1978.

67. Rizzo, J.L., Schade, R.E., "Secondary Treatment with Granular Activated Carbon," *Water & Sewage Works*, **116,** 307 (1969).

68. Hager, D.G., Reilly, P.B., "Clarification-Adsorption in the Treatment of Municipal and Industrial Wastewater," *J. Water Pollut. Control Fed.*, **42,** Part 1, 794 (1970).

69. Weber, W.J., Jr., Hopkins, C.B., Bloom, R., Jr., "Physicochemical Treatment of Wastewater," *J. Water Pollut. Control. Fed.*, **42,** 83 (1970).

70. Weber, W.J., Jr., Friedman, L.D., Bloom, R., Jr., "Biologically-Extended Physicochemical Treatment," Proceedings, Sixth Conference on Water Pollution Research, Jerusalem, June 18–24, 1973.

71. Bishop, D.F., O'Farrell, T.P., Stamberg, J.B., "Physical-Chemical Treatment of Municipal Wastewater," *J. Water Pollut. Control Fed.*, **44**, 361 (1972).

72. Atkins, P.F., Scherger, D.A., Barnes, R.A., "Ammonia Removal in a Physical-Chemical Wastewater Treatment Process," Proc., 27th Annual Industrial Waste Conference, Purdue University, Lafayette, Ind., 1972.

73. "Regional Agency Starts Up Physical-Chemical Treatment Plant," *Environ. Sci. Technol.*, **7**, 804 (1973).

74. Hager, D.G., "A Survey of Industrial Waste Treatment by Granular Activated Carbon," presented at the 4th Joint Chemical Engineering Conference, American Institute of Chemical Engineers — Canadian Society for Chemical Engineering, Vancouver, B.C., September 1973.

75. Weber, W.J., Jr., "Physicochemical Processes for Water Quality Control," Wiley-Interscience, John Wiley & Sons, Inc., New York, N.Y., 1972.

76. Culp, R.L., Culp, G.L., "Advanced Wastewater Treatment," Van Nostrand Reinhold Co., New York, N.Y., 1971.

77. Linstedt, K.D., Houck, C.P., O'Connor, J.T., "Trace Element Removals in Advanced Wastewater Treatment Processes," *J. Water Pollut. Control Fed.*, **43**, 1507 (1971).

78. Elly, C.T., Weber, W.J., Jr., "Preliminary Evaluation of the Behavior of Zinc, Copper, and Chromium in the Treatment of Wastewater by Coagulation," in "Independent Physical-Chemical Treatment Facilities for Southeastern Michigan," Southeastern Michigan Wastewater Management Study Publication, Detroit District, U.S. Army Corps of Engineers, May 1973.

79. Directo, L.S., Chen, C.-L., Kugelman, I.J., "Pilot Plant Study of Physical-Chemical Treatment," *J. Water Pollut. Control Fed.*, **49**, 2081 (1977).

80. Moss, W.H., et al., "Full-Scale Use of Physical/Chemical Treatment of Domestic Wastewater at Rocky River, Ohio," *J. Water Pollut. Control Fed.*, **49**, 2249 (1977).

81. Weber, W.J., Jr., "Fluid-Carbon Columns for Sorption of Persistent Organic Pollutants," Proceedings, 3rd International Conference on Water Pollution Research, **1**, 253 (1967).

82. Weber, W.J., Jr., Hopkins, C.B., Bloom, R., Jr., "A Comparison of Expanded-Bed and Packed-Bed Adsorption Systems," Report No. TWRC-2, U.S. Department of the Interior, Federal Water Pollution Control Administration, Cincinnati, Ohio, 1968.

83. Hopkins, C.B., Weber, W.J., Jr., Bloom, R., Jr., "Granular Carbon Treatment of Raw Sewage," Report No. ORD-17050DAL05/70, Water Pollution Control Research Series, U.S. Department of the Interior, Federal Water Pollution Control Administration, Cincinnati, Ohio, 1970.

84. Netzer, A., Norman, J.D., "Removal of Trace Metals from Wastewaters by Activated Carbon," Technical Report, Department of Chemical Engineering, McMaster University, Hamilton, Ont., Canada, 1973.

85. Koon, J.H., Kaufman, W.J., "Ammonia Removal from Municipal Wastewaters by Ion Exchange," *J. Water Pollut. Control Fed.*, **47**, 448 (1975).

86. Gilbert, R.G., et al., "Wastewater Renovation and Reuse: Virus Removal by Soil Filtration," *Science*, **192**, 1004 (1976).

87. Gerba, C.P., Wallis, C., Melnick, J.L., "Viruses in Water: The Problem, Some Solutions," *Environ. Sci. Technol.*, **9**, 1122 (1975).

88. Nebel, C., et al., "Ozone Disinfection of Industrial-Municipal Secondary Effluents," *J. Water Pollut. Control Fed.*, **45**, 2493 (1973).

89. Rosen, H.M., "Use of Ozone and Oxygen in Advanced Wastewater Treatment," *J. Water Pollut. Control Fed.*, **45**, 2521 (1973).

90. Weber, W.J., Jr., "Physicochemical Systems for Direct Wastewater Treatment," in "Applications of New Concepts of Physical-Chemical Treatment," Pergamon Press, Inc., Elmsford, N.Y., 1972.

91. Antonucci, D.C., Schaumburg, F.D., "Environmental Effects of Advanced Wastewater Treatment at South Lake Tahoe," *J. Water Pollut. Control Fed.*, **47,** 2694 (1975).

92. "1967 Census of Manufactures," Vol. 1, Summary and Subject Statistics, Bureau of the Census, U.S. Department of Commerce, Washington, D.C., January 1971.

93. "Monitoring to Detect Previously Unrecognized Pollutants in Surface Waters," prepared by the Institute for Environmental Studies, University of Illinois, for the Environmental Protection Agency, National Technical Information Service, PB 273 349, Springfield, Va., 1977.

94. *Fed. Regist.*, **42,** 6476, Feb. 2, 1977; **43**, 27736, June 26, 1978.

95. State Program Elements Necessary for Participation in the National Pollutant Discharge Elimination System, Guidelines for Acquisition of Information from Owners of Point Sources, *Fed. Regist.*, **38,** 141, Part II, 19894, July 24, 1973.

96. "The Economics of Clean Water," Vol. I, 5th Annual Report to the Congress, U.S. Environmental Protection Agency, Washington, D.C., 1972.

97. Davis, J.C., "Water Reuse: A Trickle Becomes a Torrent," *Chem. Eng.* (N.Y.), **85,** 10, 44 (1978).

98. Reiter, W.M., Stocker, W.F., "In-Plant Waste Abatement," *Chem. Eng. Prog.*, **70,** 1, 55 (1974).

99. "The Cost of Clean Water. Vol. III, Industrial Waste Profiles," U.S. Department of the Interior, Federal Water Pollution Control Administration, U.S. Government Printing Office, Washington, D.C. 1968.

100. "Chlor-Alkali Producers Shift to Diaphragm Cells," *Chem. Eng.*, (N.Y.), **81**, 4, 84 (1974).

101. "Chlor-Alkali Membrane Cell Set for Market," *Chem. Eng. News*, **56,** 12, 20 (1978).

102. Hoover, S.R., "Prevention of Food-Processing Wastes," *Science*, **183,** 824 (1974).

103. Baker, C.D., et al., "Recovering *para*-Cresol from Process Effluent," *Chem. Eng. Prog.*, **69,** 77 (1973).

104. Bennett, G.F., Lash, L., "Industrial Waste Disposal Made Profitable," *Chem. Eng. Prog.*, **70,** 75 (1974).

105. Leitz, F.B., "Electrodialysis for Industrial Water Cleanup," *Environ. Sci. Technol.*, **10,** 136 (1976).

106. "Waste Recovery," *Chem. Eng. Prog.*, **73,** 5, 45–73 (1977).

107. Schmidt, R.K., "How to Meet Water Cleanup Deadlines," *Environ. Sci. Technol.*, **10,** 140 (1976).

108. Davis, J.C., "Activated Carbon: Prime Choice to Boost Secondary Treatment," *Chem. Eng. (N.Y.)*, **84,** 8, 81 (1977).

109. Warnke, J.E., Thomas, K.G., Creason, S.C., "Wastewater Reclamation System Ups Productivity, Cuts Water Use," *Chem. Eng. (N.Y.)*, **84,** 7, 75 (1977).

110. Guy, H.P., "Sediment Problems in Urban Areas," U.S. Geol. Survey Circ. 601-E, U.S. Geological Survey, Washington, D.C., 1970.

111. "Water Policies for the Future," Final Report of the National Water Commission, U.S. Government Printing Office, Washington, D.C., 1973.

112. *Fed. Regist.*, **41,** 11458, March 18, 1976.

113. Shuyler, L.R., "National Animal Feedlot Wastes Research Program," Environmental Protection Agency, Publ. EPA-R2-73-157, U.S. Government Printing Office, Washington, D.C., 1973.

114. "Accumulation of Nitrate," National Academy of Sciences–National Research Council, Washington, D.C., 1972.

115. Goldberg, M.C., "Sources of Nitrogen in Water Supplies," in "Role of Agriculture in Clean Water," Smith, G.E., Willridge, T.H., eds., Iowa State University Press, Ames, Iowa, 1971, p. 94–124.

116. Harmeson, R.H., Larson, T.E., "Interim Report on the Presence of Nitrates in Illinois Surface Waters," Proc. 1969 Illinois Fertilizer Conference, University of Illinois, Illinois Fertilizer Industry Association, Champaign, Ill., 1969, p. 33.

117. Harmeson, R.H., Sollo, F.W., Jr., Larson, T.E., "The Nitrate Situation in Illinois," *J. Am. Water Works Assoc.*, **63,** 303 (1971).

118. Larson, T.E., Henley, L., "Occurrence of Nitrate in Well Waters," Final Report, Project 65-05G, University of Illinois Water Resources Center, Urbana, Ill., 1966.

119. "Water Pollution Caused by Inactive Ore and Mineral Mines — A National Assessment," Environmental Protection Agency, EPA-600/2-76-298, Washington, D.C., 1976.

120. "Processes, Procedures, and Methods to Control Pollution from Mining Activities," U.S. Environmental Protection Agency, EPA-430/9-73-011, U.S. Government Printing Office, Washington, D.C., 1973.

121. Barnes, H.L., Romberger, S.B., "Chemical Aspects of Acid Mine Drainage," *J. Water Pollut. Control Fed.*, **40,** 371 (1968).

122. Walsh, F., Mitchell, R., "A pH-Dependent Succession of Iron Bacteria," *Environ. Sci. Technol.*, **6,** 809 (1972).

123. Sisler, F.D., Senftle, F.E., Skinner, J., "Electrobiochemical Neutralization of Acid Mine Water," *J. Water Pollut. Control Fed.*, **49,** 369 (1977).

124. Jackivicz, T.P., Jr., Kuzminski, L.N., "A Review of Outboard Motor Effects on the Aquatic Environment," *J. Water Pollut. Control Fed.*, **45,** 1759 (1973).

125. Ard, R.W., Jr., Scarano, T.S., "Marine Waste Water Treatment Systems," *Environ. Sci. Technol.*, **8,** 219 (1974).

126. "Pollution Control in the Marine Industries," Proc., 3rd Int'l Conf., Montreal, Can., June 1973, International Association for Pollution Control, Bethesda, Md., 1973.

127. "Water Quality Criteria — 1972," National Academy of Sciences — National Academy of Engineering, Washington, D.C., 1973.

128. Stokinger, H.E., "The Spectre of Today's Environmental Pollution — USA Brand: New Perspectives from an Old Scout," *Am. Ind. Hyg. Assoc. J.*, **30,** 195 (1969).

129. Craun, G.F., McCabe, L.J., "Problems Associated with Metals in Drinking Water," *J. Am. Water Works Assoc.*, **67,** 593 (1975).

130. Ongerth, H.J., et al., "Public Health Aspects of Organics in Water," *J. Am. Water Works Assoc.*, **65,** 495 (1973).

131. Reinhardt, A.W., Spath, D.P., Jopling, W.F., "Organics, Water, and Health: A Reuse Problem," *J. Am. Water Works Assoc.*, **67,** 477 (1975).

132. Shuval, H.I., Gruener, N., "Health Considerations in Renovating Wastewater for Domestic Use," *Environ. Sci. Technol.*, **7,** 600 (1973).

133. McMillan, L.M., Stout, R.G., Willey, B.F., "Asbestos in Raw and Treated Water: An Electron Microscopy Study," *Environ. Sci. Technol.*, **11,** 390 (1977).

134. Dutka, B.J., Bell, J.B., "Isolation of Salmonellae from Moderately Polluted Waters," *J. Water Pollut. Control Fed.*, **45,** 316 (1973).

135. Ptak, D.J., Ginsburg, W., Willey, B.F., "Identification and Incidence of *Klebsiella* in Chlorinated Water Supplies," *J. Am. Water Works Assoc.*, **65,** 604 (1973).

136. Geldreich, E.E., "Applying Bacteriological Parameters to Recreational Water Quality," *J. Am. Water Works Assoc.*, **62,** 113 (1970).

137. Woodward, R.L., "Environmental Hazards: Water Pollution," *New England J. Med.*, **275,** 819 (1966).

138. Weibel, S.R., et al., "Waterborne Disease Outbreaks, 1946–60," *J. Am. Water Works Assoc.*, **56,** 947 (1964).

139. "Status of Waterborne Diseases in the U.S. and Canada," Committee Report, *J. Am. Water Works Assoc.*, **67**, 95 (1975).

140. Berg, G., "An Integrated Approach to the Problems of Viruses in Water," Proc. Natl. Specialty Conference Disinfection, American Society of Civil Engineers, New York, 1970, p. 339.

141. Sobsey, M.D., "Enteric Viruses and Drinking-Water Supplies," *J. Am. Water Works Assoc.*, **67**, 414 (1975).

142. Sorber, C.A., Malina, J.F., Jr., Sagik, B.P., "Quantitative Procedure for Evaluating the Performance of Water and Waste Water Treatment Processes at Naturally Occurring Virus Levels," *Environ. Sci. Technol.*, **6**, 438 (1972).

143. Merrell, J.C., Jr., Ward, P.C., "Virus Control at the Santee, Calif., Project," *J. Am. Water Works Assoc.*, **60**, 145 (1968).

144. Drewry, W.A., Eliassen, R., "Virus Movement in Groundwater," *J. Water Pollut. Control Fed.*, **40**, R257 (1968).

145. Sawyer, C.N., "Fertilization of Lakes by Agricultural and Urban Drainage," *J. N.E. Water Works Assoc.*, **61**, 109 (1947).

146. Sakamoto, M., "Primary Production by Phytoplankton Community in Some Japanese Lakes and Its Dependence on Lake Depth," *Arch. Hydrobiol.*, **62**, 1 (1966).

147. Dillon, P.J., Rigler, F.H., "The Phosphorus-Chlorophyll Relationship in Lakes," *Limnol. Oceanogr.*, **19**, 5, 762 (1974).

148. Schindler, D.W., "Carbon, Nitrogen and Phosphorus and the Eutrophication of Freshwater Lakes," *J. Phycol.*, **7**, 321 (1971).

149. Schindler, D.W., et al., "Atmospheric Carbon Dioxide: Its Role in Maintaining Phytoplankton Standing Crops," *Science*, **177**, 1192 (1972).

150. Morton, S.D., Sernau, R., Derse, P.H., "Natural Carbon Sources, Rates of Replenishment, and Algal Growth," in "Nutrients and Eutrophication: the Limiting Nutrient Controversy," Likens, G.E., ed., Special Symposia, Vol. 1, Amer. Soc. Limnol. Oceanogr., 1972.

151. Edmondson, W.T., "Phosphorus, Nitrogen, and Algae in Lake Washington After Diversion of Sewage," *Science*, **169**, 690 (1970).

152. Oglesby, R.T., "Effects of Controlled Nutrient Dilution on a Eutrophic Lake," in "Advances in Water Pollution Research," Jenkins, S.H., ed., Proc. 4th Int'l. Conf., Prague, Pergamon Press, New York, N.Y., 1969, pp. 747–757.

153. Sonzogni, W.C., Fitzgerald, G.P., Lee, G.F., "Effects of Wastewater Diversion on the Lower Madison Lakes," *J. Water Pollut. Control Fed.*, **47**, 535 (1975).

154. Hadeed, S.J., "Banning Detergent Phosphates — the Debate Continues," *J. Water Pollut. Contr. Fed.*, **50**, 190 (1978).

155. Yeasted, J.G., Morel, F.M.M., "Empirical Insights into Lake Response to Nutrient Loadings, with Application to Models of Phosphorus in Lakes," *Environ. Sci. Technol.*, **12**, 195 (1978).

156. "Summary Analysis of the North American (U.S. portion) OECD Eutrophication Project: Nutrient Loading–Lake Response Relationships and Trophic State Indices," Rast, W., Lee, G.F., eds., Environmental Protection Agency, Ecological Research Series, EPA-600/3-78-008, Washington, D.C., 1978.

157. Boetius, J., "Foul Taste of Fish and Oysters Caused by Chlorophenol," Medd Denmarks Fishlog Havundersdg. N.S. 1, 1 (1954).

158. Gerber, N.N., Lechevalier, H.A., "Geosmin, an Earthy-Smelling Substance Isolated from Actinomycetes," *Appl. Microbiol.*, **13**, 935 (1965).

159. Medsker, L.L., Jenkins, D., Thomas, J.F., "Odorous Compounds in Natural Waters. An Earthy-Smelling Compound Associated with Blue-Green Algae and Actinomycetes," *Environ. Sci. Technol.*, **2**, 461 (1968).

160. Safferman, R.S., et al., "Earthy-Smelling Substances from a Blue-Green Alga," *Environ. Sci. Technol.*, **1**, 429 (1967).

161. Gerber, N.N., "Geosmin, from Microorganisms, Is *Trans*-1, 10-Dimethyl-*Trans*-9-Decalol," *Tetrahedron Lett.* No. 25, 2971 (1968).

162. Henley, D.E., Glaze, W.H., Silvey, J.K.G., "Isolation and Identification of an Odor Compound Produced by a Selected Aquatic Actinomycete," *Environ. Sci. Technol.*, **3,** 268 (1969).

163. Medsker, L.L., Jenkins, D., Thomas, J.F., "Odorous Compounds in Natural Waters: 2-Exo-Hydroxy-2-Methylbornane, the Major Odorous Compound Produced by Several Actinomycetes," *Environ. Sci. Technol.*, **3,** 476 (1969).

164. Water Resource Policy Study — Task Force on Revision of Water Resources Planning and Evaluation Criteria and Procedures, 2120 L Street, N.W., Washington, D.C., Nov. 5, 1977.

165. Johnson, R., Brown, G., "Cleaning Up Europe's Waters: Economics, Management, Policies," Praeger Publishers, Inc., New York, N.Y., 1976.

166. Okun, D.A., "Water Management in England: A Regional Model," *Environ. Sci. Technol.*, **9,** 918 (1975); and "Regionalization of Water Management (A Revolution in England and Wales)," Applied Science Publishers, Ltd., Barking, Essex, England, 1977.

167. Okun, D.A., Testimony before the National Drinking Water Advisory Council, Washington, D.C., Aug. 25, 1976.

168. Andrews, C.D., "We Didn't Wait for the Rain," National Water Council (Britain), 1977.

Chapter 6

Solid Wastes

(Continued)

(Continued from page 275)

6 / Solid Wastes

INTRODUCTION

The technology used to handle and dispose of solid wastes in the United States in the late 1970's lags well behind that used to control air and water pollution. As recently as a decade ago, only 5 to 10% of the cities and towns in this country with populations of more than 2500 were disposing of community refuse by acceptable sanitary and nuisance-free methods. Municipalities resorted primarily to open dumps; relatively few used modern incinerators, resource recovery systems, or improved disposal methods such as sanitary landfills. Industry often followed a like course.

Until as recently as 1965, the Federal Government was spending only an estimated $300,000 annually on research on solid waste disposal. In the years since then, and largely under federal stimulus, solid waste management has come under increasing scrutiny, and the situation has begun slowly to change (1). Although the basic science of solid waste handling and disposal remains in relatively primitive condition, industry and government are bringing advanced technology from other fields increasingly to bear on the many problems that must be solved. The need is clear. Disposal methods that are acceptable today cannot be counted on to deal adequately with the output of solid wastes expected in this country by the year 2000 and later. Whatever the course of solid waste technology, the ultimate solution to the problem undoubtedly will be multifaceted. Important elements of that solution will be resource recovery and recycling, reuse by chemical transformation, and design of consumer products for prolonged use or at least for ease of materials recovery and recycling. And the attainment of maximum resource recovery surely will call for widespread application of chemical technology.

Legislative framework

The first federal effort to deal directly with the solid waste problem was the Solid Waste Disposal Act of 1965 (Public Law 89-272). The act had two purposes: to start and accelerate a national research, development, and demonstration program on solid wastes; and to give technical and financial support to interstate, state, and local agencies in planning, developing, and conducting solid waste disposal programs. The act authorized appropriations rising from not more than $10 million in fiscal 1966 to not more than $32.5 million in fiscal 1969. In 1970, the act was amended as the Resource Recovery Act of 1970 (Public Law 91-512). In December 1970 the U.S. Environmental Protection Agency (EPA) was formed and absorbed the functions of the Bureau of Solid Waste Management, Department of Health, Education, and Welfare. At the same time those functions were reorganized as two separate units. In 1978 the two units existed as the Office of Solid Waste and a research and development component, the Solid and Hazardous Waste Research Division of the Municipal Environmental Research Laboratory in EPA's Office of Research and Development.

The Resource Recovery Act of 1970 authorized appropriations for EPA of not more than $41.5 million in fiscal 1971 to not more than $216 million in fiscal 1974. The act authorized for the Department of the Interior not more than $8.75 million in fiscal 1971 to not more than $22.5 million in fiscal 1974. Budget constraints, however, resulted in far less being authorized by the federal Office of Management and Budget. In fiscal 1974, for example, the actual solid waste budget for EPA was less than $6 million.

For almost a decade, the federal solid waste program pursued the broad range of activities authorized in the acts of 1965 and 1970. Beginning with fiscal 1974, however, attention at the federal level was redirected in anticipation of regulation of the treatment and disposal of hazardous wastes. Legislation for regulating hazardous wastes became a reality with the amendment of the Solid Waste Disposal Act by the Resource Conservation and Recovery Act of 1976 (Public Law 94-580). The act requires EPA to identify hazardous wastes and to issue standards for generators and transporters of such wastes with regard to furnishing records and reports, labeling, and use of a manifest system. Owners and operators of treatment and storage or disposal facilities will be required to obtain permits, and penalties are provided for noncompliance. The act reemphasizes the need for resource recovery and directs the Secretary of Commerce to publish guidelines for classification of recovered materials and to perform economic analyses of impediments to resource recovery. EPA will conduct, and promote coordination of, research relating to adverse effects of solid waste on health and welfare, the operation and financing of disposal programs, and production and marketing of useful products and will explore means of reducing the generation of solid wastes. The act requires studies on the recoverability of materials, including glass and plastics, mining wastes, sludge, and tires, and on the economics of resource recovery systems.

EPA's solid waste budget for fiscal 1977 (ending Sept. 30, 1977) was about $17 million. The total included funds reprogrammed by the agency to begin implementation of the Resource Conservation and Recovery Act. The solid waste budget for 1978, the first full fiscal year under the act, was expected to total about $40 million (2).

Magnitude of the solid waste problem

The complexity of the solid waste problem is due in part to lack of firm data on waste generation rates. The difficulty was especially acute in the 1960's and persists in the 1970's albeit to a lesser degree. Consequently, the literature contains a wide range of estimates for the various types of wastes. The absence of firm data stems partly from the relative lateness of the interest in solid waste

Table 1

Solid wastes originated from certain resource categories

Source	Million tons/year	Percent
Municipal	230	5.2
Industrial	140	3.1
Mineral	1700	38.2
Animal wastes	1740	39.1
Crop wastes	640	14.4
Total	**4450**	**100.0**

Source: "First Report to Congress — Resource Recovery and Source Reduction," U.S. Environmental Protection Agency, Report SW-118, Washington, D.C., February 1973.

management and the resulting failure of public and private agencies to keep accurate records of wastes handled, or, indeed, to keep records at all. The numbers quoted here reflect the data problem at some points.

The variation in data is especially apparent in past, current, and projected estimates for the production of solid wastes nationwide. One prediction, based on data available in 1968 (3), said that by 1980 some 235 million people would be generating 8 lb of waste per person per day, or more than 340 million tons per year, including only those wastes handled by collection agencies. A second estimate put the current overall rate at 10 lb of household, commercial, and industrial wastes per person per day or more than 360 million tons per year. Yet a third estimate put production of solid wastes of all kinds in the United States in 1969 at about 4.5 billion tons (Table 1).

SOURCES OF SOLID WASTES

Municipal Refuse

Although "municipal refuse" accounts for only about 5% of the total solid wastes generated, it is the primary concern of the nation's solid waste effort. The reason is that the solid waste problem is largely one of collection and disposal, and municipal refuse is the major solid waste in areas of high population density, where disposal and collection for disposal comprise the biggest, most visible problem.

In 1968 it was estimated that public and private agencies in the U.S. collected 4.73 lb/person/day of solid wastes, nationwide, or more than 190 million tons per year exclusive of industrial and agricultural wastes (Table 2). Recent calculations, based on measured values rather than on estimates, however, suggest that the rate today for collected municipal wastes averages more like 3 lb/person/day (4). If the latter estimate is valid, and it probably is, then the earlier projections of the total production of wastes — nonurban plus urban — must be revised downward significantly. The economic impacts of the energy shortage and inflation, moreover, will help materially to reduce the rate of generation of municipal wastes, thus upsetting the earlier projections even further. EPA estimates for 1971–75, based on typical municipal collections of solid wastes, appear in Table 3. The composition of municipal refuse has been analyzed in many studies (5). Predictably, it varies from region to region and from season to season within a region (Table 4).

Abandoned automobiles. The problem of the abandoned automobile, strictly speaking, transcends that of municipal wastes, but it is considered here in that category because it is more acute in urban areas. Despite the backlog of junked automobiles that has accumulated over the years, industry actually processes a large number of them. Estimates of the numbers involved are subject to considerable uncertainty (6, 7). They indicate, however, that an average of perhaps 7 million cars were retired annually during 1970–75, and that all but about 1.2 million annually were recycled. The accumulated inventory of unprocessed automobiles in 1975 was an estimated 17 million or 21 million tons of steel. Some 14 million of the 17 million were in the scrap cycle. The remaining 3 million cars were abandoned and are both a real disposal problem and a waste of steel. As with municipal wastes, estimates of the numbers of abandoned automobiles must be revised downward because of the difficult economic situation of the early 1970's. The number of reclaimed auto hulks, on the other hand, should be revised upward. In 1974 on the West Coast, for example, most of the abandoned automobiles were being collected and processed to meet a sharply increased demand for scrap. In addition, the automobile industry has been reducing the weight of its products to improve their gas mileage as mandated by the Energy Policy and Conservation Act of 1975 (Public Law 94-163). The amount of solid waste represented by scrapped automobiles will decline correspondingly.

Industrial Solid Wastes

Total solid waste production in the U.S. in the past usually has been put at 100 lb/person/day. Of the 100 lb, 10 were attributed to residential, commercial, and industrial activities; of these 10 lb, 3 were considered industrial solid wastes. Studies of the early 1970's indicate, however, that the industrial contribution shortly will reach 6 lb/person/day. If this projection is correct, industrial solid waste production per person will surpass the residential and commercial output.

Industrial solid wastes can be divided into nonprocess and process wastes. Nonprocess wastes are those common to most industries, such as packaging and shipping, office, and cafeteria wastes; their composition is similar to that of commercial wastes in that they contain a high percentage of paper. Processing wastes are specific to the industrial plant, and their composition depends on the product or products produced.

Nonprocess solid wastes generally are handled like residential and commercial wastes. Industrial process wastes collectively represent different and greater problems than nonprocess wastes in that they are more varied both chemically and physically. They are more likely, moreover, to contain hazardous substances that might create environmental problems when incinerated or buried. Process wastes nevertheless are disposed of commonly by incineration and landfill and variations thereof.

Because industry is a producer, not solely a consumer, it has strong incentives to reuse material that otherwise might require disposal. Not surprisingly, economics determine in the end whether a material will be reused or discarded. For this reason, the prognosis for sharply intensified efforts at reclamation has

Table 2

Average of solid wastes collected (lb/person/day)

Solid wastes	Urban	Rural	National
Household and commercial	4.35	3.43	4.15
Demolition and construction	0.23	0.02	0.18
Street and alley	0.11	0.03	0.09
Miscellaneous	0.38	0.08	0.31
Total	**5.07**	**3.56**	**4.73**

Source: Black, R.J., et al, "An Interim Report. 1968 National Survey of Community Solid Waste Practices," presented at the 1968 annual meeting of the Institute for Solid Waste, American Public Works Association, Miami Beach, Fla., October 1968.

improved materially as a side effect of the petroleum and natural gas situation. The sudden realization that supplies of cheap oil and gas are not endless has dramatized the need to conserve all resources so as to extend supplies and minimize reliance on foreign sources.

Process wastes ideally would be recycled within the plant and usually are when it is technically and economically feasible to do so. Material that cannot be recycled internally is sold as a rule to salvage dealers, members of the well established secondary materials industry. In the U.S. in 1967, the industry was estimated to include some 7900 establishments with annual sales exceeding $4.6 billion (8). Many of these establishments were quite small, although the trend was toward fewer, larger organizations. The National Association of Recycling Industries estimated that in 1975 its 800 member companies sold some $9 billion worth of secondary materials, principally nonferrous metals, paper, and textiles. Although metals comprise three quarters of the secondary materials industry's business, it handles virtually every kind of used resource. Thus the industry provides an institutional framework for returning potential solid waste to productive use. Historically, the industry's ability to perform has been limited by a weak or highly fluctuating market for secondary materials. The growing demand to conserve resources, however, should change this situation for the better.

Table 3

Postconsumer solid waste generated and disposed of[a] (millions of tons, wet weight)

Materials	1971	1972	1973	1974	1975
Paper	39.1	42.5	44.2	43.4	37.2
Glass	12.0	12.7	13.2	12.9	13.3
Metal	11.8	12.1	12.4	13.0	12.2
Ferrous	(10.6)	(10.8)	(11.0)	(11.5)	(10.8)
Aluminum	(0.8)	(0.9)	(1.0)	(1.0)	(0.9)
Other	(0.4)	(0.4)	(0.4)	(0.4)	(0.4)
Plastics	4.2	4.7	5.0	4.5	4.4
Rubber and leather	3.3	3.4	3.6	4.1	3.3
Textiles	1.8	1.8	1.9	2.1	2.1
Wood	4.6	4.7	4.9	4.8	4.9
Food waste	22.0	22.2	22.4	22.6	22.8
Yard waste	24.1	24.5	25.0	25.5	26.0
Miscellaneous inorganics	1.8	1.8	1.9	1.9	1.9
Total disposed of	**124.8**	**130.5**	**134.6**	**134.8**	**128.2**
Total recycled	**8.1**	**8.8**	**9.6**	**9.4**	**8.0**
Total generated	**132.9**	**139.3**	**144.2**	**144.1**	**136.1**

a. Includes residential and commercial wastes that comprise the bulk of typical municipal collections. Does not include mining, agricultural, industrial processing, and demolition and construction wastes, sewage sludge, and junked autos and other obsolete equipment wastes.

Source: "Fourth Report to Congress — Resource Recovery and Waste Reduction," Environmental Protection Agency, Report SW-600, Washington, D.C., August 1977.

Mining wastes

The mining industry produces immense amounts of wastes; the total accumulated by 1968 in this country was an estimated 21 billion tons (9). By 1978, the total was rising by nearly 2 billion tons annually of mine waste, mill tailings, washing plant rejects, processing plant wastes, and smelter slags and rejects. About 60% of these wastes have economic value either for their mineral content or for use as aggregate and for similar structural purposes. About 37% of mining and processing wastes have no known economic value, but are hazards to public health and safety, prevent the use of the land they occupy for other purposes, or are an aesthetic blight. The remaining 3% of mineral wastes are of small volume, have no value, and are located remotely and thus constitute no real problem.

One reason for the mining industry's huge output of wastes is the high ratio of rock to ore. In addition, large volumes of solid waste are produced in the milling, beneficiation, and processing of the ore. In 1974, for example, the copper ore processed in this country yielded an average of only 0.49% copper. Consequently, almost a ton of waste was produced for every 10 lb of copper obtained.

Copper mining. The copper mining industry in the western U.S. recovers about 450 tons of copper daily by leaching mine waste that in 1976 was accumulating at a rate of about 1 million tons daily. In the process, the industry uses some 275,000 tons per year of incinerated, detinned, and shredded tin plate scrap and some 100,000 tons per year of shredded auto hulks.

To recover copper, leach solutions containing a small amount of sulfuric acid (pH 1.9–3.5) are percolated through the waste dumps. The copper-bearing effluent from the dumps is then passed through beds of shredded iron contained in cementation launders or precipitation reactors. The iron enters solution as iron

Table 4

Composition of municipal refuse

Component	Composition (% of dry weight)[a] Range	Nomimal
Metallics	7 to 10	9.0
Ferrous	6 to 8	7.5
Nonferrous	1 to 2	1.5
Glass	6 to 12	9.0
Paper	37 to 60	55.0
Newsprint	7 to 15	12.0
Cardboard	4 to 18	11.0
Other	26 to 37	32.0
Food	12 to 18	14.0
Yard	4 to 10	5.0
Wood	1 to 4	4.0
Plastic	1 to 3	1.0
Miscellaneous	<5	3.0

a. Moisture content: range, 20 to 40%; nominal, 30%.

Source: Abert, J.G., Alter, H., Bernheisel, J.F., "The Economics of Resource Recovery from Municipal Solid Waste," *Science*, **183**, 1052 (1974). Copyright 1974 by the American Association for the Advancement of Science.

(ferrous) sulfate, displacing the dissolved copper, which precipitates in metallic form and is recovered.

Copper mining wastes also may be a source of other metals. At Kennecott Copper's Bingham mine in Utah, for example, about 65 million gallons of leach liquor per day was available in 1976. In addition to copper, the liquor contained 4 to 10 mg/l of uranium oxide, 6 g/l of aluminum, and 8 to 10 mg/l of yttrium. Research has found that 85 to 90% of the uranium can be recovered from solution by ion exchange at costs competitive with those of uranium extraction processes already in use. Early in 1978, a subsidiary of Westinghouse Electric began operating a commercial plant that applies this process to Kennecott leach liquor.

Bauxite processing. More than 10 million tons of process wastes (commonly called "red mud") are produced annually in this country by plants that convert bauxite into aluminum oxide (alumina). Red muds vary in composition with the type of bauxite and the method of processing. A typical mud, however, would include 10 to 20% aluminum oxide, 40 to 60% iron (ferric) oxide, 10 to 20% silicon dioxide, 5 to 30% calcium oxide, and lesser amounts of compounds of other elements, including sodium, titanium, phosphorus, and sulfur. Red muds traditionally have been disposed of in "red mud lakes": impervious impoundments of 20 to 250 acres in which the mud settles until its water content (again depending on the particular bauxite and process) is about 50%. The surface water from these lakes is decanted and returned to the refining process for its residual aluminum and sodium content; the red mud is held in impoundments.

Potential uses for red muds have been investigated for many years (10). Uses that have been proposed include: a source of iron ore; raw material for bricks or ceramics; aggregate for cement and concrete; fill material; and extender for various rubber products. None of these alternatives to the present method of disposal has proved economically feasible, primarily because of the large amounts of energy required to dry the red mud and the distance of existing plants from suitable market areas. Should the energy and transportation situations change in the near future, it is still doubtful that present production of red muds could be consumed, much less the large volumes of muds stored in existing impoundments. Where evaporation rates permit, impounded red muds may be consolidated to 60 to 90% solids by a combination of decantation and evaporation. Where rainfall exceeds evaporation, however, other means of consolidation may have to be devised. At least one aluminum company in 1976 was operating an impoundment with an "underdrain system" (essentially a sand filter and tile field) that consolidates red mud by natural drainage and evaporation. Both methods of disposal provide equally suitable raw material for whatever uses may be developed for the muds.

Phosphate mining. A mineral waste problem of very large proportions that awaits improved solutions is the disposal and utilization of the colloidal clay wastes or slimes produced in phosphate mining operations in Florida (11). The slimes leave the processing plant containing 2 to 6% solids in water and about one third of the total phosphate mined, measured as bone phosphate of lime equivalent. They are pumped to huge settling ponds and, as the solids settle, clear water is drawn from the tops of the ponds and returned to the process. The solids settle very slowly, and after a decade the slimes may still have been concentrated to only 20% solids. The settling properties of the slimes impede the reclamation of land occupied by the ponds and, furthermore, there is no economical means of recovering the phosphate in the slimes. Efforts to find means of improving the settling of slimes and of using them in some way or recovering the phosphate they contain have produced some progress, but no fully satisfactory solution to the problem.

Hazardous and toxic wastes

The Resource Recovery Act of 1970 required in part that the Environmental Protection Agency prepare a comprehensive report for Congress on the storage and disposal of hazardous wastes (12). The act defines "hazardous wastes" as any waste or combination of wastes that pose a substantial present or potential hazard to human health or living organisms. A given waste is regarded as hazardous if it is lethal, nondegradable, persistent in the environment, can be biologically magnified, or otherwise causes or tends to cause detrimental cumulative effects. General categories of hazardous wastes are toxic chemicals and flammable, radioactive, or biological substances. (Biological substances include pathological and warfare agents.) These wastes can be solids, sludges, liquids, or gases, and their sources are numerous and widely scattered. Prime producers are industry, the Federal Government (mainly the Department of Energy and the Department of Defense), agriculture, and institutions such as hospitals and laboratories. The Resource Conservation and Recovery Act gave the Government the primary responsibility for controlling hazardous waste from generation to ultimate disposal. The act provides for state participation, and early in 1978 EPA proposed guidelines for state hazardous waste programs (13).

Radioactive wastes. Most radioactive wastes are nonradioactive materials contaminated with gaseous, liquid, or solid radionuclides. The concentration of radionuclides can range from a few parts per billion to as high as 50% of the total waste. A given waste often contains many radionuclides, whose nature changes as they decay. Solid radioactive wastes normally are not important potential contaminants of the biosphere until they become airborne, usually as particles, or waterborne by leaching. Existing regulatory limits on radioactivity, therefore, are based primarily on concentrations in air and water. Specific sources of radioactive solid wastes include uranium mining, nuclear power generation and fuel processing facilities, and operations that involve work with radioisotopes or devices or materials containing them.

The biological hazard of radioactive wastes is due primarily to penetrating and ionizing radiation rather than to chemical toxicity. On a weight basis, certain radionuclides are far more toxic than many of the more common toxic chemicals. The only practical way to neutralize a radionuclide currently is to allow it to decay to stable form; wastes that contain decaying radionuclides must be stored under carefully controlled conditions to assure their containment and isolation (14).

No satisfactory method has yet been demonstrated for disposing permanently of radioactive wastes (as opposed to storing them in controlled conditions), and the problem is exceptionally difficult. Uranium mill tailings, for example, present a unique disposal problem, similar in magnitude to the disposal of all other industrial hazardous wastes together. The several federal agencies at work on the problem have been unable as yet to devise sound means either of disposing of the tailings or of recovering the radionuclides from them. Aside from uranium mill tailings, the commercial nuclear electric power industry and other private sources in the U.S. were estimated in the mid-1970's to be generating about 24,000 tons of radioactive material per year. This amount is less than 1% of the total hazardous wastes from all industry, but the percentage will increase substantially if the nuclear power industry expands as expected. (See Chapter 8, Radiation in the Environment.)

Nonradioactive wastes. Data on nonradioactive hazardous wastes, most of which are generated by industry, are approximate at best. In 1970, industry generated an estimated 10 million tons of nonradioactive hazardous wastes. That amount was about 10% of industry's total solid waste output of about 110 million tons. In 1977, according to EPA estimates, 14 categories of industry generated more than 30 million tons of potentially hazardous nonradioactive solid wastes (15); the agency estimates that about 10% of all industrial waste is hazardous (2). Practically all of the nonradioactive hazardous wastes fall into one or more of four categories: inorganic toxic metals, salts, acids, or bases; synthetic organics; flammables; and explosives.

About one in four of the metals in common use today is toxic enough to be considered hazardous even at low concentrations. The largest amounts of toxic metal wastes are produced in mining and metallurgy and in electroplating and metal finishing. Arsenic-containing flue dusts collected from the smelting of copper, lead, zinc, and other arsenic-bearing ores, for example, amount to 40,000 tons per year. Approximately 30,000 tons of chromium-bearing wastes are discharged from the metal finishing industry annually.

Hazardous synthetic organic compounds include the halogenated hydrocarbon pesticides, the polychlorinated biphenyls (PCB's), and the phenols. Approximately 5000 tons of synthetic organic pesticide wastes were produced in 1970. The Department of Defense in 1974 had 850 tons of dry pesticides and 15,000 tons in liquid form requiring disposal. The problem involves not only the pesticides themselves, but also their containers. It is estimated, for example, that more than 250 million pesticide containers of all types were used in this country in 1973.

Flammable wastes consist mainly of contaminated organic solvents, but also may include oils, pesticides, plasticizers, complex organic sludges, and off-specification chemicals. Highly flammable wastes pose acute handling and chronic disposal hazards.

Explosive wastes are mainly obsolete ordnance, explosives manufacturing wastes, and contaminated industrial gases. The largest amount of explosive wastes, not surprisingly, is generated by the Department of Defense, which in 1974 had on hand some 150,000 tons of obsolete conventional ammunition. Disposal of obsolete munitions has become complex because the practice of loading them on old ships and sinking ship and cargo in the ocean has been discontinued. Now, final disposal must be postponed until a more suitable method becomes available.

Treatment processes for hazardous wastes ideally should achieve four general ends: reduce the volume of the waste where required; separate the components; detoxify the waste; and recover useful materials. Since no single process can perform all four functions, adequate treatment usually involves several processes linked in series. Residues from these processes, as well as untreated hazardous wastes, require ultimate disposal. Treatment technology is generally available for most hazardous wastes. Several governmental agencies are supporting research on improved disposal methods.

Recommendation S1: *The appropriate federal, state, and local government agencies should press their efforts to define the nature and magnitude of the solid waste problem both now and in the future. Augmented programs are* *necessary in education, research, development and demonstration, and local and regional planning for solid waste management, utilization, and disposal.*

Recommendation S2: *Research and development on processes for recovering various minerals from mining and processing wastes should be maintained at an adequate level against the day when changing economics warrant the recovery of such minerals. Work on other means of utilizing or disposing of these wastes should be maintained at a steady level.*

Recommendation S3: *Increased disposal problems associated with disposal of sludges and hazardous wastes on land should receive the additional research and development required in the areas of health effects, environmental effects, control technology, management systems, and economic/legislative efforts.*

METHODS OF SOLID WASTE DISPOSAL

Of the methods used in the United States to dispose of solid wastes, disposal on land is by far the most common. In the late 1970's, landfill probably accounts for more than 90% of the nation's municipal refuse; incineration accounts for most of the remainder, and composting is an insignificant factor. Although these traditional techniques very often are not applied properly, modern practice is slowly taking hold. In addition, a trend has developed toward recovery of heat and materials from refuse. The selection of the solid waste management method to be used depends strongly on cost, which in turn may reflect local circumstances. Sanitary landfill, for example, is the cheapest satisfactory means of disposal, but only if suitable land is within economic range of the source of the wastes. Estimates of the costs of technically feasible methods appear in Table 5.

It should be noted that collection and transportation typically account for perhaps 75% of the total cost of solid waste management. Studies during the past decade have steadily upgraded the information available to local authorities who must deal with this aspect of the solid-waste problem as well as with processing and disposal (16).

Sanitary Landfill

Modern practice in disposing of solid wastes by the sanitary landfill method has been extensively described (17–19) and is well exemplified at several operations in the U.S. (20). Even so, the method still is not used to the degree that it ought to be. EPA estimated early in 1976 that fewer than one third of the nation's 18,500 known municipal land disposal sites met state standards. It was not certain that prevailing landfill standards were always high enough to protect groundwater from damage by leachate. And besides the known land disposal sites, an unknown number of illegal open dumps were operating. Elimination of open dumps in the U.S. is one goal of the Resource Conservation and Recovery Act. Early in 1978, EPA took the first step by proposing criteria for the states to use in evaluating solid waste disposal facilities and identifying open dumps (21).

Of the several definitions of "sanitary landfill," one by the Environmental Protection Agency (22) is as appropriate as any: ". . . an engineered method of disposing of solid wastes on land in a manner that minimizes environmental hazards by spreading the solid wastes in thin layers, compacting the solid wastes to the smallest practical volume, and applying and compacting cover material at the end of each operating day." The key elements of the definition are compaction and daily covering of the wastes with soil. Well planned, designed, and operated sanitary landfills are a necessary tool in the management of municipal solid wastes and of certain other wastes as well. The method does require relatively large areas and large volumes of cover material. The economics of sanitary landfilling at a given location, therefore, depend in significant degree on the availability of suitable sites.

In a modern sanitary landfill (Fig. 1), refuse is spread in thin layers, each compacted by a bulldozer before the next is spread. When about 10 ft of refuse has been laid down in this manner, it is covered by a thin layer of clean earth, which also is compacted. At the end of each working day, the landfill is sealed with a thin layer of compacted earth. The process is repeated daily, with the result that the 10-ft layers or lifts make up a series of earth-enclosed cells. When a landfill site is filled completely, a final 1 to 2 ft of soil is added to the topmost layer to bring it to a depth of 2 to 3 ft. No burning should be allowed at the site. A properly operated sanitary landfill is free of odors, flies, and rodents. Pollution of surface water is minimized by suitably compacting and planting the cover, contouring the fill, diverting upland drainage, selecting a proper soil type for final cover, and by not depositing wastes in sites subject to inundation by flooding or high groundwater. In unique site conditions it may be necessary to

Table 5

Net operating costs for refuse disposal and resource recovery systems[a]

Process concept	Municipal owner[b]	Private owner
	$/wet ton[c]	
Sanitary landfill (close-in)	2.57	3.04
Fuel recovery	2.70	4.30
Materials recovery	4.77	7.21
Pyrolysis	5.42	8.02
Sanitary landfill (remote)	5.94	6.47
Composting (mechanical)	6.28	9.91
Incineration, steam and residue recovery	6.57	9.27
Incineration and steam recovery	7.05	9.50
Incineration and residue recovery	7.18	9.43
Incineration only	7.68	9.64
Incineration and energy recovery	8.97	12.72

a. For 1000 tons per day of raw refuse, 300 operating days per year, and 20-year life. Economic analysis based on generalized and "national average" data for 1971; results cannot be considered typical for any geographic location and conditions. The costs shown reflect capitalization, operating costs, and credits for resources recovered, which were assumed to be salable in the amounts recovered at the prices of comparable competing commodities.

b. Most systems have high capital costs per ton of capacity, so that municipally owned plants, which can be financed with tax-free municipal bonds, cost significantly less than privately owned plants.

c. Moisture content assumed to average 27%; 1000 tons of raw refuse contains 700 to 800 tons of potentially recoverable resources.

Source: "Resource Recovery: The State of the Technology," a report to the Council on Environmental Quality, Midwest Research Institute, U.S. Government Printing Office, Washington, D.C., February 1973.

protect groundwater from leachate — rainwater that passes through the landfill and picks up dissolved, colloidal, and suspended solids in concentrations that vary widely with local conditions (Table 6). Leachate can become highly contaminated and, where indicated, is collected and treated before it is discharged to the environment.

Decomposition in a sanitary landfill is very largely anaerobic. When a load of wastes is deposited, it is attacked first by aerobic organisms that soon consume the available oxygen trapped in the interstices in the compacted wastes. This initial aerobic decomposition causes the temperature of the buried mass to rise somewhat, but not for long. The temperature declines to ambient when the oxygen is consumed, and anaerobic degradation sets in. The anaerobic phase yields substantial amounts of methane and carbon dioxide, some hydrogen sulfide, and traces of hydrogen. Where these gases may be a problem, it can be eased by providing suitable vents. Landfill gases will move laterally because of relative differences in soil permeabilities; diffusion apparently is the primary force in such migration. Migrating gases have been noted to cause vegetative kills, localized mineralization of groundwater, and explosion and odor problems in nearby structures. Methane in itself does not appear to be an environmental hazard, but mixtures of as little as 5% of the gas in air are explosive when ignited. Thus buildings should not be located on or near landfills unless measures are taken to prevent the gas from accumulating. Carbon dioxide can be a problem because it dissolves in groundwater to form a carbonic acid complex. Because of the water's resulting higher acidity, it will likely attack its surround-

FIGURE 1

Schematic diagram of a sanitary landfill

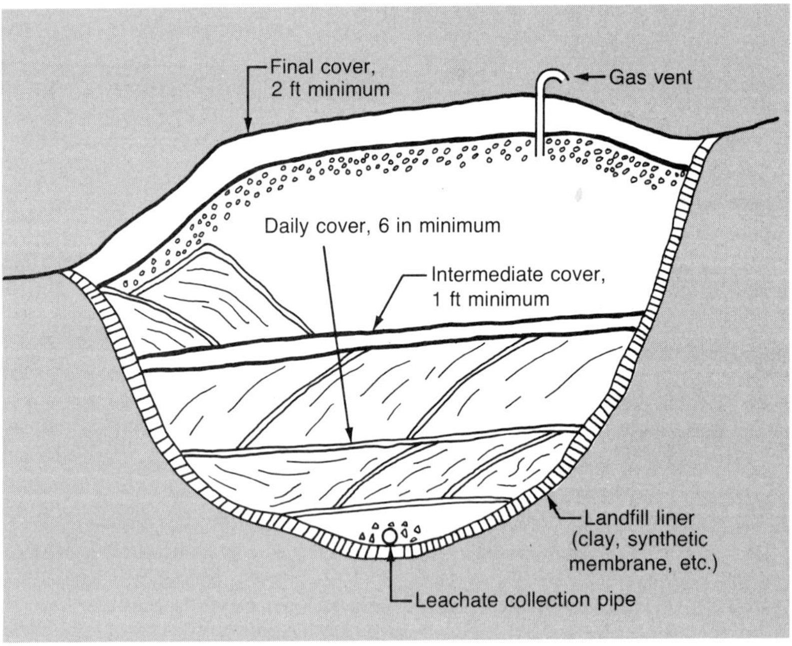

ing aquifer more actively, acquiring a higher concentration of dissolved minerals and generally deteriorating in quality.

Landfill uses and cost

Landfills have been used to reclaim land for conversion into golf courses and for other recreational uses. With the proper precautions and design of foundations and utility corridors, light commercial buildings and possibly industrial centers may be constructed on completed sanitary landfills. The prospective use of the completed landfill should be decided before operations begin, however, so that the fill can be designed accordingly. Specific design criteria are available for uses that range from low-density recreation to building construction (23). The criteria are oriented toward settlement and control of gas migration; they include the reservation of areas of original, unfilled earth to provide structural support for buildings. Settlement, whether uniform or not, generally limits the usefulness of recently completed fills.

A most important potential use for sanitary landfilling, the reclamation of strip mine areas (24, 25), has been blocked by public opposition. Maximum

Table 6

Composition of leachate from landfill

Component	Range of all values (mg/l)
Alkalinity ($CaCO_3$)	0–20850
BOD (five day)	9–54610
Calcium	5–4080
COD	0–89520
Copper	0–9.9
Chloride	34–2800
Hardness ($CaCO_3$)	0–22800
Iron — total	0.2–5500
Lead	0–5.0
Magnesium	16.5–15600
Manganese	.06–1400
Nitrogen (NH_3)	0–1106
Nitrogen (Kjeldahl)	0–1416
Nitrogen (NO_3)	0–1300
Potassium	2.8–3770
Sodium	0–7700
Sulfate	1–1826
Total dissolved solids	0–42276
Total suspended solids	6–2685
Total phosphate	0–154
Zinc	0–1000
pH	3.7–8.5

Source: Bruner, D., "An Environmental Assessment of Potential Gas and Leachate Problems at Land Disposal Sites," Open-file Report SW-110, U.S. Environmental Protection Agency, 1973.

reclamation requires that solid wastes be brought from cities outside the strip mine areas, whose citizens object strongly to the imports. While such objections exist, strip mines are unlikely to be reclaimed on a large scale by sanitary landfilling.

Citizen opposition to sanitary landfills is not confined to strip mine areas. The public tends generally to downgrade such operations — probably confusing them with the traditional city dump — despite the many improvements that have occurred in practice. EPA, in fact, has assembled a series of pamphlets, speeches, films, and other education aids designed to overcome the general antagonism to sanitary landfilling (26).

The cost of sanitary landfilling varies widely with the size of the operation, differences in specific site development, operating requirements, and the quality of the operation. Excluding land and hauling, the cost ranges generally from $1.00 to $5.00 per ton of solid waste. The costs of land and hauling are such that wastes sometimes can be transported long distances — up to 50 miles has proved economically feasible — to a landfill site. The cost of hauling can be minimized by using high-capacity transfer vehicles operating out of transfer stations near the centers of waste generation.

Landfill research

Current research on sanitary landfilling is concerned mostly with municipal wastes. The research is directed primarily at identifying the potential leachate problem, and providing appropriate solutions to it, and to extending the life of the fill by pretreating the solid wastes. In addition, work is under way on recovery of methane from landfills for use as fuel.

Studies of leachate ranging from laboratory to field scale indicate that the concentrations of its constituents vary widely. Reported values for chemical oxygen demand (COD), for example, range from 8000 to as high as 89,520 mg/l; total organic carbon ranges from 256 to 28,000 mg/l; total dissolved solids range from 1630 to 59,200 mg/l (27). The wide spread in reported values is due no doubt to a combination of variables that include operating parameters, waste density and composition, type and thickness of cover material, available moisture, temperature regimen, and the age of the landfill. Variations also may result from differences in the methods used to determine the reported values.

The potentially adverse effect of high concentrations of contaminants in landfill leachate are perhaps lessened by the relatively small volume of leachate that normally is formed. Even a small amount of leachate can pose an environmental hazard, however, and leachate often cannot readily be contained within the fill. Thus several studies have sought suitable treatment methods for leachates. Results obtained at the University of Kentucky (28) indicate that anaerobic digestion and anaerobic filters may remove as much as 96% of the COD. Aerobic treatment can achieve slightly better COD removals (29). In both cases, the treated leachate retained a COD and biochemical oxygen demand comparable to those of raw domestic sewage. Scientists at the University of Wisconsin (30) and the University of Illinois (27) found that a large fraction of the organic matter can be removed from leachate. They concluded that anaerobic treatment should be used to remove most of the organic matter. They recommended that the anaerobic effluent be "polished" by aerobic oxidation, followed by physical/chemical treatment when a high-quality effluent is required.

Work at Oceanside, Calif., indicates that wastewater sludges and septic tank pumpings can be mixed with solid wastes and deposited in sanitary landfills in arid climates without producing a leachate (31, 32). Another sucn study is being conducted at Sonoma County, Calif. (33). Experience in both studies indicates good potential for the method, provided that the moisture regimen in a specific landfill site is evaluated carefully before mixing sludge with solid wastes on a large scale.

Attempts to accelerate decomposition and to provide some control over the sanitary landfill after deposition have involved two different approaches — aerobic stabilization and leachate recirculation. Aerobic stabilization was evaluated on a pilot scale (34) and then on a demonstration scale at Santa Clara, Calif. (35). The process involves pumping air to manifolds in the bottoms of prepared trenches where the wastes are placed for "underground composting." Following degradation, the solid waste must be removed for permanent burial elsewhere or the trench must be abandoned. The estimated cost is 20% higher than that of conventional sanitary landfilling. Leachate recirculation has been used to accelerate decomposition by the anaerobic process that ordinarily occurs in a landfill (33, 36). In this approach, leachate production is encouraged by adding water, which is recirculated through the fill in an attempt to build up a pH buffer capacity, replenish nutrients, and increase the population of those organisms active in stabilizing the wastes. Pilot-scale data indicate that initial decomposition is completed within three to six months if pH is controlled. An independent field study is required to develop ways of managing a large landfill in which leachate is recycled. In the long run, excess leachate would be expected in the system from rainfall.

Pre-landfill processing — shredding and baling — of municipal waste has received considerable attention in recent years (37–39). Such processing would be done at centralized transfer stations and would require extra truck-to-truck transfers. The added costs involved would be offset by benefits that include: higher densities in the landfill and thus greater disposal capacity per acre; easier handling; more economical use of collection vehicles by reducing the number of trips to the disposal site; and fewer rodent and fly problems at the site. Shredding would cost about $3.00 per ton (1972 dollars) and baling about $2.50 per ton. Baling may reduce the rate of leaching to an acceptable level, and thus remove the need to collect and treat leachate (40). Shredding tends to accelerate the rate of contaminant removal, thereby giving the landfill designer some control of the leachate problem (41).

Methane recovery. Methane, the primary constituent of natural gas, has been collected from a sanitary landfill in southern California for use as fuel since mid-1975 (42). In mid-1974, an extensive study of the treatment and use of landfill methane was started by the city of Mountain View, Calif., near San Francisco, with EPA support (43, 44). Later, with support from EPA and Pacific Gas & Electric Co., the project was expanded to include a molecular sieve gas treatment plant with a capacity of 1 million cubic feet of gas daily. The expanded project began shakedown runs in mid-1978.

Incineration

Incinerator technology in this country was largely empirical for many years, but was evolving rapidly by 1970 (45). Modern air pollution control requirements, for example, have made control devices such as scrubbers and electrostatic precipitators mandatory for incinerators. Other developments include more sophisticated designs, closer examination of the technology and economics of recovering waste heat, and an incipient trend toward large regional incinerators rather than small local units. U.S. practitioners also have been looking more closely at incinerator and other solid waste management practices in Europe (46–48). Urban pressures there have demanded the application of technology that, while generally available, is not yet used widely in this country to dispose of municipal refuse.

In modern U.S. incinerators of conventional design, refuse is burned on moving grates in refractory-lined chambers, while combustible gases and their entrained solids are burned in secondary combustion chambers or zones. Combustion is 85 to 90% complete for the combustible materials. Up to three times as much air is introduced into the incinerator as would be needed to supply enough oxygen to oxidize the refuse completely. The temperature in the bed of burning refuse may reach 2500°F or more, and the excess air is required mainly to hold the overall temperature of the furnace at 1400° to 1800°F. At temperatures higher than 1800°F, slag formation becomes a problem. The products of incineration include, in addition to heat, the normal primary products of combustion — carbon dioxide and water — as well as oxides of sulfur and nitrogen and whatever other gaseous air pollutants might be generated; nongaseous products are fly ash and unburned solid residue (Table 7).

A large body of technology has grown up around the incineration process, and a variety of mechanical designs is available. The scientific principles that underlie this technology, however, often are not well defined (49), partly because the heterogeneous nature of refuse and its variable moisture content make analysis of the combustion process extremely complex. Although useful combustion and heat calculations can be made, for example, a sound scientific basis is not yet available for estimating a refuse incinerator's maximum capacity — the maximum weight of refuse that it can burn per hour (50, 51). One study aimed at the lack of scientific information focused on fuel-bed height, combustion temperature profiles, fractions of combustion air used in overfiring and underfiring, and analytical modeling of the process (52). A second study was concerned with the benefits of increased turbulence in overfire combustion air in reducing the amount of entrained, unburned particles (53).

Air pollution

The most immediate problem in municipal incineration in the 1970's is control of air pollutants, chiefly fly ash (54). Odors caused by gaseous air pollutants normally can be prevented by proper furnace design and operation. Emissions of noxious gases, such as oxides of sulfur and nitrogen, are quite low compared with those from other sources and do not appear to pose serious difficulty. Questions have been raised about incinerating plastics, whose concentration in municipal refuse now is estimated to be only about 3.8%, but is growing (55–57). A case in point is the widely used packaging material polyvinyl chloride. The plastic is about 50% chlorine, and upon incineration a significant portion of the element is emitted as hydrogen chloride.

Control of the emission of fly ash and other particles is relatively clear-cut. EPA's particle emission standard for new incinerators is 0.18 g/m^3 (0.08 grain per standard cubic foot) of stack gas (adjusted to 12% CO_2). Wet scrubbers and electrostatic precipitators of 99% collection efficiency can meet this standard, although such equipment is expensive. Some states and cities have imposed emission standards of 0.11 g/m^3 (0.05 grain per standard cubic foot) and even lower, for both new and existing incinerators. Such limits, in effect, prohibit incineration. The growing stringency of particle emission limits has caused many conventional incinerators to close down. EPA has estimated that about 300 conventional municipal incinerators were operating in the U.S. in 1968, but only about 190 in 1972 and roughly 125 in 1974.

The trend in particle emission control for incinerators in the 1970's has been toward electrostatic precipitators (58). The water used in wet scrubbers must be condensed and removed to avoid a stack plume; this is done by cooling the stack gases, which then must be reheated to make them sufficiently buoyant. The scrubber, moreover, uses a large amount of water, which can create a water pollution problem. With an electrostatic precipitator, the gas to be treated must be cooler than about 600°F; uncooled gas from a conventional refractory-lined incinerator may range from 1200° to 1500°F or higher. The gases can be cooled by water sprays, the most common method, or by using a waste heat boiler, but need not be reheated. Air cooling is uneconomical because of the large volume of air required. A third device, the bag filter, has been installed on a few municipal incinerators (59). Bag filters display the necessary collection efficiency, but have a relatively high first cost and space requirement. The life of the available filter fabrics, furthermore, has been too short at the prevailing operating temperatures, although new materials could solve this problem. Glass fabric filter bags, for example, have a long life at 500°F, only about 100°F below the temperature to which gases from refractory-lined furnaces can be cooled by normal means; operating temperatures of 600°F and higher are claimed for some newer fabrics.

Table 7

Typical products of incineration of municipal refuse.

Stack gases	Lb/ton refuse	Fraction by volume (dry)
Carbon dioxide	1,738	6.05%
Sulfur dioxide	1	22 ppm
Hydrogen chloride	—	5–500 ppm
Carbon monoxide	10	0.06%
Oxygen	2,980	14.32%
Nitrogen oxides	3	93 ppm
Nitrogen	14,557	79.57%
Total dry gas	**19,289**	**100%**
Water vapor	1,400	
Total	**20,689**	
Solids, dry basis:		
Grate residue	471	
Fly ash	20	
Total, lb/ton refuse	**21,180**	

Source: Kaiser, E.R., "Refuse Reduction Processes," in "Proceedings, The Surgeon General's Conference on Solid Waste Management for Metropolitan Washington, D.C., U.S. Public Health Service Publication No. 1729, U.S. Government Printing Office, Washington, D.C., July 1967, p. 93.

On the whole, however, bag filters require further research and development to fulfill their promise for use on incinerators.

Water-wall incinerators. Although electrostatic precipitators are effective on refractory-lined incinerators, they are better adapted to the water-wall design. Water-wall incinerators precondition the stack gases automatically for cleaning and also can recover the heat of combustion in the form of steam for district heating or generating electricity. Large, steam-generating, water-wall incinerators are common in Europe (60). The relatively high cost of fuel there has helped to justify the recovery of heat from incinerators, and urban pressures have intensified the need to convert refuse to the lowest-volume, most-readily-disposable form. This need as a rule is best filled by incineration; the auxiliary fuel used typically in heat recovery incinerators, moreover, can increase the efficiency of combustion of refuse to well over 90%. Waste heat can be recovered from refractory-lined incinerators, but the process is relatively ineffective and has seen only limited use in this country. In any event, urban pressures and the cost of fuel in parts of the U.S. and Canada are approaching conditions in Europe, and so European incineration practice is taking root in both countries.

Commercial buildings, apartments. The relatively small incinerators used in commercial buildings and apartment houses present a different problem than do big municipal units. Attempts to use dry collectors such as filters or cyclones on the small units have failed because the equipment is fouled rapidly by condensation of fats. A wet scrubber on a properly designed incinerator, however, can reduce particle emissions by at least 90% and noxious gas emissions by up to 60%. All in all, the problem is more economic than technical, particularly where old incinerators must be modernized. The capital cost of modernizing old equipment in New York City in the late 1960's was estimated at $1500 to $5000 per unit, depending on size.

Composting

Composting as a waste treatment process may be defined as the biological stabilization of the organic content of solid wastes under controlled conditions. The critical word is "controlled" — without that qualification "composting" could describe any biological decomposition, including that in a poorly operated open dump. The biological nature of composting subjects the process to certain needs and limitations: a microbial population and a nutritionally balanced substrate must be present; the rate of processing depends on the rate of bacterial activity; the capacity of a given composting operation depends on the size and composition of the microbial population; and the process is affected by environmental factors like temperature, pH, and degree of aeration. Two further, and very important, limitations are those imposed by the genetic traits of the microbes involved and by the interactions, both stimulatory and inhibitory, among the many groups of microorganisms necessarily present. These two biological limitations together place an upper limit on the extent to which the composting process can be optimized by mechanical means.

Composting operations in the U.S. include three basic steps: refuse preparation; stabilization of organic matter by aerobic microorganisms; and product upgrading (61). Refuse is presorted to remove noncompostable materials or those that might have salvage value. The refuse is then ground to improve the efficiency of the aerobic decomposition process. The ground material is stacked in windrows on the ground or placed in mechanical systems where it is degraded biologically to a humus that has a total nitrogen, phosphorus, and potassium content ranging from 1 to 3%, depending on the nature of the material composted. The finished product may be cured, ground, and bagged.

Windrow composting costs less than mechanical systems, and the total time required is about the same with both processes if the windrow program developed at the University of California (62) is followed. The literature says generally that refuse in windrows requires six weeks or more to degrade to the proper level of stabilization, compared with only about six days for mechanical composters. Actually, composting does not progress far enough in six days in a mechanical digester to permit the material to be stored without creating a severe odor nuisance. That is why specifications for a mechanized operation usually call for six to nine weeks of "maturation," which amounts to windrowing the processed material. The University of California windrow method, however, allows most wastes to be composted within three weeks or less. The development of efficient machines for "turning" (aerating) windrows (63) has done much to make the windrow system more attractive than the fully mechanized system operationally and certainly economically.

Almost every type of organic waste has an indigenous microbial population more than large enough to carry on the composting process from beginning to end. Consequently, the use of additives — inoculums, enzymes, hormones — is superfluous.

Almost any organic material can be composted, but it must contain the proper proportions of essential nutrients if the process is to proceed rapidly enough to be practical. The necessary trace elements generally are present in all wastes. The limitation, if any, is usually an imbalance between carbon and nitrogen. Too high a carbon-to-nitrogen ratio (C/N greater than 25:1) slows the process; too low a ratio (C/N less than 16:1) leads to a loss of nitrogen in the form of volatilized ammonia-nitrogen. In the U.S. the carbon-to-nitrogen ratio of the average municipal refuse as it reaches the disposal site is too high for efficient operation. The problem can be rectified by adding nitrogen-rich wastes such as manures or digested sewage sludge. Other restrictions on the suitability of a waste for composting arise from the need to maintain aerobic conditions. The degree of aerobiosis depends on the amount of air trapped in the voids in the waste. If the total volume of the voids is too small, or if they are filled with water, the lack of atmospheric oxygen will cause the mass to become anaerobic, with all of the attendant problems, including odors. Material of "weak structure," such as saturated paper, or material that contains too much moisture will not compost satisfactorily unless remedial measures are taken. For example, fresh manure should be mixed with bedding material (straw, sawdust, dry leaves, etc.); refuse consisting largely of paper should contain less than 60% moisture.

Compost is used primarily as a soil conditioner and only secondarily as a source of plant macronutrients (nitrogen, phosphorus, potassium). Materials rich in nitrogen and other nutrients before composting, however, usually yield a compost suitable for fertilizing as well as conditioning soil. A potential large-scale use for compost is in renovating lands damaged by strip mining or other human activities.

Status of composting

The status of composting in treating municipal wastes remains poor in the U.S. The number of municipal-scale plants operating at any one time has never exceeded two or three. The reason for the poor showing is largely economic: there is no significant market for compost and, less importantly, the cost of composting is relatively high. A market large enough to absorb all of the compost produced, were the process used by every municipality, would necessarily include large-scale agriculture as well as small growers. The fact is that large-scale agriculture ("agribusiness") is not interested in compost as a soil conditioner, much less as a low-grade fertilizer. The bulkiness of the product and its low cost-benefit ratio make it unattractive to agribusiness. Conceivably this aversion would be eased by fertilizer shortages, which some have predicted will become permanent (64). Another factor in the poor record of composting in this country is the fact that any such undertaking is judged on an economic basis — "Is the venture making money?" If the compost cannot be sold, and expenditures exceed income, the operation is a "failure," and technical performance has little or nothing to do with "success." This criterion is not applied to other solid waste disposal methods. Were it applied to incineration, for example, all incineration operations would be "failures."

Recommendation S4: *Continuing attention should be paid to collection and transportation of municipal refuse, both in the development of improved technology and in mechanisms for promoting the application of such technology by local agencies who are responsible for handling and disposing of municipal refuse.*

Recommendation S5: *The use of known science and technology peripheral to solid waste management in developing improved methods for sanitary landfill and incineration should be encouraged and supported. Specific efforts should be directed toward pollutant identification and the potential for pollutant migration into ground and surface waters from disposal of waste materials on land.*

RESOURCE RECOVERY

The desirability of reclaiming materials from solid wastes has long been recognized, but in the past decade or two only limited amounts have been recovered relative to the consumption of new or "virgin" materials. This has not always been the case; during World War II and the years leading up to it, materials were recycled extensively. In the period 1935–39, for example, recycled metal accounted for 44% of the copper shipped to fabricators, 39% of the lead, and 28% of the aluminum. During the same period scrap supplied about 25% of the ferrous metals and 30% of total paper production. After the war the use of recycled resources began to decline and had dropped precipitously by the late 1950's. The reasons for the decline were a rapid rise in the cost of processing recycled or secondary materials and an over-optimistic impression of the abundance of virgin materials.

The Resource Recovery Act of 1970 in part required EPA to study and report to Congress the status of resource recovery as a means of managing solid wastes (65). "Resource" was taken to include energy as well as materials. In this vein it should be noted that recovery of energy involves, in addition to combustibles, the "energy content" of noncombustible materials.

Oak Ridge National Laboratory has investigated the energy requirements for recycling selected metals (66). One comparison involved the production of steel bars using metal from the basic oxygen process, which consumes primarily iron ore, and from the electric furnace, which uses ferrous scrap almost entirely. The results indicated that raw steel can be made from scrap for a little more than a quarter of the energy required to make it from ore, and finished steel for about half that required with ore. Aluminum can be made from scrap at 4 to 5% of the energy cost of making it from bauxite. Products of recycled copper require only about 10% of the energy needed to make them from ore. Recycling magnesium would require less than 2% of the energy needed to make the metal from seawater. These analyses do not include the energy consumed in separating the scrap from the solid waste and collecting and transporting it to a processing site, but the inclusion of such energy should not substantially affect the overall results.

By 1973–74, the portents for an upswing in resource recovery had become especially good. The reasons were a growing awareness of this country's profligate use of natural resources and uneasiness about depending too heavily on foreign sources for materials (67) and fuels. Nevertheless, a considerable effort will be required to expand recycling activities to a satisfactory level. Trends in materials recovery from municipal wastes appear in Table 8. By 1978, a number of resource-recovery projects were operating, under construction, or being planned in the U.S. (68); for 1976, EPA listed 21 operational facilities, including pilot and demonstration units; 10 units in various stages of construction, startup, or modification; and 87 projects in the planning stage (44). These operations extended from the extraction of heat energy from the combustible fraction of solid wastes to the reclamation of metals, fibers, and other materials for reuse by industry. Energy and materials recovery are treated separately here, although, as will be evident, they may be parts of the same process. As will also be evident, resource-recovery technologies, excepting perhaps the water-wall incinerator, are not yet highly developed in this country.

Thermal Energy Recovery

Of the nearly 4.45 billion tons of municipal, industrial, mineral, and agricultural solid wastes generated in 1970, about 13% or 570 million tons was dry combustible material having a net value of about 8.6×10^{15} British thermal units (Btu) (69). The latter is equivalent to about 12.4% of this country's energy consumption — 68.8×10^{15} Btu — for that year (70). Not all combustible wastes, however, are available in lots of sufficient size to justify their use in energy recovery systems. In summary, about 225×10^{9} lb of dry combustible solid waste representing 20×10^{14} Btu were available in 1970 in 100 ton/day lots for use as fuel. In lots of 500 ton/day of dry combustibles, about 138×10^{9} lb/year representing about 11×10^{14} Btu/year were available. These represent 2.9% and 1.6% respectively of this country's energy requirements in 1970.

Numerous thermal processes in various stages of development recover energy in one form or another from solid waste. (Recovery of methane from landfills, described earlier, represents energy recovery, but not directly by a thermal process.) The only such systems actually in use in this country, however, are those that generate steam. These systems, including those in development, fall into two groups: combustion processes and pyrolysis processes. Detailed information on individual processes is available (71–73). Economic projections for processes described in this section are largely those of their developers and so are probably favorably biased; for this and other reasons, including differences in local conditions, the projections are not necessarily comparable directly or with the generalized data of Table 5. Costs in general are based on conditions of the mid-1970's.

Combustion processes

Industry for some years has used materials such as sawdust and sugarcane wastes (bagasse) as fuels. By the mid-1970's, a number of companies were burning in-plant wastes in conventional incinerators to produce steam, and others were exploring such processes (74). A few municipalities produce steam in water-wall incinerators operated at conventional temperatures; the steam may be used for heating or to drive a turbine to generate electricity. Other types of steam-producing systems are based on high-temperature incineration. The feasibility of substituting mixed urban refuse for conventional fuels in power plants and industrial furnaces also has been under study in the U.S. (75–77). Still another process bypasses steam production and instead uses hot furnace gases to drive a turbine.

Table 8

Materials recovery from postconsumer municipal waste[a] (thousands of tons)

Material recycled	1971	1972	1973	1974	1975
Paper and paperboard	7495	8075	8730	8430	6830
Percentage of gross paper and board discards	15.9	16.0	16.5	16.3	15.5
Aluminum	20	30	35	52	87
Percentage of gross aluminum discards	2.4	3.2	3.4	5.0	8.7
Ferrous metals[b]	140	200	300	400	500
Percentage of gross ferrous discards	1.3	1.4	2.4	3.4	4.4
Glass	221	273	306	327	368
Percentage of gross glass discards	1.8	2.1	2.3	2.5	2.7
Rubber (including tires and other)	257	245	219	194	189
Percentage of gross rubber discards	8.9	7.9	6.8	6.1	6.9
Total materials	**8133**	**8825**	**9590**	**9400**	**7975**
Percentage of gross nonfood product waste	9.5	9.6	10.1	10.0	9.3
Percentage of total postconsumer waste	6.1	6.2	6.7	6.5	5.9

a. Includes residential and commercial wastes that comprise the bulk of typical municipal collections. Does not include mining, agriculture, industrial processing, and demolition and construction wastes, sewage sludge, and junked autos and other obsolete equipment wastes.

b. Data uncertain. No statistics are collected regularly.

Source: "Fourth Report to Congress — Resource Recovery and Waste Reduction," Environmental Protection Agency, Report SW-600, Washington, D.C., August 1977.

The U.S. Naval Station, Norfolk, Va., started to operate a water-wall incinerator, the first in this country, early in 1967 (78) (Fig. 2). The plant has a refuse capacity of 360 tons per day, and the boilers can be fired with oil when the supply is inadequate. Steam capacity is 100,000 to 120,000 lb/hour. The water-wall incinerator in Montreal, the first to be built by a municipality in North America, began operating in 1971. It is rated at 1200 tons of refuse per day and 400,000 lb/hour of steam. By mid-1976, water-wall units were operating in six U.S. municipalities: Braintree and Saugus, Mass.; Chicago, Ill.; Harrisburg, Pa.; Nashville, Tenn.; and Hempstead (Oceanside), Long Island, N.Y. (Table 9). Others were in the planning and construction stage.

The Torrax system of Carborundum Environmental Systems, Inc., is an example of high-temperature incineration (and pyrolysis) with steam production. The Torrax High Temperature Incinerator uses a vertical-shaft furnace; solid waste is charged at the top. Combustion air is preheated in a "Super Blast Heater" by passing it through silicon carbide tubes around which flow the hot combustion products of a fuel, ordinarily gas or oil. The combustion air, at about 2000°F, then enters the shaft furnace near the bottom, passes upward through the solid

FIGURE 2

Water-wall incinerator at U.S. Naval Station, Norfolk, Va.

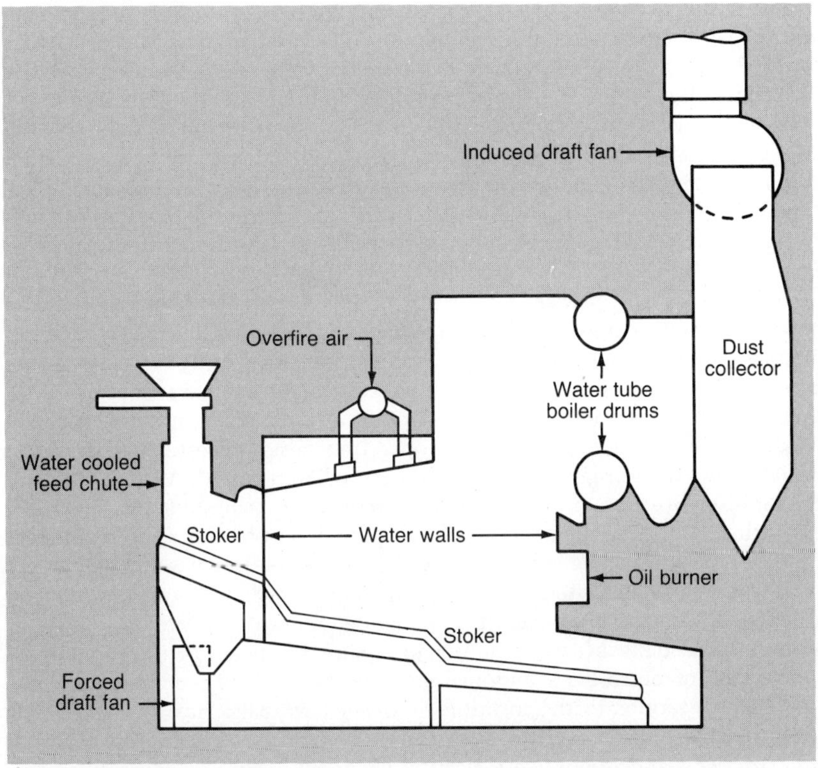

Source: Moore, H.C., "Refuse Fired Steam Generator at Navy Base, Norfolk, Va.," in "Incineration of Solid Wastes," Metropolitan Engineers Council on Air Resources, New York, N.Y., March 1967, p. 10.

wastes, and exits below the top of the refuse charge. As the combustion air passes upward through the refuse, its oxygen is depleted, and solid-waste pyrolysis products, primarily carbon monoxide and unburned hydrocarbons, are carried out the exhaust where they are mixed with air and burned at above 2000°F in a tangentially fired igniter. Noncombustible materials in the exhaust gas fuse and flow down the refractory wall of the igniter into a quench tank. From the igniter, the gas passes through a gas cooler where steam is generated and then through a bag filter for particle removal. The solid waste passing downward through the shaft furnace loses its readily combustible material in the pyrolysis zone, and the difficult-to-oxidize and noncombustible wastes descend slowly into the combustion zone in the lower portion of the furnace. A liquid slag in the range of 2600° to 3000°F is removed from the bottom of the furnace and quenched with water.

The Torrax system was under development during 1969–74 with the help of an EPA demonstration grant to Erie County, N.Y., where a 75-ton/day plant was built and tested. In mid-1974 the operating costs for a 300-ton/day plant were estimated at $6.00 to $8.00/ton of refuse. A 300-ton/day plant would generate about 600 ton/day of steam. A market value for the steam had not been established, but it could be the equivalent of a disposal credit of up to $2.00/ton of solid waste. The Torrax process is marketed by the Andco Torrax Co. A 200 ton/day unit began operating in Luxembourg in 1976, and two others were being built in Europe.

The substitution of urban refuse for conventional fuels is exemplified by the Horner & Shifrin Fuel Recovery Process, in which solid waste is prepared for firing in a suspension-fired utility boiler. In such a system, solid waste provides 10 to 20% of the boiler's heat energy, and powdered coal or other fuel provides the balance. By weight, solid waste provides up to one half the fuel input.

The solid waste processing facility in this system includes a shredder to reduce the waste to a relatively uniform size range, followed by an air classifier to separate the light, combustible material from the heavy noncombustibles. After air classification, the combustible fraction is transferred to a storage facility. The heavy noncombustibles may be processed further to reclaim steel, aluminum, copper, or other materials.

If the solid waste processing facility is adjacent to the utility boiler, the prepared waste is conveyed from the storage bin to a rotary airlock feeder and pneumatically injected into the boiler. If the processing facility is not next to the boiler, truck transportation and additional storage facilities are required.

Capital costs for a 1500-ton/day Horner & Shifrin plant are estimated at $6000 to $8000 per ton of daily capacity; operating costs are estimated at about $5.00/ton, not including waste transportation when required. At a fuel value of $0.45/million Btu, a solid waste fuel value credit of $2.00 to $3.00/ton is possible. In addition, $2.00 to $6.00/ton of solid waste credit may be realized from the sale of recovered metals and glass. A system employing the Horner & Shifrin process was developed by the Union Electric Co. and the city of St. Louis during 1968–76 under an EPA demonstration grant and operated intermittently for four years. In mid-1976, Union Electric was planning to expand the operation to accommodate 8000 ton/day of refuse, but abandoned the project because of economic problems and local objections to the location of waste terminals. However, a similar system, with a refuse capacity of 200 ton/day, was started up by the city of Ames, Iowa, in 1975, and others were starting shakedown runs in Milwaukee and Chicago early in 1978.

Steam production is bypassed by the CPU-400 system, which burns prepared solid waste to provide heated gas for injection into a conventional gas turbine for generating electricity. As in the Horner & Shifrin system, the refuse is shredded, air classified, and placed in storage before use. The combustion system consists of a fluidized bed combustor and three stages of cyclone separators to remove particulates from the gas before it enters the turbine. The turbine compressor is used to provide combustion air, and the whole system operates at a pressure of four atmospheres. Solid waste is fed into the fluidized bed combustor through two rotary airlock feeder valves that discharge into a pneumatic transport system.

The capital costs of the CPU-400 process are estimated to be $12,000 to $15,000/ton daily capacity for a 600-ton/day, three-module power system to burn both solid and liquid wastes. Operating costs are estimated to be $4.00 to $6.00/ton. Income from electricity generated at $0.008/kilowatt-hour (Kwh) would be equivalent to about a $4.00/ton solid waste credit, and sewage sludge disposal income at $25.00/ton of solids would be equivalent to about $0.75/ton of solid waste credit. In addition, $2.00 to $6.00/ton of solid waste credit may be realized from the sale of metals and glass.

The CPU-400 process has been under development (79) since 1967 through an EPA research contract with the Combustion Power Co. of Menlo Park, Calif. By early 1976, a pilot facility, capable of converting about 100 ton/day of solid waste into energy, had been under test and modification for more than a year. The system has encountered severe technical difficulties, principally damage to the turbines caused by entrained materials in the combustion gases, and its future is uncertain (44).

Pyrolysis processes

The pyrolysis approach to energy recovery is exemplified by three processes: the Garrett Pyrolysis Process for producing oil; the Union Carbide Oxygen

Table 9

Water-wall incinerators operating in the United States

City	Year in service	Refuse capacity tons/day	Steam use
Braintree, Mass.	1971	240	Plant auxiliaries
Chicago, Ill. (Northwest)	1970	1600	No market
Harrisburg, Pa.	1972	720	No market
Hempstead, N.Y. (Oceanside)	1974	750	Plant auxiliaries
Nashville, Tenn.	1974	720	Urban heating and cooling
Norfolk, Va. (Naval Station)	1967	360	District heating, plant auxiliaries
Saugus, Mass.	1976	1200	Plant auxiliaries

Source: "Fourth Report to Congress — Resource Recovery and Waste Reduction," Environmental Protection Agency, Report SW-600, Washington, D.C., 1977.

Refuse Converter System for producing fuel gas; and the Monsanto Landgard System for producing steam.

The Garrett Pyrolysis Process of Occidental Research Corp. uses low-temperature flash pyrolysis to produce a char and a high-viscosity, highly oxygenated fuel oil having a heating value of about 10,000 Btu/lb (about 75% of that of No. 6 fuel oil). Two stages of shredding, an air classification operation, and a drying operation are employed to produce a pyrolysis-reactor fuel of minus 14 mesh (80% of the particles will pass a screen with 14 openings to the inch). The heat required for the pyrolysis operation is derived from the combustion of the pyrolysis off-gas and a fraction of the char product. The heat energy is transferred by means of a heat exchanger of proprietary design. From the pyrolysis unit the gases are exhausted through a cyclone to remove the char and then scrubbed to remove the oil, water, and other solids and liquids. Part of the cleaned gas, which has a heating value of about 550 Btu/cu ft (about half that of natural gas), is used to heat the pyrolysis reactor.

The Garrett process has been tested in a 4-ton/day pilot plant. A 200-ton/day plant, supported partly by an EPA demonstration grant, was starting up early in 1978 in San Diego. It will be used to verify predicted process costs and product yields. Capital costs for a 1000-ton/day plant are estimated to be $12,000 to $15,000/ton daily capacity, and operating costs are estimated to be $5.00 to $7.00/ton. At $0.60/million Btu, a credit of about $3.00/ton solid waste could be claimed from the sale of oil. An additional $2.00 to $6.00/ton of solid waste credit may be realized from the sale of recovered metals and glass.

The Union Carbide Oxygen Refuse Converter System consists of a shaft furnace into which unprepared municipal refuse is charged at the top and pyrolyzed as it passes downward through the furnace. Oxygen enters the shaft through tuyeres near the bottom of the furnace, passes upward through a 2600°-to-3000°F combustion zone and a pyrolysis zone, and exits at about 200°F. (A cryogenic gas-separation unit at the site produces the gaseous oxygen at 95% purity.) The off-gas then passes through an electrostatic precipitator to remove fly ash and the oil formed during pyrolysis; both are recycled to the furnace combustion zone. From the precipitator the gas then passes through an acid absorber and condenser. The clean fuel gas has a heating value of about 300 Btu/cu ft and a flame temperature equivalent to that of natural gas. The solid waste passing downward through the furnace traps part of the oil and fly ash from pyrolysis while the gas itself loses moisture. The solids that remain after pyrolysis and combustion are removed as a slag from the bottom of the furnace.

The Union Carbide system was tested in a 5-ton/day pilot plant beginning in 1971. In 1974, a 200-ton/day plant was completed in South Charleston, W.Va., and extended runs have been made with it using municipal refuse alone and mixed with sewage sludge. For a 1000-ton/day system, capital costs are estimated to be $10,000 to $13,000/ton daily capacity. Operating costs are estimated to be about $5.00/ton. About 7 million Btu of gas are produced from each ton of solid waste. At $0.75 per million Btu, a credit of about $5.00/ton would be realized from the sale of the gas. Union Carbide offers the system commercially as the Purox System.

The Monsanto Landgard System consists of a rotary kiln pyrolyzer that is fed with shredded solid waste at one end and with air and fuel at the other end. As the hot products of combustion contact the solid wastes, pyrolysis gases and char are formed. The char and solid waste inerts exit from the kiln at one end into a quench tank, while the pyrolysis gas exits from the other end into an afterburner. Hot exhaust gases from the afterburner pass into a waste heat boiler to produce steam and finally through a wet scrubber and exhaust fan. The process uses about 47 Kwh of electricity and 1 million Btu of fuel to produce 4800 lb of steam and 1 million Btu of char from each ton of solid waste. Estimated capital cost for a 1000-ton/day system is $12,000 to $14,000/ton daily capacity, and operating costs are estimated to be about $6.00/ton. Revenue from the sale of steam is yet to be established, but probably will be about $2.00/ton solid waste. An additional credit of $2.00 to $6.00/ton of solid waste may be available from the sale of reclaimed metals and glass.

A 50-ton/day Landgard pilot plant has been operated. A 1000-ton/day plant was completed in 1976 in Baltimore, Md., with support from an EPA demonstration grant. A number of operating problems were encountered, however, and extensive modification failed to solve them satisfactorily. Early in 1977 Monsanto withdrew from the project and recommended that the unit be operated as a conventional incinerator. The city of Baltimore continued to operate it as a pyrolysis unit, but at reduced capacity.

Materials Recovery

The upsurge in interest in recovering materials from solid wastes in the 1970's has been focused primarily on the residential, commercial, and institutional discards that constitute municipal refuse. These "postconsumer" wastes, as noted earlier, make up only about 5% of the nation's total solid waste load, but have the most immediate impact because they are concentrated in urban areas. Also, postconsumer wastes contain substantial amounts of potentially useful materials (Table 3).

Municipal refuse is heterogeneous and highly variable in composition (Table 4), and hand sorting probably is the most common means of recovering salable materials from it. Still, by 1978 mechanized systems were planned, under construction, or operating at a number of sites, often as parts of energy recovery schemes like those mentioned earlier (68). Attention was being devoted also to problems and processes that affect the recyclability of individual materials.

Systems for recovering materials from raw municipal refuse are typified by the pilot plant that the U.S. Bureau of Mines began operating in 1973 at its Metallurgy Research Center in College Park, Md. The steps in the process include shredding, magnetic separation, air classification, screening, and washing (80) (Fig. 3). The pilot plant separates ferrous and nonferrous metals, glass, and combustibles from 5 ton/hour of raw municipal refuse. By early 1976, several municipalities in this country and governments abroad planned to install adaptations of the Bureau of Mines process.

A different approach to raw municipal refuse is the wet pulping process developed by Black Clawson Co. A facility of this type, supported partly by an EPA grant, has been operating continuously in Franklin, Ohio, since 1971 (81). Its capacity is 150 ton/day of raw refuse. Incoming refuse is mixed with water and ground to a slurry in the wet pulper, which resembles a large kitchen disposal unit. Large pieces of metal and other nonpulpable materials are thrown out the side of the pulper and conveyed to a magnetic separator which removes ferrous materials; the residue is landfilled. The slurry from the pulper goes to a liquid cyclone, where centrifugal force separates heavier noncombustibles such as glass, metals, and ceramics. The heavy fraction goes to a glass and aluminum recovery system; the light fraction goes to a paper fiber recovery system that rejects the shorter, less valuable fibers. Combustible residues from both systems are mixed with sewage sludge, and the mixture is dewatered mechanically and incinerated in a fluid bed unit. Noncombustible residues are landfilled. A Black Clawson facility with a capacity of 2000 tons of refuse daily was expected to

FIGURE 3

Raw refuse separation flow sheet, U.S. Bureau of Mines

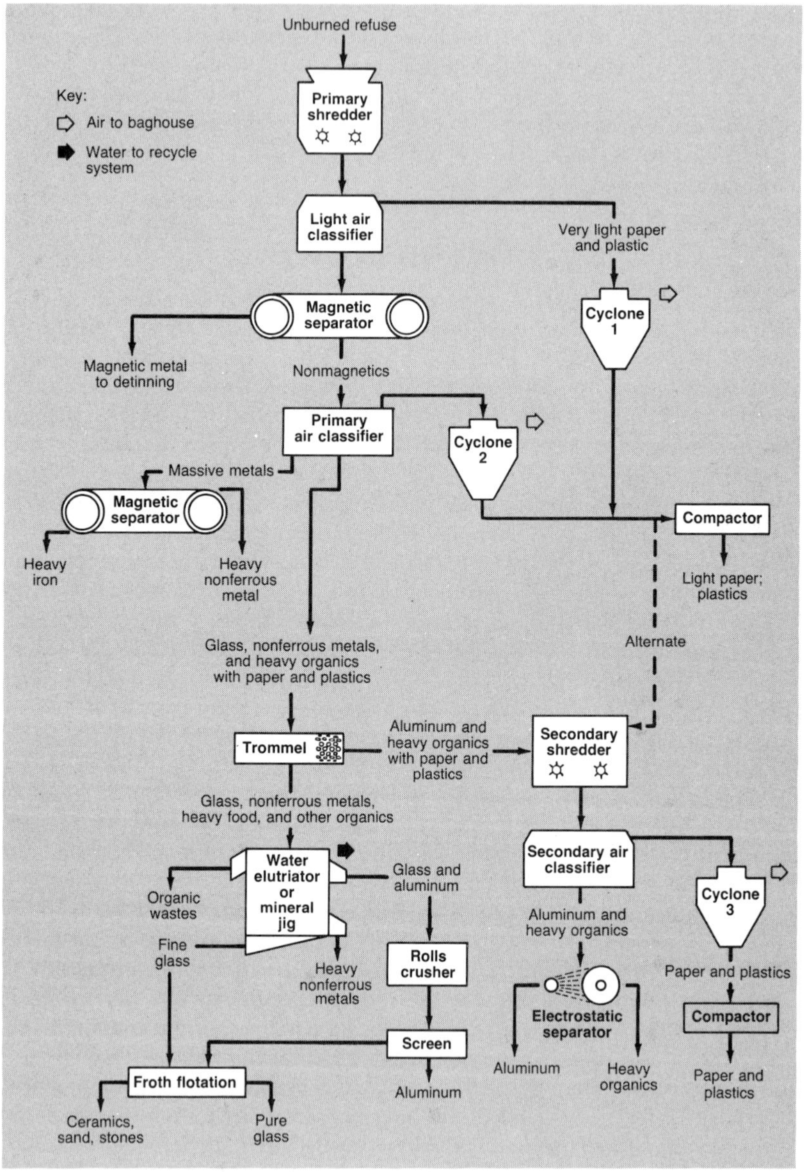

Source: Sullivan, P.M., Makar, H.V., "Bureau of Mines Process for Recovering Resources from Raw Refuse," Proceedings, Fourth Mineral Waste Utilization Symposium — IIT Research Institute, Chicago, Ill., 1974.

begin operating in 1978 in Hempstead, N.Y. Combustible residues will be burned in a bark-type boiler to generate steam for producing electricity at the site.

Still another approach to municipal refuse is to recover materials from incinerator residue. A ton of such residue contains on the order of 400 lb of ferrous metals, 35 lb of aluminum, 15 lb of copper, lead, and zinc, and 600 lb of glass. The Bureau of Mines has developed and tested a recovery process in a pilot plant at its College Park research center that can treat about 12 tons of incinerator residue daily (82).

Metals

Metals are important to solid waste management because they degrade relatively slowly and because they are nonrenewable resources. Scrap metal produced by industrial operations is comparatively easy to handle; often it is readily recycled and may never leave the plant where it is generated. Metals that enter the waste stream, however, such as those in obsolete automobiles and municipal refuse, are especially difficult to manage. All of the easy work has been done — industrial wastes, for example, have already been picked over for valuable or easily recovered materials.

Nonferrous metals. Nonferrous metals constitute only about 1% of the municipal solid waste stream, by weight, compared with 6.5 to 10% for ferrous metals. Aluminum accounts for only 0.5 to 0.7% of municipal wastes, but its high resale value, up to $300/ton in the late 1970's, has led to greater effort to reclaim it than in the past. In municipal waste the largest source of aluminum for recycle is beverage containers; collections in 1977, for example, totaled a record 6 billion cans, about one of every four sold. Despite the strong motivation for reclaiming aluminum, however, only about 20% is recovered from all sources. Recovery of scrap aluminum, other than that in containers, is impeded by the high cost of identification and sorting, by contaminants in the metal, and by competition with natural sources.

Certain other nonferrous metals, such as chromium, copper, nickel, tin, and lead, are recovered in greater proportions than aluminum because of their high value. The absolute amounts of these metals in the solid waste flow, however, are much smaller than those of aluminum and ferrous materials.

Ferrous metals. Overall, about 20 million tons of postconsumer ferrous scrap are recycled each year, while the amount of ferrous solid waste not being utilized is increasing at a rate of about 140,000 tons annually. During the five years ending with 1974, the fraction of available ferrous scrap recycled dropped from about 60% to about 40%. The raw steel and foundry industries provide the main domestic demand for obsolete ferrous scrap, and much of the decline in demand has been due to the rise of the basic oxygen furnace for making raw steel. The BOF is essentially self-sufficient in scrap, since it generates about as much as it consumes. Various means of increasing the fraction of scrap in the charge have been explored, but in general they cost too much to be practical. The BOF process, meanwhile, has been steadily displacing the open hearth furnace, long a significant user of scrap, both in the U.S. and abroad. The BOF accounted for 62% of this country's steel output in 1977, compared with 16% for the open hearth (including a small amount of Bessemer steel). The third common process, the electric furnace, can use up to 99% scrap in its charge, but accounted for only 22% of the raw steel produced in 1977.

The demand for obsolete ferrous scrap is limited also because the scrap is often contaminated with elements like chrome, nickel, copper, aluminum, and tin. The use of such scrap in making raw steel involves the risk of adding elements in undesirable proportions. The amounts and kinds of contaminants that can be tolerated depend to some extent on the ultimate end products. The long-term effects of using contaminated ferrous scrap at current levels or higher apparently are not known. Some industry experts suggest the possibility of a build-up of residuals such as tin, while others have noted the possibility of an undesirable accumulation of lead in the furnaces (83).

Scrap in the form of consumer durables, especially the so-called "white goods" — steel sheet coated with porcelain — is difficult to market. Also, the insulating material in these items poses difficulties, while the presence of tin, nickel, chrome, and copper results in a less attractive scrap. The white-goods problem is significant. In 1971, discards of the nine major appliances added nearly 1.7 million tons to the ferrous solid waste stream (84); gross discards of major appliances in 1975 came to an estimated 2.4 million tons (44).

Automobile scrap. The discarded automobile typically goes first to the auto wrecker, who removes those parts that can be sold for further use. The remainder goes to the scrap processor, who may sell the battery for its lead content, the radiator, electric motors, and some other parts for their copper content, the engine block and other cast-iron components for use in foundries, and heavy components such as the frame for use as heavy melting stock. The remaining hulk, weighing from 1000 to 1500 pounds, is cleaned of combustible or other nonmetallic material by hand-stripping or burning and converted into scrap by baling, shearing, or shredding.

The auto scrapping industry is plagued by transportation costs ranging from those for moving the hulk to the processing plant to those for hauling the scrap to the steel producer or other user. Another problem is ridding the hulk of combustibles. Hand-stripping is expensive, and increasingly stringent air pollution laws have limited the use of open burning. Also troublesome is the generally lowered demand for ferrous scrap brought about by the steel industry's shift to the basic oxygen furnace. Nevertheless, improvements in automotive scrap recycling are resulting from new processing techniques. Smokeless incineration technology has been developed (85, 86) and is slowly coming into use. And the number of operating automobile shredders, which produce a relatively high grade of scrap, increased from 69 in 1969 to more than 100 in 1972 (7).

The U.S. Bureau of Mines has investigated various chemical, metallurgical, mechanical, and cryogenic methods for processing automotive scrap (87, 88). At its Salt Lake City installation, the Bureau has worked in two complementary areas: dismantling and chemically analyzing selected cars; and developing methods for making clean scrap from cars or other sources. Methods of upgrading automotive scrap have been studied also at Iowa State University. Of those investigated, electroslag remelting, vacuum induction melting, and electron beam melting showed the greatest potential (89).

Rubber

The resilience, impermeability, and virtual indestructibility that make rubber ideal for its intended uses also make it extremely difficult to dispose of or to reclaim. Furthermore, consumption of natural and synthetic rubber in the U.S. is rising. In 1970 consumption totaled some 2.5 million metric tons; it climbed

25% by 1973, fell sharply in the recession years 1974–75, but in 1977 exceeded 3.2 million metric tons.

Despite the rising amount of waste rubber, the future for reclaiming more of it by the standard chemical means is not promising. The major rubber producers maintain very limited reclamation facilities, and independent rubber reclaimers are few. Reclaimed rubber accounted for only about 3.5% of the rubber consumed in the U.S. in 1977. The poor prospects result in part from a large increase in the cost of manufacturing reclaimed rubber relative to that of new rubber; the problem has been aggravated by restrictions on air and water pollution.

Vehicular tires account for some 65% of the rubber consumed in the U.S. On the order of 225 million tires are discarded annually in this country, and they constitute one of the major headaches in the disposal of solid wastes. The majority of scrap tires are landfilled. Other options include retreading, building "artificial reefs" (fish shelters), use as crash barriers, use in ground form in asphalt, and conversion to energy either directly or through destructive distillation. Experimentally, tires have been chilled at cryogenic temperatures and processed in a hammer mill; the rubber fragments instantaneously and separates almost completely from wire and cords (90). Although retreading is the most economically attractive option, it only delays the disposal problem because each retreaded tire ultimately must be discarded. Retreading does, however, reduce the amount of tire waste produced in any given period; without it, some 22% more passenger tires would be discarded per year. There is persuasive evidence that retreaded tires can perform as well as new tires on nonpassenger vehicles. In 1968, retreads comprised 28% of tire sales in the truck and bus market and 17% in the passenger vehicle market. Tires for trucks and buses are retreaded 1.8 times, on the average, and many are retreaded four or five times. In the late 1970's some 46 million auto and truck tires were retreaded annually (91).

The heating value of waste rubber is 50% greater than that of coal and equal to that of oil. Direct combustion of rubber (especially tires) has been hampered by the attendant air pollution potential and by the fouling of the combustion chamber inherent in burning rubber products. In 1973, Goodyear Tire & Rubber Co. installed a novel cyclonic furnace with a capacity of 3000 tires daily at Jackson, Mich. (92). The device is reported to be essentially pollution-free, but has suffered intermittently from mechanical problems.

Fractional distillation of tires probably holds greater promise than direct combustion. Fractional, destructive distillation converts the waste rubber to oil, gas, carbon black, char, and miscellaneous petrochemicals through processes that take place at high temperatures in the partial or complete absence of oxygen. Although tire distillation is expensive, several research projects have shown and are showing now that it is technically feasible (93). In mid-1976, Goodyear and Oil Shale Corp. (later Tosco Corp.) completed a test facility in Rocky Flats, Colo., that was designed to recover liquid hydrocarbons, carbon black, and steel from 15 tons of tires daily. Depending on results with this pilot plant, Goodyear was contemplating a full-scale facility to consume more than 10 million tires annually. It would produce 57 million pounds per year of carbon black, 14 million gallons of fuel oil, and 9.6 million pounds of steel (91).

Paper

The amount of wastepaper recycled in the U.S., as a percent of total paper and paperboard consumption, fell from 27.4% in 1950 to 18.3% in 1974. Paper and paperboard consumption in 1974 was some 65 million tons, of which 81.7% or some 53 million tons entered the solid waste stream. The recycle rate has remained essentially level since 1968 (65); estimates for 1977 show paper and paperboard consumption of 66 million tons and a recycle rate of about 18% (94).

Paper that is recycled is pulped to supply cellulose fiber for the paper industry. About 40% of the paper recycled is "converting" wastes (in-house wastes); the remaining 60% is postconsumer waste (95). Of the postconsumer waste, about 35% is corrugated boxes, 35% is mixed office and high-grade paper, and 30% is newspaper. As of 1970, 3.1% of the total wastepaper consumed was used in newsprint, 5% in printing, writing, and related paper, and 69% in paperboard. About 12.2% was used in construction paper and board, molded pulp, and similar materials, and the remainder for miscellaneous purposes.

The major reason for the limited recycling of wastepaper is that, in general, it competes with virgin fiber in papermaking. De-inking and the removal of other additives from wastepaper present problems — although the technology is steadily being improved — and the properties of recycled fiber are somewhat inferior to those of virgin fiber. The virgin fiber, as a result, is strongly favored by consumer preferences for paper products and by the economics and technology of paper manufacture. Mills built to use virgin fiber do so as a rule for the greater part of their output; most of the recycled wastepaper is used in secondary mills that are built for the purpose and are far outnumbered by those designed to use virgin fiber. Also, no shortage of timber for virgin fiber seems imminent.

The long-term demand for wastepaper will depend to a great extent on the price and availability of virgin wood pulp. Factors like increasing labor and equipment costs could raise the cost of virgin pulp to the point where wastepaper would be economically competitive. Nevertheless, the full potential of wastepaper recycling most likely will not be reached in the foreseeable future because of the economic and technical advantages of using virgin pulp and the adequate supply of timber. A fiscal incentive program could increase the fraction of wastepaper recycled. However, such a program may be inappropriate in view of the availability of other uses for the material. The use of wastepaper in cellulosic insulation, for example, was growing rapidly by 1978 because of the rising cost of energy and the effort to conserve it.

Cellulose

The nation's solid waste stream contains cellulose not only as paper, but also in the large quantities of cellulosic wastes produced by agriculture and the timber industry. Some such wastes serve as fuel, but the waste cellulose that is used in any way, including the recycling of paper, remains a small fraction of the total available. Yet the characteristics of this very common chemical would permit it to be put to use by diverse chemical means. The same technology might be used to recover other chemicals, such as the lignins and pentosans that may be associated with cellulose in wood and straw (96).

Cellulose comprises about half the organic matter created by the photosynthetic process in green plants and is, therefore, replenishable annually in immense amounts. The compound is a polysaccharide, a polymer of the simple sugar glucose; when hydrolyzed — broken down in the presence of water — it yields only glucose. Cellulose also can be degraded thermally to yield carbon, carbon dioxide, water, and a number of other compounds that could be recovered for use (Table 10). The many chemical reactions of the polymer involve primarily its hydroxyl (OH) groups and are exploited commonly in industry.

Cellulose in straw, a by-product of the cereal and grass seed industries, is used extensively to make paper abroad, but not in this country. Paper of relatively high quality can be made from straw by most pulping processes, but particularly by nitric acid or soda pulping. Chopped straw may be pressed into finished products such as door-frames and doors, and cellulose as municipal refuse or straw can be converted to various types of particle board. In both kinds of products an appropriate resin must be used. The urea-formaldehyde or phenol-formaldehyde resins used to make plywood are used often in particle board. Research has indicated, however, that a better board results from the use of cross-linking chemicals like diisocyanates, which tie the cellulose chains together by reacting with their hydroxyl groups. The tubular structure of the straw, by exposing more surface, increases the amount of cross-linking that can occur and thus increases the internal strength of the finished product. High density hardboard made in this way is of high quality; its natural appearance is attractive, and it will accept most finishes readily.

A number of products are made commercially from cellulose (Table 11). Familiar examples include rayon, cellophane, a variety of lacquers, and smokeless powder.

Cellulose (in wood) almost certainly was man's first fuel. Its heating value is only about half that of coal or oil, however, and as wood or other plant matter it is bulky and uneconomic to transport. A number of investigators over the years have demonstrated that cellulose can be hydrogenated to produce hydrocarbons (97–100), and the resulting liquid, which also includes oxygen-containing compounds, has been suggested as a fuel. The economics of such a scheme have not been attractive, but the latent shortage of petroleum has aroused new interest in cellulose as a source of easily transported liquid fuel. The polymer can be hydrogenated by a number of reactions. In particular, a great deal of work has been done in recent years on hydrogenating cellulose by a process using water

Table 10

Products of thermal decomposition of cellulose

	Cotton weight %	Pine pulp weight %	Spruce pulp weight %	Birch pulp weight %
Carbon	38.8	36.9	34.9	33.4
Carbon dioxide	10.4	12.8	11.9	11.1
Ethene	0.2	0.2	0.2	0.4
Carbon monoxide	4.2	3.4	3.9	3.5
Methane	0.3	0.3	0.2	0.5
Methanol	0.0	0.0	0.1	0.0
Acetone	0.1	0.1	0.1	0.2
Acetic acid	1.4	2.2	2.8	3.9
Other organic compounds	5.1	4.2	8.5	7.7
Tar	4.2	4.8	6.3	9.6
Water	34.5	34.2	30.0	29.4

Source: Barbour, J.F., Groner, R.R., Freed, V.H., "The Chemical Conversion of Solid Wastes to Useful Products," Environmental Protection Agency, National Technical Information Service, PB 233 178, Springfield, Va., April 1974.

and carbon monoxide (97). Promising studies of this approach were in the pilot plant stage by 1974; developments in the use of metal catalysts to increase the efficiency of the reaction were proving most encouraging.

Waste cellulose and related products are valued highly as organic matter for application to soil, but they improve primarily the structure of the soil and contribute little in the way of plant nutrients. Cellulose may be converted, however, to a material that contains appreciable amounts of nitrogen, phosphorus, potassium, and other plant nutrients. Such a product not only combines organic matter with the desired nutrients; unlike most commercial fertilizers, it releases the nutrients over a period of time, reducing losses by leaching and other mechanisms. Cellulose can be enriched with nitrogen by reacting it under high pressure with ammonia or, even better, with urea. The reaction of urea forms a cellulose carbamate, and nitrogen contents of as high as 10% are readily attainable. The same reaction carried out in the presence of phosphoric acid yields a cellulose amide phosphate. To the latter can be added potassium to give an organic fertilizer containing the three major plant nutrients — nitrogen, potassium, and phosphorus. Experiments with this product have shown it to be quite acceptable as a soil amendment.

Waste cellulose also may be converted to useful products biochemically (101). In one such process, bacteria convert cellulose to single-cell protein (102). Milled or shredded cellulose, slurried in water with appropriate nutrient salts, serves as a growth medium for a bacterium of the *Cellulomonas* species or for mixed cultures of microorganisms. The organisms use the cellulose as a source of carbon and incorporate the nutrient salts into protein. They multiply to a bacterial mass that can be separated from the residue of the fermentation and processed to yield a high-protein material that has potential as a food for animals

Table 11

Cellulose derivatives and their uses

Compound	Uses
Ethyl cellulose	plastics, lacquers, sheeting, varnishes, adhesives
Methyl cellulose	adhesives, latexes, emulsions, foods, sheeting, cosmetics, pharmaceuticals
Sodium carboxymethyl cellulose	colloid, thickener
Benzyl cellulose	coatings, plastics, lacquers
Ethyl hydroxyethyl cellulose	emulsifier, thickener, stabilizer
Cellulose nitrate	lacquer, plastic, explosives
Sodium cellulose sulfate	gelatin films, glues, paints, textiles, paper coating
Cellulose acetate	yarn, photographic films, sheeting, plastics, coatings, membrane
Cellulose xanthate	fiber, cellophane, plastics, sponges

Source: Barbour, J.F., Groner, R.R., Freed, V.H., "The Chemical Conversion of Solid Wastes to Useful Products," Environmental Protection Agency, National Technical Information Service, PB 223 178, Springfield, Va., April 1974.

and even for humans. A second microbial process converts cellulose to glucose (103). The latter could serve as a food or be converted microbially to single-cell protein or chemicals like ethyl alcohol and the solvent acetone. The glucose is produced by enzymic hydrolysis of cellulose by the mutant fungus *Trichoderma viride*, developed by the U.S. Army Natick Laboratories, Natick, Mass.

Glass

The total amount of glass in municipal wastes in 1975 was about 13 million tons, mostly in the form of containers and packaging. Of this amount, somewhat less than 3% was recycled. The considerable amount of waste glass (cullet) recycled by the glass industry was primarily in-plant or process wastes.

There is no lack of a market for clean, color-sorted cullet. The problem is to produce such a cullet from glass in the condition in which it occurs in municipal wastes. Thus glass recycling is limited by supply rather than by demand. The best hope for producing a suitable cullet is mechanical separation and color-sorting in resource recovery facilities where wastes are being processed and classified; technology of this kind is in development (104). New methods of utilizing waste glass are also in development (105).

Plastics

Recovery of plastics from mixed wastes is practically nonexistent, largely for two reasons: the difficulty of extracting plastics in the purity needed for reuse; and the low costs, at least today, of the virgin raw materials. Discarded plastics could, however, become a significant source of energy. The heat content of plastics averages about 11,000 Btu/lb, about the same as coal. A potential drawback to such an approach is the formation of the corrosive hydrogen chloride in the incineration of polyvinyl chloride (PVC) that might be in the plastic mix. PVC constitutes about 13% of the total production of plastics.

The Product Charge Concept

In recycling, as in any economic activity, there is a social optimum. For most recoverable materials, the optimum probably is neither 100% nor 0%, but somewhere between those extremes. For private industry, the optimum level of recycling, without government intervention, is the point at which the marginal cost of recycling equals the marginal revenue from the sale of materials. The market system, however, does not take into account the social cost of not recycling. A specific example is the manner in which manufacturers design their products.

The tractability of solid wastes is governed in part by product design, including materials selection (7). In this light, a well designed product would be durable; relatively easy to disassemble, chemically or physically, for recycling; and readily degradable in the environment. The designer, however, must consider additional parameters, including consumer preferences, product safety, and the costs of materials and manufacture. Some design parameters, such as durability and degradability, may be in varying degree incompatible. More importantly, products are designed in an economic framework that does not reflect the design-related costs of solid waste handling and disposal, recovering or not recovering depletable resources, and environmental impact. These costs might be built into the design process by various means, including a "product charge" imposed at the point of manufacture.

Conceptually, the product charge is simply a mechanism for building social costs into the market system. In the context of solid wastes it need not necessarily be imposed on the manufacturer. A charge could be levied at the point of wholesale or retail sale or at the point of waste discharge, although the net effect might vary with the point of imposition. In any event, the creation of such a mechanism poses problems. Among them are the difficulty of assigning equitable costs to, say, environmental impact; the danger of erecting yet another unwieldy bureaucracy; and the possibility of channeling resources inadvertently into socially undesirable uses. Nevertheless, the prospective benefits of the product charge and related measures in solid waste management have seemed sufficient to warrant continued investigation of the approach. The Resource Recovery Act of 1970 directed EPA to conduct such studies, and the results through about mid-1976 have been summarized (44).

Recommendation S6: *Research and development on utilization and recycle of components of solid wastes, including their energy content, should be maintained at a level that will insure that radically new approaches are not overlooked or inadequately investigated.*

Recommendation S7: *Efforts by private industry to improve the economics of the automobile scrap processing industry should be stimulated. The improvement of present scrapping methods or the development of radically new scrapping methods that would permit the use of less costly equipment should be pursued at all levels.*

Recommendation S8: *Economic mechanisms that will encourage a more rational approach to solid waste management should continue to be developed and evaluated. Such measures should be directed at product design as well as at waste handling and disposal and resource recovery.*

EFFECTS OF SOLID WASTES

The primary purpose of controlling environmental pollution is to protect human health and the ecological integrity of the biosphere on which humans and other organisms depend. Environmental protection can produce many other benefits — production economies, higher efficiency, resource conservation, improved aesthetics — but the fundamental concerns are the potential threats to health and life. Numerous incidents of death, injury, or illness have arisen from the migration of gaseous and waterborne pollutants from disposal sites, from direct contact with wastes, and from disease vectors harbored by wastes.

For these reasons, deeper understanding is needed in several areas: the effects of environmental pollutants arising from the recovery of resources from wastes; the ultimate disposal of wastes; the mechanisms by which waste materials are transformed into pollutants not present in the original wastes; the mechanisms by which pollutants migrate from disposal sites; and the routes by which pollutants come into contact with the organisms they affect. Only within the framework of such knowledge can effective and economic solutions to solid waste recycle and disposal problems be developed (106).

Much is already known of the effects of environmental pollutants, and research is in progress to develop and refine data on effects on humans and animals, aquatic organisms, and, to some degree, plant life. Yet research on the effects of solid-wastes pollution is a complex area, and some aspects of it take considerable time, often as much as decades. This is because the effects of pollutants on health often are subtle and may become apparent only when corrective measures are no longer possible. Indeed, the results of contamination often become known so belatedly that few recall any relationship between cause and effect. This slow response makes it difficult to determine and evaluate the impact of environmental degradation on receptor organisms, both individually and collectively. What each pollutant does to a human depends on its physical and chemical properties; the time, intensity, and route of exposure; and the individual's tolerance for the pollutant. Some people — notably the very young, the old, and the sick — are especially susceptible. Still, resistance to pollution is an individual trait, governed by factors such as age, heredity, general health, climate, occupation, residence, and smoking and dietary habits. It is thus extremely difficult to assess precisely the general effects on a large population.

All analyses of the environmental effects arising from the ultimate disposal of solid wastes on land should consider: whether the pollutant is toxic, radioactive, carcinogenic, mutagenic, teratogenic, pathogenic, explosive, flammable, corrosive, etc.; the chemical and microbiological transformation of waste components into pollutants not originally present and the effects of these transformation products; the route by which organisms are exposed to pollutants, including consideration of transport of pollutants to air or water (intermedia transport), transport by vectors, accumulation in and transport through food chains, and direct exposure of humans and animals; soil contamination and damage, use of the adsorptive or self-purification properties of soil for pollution control, land reclamation and reuse, and similar factors; and, finally, aesthetics. In performing such analyses, it is imperative also to consider other properties of the pollutants, such as degradability and environmental persistence, environmental mobility, biomagnification, and bioaccumulation.

When considering methods of resource recovery and disposal of wastes on land, the potential impact of alternative technologies on human health and the biosphere can be evaluated during the research and development stages to eliminate environmentally undesirable technologies from consideration. Data are needed on the effects of waste disposal on aesthetics and on soil contamination, damage, and reuse so that cost/benefit assessments can be made as a measure of waste management and disposal efficiency. Basic data on effects on human health and ecological impact are also required for cost/benefit assessments. However, these data are being developed in programs to control air and water pollution and insure safe drinking water, so that no separate, overall effort should be required to support waste management activities for land-destined wastes.

Recommendation S9: *Assessments of the environmental effects of pollutants must be made and documented as an integral part of any activity dealing with resource recovery and the ultimate disposal of wastes on land. Research to determine the levels at which effects occur and assessments to determine risk and cost/benefit levels are both required.*

LITERATURE CITED

1. Golueke, C.G., et al., "Comprehensive Studies of Solid Waste Management," SERL Reports 67-7 (May 1967), 69-1 (January 1969), 70-2 (August 1970), 72-3 (May 1972), Sanitary Engineering Research Laboratory, University of California, Berkeley, Calif.
2. "EPA Activities under the Resource Conservation and Recovery Act of 1976 — Annual Report to the President and the Congress, Fiscal Year 1977," Environmental Protection Agency, Report SW-663, Washington, D.C., February 1978.
3. Black, R.J., et al., "An Interim Report. 1968 National Survey of Community Solid Waste Practices," presented at the 1968 Annual Meeting of the Institute for Solid Waste, American Public Works Association, Miami Beach, Fla., October 1968.
4. "Municipal Solid Waste . . . Its Volume, Composition, and Value," NCRR *Bulletin*, Vol. III, No. 2, National Center for Resource Recovery, Washington, D.C., 1973.
5. "Analysis of Composition of Rubbish in the United States," *Solid Wastes Management — Refuse Removal J.*, **15**, 74, September 1972.
6. "Automobile Disposal — A National Problem," U.S. Bureau of Mines, Department of the Interior, U.S. Government Printing Office, Washington, D.C., 1967.
7. "Second Report to Congress — Resource Recovery and Source Reduction," Publication SW-122, U.S. Environmental Protection Agency, Washington, D.C., 1974.
8. Darnay, A., Franklin, W.E., "Salvage Markets for Materials in Solid Wastes," U.S. Environmental Protection Agency, Report SW-29c, Washington, D.C., 1972.
9. Proceedings of the Symposium: "Mineral Waste Utilization," IIT Research Institute, Chicago, Ill., March 1968.
10. Parekh, B.K., Goldberger, W.M., "An Assessment of Technology for Possible Utilization of Bayer Process Muds," Battelle Columbus Laboratories, Columbus, Ohio, draft dated Jan. 30, 1976, EPA Grant R-803760-01.
11. "The Florida Phosphate Slimes Problem," Bureau of Mines Information Circular 8668, U.S. Department of the Interior, Washington, D.C., 1975.
12. "Report to Congress: Disposal of Hazardous Wastes," Environmental Protection Agency, Publication SW-115, U.S. Government Printing Office, Washington, D.C., 1974.
13. *Fed. Regist.*, **43**, 4366, Feb. 1, 1978.
14. Schneider, K.J., "Solidification and Disposal of High-Level Radioactive Wastes in the United States," *React. Technol.*, **13**, 4, 387 (Winter 1970–71).
15. "Environmental Quality — 1977," eighth annual report of the Council on Environmental Quality, U.S. Government Printing Office, Washington, D.C., December 1977.
16. "Decision-Makers Guide in Solid Waste Management," Environmental Protection Agency, Publication SW-500, U.S. Government Printing Office, Washington, D.C., 1976.
17. Brunner, D.R., Keller, D.J., "Sanitary Landfill Design and Operation," U.S. Environmental Protection Agency, Report SW-65ts, Washington, D.C., 1972.
18. "Sanitary Landfill," ASCE "Manuals of Engineering Practice," No. 39, Sanitary Engineering Div., Am. Soc. Civil Engineers, New York, N.Y., 1959.
19. "Sanitary Landfills," in "Municipal Refuse Disposal," Chap. 4, p. 91, Institute for Solid Wastes, American Public Works Association, Washington, D.C., 1970.
20. Bowerman, F.R., "A Summary of Current Sanitary Landfill Practices," testimony before Subcommittees on Business and Commerce and on Public Health, Education, and Welfare, and Safety, of the Committee on the District of Columbia, U.S. Senate, 90th Congress, 1st Session, U.S. Government Printing Office, Washington, D.C., 1967.
21. *Fed. Regist.*, **43**, 4942, Feb. 6, 1978.
22. Environmental Protection Agency, "Thermal Processing and Land Disposal of Solid Wastes — Guidelines," *Fed. Regist.*, **39**, 158, Part III, Aug. 14, 1974.

23. "Construction and Use Criteria for Sanitary Landfills," Final Report submitted to Office of Solid Waste Management, Environmental Protection Agency, Washington, D.C., 1971.

24. Lion, L., Hubbard, S.J., "Solid Waste Disposal in Abandoned Coal Strip Mines," U.S. Environmental Protection Agency, Washington, D.C., 1973.

25. "Use of Abandoned Strip Mines for Disposal of Solid Waste in Maryland," Final Report, Maryland Department of Health and Mental Hygiene, Bureau of Environmental Services, Division of Solid Wastes, Baltimore, Md., 1970.

26. "Solid Waste Management: Available Information Materials," U.S. Environmental Protection Agency, Report SW-58.27, Washington, D.C., December 1977.

27. Chian, E.S.K., DeWalle, F.B., "Sanitary Landfill Leachates and Their Treatment," *J. Environ.*, Eng. Div., American Society of Civil Engineers, **102** (EE2), 411 (1976).

28. Foree, E., Reid, V.M., "Anaerobic Biological Stabilization of Sanitary Landfill Leachate," Office of Research and Engineering Services, University of Kentucky, Lexington, Ky., 1973.

29. Cook, E.N., Foree, E.G., "Aerobic Biostabilization of Sanitary Landfill Leachate," *J. Water Pollut. Control Fed.*, **46**, 380 (1974).

30. Boyle, W.C., Ham, R.K., "Treatability of Leachate from Sanitary Landfills," *J. Water Pollut. Control Fed.*, **46**, 860 (1974).

31. Stone, R., "Disposal of Sewage Sludge into a Sanitary Landfill," Final Report to the U.S. Environmental Protection Agency, SW-71d, Washington, D.C., 1974.

32. Stone, R., Kahle, R.L., "Water and Sewage Sludge Adsorption by Solid Waste," *J. of Sanitary Engineering Div., Proceedings ASCE*, **98**, 731 (1972).

33. Conway, J., Pacey, J., "Second Interim Annual Report: Sonoma County Solid Waste Test Cells," Report to Office of Solid Waste Management, U.S. Environmental Protection Agency, Washington, D.C., May 30, 1973.

34. Merz, R.C., Stone, R., "Special Studies of a Sanitary Landfill," Final Summary Report and Third Progress Report to the Bureau of Solid Waste Management, U.S. Department of Health, Education, and Welfare, Washington, D.C., 1969.

35. Stone, R., "Aerobic Landfill Stabilization," Final Report to the Bureau of Solid Waste Management, U.S. Department of Health, Education, and Welfare, Washington, D.C., May 1969.

36. Pohland, F.G., "Sanitary Landfill Stabilization with Leachate Recycle and Residual Treatment," U.S. Environmental Protection Agency, EPA-600/2-75-043, October 1975, National Technical Information Service, PB 248 524, Springfield, Va.

37. "Baling Solid Waste to Conserve Sanitary Landfill Space: San Diego," Environmental Protection Agency, National Technical Information Service PB 214 960, Springfield, Va., 1972.

38. Reinhardt, J.J., Ham, R.K., "Milling of Solid Waste at Madison, Wisconsin," Final Report to Office of Solid Waste Management Programs, U.S. Environmental Protection Agency, Washington, D.C., March 1973.

39. Wold, K.W., Sosnovsky, C.H., "High Pressure Compaction and Baling of Solid Wastes," U.S. Environmental Protection Agency, Publication SW-32d, Washington, D.C., 1972.

40. "Evaluation of Solid Waste Baling and Balefills," draft report prepared by Ralph Stone and Co., Inc., for the U.S. Environmental Protection Agency under Contract No. 68-03-0332, December 1974.

41. Caffrey, R.P., Ham, R.K., "The Role of Evaporation in Determining Leachate Production from Milled Refuse Landfills," *Compost Sci.*, **15**, 2, 11 (1974).

42. "First Landfill Methane Plant Dedicated: Shredding Approach Also Gets Nod," *Solid Waste Systems*, July/August 1975, p. 24.

43. "Treatment and Utilization of Landfill Gas — Mountain View Project Feasibility Study," Environmental Protection Agency, Report SW-583, Cincinnati, Ohio, 1977.

44. "Fourth Report to Congress — Resource Recovery and Waste Reduction," U.S. Environmental Protection Agency, Report SW-600, U.S. Government Printing Office, Washington, D.C., August 1977.

45. "Incineration of Solid Wastes," Rubel, F.N., Noyes Data Corp., Park Ridge, N.J., 1974, 246 pp.

46. Eberhardt, H., Mayer, W., "Experience with Refuse Incinerators in Europe: Prevention of Air and Water Pollution, Operation of Refuse Incineration Plants Combined with Steam Boilers, Design and Planning," in "Proceedings of 1968 National Incinerator Conference," Am. Society Mechanical Engineers, New York, N.Y., May 1968, p. 73.

47. Sebastian, F.P., Arley, A.F., Garretson, B.B., "Modern Refuse Incineration in Dusseldorf — A Composite of the Best European Practices," Paper 68-PWR-3, Inst. of Electrical and Electronic Engineers Joint Power Generation Conference, San Francisco, Calif., September 1968.

48. "Solid Waste Management in Germany, Report of the Solid Waste Study Team Visit, June 25–July 8, 1967," U.S. Public Health Service Publication No. 1812, U.S. Government Printing Office, Washington, D.C., 1968.

49. Essenhigh, R.H., "Incineration — A Practical and Scientific Approach," *Environ. Sci. Technol.*, **2,** 524 (1968).

50. Kaiser, E.R., "Combustion and Heat Calculations for Incinerators," in "Proceedings, 1964 National Incinerator Conference," Am. Society Mechanical Engineers, New York, N.Y., May 1964, p. 81.

51. Essenhigh, R.H., Shieh, W., "Combustion of Computer Cards in a Continuous Test Incinerator: A Comparison of Theory and Experiment," Proceedings, 1972 National Incinerator Conference, Am. Society of Mechanical Engineers, New York, N.Y., June 1972, p. 120.

52. Sarofim, A.F., et al., "Design and Control of Incinerators," Vol. I, Environmental Protection Agency, EPA-670/2-73-089A, National Technical Information Service, PB 223 626, Springfield, Va., September 1973.

53. "Study of Reduction of Combustible Pollutant Emissions Using Overfire Air Mixing," A.D. Little, Inc., Environmental Protection Agency Research Contract No. 68-02-0204 (no date).

54. Walker, A.B., "Air Pollution Control Equipment for Incinerators," in "Incineration of Solid Wastes," Metropolitan Engineers Council on Air Resources, New York, N.Y., March 1967, p. 73.

55. Kaiser, E.R., Carotti, A.A., "Municipal Incineration of Refuse with 2 Per cent and 4 Per cent Additions of Four Plastics: Polyethylene, Polyurethane, Polystyrene, and Polyvinyl Chloride," Proceedings, 1972 National Incinerator Conference, American Society of Mechanical Engineers, New York, N.Y., June 1972, p. 230.

56. Vaughan, D.A., Anastas, M.Y., Krause, H.H., "An Analysis of the Current Impact of Plastic Refuse Disposal Upon the Environment," Environmental Protection Agency, EPA-670/2-74-083, National Technical Information Service, PB 238 654, Springfield, Va., December 1974.

57. Vaughan, D.A., et al., "Environmental Assessment of Future Disposal Methods for Plastics in Municipal Solid Waste," EPA-670/2-75-058, Environmental Protection Agency, Cincinnati, Ohio, June 1975.

58. White, H.J., "Role of Electrostatic Precipitators in Particulate Control: A Retrospective and Prospective View," *J. Air Pollut. Control Fed.*, **25,** 102 (1975).

59. Bergmann, L., "New Fabrics and Their Potential Application," *J. Air Pollut. Control Fed.*, **24,** 1187 (1974).

60. Astrom, L., et al., "A Comparative Study of European and North American Steam Producing Incinerators," Proceedings, 1974 National Incinerator Conference, American Society of Mechanical Engineers, New York, N.Y., May 1974, p. 255.

61. Golueke, C.G., "Composting: A Study of the Process and Its Principles," Rodale Press, Inc., Emmaus, Pa., 1972, 110 pp.

62. McGauhey, P.H., Golueke, C.G., "Reclamation of Municipal Refuse by Composting," Technical Bulletin No. 9, Sanitary Engineering Research Laboratory, University of California, Berkeley, Calif., June 1953.

63. Franz, M., "General Motors Enters Composting Field," *Compost Sci.*, **13** (3), 16 (May–June 1973).

64. Goldstein, J., "Manure Gets 'Rediscovered'," *Compost Sci.*, **15** (2), 24 (March–April 1974).

65. "Third Report to Congress — Resource Recovery and Waste Reduction," Publication SW-161, U.S. Environmental Protection Agency, Washington, D.C., 1975.

66. Bravard, J.C., Flora, H.B., II, Portal, C., "Energy Expenditures Associated with the Production and Recycle of Metals," Oak Ridge National Laboratory Report ORNL-NSF-EP-24, Oak Ridge, Tenn., 1972.

67. "Raw Materials: U.S. Grows More Vulnerable to Third World Cartels," *Science*, **183**, 185 (1974).

68. "Resource Recovery Briefs," (a periodical status report on resource recovery activities), National Center for Resource Recovery, Washington, D.C.

69. Chapman, R.A., "Distribution and Availability of Combustible Solid Waste in the United States," U.S. Environmental Protection Agency (unpublished data).

70. Hottel, H.D., Howard, J.B., "New Energy Technology — Some Facts and Assessments," MIT Press, Cambridge, Mass., 1971, 364 pp.

71. Franklin, W.E., et al., "Resource Recovery: Catalogue of Processes," Council on Environmental Quality, 151 pp., National Technical Information Service, PB 214 148, Springfield, Va., 1973.

72. "Resource Recovery from Municipal Solid Waste," National Center for Resource Recovery Inc., Lexington Books, D.C. Heath & Co., Lexington, Mass., 1974, p. 2.

73. "Problems and Opportunities in the Management of Combustible Solid Waste," International Research and Technology Corp. Contract No. 68-03-0060, Final Report to Solid and Hazardous Waste Research Laboratory, National Environmental Research Center, U.S. Environmental Protection Agency, Cincinnati, Ohio, 1972, 509 pp.

74. NCRR *Bulletin*, Vol. V, No. 2, National Center for Resource Recovery, Washington, D.C., 1975.

75. "Chicago to Build Energy Recovery Plant," *Waste Age*, **5** (2), 14 (March–April 1974).

76. Nydick, S.E., Hurley, J.R., Oberacker, D.A., "Study Program to Investigate Use of Solid Waste as a Supplementary Fuel in Industrial Boilers," U.S. Environmental Protection Agency report (unpublished), Contract 68-03-0355, May 15, 1975.

77. Oberacker, D.A., "Processed Municipal Refuse — A Fuel for Small Power Plant Boilers," U.S. Environmental Protection Agency, Municipal Environmental Research Laboratory, Cincinnati, Ohio, Newsletter on Environmental Research in Cincinnati, Nov. 15, 1976.

78. Moore, H.C., "Refuse Fired Steam Generator at Navy Base, Norfolk, Va.," in "Incineration of Solid Wastes," Metropolitan Engineers Council on Air Resources, New York, N.Y., March 1967.

79. Chapman, R.A., Wocasek, F.R., "CPU-400 Solid Waste-Fired Gas Turbine Development," Proceedings, 1974 National Incinerator Conference, Miami, Fla., May 12–15, 1974, American Society of Mechanical Engineers, New York, N.Y.

80. Sullivan, P.M., Makar, H.V., "Bureau of Mines Process for Recovering Resources from Raw Refuse," Proceedings, Fourth Mineral Waste Utilization Symposium, IIT Research Institute, Chicago, Ill., 1974.

81. "Recovering Resources from Solid Waste Using Wet-Processing," Summary Report SW-47D, U.S. Environmental Protection Agency, Washington, D.C., 1974.

82. Henn, J.J., Peters, F.A., "Cost Evaluation of a Metal and Mineral Recovery Process for Treating Municipal Incinerator Residues," Information Circular 8533, U.S. Bureau of Mines, Washington, D.C., 1972.

83. Albrecht, O.W., McDermott, R.G., "Economic and Technological Impediments to Recycling Obsolete Ferrous Solid Waste," U.S. Department of Commerce, National Technical Information Service, PB 223 034, Springfield, Va., October 1973.

84. "The Disposal of Major Appliances," Report to the U.S. Department of Commerce by the National Industrial Pollution Control Council, 1971, p. 32.

85. Chindgren, C.J., Dean, K.C., Sterner, J.W., "Construction and Testing of a Junk Auto Incinerator," Technical Progress Report No. 21, Bureau of Mines Solid Waste Program, U.S. Department of the Interior, Washington, D.C., February 1970.

86. Kaiser, E.R., Tolciso, J., "Smokeless Burning of Automobile Bodies," Technical Report 7642, College of Engineering, New York University, New York, N.Y., June 1961.

87. Dean, K.C., Chindgren, C.J., "Advances in Technology for Recycling Obsolete Cars," Proceedings, Third Mineral Waste Utilization Symposium, IIT Research Institute, Chicago, Ill., 1972.

88. Bilbrey, J.H., Jr., "Use of Cryogenics in Scrap Processing," Proceedings, Fourth Mineral Waste Utilization Symposium, ITT Research Institute, Chicago, Ill., 1974.

89. Carlson, O.N., Schmidt, F.A., "The Metallurgical Upgrading of Automotive Scrap Steel," Final Report, Metallurgy Division, Ames Laboratory, U.S. Atomic Energy Commission, Iowa State University, U.S. Environmental Protection Agency, Washington, D.C., June 1973.

90. Braton, N.R., "Cryogenic Recycling," Proceedings, Fourth Mineral Waste Utilization Symposium, IIT Research Institute, Chicago, Ill., 1974.

91. *Rubber World,* **176,** 4, 49 (1977).

92. "Waste Rubber," NCRR *Bulletin,* Vol. IV, No. 1, National Center for Resource Recovery, Washington, D.C., Winter 1974.

93. Beckman, J.A., Crane, G., Laman, J.R., "The Destructive Distillation of Used Tires in a Continuous Pilot Plant," *Rubber Age* (N.Y.), April 1973.

94. "Pulp, Paper, and Board," Vol. XXXIII, No. 4, Quarterly Industry Report, U.S. Department of Commerce, Washington, D.C., Winter 1977/78.

95. "First Report to Congress — Resource Recovery and Source Reduction," U.S. Environmental Protection Agency, Washington, D.C., February 1973.

96. Barbour, J.F., Groner, R.R., Freed, V.H., "The Chemical Conversion of Solid Wastes to Useful Products," U.S. Environmental Protection Agency, National Technical Information Service, PB 233 178, Springfield, Va., April 1974.

97. Appell, H.R., Wender, I., Miller, R.D., "Conversion of Urban Refuse to Oil," U.S. Bureau of Mines Technical Progress Report No. 25, Washington, D.C., May 1970.

98. Balandin, A.A., et al., "Hydrolytic Hydrogenation of Cellulose," *Chem. Abstr.,* **54,** 7140f, 1960.

99. Berl, E., "Cellulose als Grundstoffe der Steinkohle und Erdolbildung," *Papier-Fabrikant,* **31,** 141 (1933).

100. Appell, H.R., et al., "Conversion of Cellulosic Wastes to Oil," Report of Investigations 8013, U.S. Bureau of Mines, Washington, D.C., 1975.

101. "Enzymatic Conversion of Cellulosic Materials: Technology and Applications," Proceedings of a Symposium, National Research Council, U.S. Army Natick Research and Development Command, "Biotechnology and Bioengineering," Symposium No. 6, John Wiley & Sons (Interscience), New York, N.Y., 1976, 326 pp.

102. Callihan, C.D., Dunlap, C.E., "Construction of a Chemical — Microbial Pilot Plant for Production of Single Cell Protein from Cellulosic Wastes," U.S. Environmental Protection Agency, Report SW-24C, Washington, D.C., 1971.

103. "Some Trash Can Really Be Sweet!" *Environ. Sci. Technol.,* **9,** 1100 (1975).

104. Cummings, J.P., "Glass and Aluminum Recovery Subsystem — Franklin, Ohio," Proceedings, Fourth Mineral Waste Utilization Symposium, IIT Research Institute, Chicago, Ill., 1974.

105. Cutler, I., "Foam Glass Insulation from Waste Glass," EPA Research Grant R-00937-02, 1973–75, University of Utah, Salt Lake City, Utah.

106. Kupchik, G.J., Franz, G.J., "Solid Waste, Air Pollution, and Health," *J. Air Pollut. Control Assoc.*, **26**, 116 (1976).

Chapter 7

Pesticides in the Environment

7 / Pesticides in the Environment

INTRODUCTION

Great concern continues to be expressed over the effects of pesticides in the environment. The principal interest has centered on the widespread presence of residues of the chlorinated hydrocarbon insecticides, such as DDT, although fears have been raised also by incidents of pesticide mishandling. A notable example of the latter was the Kepone episode of 1974–75 which involved severe damage to occupationally exposed workers in Hopewell, Va., and extensive contamination of the surrounding area (1, 2). Such problems have produced increasingly rigid legal controls on the composition and use of the materials and on the amounts of their residues permitted in environmental samples. (They have also produced the cancellation of most major uses of certain chlorinated hydrocarbon insecticides, starting in 1972 with DDT and including aldrin, dieldrin, chlordane, and heptachlor.) However, pests still exist, and more satisfactory solutions are needed.

Paralleling and in part supporting the effort to limit the environmental presence of pesticides has been the development of more and more powerful analytical and toxicological tools. Recent years also have seen careful and continuing examination of pest control practices (3) and growing interest in a more balanced pest control arsenal, including cultural and biological as well as chemical weapons. Pesticide chemists have synthesized new control chemicals, such as insect sex attractants or pheromones and the growth-regulating juvenile hormone analogs. These and related developments may well mark the beginning of a new era in pesticide chemistry and in pest control generally (4).

Whatever changes occur in pest-control measures, chemicals will remain an important tool, and concern over their environmental impact remains justifiably high. If that concern is to continue to yield positive results, it must be accompanied by full realization of the nature, uses, and value of pesticides as well as of the dangers they present. Within the framework of current practice and knowledge, certain basic questions must be answered if the position of pesticides as environmental contaminants is to be seen clearly:

- What is known about pesticide residues in the environment?
- Are the existing data adequate to support valid conclusions on the extent of contamination by such residues?
- How can contamination of the environment by pesticide residues be reduced?
- What is the effect of pesticide residues on wildlife?
- What is the effect of pesticide residues on human health?
- What additional efforts should be undertaken to maximize the advantages of pesticide use as well as the safety of ourselves and our environment?

Background

Long ago, the Greeks and Romans used sulfur to control plant diseases and arsenic to control insects. The birth of chemistry as a science provoked a more systematic search for pest-control chemicals; the diverse products of the organic chemist became the center of attention and, eventually, the mainstay of the modern era of chemical pest control. That era, ushered in by World War II, began with the discovery and development of DDT, the organophosphorus insecticides, and the phenoxy herbicides. These were the forerunners of a veritable avalanche of synthetic organic pesticides of unprecedented versatility and effectiveness.

Pest control chemicals — the pesticides — often are called "economic poisons." The term includes all preparations intended for use as insecticides, rodenticides, nematicides, fungicides, herbicides, repellents, plant growth regulators, defoliants, desiccants, and for many other purposes. The utility of these materials is attested to by countless examples of abatement of disease and increased production of crops. Their value also is evident in the immense amounts of them that are used.

The production and use of synthetic organic pesticides has increased dramatically over the three decades since World War II. By 1960, the U.S. alone was producing more than 500 million pounds of pesticides annually; in the succeeding 15 years, annual production tripled, to about 1.6 billion pounds (Table 1). More than half of all U.S. farmers use chemicals to control insects, weeds, and plant diseases, and the acreage treated has increased steadily. In this country in 1969, some 90 million acres were treated with herbicides, more than 40 million acres with insecticides, and about 4 million acres with fungicides (5). In 1971, herbicides were applied to 95% of the rice acreage, 82% of the cotton, 79% of the corn, and 68% of the soybeans; insecticides were applied to 77% of the Irish potatoes, 61% of the cotton, and 28% of the corn (3). In 1969, about 1000 different chemical substances (active ingredients) and more than 30,000 different pesticide products were reported to be in regular use as "economic poisons" (6); in October 1977, the Environmental Protection Agency listed some 1850 substances and more than 40,000 pesticide products (7). Relatively few of these products are applied on the massive scale that we have come to associate with DDT and 2,4-D. The continued, *correct* use of chemical pest control is amply justified by the facts: two out of three persons on the earth go to bed hungry every night; 80 million new mouths to be fed appear each year; other millions suffer and die of diseases transmitted by insects and mollusks. In the struggle to ease such afflictions — to provide food and fiber and suppress disease — pesticides rank in importance with tractors and antibiotics as tools of agriculture and medicine. Still, the only valid reason for using *any* pesticide is its pronounced effect on some form of life, and that in itself should provide sufficient basis for curiosity, caution, and respect.

Toxicity

The world we live in is essentially a chemical one. Aspirin, salt, gasoline, clothing, and the many other substances that people use in their daily lives are made up of chemical elements or compounds. So, too, are the air we breathe, the water we drink, the food we eat. People tend to believe that only certain types of chemicals are inherently "poisons." But one of the very important principles of toxicology is the fact that *any chemical in sufficient amount will harm an organism*. Thus even vitamins, minerals, and amino acids — all essential nutrients — are dangerous when taken in too great a dose.

Dose-response relationships. An animal's response to a specified dose of a given chemical depends on its species and, to a lesser degree, on its individual peculiarities. The quantitative response of a species to different chemicals can be quite variable (Table 2), but the response to a single chemical is generally considered to increase in a predictable way as the exposure or "dose" increases.

The dose-response relationship for a species is determined by feeding varying doses of the chemical to groups of animals. Individual animals will respond differently to the same dose, but if death, for example, is the response of interest, the most statistically accurate estimate of the minimum lethal dose for the species is the amount that kills 50% of the test animals. That amount is called the median lethal dose (LD50). Usually, the LD50 is expressed as mg/kg, milligrams of toxicant administered per kilogram of body weight of the victim. LD50 values

Table 1

Production of synthetic organic pesticidal chemicals, United States

Chemical	1970 1000 pounds	1975[a] 1000 pounds
Fungicides:		
Copper naphthenate	1,730	1,078
Dithiocarbamic acid salts	39,381	40,666
Mercury fungicides	1,114	[b]
Pentachlorophenol (PCP)	47,170	39,447
All other	50,307	74,027
Total	**139,702**	**155,218**
Herbicides:		
2,4-D acid	43,576[c]	[b]
Maleic hydrazide	3,271	[b]
Methanearsonic acid salts	30,454	[b]
Phenylmercuric acetate (PMA)	457	147
Silvex	2,016	[b]
2,4,5-T acid, esters, and salts	12,335	[b]
All other	312,132	788,030
Total	**404,241**	**788,177**
Insecticides, fumigants, rodenticides:		
Aldrin-toxaphene group	88,641	[b]
Chloropicrin	[b]	5,698
DDT	59,316	[b]
Methoxychlor	[b]	5,504
Methyl bromide	21,047	36,048
Organophosphorus insecticides	132,496[c]	209,795[c]
Methyl parathion	41,353	53,668
Parathion	15,259	[b]
Other	75,884	156,127
Toxaphene	N.A.	59,336
All other	188,632	349,345
Total	**490,132**	**665,726**
Grand total	**1,034,075**	**1,609,121**

a. Preliminary.
b. Witheld to avoid disclosure; figure included in "All other" and "Total."
c. Represents duplication, but is included in totals.

Source: Fowler, D.L., Mahan, J.N., "The Pesticide Review — 1976," Agricultural Stabilization and Conservation Service, U.S. Department of Agriculture, Washington, D.C., July 1977.

as a rule are reproducible only within a fairly broad range (Table 2). The accuracy and precision of the data improve as the test organisms and experimental conditions — temperature, diet, etc. — are standardized more rigidly. The more rigid the experiment, however, the less likely is it to reflect environmental reality.

Although LD50 values provide a useful measure of toxicity, death is a crude, uninformative, and often inappropriate indicator of organic damage. For one thing, not all harmful effects are lethal. Continued exposure to certain coal-tar hydrocarbons, for example, eventually causes tumors in animals: the tumors may be harmful, even though the hydrocarbons themselves are "nontoxic" in the usual sense of poisoning. So-called "dioxin" (2,3,7,8-tetrachlorodibenzo-p-dioxin, or TCDD) causes serious skin eruptions in doses far below those that are lethal. In each of these instances, the direct dose-response relationship seems to hold, although physiological complexity may obscure the details. Because toxic effects other than death are not in general well understood, they must often be measured under quite unrealistic conditions and within wide statistical limits. Although such toxicologic studies are helpful in assessing the potential hazards of pesticides, they do not provide extremely precise information.

Certain chemical agents, as noted, are known to be carcinogenic or tumorigenic in that they cause tumors or other localized tissue abnormalities. Examples of such agents are the polynuclear hydrocarbons, which cause tumors in people exposed continuously to them in their work or in certain foods (such as smoked fish). Laboratory assays have implied that DDT is tumorigenic to mice, but more than 25 years of field experience with the insecticide has not revealed the effect in man or other animals under practical conditions. Other chemicals, pesticides among them, are mutagenic in laboratory tests in that they cause irreversible genetic damage; for example, such tests — necessarily artificial — show that caffeine causes chromosome breakage in both plant and animal cells, but centuries of experience with caffeine-containing coffee and tea have not yet

Table 2

Variability in toxicity measurements in rats

Pesticide	Average LD50[a]	Range of values
DDT	113mg/kg	87–147
Dieldrin	46	41–51
Toxaphene	90	67–122
Parathion	13	10–17
Dichlorvos (DDVP)	80	62–104
Malathion	1375	1206–1568
Carbaryl (Sevin)	850	733–986

Data show, for various pesticides, the variability in the dose that is lethal among individuals of the same species. Among the rats tested, for example, the lethal oral dose of DDT might vary between 87 and 147 milligrams per kilogram of body weight, and the LD50 or median lethal dose — the amount calculated to kill half the rats in the group — was 113 mg/kg.

a. Acute male rat oral, mg/kg.

Source: Gaines, T., *Toxicol. Appl. Pharmacol.* **2,** 88 (1960).

demonstrated caffeine-derived genetic abnormalities in humans. Still other compounds, including the drug Thalidomide and the plant alkaloid cyclopamine, are teratogenic in that they can cause congenital malformations in unborn animals. Present test methods for teratogenicity have proven to be difficult to use or are unreliable, and attempts to relate the results of such tests on pesticides to actual environmental problems have been unsuccessful. None of this is to suggest that any pesticide, as an environmental contaminant, will never exhibit toxic properties of these types. The purpose, rather, is to point out our present inability to predict such effects with extreme accuracy.

Species differ, moreover, in their responses to toxicants. A rodenticide is not necessarily very toxic to creatures other than the rat, a common test animal. The insecticide methoxychlor is only slightly toxic to mammals, but is highly toxic to insects and fish. Powerful herbicides are highly toxic to plants, but often are only slightly toxic to animals. This selective toxicity ot pesticides, which often is associated with an organism's ability to metabolize (decompose) a compound to more toxic or less toxic products (8), undoubtedly is their most significant characteristic.

Table 3

Some important pesticide classifications[a]

Name	Use	Classification
Aldrin	Insecticide	Chlorinated hydrocarbon
Atrazine	Herbicide	Triazine
BHC	Insecticide	Chlorinated hydrocarbon
Carbaryl	Insecticide	Carbamate
Chlordane	Insecticide	Chlorinated hydrocarbon
2,4-D	Herbicide	Phenoxy
DDT	Insecticide	Chlorinated hydrocarbon
Dibromochloropropane	Nematicide	Halogenated alkane
Dichlorvos	Insecticide	Organophosphate
Dicofol	Acaricide	Chlorinated hydrocarbon
Dieldrin	Insecticide	Chlorinated hydrocarbon
Dinoseb	Herbicide	Phenol
Endrin	Insecticide	Chlorinated hydrocarbon
Heptachlor	Insecticide	Chlorinated hydrocarbon
Malathion	Insecticide	Organophosphate
Methyl bromide	Nematicide (fumigant)	Halogenated alkane
Nabam	Fungicide	Dithiocarbamate
Paraquat	Herbicide	Pyridinium
Parathion	Insecticide	Organosphosphate
Pentachlorophenol	Herbicide, fungicide	Phenol
Phenylmercuric acetate	Herbicide, fungicide	Mercurial
2,4,5-T	Herbicide	Phenoxy
Toxaphene	Insecticide	Chlorinated hydrocarbon
Trifluralin	Herbicide	Dinitroaniline
Zineb	Fungicide	Dithiocarbamate

a. See Figure 1 for representative structures and Table 4 for representative physical properties.

Hazard

Some pesticides are readily detected in soil and water. This fact has aroused concern that the environment might act as a reservoir, subjecting various organisms in the food web to continuing exposure to persistent chemicals, and that continued accumulation of toxic substances in the environment could end in a biological catastrophe. The issue actually is not whether detectable amounts of toxic chemicals remain, but whether the residues are at concentrations and locations that will result in harm. Hazard — the probability of harm — is the key question.

"Harm" depends on the organism's exposure to, uptake, and retention of the chemical. A pesticide must enter the organism by absorption through an external or internal surface (skin, stomach, etc.) and remain there long enough to exert an effect despite metabolic detoxication and excretion. In other words, it must be available for toxic action. In laboratory tests, for example, the herbicide trifluralin, when dissolved in water, is highly toxic to fish; when tightly bound to mud, however, it is almost nontoxic (9), and its environmental hazard is low.

The apprehension over persistent pesticides has rested in part on the premise that organic pesticides can exist indefinitely in the environment, but experience has shown that this is a misconception. Wherever the persistence of pesticides has been measured, the concentration of the parent chemical has been found to diminish with time at a rate dependent on its physical and chemical characteristics. Wind and water disperse the chemicals physically. Within a finite time, usually measured in days or weeks, their concentrations are reduced by degradation — chemical, photochemical, or microbiological — to below the levels of recognizable biological significance (10). Still, there are exceptions. A degradation product (of DDT) such as DDE, which is slow to break down and becomes concentrated via fat in the food chain, can exert a strong effect on certain birds despite a low mammalian LD50. That is, the hazard from the formation of DDE is unexpectedly great.

PESTICIDE CHEMICALS

Classification

Pesticide chemicals are used to control a variety of pests, and it has long been common practice to classify the chemicals accordingly: insecticides act against insects, herbicides against weeds, fungicides against fungi, and so on (Table 3). Such classifications bear little relationship to the chemical types or structures of the compounds involved (Fig. 1). Pesticides represent almost all major classes of organic compounds. No mysterious property determines that a chemical will affect one type of organism but not another; effectiveness against a given class of pests often is only relative and can be accompanied by toxicity to other classes as well. Most insecticides, for example, are distinctly toxic to mammals, although in widely varying degree.

Specific chemicals *can* be categorized by broad structural types, an accepted scheme that has evolved over the years. Typical categories include the chlorinated (or halogenated) hydrocarbons, the organophosphorus compounds, the triazines, and the carbamates (Table 3). The physical and biological properties of the chemicals in a given category may differ markedly; the carbamates, for example, include herbicides, insecticides, and fungicides. These groupings serve nonetheless to characterize chemicals roughly by the parts of their molecu-

FIGURE 1

Chemical structures of representative pesticides

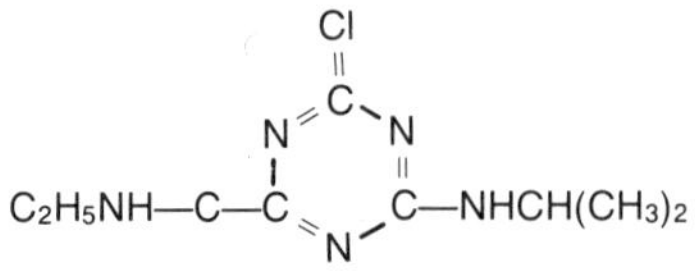

Atrazine

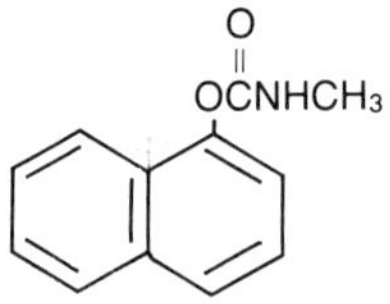

Carbaryl

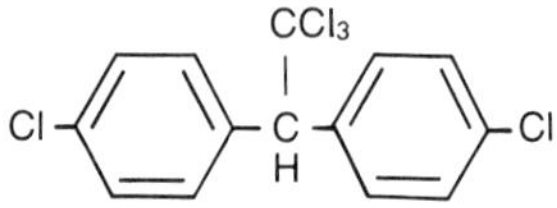

DDT

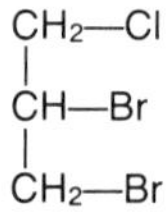

Dibromochloropropane

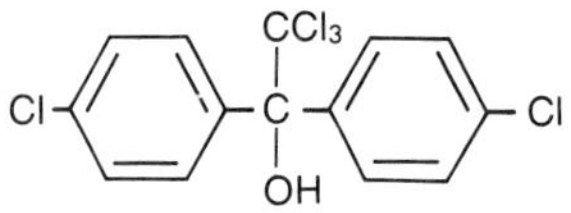

Dicofol

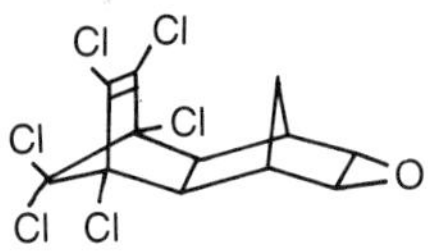

Dieldrin

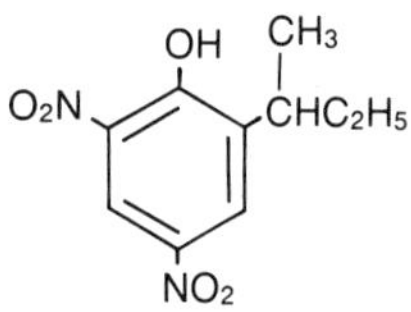

Dinoseb

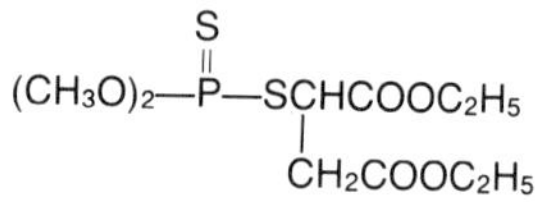

Malathion

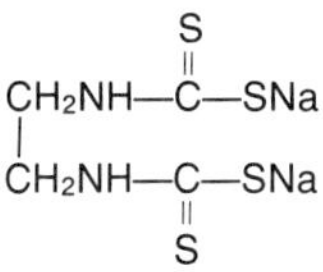

Nabam

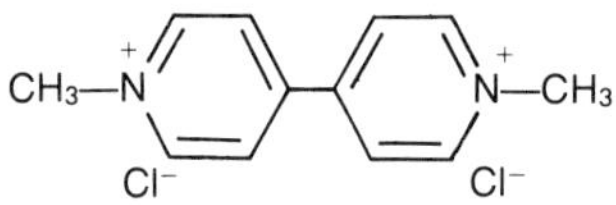

Paraquat

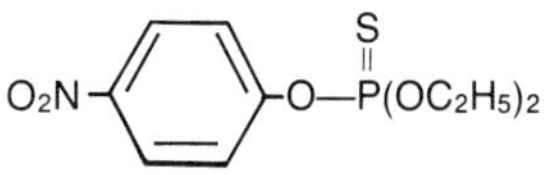

Parathion

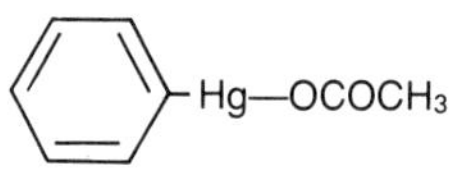

Phenylmercuric acetate

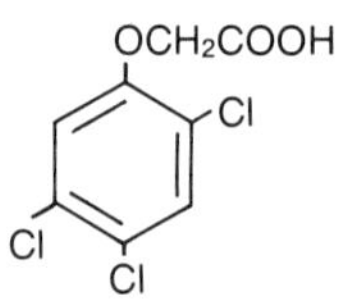

2,4,5-T

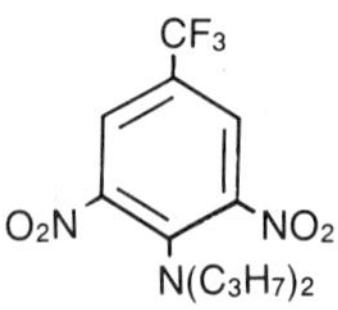

Trifluralin

Note: See also Tables 3 and 4.

lar strucures that determine their biological activity and environmental transformations. Detailed descriptions of the major pesticides have been compiled (11).

Reactivity and persistence

Pesticides, like any other chemicals, undergo many different reactions. The reactivity of a given compound depends on its specific chemical composition and on the environmental conditions. The more reactive the pesticide, as a rule, the less persistent it is; some are so reactive that they exist only briefly in the environment. Nonpersistent pesticides generally are considered environmentally desirable and persistent materials undesirable; it should be recognized, however, that pesticide breakdown products sometimes may be more toxic than the original compound. A case in point is the carbamate insecticide carbaryl, whose principal decomposition product, 1-naphthol, is more toxic to clams than is the parent compound while being less toxic to insects (12). Nonpersistent pesticides, such as the organophosphorus insecticide disulfoton, also may be metabolized to products more toxic than the parent compound (13). Fortunately, however, most reactions of pesticides lead to progressively less toxic products, and many such reactions offer practical means of intentionally destroying pesticides.

Many factors influence the rate at which a pesticide breaks down. Among them are temperature, moisture, biotic activity, the intensity of light, and the nature of the reaction medium. Therefore, the environmental persistence of a given compound — often expressed as "half-life," the time required for half of the chemical to dissipate — is not an absolute value but may vary widely with the prevailing climatic and biotic conditions (10). DDT, for example, may dissipate in a matter of weeks where temperature and moisture favor rapid volatilization and chemical breakdown; in less favorable climates it may persist for years. Other physical sources of "nonpersistence," such as volatilization, soil absorption, leaching, and surface runoff into water, may only shift the environmental burden from one place to another. The toxicity, persistence, and environmental fate of pesticides, then, are determined ultimately by their susceptibility to several key types of chemical reaction, including oxidation, reduction, and hydrolysis, whether carried out by living creatures or by the nonliving environment (10).

Physical properties and environmental behavior

The chemical diversity of pesticides is paralleled by their widely varying physical properties (Table 4). The compounds range from gases and volatile, low-boiling liquids to solids with high melting points. Some pesticides, such as the herbicide paraquat, are very soluble in water; others, such as DDT, are almost insoluble. A compound's physical properties strongly influence its behavior in the environment; they determine its mobility in soil, the amount of it that may vaporize into the air, and its distribution between water and fat. Physical properties combine with chemical reactivity to determine the chemical's persistence in the environment (14). Thus a relatively nonreactive compound that also is volatile and fat-soluble is likely to persist in the environment for substantial periods and to have the opportunity to become widely distributed. A notable example is DDE, a breakdown product of DDT.

Formulations and mixtures

Pesticides normally are applied in dilute form as formulations. Common formulations are dusts, wettable powders, emulsifiable concentrates, and aqueous concentrates. Formulations are used partly to modify the physical and chemical properties of pesticides and partly to increase pesticidal effectiveness. Many pesticides can be effective at rates of a few ounces of active ingredient per acre; proper formulation eases the difficulty of applying such a small amount of active chemical evenly over large areas. Appropriate choice of formulating agents, moreover, makes a pesticide safer for applicators and nontarget organisms, increases its effectiveness, helps to control environmental dispersion, and reduces cost.

Formulated pesticide concentrates often are mixed into relatively large proportions of water and applied as a spray by hand-operated equipment, powered ground-rigs, or aircraft. Considerable interest has developed in ultralow volume (ULV) application, in which a highly concentrated formulation or even the undiluted pesticide is sprayed. ULV application is a means of reducing the amount of pesticide needed to control a given pest. Pesticides that are extremely volatile, such as the fumigant Telone, or that are unstable to sunlight, such as the herbicide trifluralin, are incorporated directly into the soil. Others that are applied directly to soil are the systemics, which move through the plant after entering through the roots, and preemergence herbicides, which affect plants before they emerge from the ground.

DETECTION AND MEASUREMENT

Pesticide analysis

The ability to detect and measure pesticides as they exist in crops, soils, natural waters, air, and aquatic and terrestrial animals is essential to understanding the impact of these biologically active chemicals on the environment. That ability also is needed in devising ways to apply pesticides so as to minimize contamination of nontarget elements of the environment, both animate and inanimate. The capacity to demonstrate that some critical concentration of a pesticide must be reached before it is detrimental to one or another of those elements can do much to avoid or at least to calm the acrimonious, often meaningless debates that have erupted over the cause and remedy of many pesticide-caused environmental problems.

Appropriate use of analytical chemistry can and has resolved such debates. In one case (15), analysis demonstrated that the organophosphorus insecticide Azodrin (monocrotophos) was responsible for the death of numerous pheasants in cotton fields and permitted corrective action to be taken promptly. Another instance was the great 1963 fish-kill in the lower Mississippi River basin. Early reports implicated normal agricultural use of the insecticide endrin as the principal source of the pollution; later, analytical monitoring showed that the problem indeed had been endrin, but that it had entered the water accidentally as an industrial waste and not as a result of normal use (16). It is clear from such examples that the effective use of pesticides and the control of pesticide contamination of our food supply and environment depend heavily on the ability to detect, identify, and measure chemicals, often in only trace amounts. [Also, chemical analysis is important for quality control in the manufacture of concentrated (technical) pesticides and their formulations.]

Detection

The pesticide analyst often faces the challenge of detecting, identifying, and measuring any of the hundreds of registered pesticide chemicals and their metabolites or other degradation products, either alone or in mixtures with other substances. He must deal with these compounds, moreover, at extremely low concentrations and in the presence of massive proportions of the various chemical components of the environment. Only since about 1955 have the minute concentrations involved become generally accessible to chemical analysis. The relevant units of measure* have become realistic so recently that in 1978 much of the terminology is not yet common language.

In residue work today, the analyst must devise a procedure that will measure the pesticide, its significant degradation products, or both in the sample of interest in very low concentrations: 0.1 to 0.01 part per million (ppm) or even lower.** The procedure must be selective; that is, it must respond primarily to the compounds being sought. In addition, the procedure usually incorporates isolation or "cleanup" methods that effectively eliminate all or most of the natural bulk of the sample that might interfere with the detection system without

*Micrograms ($1 \mu g = 10^{-6} g$); nanograms ($1\ ng = 10^{-9} g$); picograms ($1\ pg = 10^{-12} g$).

**One ppm is equivalent to 1 mg of chemical in 1 kg of a substrate or 1 μg/g; one part per billion (ppb) is equivalent to 1 μg/kg. Because of the widespread common use of the terms ppm and ppb in pesticide chemistry, they will generally be used in preference or in addition to their metric equivalents in this chapter.

Table 4

Physical properties of representative pesticides[a]

Name	Melting point, °C	Boiling point, °C	Vapor pressure torr (°C)	Solubility in water, mg/l
Atrazine	173–175		3×10^{-7} (20°)	33 (20°)
Carbaryl	142		$>5 \times 10^{-3}$ (26°)	40 (30°)
DDT	108.5		1.0×10^{-7} (20°)	0.001 (25°)
Dibromochloropropane		196	0.8 (21°)	1000
Dicofol	79	decomp		
Dieldrin	175		7.8×10^{-7} (25°)	0.5 (35°)
Dinoseb	42			50
Malathion		156–157 (0.7 torr)	4×10^{-5} (30°)	145
Nabam	decomp		negligible	3×10^{5}
Paraquat	>300 decomp		negligible	very high
Parathion	6	157–162 (0.6 torr)	3.8×10^{-5} (20°)	24
Phenylmercuric acetate	149–153	sublimes	9×10^{-6} (35°)	438
2,4,5-T	158		$<6.46 \times 10^{-6}$ (25°)	278
Trifluralin	49	96–97 (0.18 torr)	2×10^{-4} (29.5°)	24

a. See also Table 3 and Figure 1.

significant loss of the compound of interest. And a practical residue analysis method requires maximum response from the pesticide or metabolite and minimum response from whatever interfering constituents remain in the sample.

During the early years of the synthetic organic pesticides, analytical methods for residues relied on color-forming procedures which at the time were considered sensitive and reliable in the microgram range. These procedures usually were selective for a part of the molecule (a functional group) considered unique to the pesticide to be measured. However, the burgeoning arsenal of new pesticides and the growing awareness of environmental contamination in the parts-per-billion range called for analytical techniques that were much more sensitive and selective and that could handle more than one type of residue. The development of the several forms of chromatography—paper, thin-layer, liquid, and gas — provided this capability (17); gas chromatography with electron-capture, flame-photometric, alkali-flame, or other sensitive and selective detectors has achieved particularly widespread use and importance (18).

Even with these advantages, inaccurate results still arise from impurities in the original pesticide, breakdown products, and nonpesticide contaminants in the environment. For example, the phthalate plasticizers used commonly in plastics produce responses on the gas chromatograph that are almost identical to those of DDT (19); it is essential, therefore, that the presence of the pesticide be confirmed by independent tests.

Confirmation

Considering the complexities, it should not be surprising, perhaps, that misinformation about pesticide residues enters the news media and the scientific literature. This is regrettable, because it is axiomatic that the results of a pesticide residue analysis that has scientific or regulatory significance should never be reported until the identity of the responding substance has been confirmed. Gas chromatography is one of the more frequently abused techniques in this regard. Despite the high selectivity of the detection system for a particular pesticide whose presence is suspected, the method can — and often does — respond quite similarly to other compounds: natural components of the sample, industrial pollutants, or other pesticides exhibiting similar chemical and physical properties.

In addition to the phthalate esters, the ubiquitous industrial compounds called polychlorinated biphenyls (PCB's) provide an example (20). For many years, these compounds (which are not pesticides at all) were interpreted as DDT and its metabolites in numerous environmental samples. The PCB's, as it happens, also are highly persistent organochlorine compounds; their confusion with DDT delayed overlong the realization of the need for the controls that since have been placed on PCB distribution.

Analyses of biological samples predating the use of chlorinated hydrocarbon insecticides and maintained in sealed containers in museums have been reported to contain these pesticides, although contamination of the samples with pesticides from any source before or during analysis certainly was excluded (21). Confirmation of identity might have corrected these and the many other misinterpreted analyses that have entered the scientific literature and influenced public decisions. In the period when analysts relied on colorimetric methods, means of confirming the identity of a compound were not readily available.

Now, however, reliable and convenient confirmation techniques are available to all pesticide residue analysts.

Residue analysts face many challenges in addition to confirmation of identity. They include securing a representative sample for analysis, completely extracting the pesticide from the sample, accurately measuring microliter portions of extract, preventing breakdown of the pesticide during analysis, and achieving consistent response by the analytical detector. Despite such difficulties, analytical chemists today can estimate reliably the concentrations of many pesticides in environmental samples at the part-per-billion level. Still, the utility of even the most accurate analysis is limited by the quality of the original sample: a pesticide detected in a sample of water may have been adsorbed on suspended sediment; measurement of DDT in a dead bird may be meaningless in diagnosis if the bird died of disease; a sample of muscle from the carcass may yield an unrealistically low value for DDT, while a sample of fat may yield an unrealistically high value. For these and similar reasons, analytical data commonly are endowed with significance that they do not possess. Residue analysis, in short, is a vital tool, but a tool to be used with discretion and thorough understanding.

PESTICIDE RESIDUES: OCCURRENCE AND SIGNIFICANCE

Pesticides are unique among environmental pollutants in that their function requires that they be released intentionally into the environment. Of the millions of pounds of organic pesticides used annually in this country, about one third is applied by aircraft and the remainder by ground equipment. (Aircraft apply more than 60% of the pesticides used in agriculture, but these amount to only about 60% of the total volume of pesticides used.) The impact of a pesticide on the environment depends on factors that include in particular the compound's toxicity, persistence, and mobility and its tendency to be biologically concentrated by food chains. The most troublesome environmental problems with pesticides have involved persistent, mobile compounds that are lipophilic (have an affinity for fatty materials). Such compounds can move readily from the point of application and eventually may bioconcentrate in the fatty tissue of organisms in various aquatic and terrestrial food chains.

Research and monitoring efforts on the environmental aspects of pesticides themselves have been paralleled by increasing attention to pesticide breakdown products (metabolites). A most important feature in such investigations overall is the sheer number of compounds involved. The hundreds of registered herbicides, insecticides, and fungicides include a number that may be used in more than one form, such as various salts or esters of the parent compound. Many pesticides, moreover, are registered in more than one formulation, and altogether, as noted, more than 40,000 different products were registered for use in 1977. Each pesticide can yield one or more primary metabolites; many of these, in turn, may conjugate with natural products such as amino acids, sugars, or other materials in plants and animals to yield still other chemicals. Most pesticides also break down nonbiologically. Thus the total number of compounds that can enter the environment as a result of the use of pesticides is potentially tremendous.

A great deal of data on pesticides in the environment has been gathered over the years. This has been true especially since the advent of highly selective and sensitive analytical techniques such as gas chromatography. For example, various agencies of the Federal Government have conducted nationwide monitoring programs for pesticides, coordinated by the Federal Working Group on Pest Management, which answers directly to the President's Council on Environmental Quality. At present, the Environmental Protection Agency, the Food and Drug Administration, and the Department of Agriculture are responsible for much of the pesticide-monitoring activity, although chemical manufacturers, food processors, and academic institutions contribute significantly. The results of this extensive effort are published periodically in the *Pesticide Monitoring Journal* (22) and other technical journals as well as in reference books (23, 24). The following sections will provide only a few illustrative examples of the occurrence of pesticide residues

Residues in soil

Soils are the ultimate repository for many pesticides. The processes that act on pesticides in and on soils include erosion, photodecomposition, and volatilization at the surface of the soil as well as microbial metabolism, chemical reaction, leaching, plant uptake, and adsorption below the surface. Erosion, volatilization, and leaching are transport mechanisms that eventually move the pesticide into other phases of the environment; photodecomposition, microbial metabolism, and purely chemical reactions are degradative processes that may or may not convert the pesticide to less toxic substances. The degradation products of pesticides in soils tend as a rule to be more soluble (more polar) than the parent compounds.

The persistence of a pesticide in soil depends on a number of chemical, edaphic (soil-related), and climatic factors (25). The chemical factors depend largely on the chemistry of the pesticide molecule itself. Highly chlorinated compounds of low solubility in water, such as the chlorinated hydrocarbon (organochlorine) insecticides, are among the most persistent pesticides. Water-soluble pesticides that contain easily biodegradable or hydrolyzable chemical linkages generally do not persist as long in soils. Such compounds include the methylcarbamate and organophosphorus insecticides and most organic herbicides and fungicides. The other factors that influence pesticide persistence in soils include cultivation, soil type, wind velocity, rainfall, soil moisture and temperature, and light intensity. The persistence of the major types of insecticides and herbicides in soil is shown in Figure 2 (26).

In the numerous pesticide monitoring programs conducted over the past decade, the residues encountered most commonly have been those of the chlorinated hydrocarbon insecticides — the DDT group (including DDE and DDD), dieldrin, endrin, toxaphene, and chlordane. A typical survey (27) determined soil residues at 51 locations, nationwide, in the three years 1965–67. Samples were collected from 17 areas where pesticides were used regularly, from 16 areas where they had been used infrequently, and from 18 areas with no history of use. In regular-use areas, DDT and dieldrin were the most consistently detected pesticides in soils from vegetable- and cotton-growing sites. DDT was found in amounts ranging from 0.29 ppm to 15.63 ppm (0.29 to 15.63 mg/kg); dieldrin was found in 80% of the samples, in amounts ranging from 0.02 ppm to 3.08 ppm. Among regular-use areas, tree-fruit sites had the highest residues; DDT in orchard soils ranged from 0.07 ppm to 245.4 ppm and was highest in apple orchards in Washington, Michigan, and Pennsylvania where it had been applied heavily for some years. Dieldrin was found in 70% of the orchards sampled and ranged from 0.02 ppm to 2.84 ppm. DDT residues were detected in soils from all of the limited-use locations, but the levels averaged only 0.001 ppm to 0.99 ppm. DDT was found, at a level of 0.001 ppm, in soil from only one nonuse area, and no other pesticide residues were detected in the nonuse areas. Another national monitoring program (28) involved locales in six states: Georgia, Idaho, Maine, Nebraska, Virginia, and Washington. Again, the most commonly encountered pesticides were members of the DDT group, followed by dieldrin and chlordane. The mean pesticide residues from cropland areas in the six states were: chlordane, 0.01 ppm; dieldrin, 0.02 ppm; DDT group, 0.38 ppm; and toxaphene, 0.13 ppm. In a third extensive survey (29), scientists looked for 35 pesticides and their relatives in 43 states; mean residues generally were at or below 0.01 ppm in both cropland and noncrop areas except for arsenic, which existed at more than 5 ppm in both.

Residues in air

Air is a global transport medium for pesticides. The results of worldwide atmospheric monitoring provide good evidence that residues of certain pesticides are ubiquitous and travel great distances in the atmosphere. The amounts detected, however, are minute and usually represent only nanograms of pesticide per cubic meter (ng/m^3) of air (30). Innovations in instrumentation have made it possible to detect pesticides in air at levels below 0.1 ng/m^3 (31).

Pesticides occur in air as vapors, aerosols, and adsorbed on dust particles. A panel of the National Academy of Sciences concluded that as much as 25% of the DDT produced in the world may have moved through the atmosphere to the oceans as a result of aerial drift during application and vaporization from the

FIGURE 2

Persistence of pesticides in soils

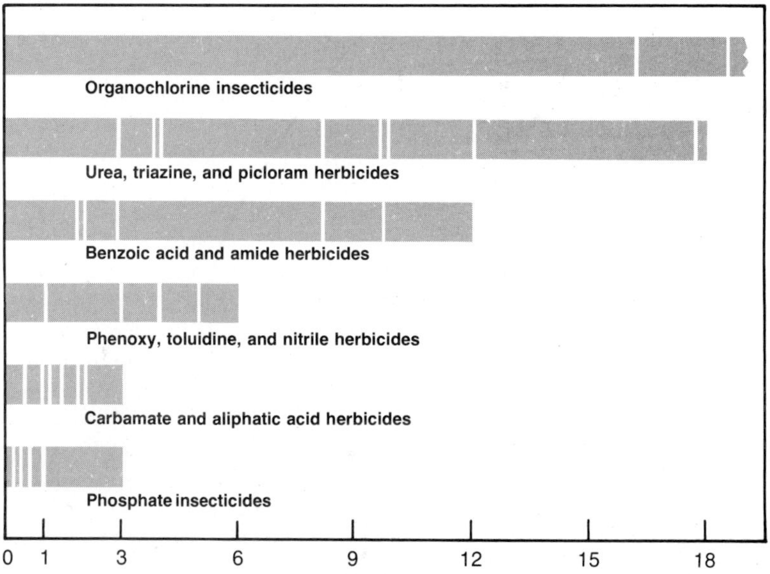

Source: Kearney, P.C., et al, *Residue Rev.*, **29**, 137 (1969).

surfaces of plants and soils (32). The panel worked with very limited data on the concentrations of pesticides in rainwater and dust over oceans, and its conclusion was controversial, but there is no doubt that particulate "drift" during application is an important source of atmospheric pesticides (33). Another investigation incriminated dust from the southern high plains area of Texas as a major airborne carrier of chlordane, DDT, DDE, BHC, heptachlor epoxide, aldrin, and 2,4,5-T isooctyl ester; the contaminated dust was washed from the air by rainfall over Cincinnati and backtracked to its origin in Texas (34). More recent work (35) shows that pesticides in air are largely associated with particles.

A major fraction of many pesticides on environmental surfaces enters the air through volatilization. A pesticide may evaporate from soil even after it is applied and worked in, or fine spray particles may vaporize during application and never reach the soil. Field observations show that vapor losses of dieldrin range from 30% to more than 60% when the compound is sprayed on soil as an aqueous emulsion (36). While atmospheric residues remain small — typically in the ng/m^3 range — even in agricultural communities (35), residues close to the point of application or in enclosed spaces can become significant for short periods: parathion in a sprayed orchard averaged 130 $\mu g/m^3$ (37), methoxychlor in a barn averaged 7.68 mg/m^3, and DDT applied by air to a forest produced a concentration of 0.17 g/m^3 (38).

Residues in water

Pesticides enter surface waters and groundwater as a result of runoff from treated soils, leaching, aerial transport, misapplication, accidental spillage, and faulty disposal of wastes and containers. Few data are available, however, on the relative contributions of each process to water pollution. The results of many nationwide surveys indicate clearly that very low but detectable amounts of the chlorinated hydrocarbon insecticides are present in most of the surface waters of the United States. (For most chlorinated hydrocarbon insecticides in water, electron-capture gas chromatography can achieve a sensitivity of less than 0.01 ng).

Incidents of water pollution by pesticides have been summarized (25); government agencies have conducted extensive monitoring surveys for more than a decade. The amounts of pesticide residues detected in water usually are small, in the range of parts per billion or parts per trillion (micrograms or nanograms per liter).

An extensive 1967 survey of selected streams in the western U.S., for example, detected maximum concentrations of 0.12 ppb or $\mu g/l$ of DDT (average 0.01) and 0.35 ppb of 2,4-D (39). The vast majority of the samples, however, contained no measurable residues at a limit of detection of 0.005 ppb, and the concentrations of the residues that were detected ranged between 0.01 and 0.10 ppb. Most data also showed a direct correlation between the amount of suspended sediment in the water and the level of pesticide residues.

There are indications that the amounts of pesticide residues in water are declining. A 1970 report (40) on a five-year synoptic survey of pesticides in the nation's surface waters made the following main points:

- Chlorinated hydrocarbon insecticides were detected widely in surface waters.
- Dieldrin and the DDT group (including DDE and DDD) were the compounds detected most frequently throughout the period of the survey.

- The number of occurrences of residues reached a peak in 1966 and declined sharply in 1967 and 1968.

The decline in 1967–68 was consistent with reports by the U.S. Departments of Agriculture and Interior which showed declining use of the persistent chlorinated hydrocarbon insecticides and rising use of the less-persistent organophosphorus and carbamate compounds. Studies of estuarine mollusks in the period 1965–72 also showed a clear-cut downward trend in organochlorine residues (41).

Detection of pesticide residues in the sea has been restricted almost entirely to minute levels of chlorinated hydrocarbon insecticides or their derivatives (32). For example, DDE has been observed in Pacific coastal waters at rather uniform levels of 0.01 ppb (42), and DDE and dieldrin have been detected at mean concentrations of 3.8 ppt and 5.8 ppt, respectively, in the western North Atlantic (43). Except where heavy pesticide runoff occurs from urban and industrial activities, as at Los Angeles, Calif., the principal sources of oceanic pesticides must be windborne dust (44) and rain (34).

Recommendation P1: *Pesticide monitoring in all phases of the environment should be continued for specific substances recognized to be harmful, including the parent pesticides, their environmental transformation products, and their manufacturing by-products, and the identity of each substance should be confirmed by appropriate analytical chemical methods.*

Residues and Living Receptors

While pesticides most often are degraded to innocuous form in soil, air, and water, some may also move through these phases of the environment to living receptors or sequences of receptors, including plant and animal life, foods and feeds, and humans (Table 5). Pesticides or their breakdown products in soils may affect crops and soil organisms and their predators in unintended ways. Much remains to be learned of the behavior of pesticide residues in air, particularly their interactions with other atmospheric contaminants. Airborne residues are known, however, to reach receptors, either directly or after depositing in soils or water. Receptors on occasion may be exposed to residues originating in all three phases of the environment. One example is the aquatic organisms in estuaries, where residues can accumulate as a result of runoff from soil, aerial drift, and deposition of river silt carrying adsorbed pesticides. Exposure of animals, where it occurs, appears to be primarily through food; exposure through air and water appears to be minor, although aquatic animals can be exposed through water.

Residues in plants and foods

Plants can become contaminated with pesticide residues by direct application of the chemical and by at least four indirect methods: aerial drift, soil dust or splashing, volatilization from surfaces, and root uptake. The timing and rates of application generally are designed to protect the plant during crucial growing periods and to avoid excessive residues on the commodity ultimately produced from the crop. Contamination by aerial drift can be reduced by careful, rigorously controlled application procedures (33); that from other indirect sources results, as a rule, from contaminated soil and can be controlled by adequate crop management methods. Normally, the indirect residues are small in comparison with those intended for the plant surface.

Plant root uptake can be a source of low level contamination of plants by chlorinated hydrocarbon insecticides, and crops have been damaged by root uptake of persistent herbicides. One study (45) has shown that dieldrin, heptachlor, endrin, and DDT can contaminate plants upon volatilizing from soil (Table 6). For DDT, volatilization was a more important contamination pathway than root uptake; for dieldrin, the two pathways were about equivalent; for endrin and heptachlor, root uptake was the more important pathway. Volatilization, as expected, was greater when the soil was moist than when it was dry.

Some pesticides are "systemic"; that is, they are readily absorbed and translocated throughout the plant. Use of systemic materials such as disulfoton can cause concern; although the parent compound dissipates quickly enough from treated plants (13), its toxic metabolites build up and can remain intact for long periods. Also, a few pesticides, most notably the chlorinated hydrocarbon insecticides, have an affinity for the surface of root crops such as carrots and tend to concentrate there (46).

If pesticides are to be used safely in agriculture, the foods, fiber, and feeds produced must be monitored systematically for residues. This is done by the food industry, the states, and the federal agencies that implement the national monitoring program coordinated by the Environmental Protection Agency (47).

Assessing the significance of pesticide residues in foods is difficult. The many commodities that enter the nation's food supply are produced in numerous geographical areas; because of the rapid and flexible distribution system that moves these commodities to market, their geographical origins shift constantly. Uniformity cannot be expected, even in a single food item. because of extreme variations in local growing, harvesting, and processing methods as well as geographical and environmental variables. Nonetheless, sampling procedures

Table 5

DDT and related compounds in the environment, 1960–66

Substrate	DDT (ppm; mg/kg)	Remarks
Apple	0.20	Typical, ready-to-eat
Beef steak	0.10	Typical, ready-to-eat
Milk (cow)	0.02	Typical, ready-to-eat
Milk (human)	0.37	San Francisco
Alfalfa hay	0.10	
Wheat-grass	211.	Salmon Nat. For., Idaho
Pheasant (fat)	8.	Richvale, Calif.
San Joachin River	0.000066	Vernalis, Calif.
Rainwater	0.00015	Ripley, Ohio
Ice	0.0003	Mt. Olympia, Wash.
Abalone	0.0087	Ensenada, Mexico
English sole	0.57	San Francisco Bay
Adelie penguin (fat)	0.024	Antarctica
Pear-orchard soil	38.	Medford, Ore., top 6 in.
Bald eagle (breast muscle)	5.	Canada (average)
Human (fat)	6.7	U.S. (average, 1963)
Human (fat)	27.2	India (average, 1964)
Human (fat)	1131.	U.S. (occupational)
Human (fat)	3.0	Eskimo

Source: Various.

and highly sensitive analytical chemical techniques have been established for the Total Diet Program or Market Basket study conducted annually by the Food and Drug Administration since 1964.

Residues are detected regularly in the Market Basket surveys (Table 7). Only rarely, however, do they exceed the legal tolerances — which incorporate large safety margins — established by the U.S. and international authorities. Food shipments or sources found by regulatory monitoring to contain excessive residues are removed from market or production as soon as possible; follow-up procedures prevent further consignments of similarly contaminated food from reaching the consumer, and import or export of food containing excessive residues is not permitted unless it complies with the laws of the country involved. [It should be recognized that some findings of the Market Basket surveys may be questionable because the identity of the compounds detected has not been confirmed (see earlier section on Confirmation).]

The raw foods sampled in these programs generally undergo further processing or preparation before they are consumed. This usually reduces residue levels even further, and so the intake of pesticide residues by the U.S. public is at or below the acceptable daily levels established by the World Health Organization–Food and Agriculture Organization (Table 8). Meat, fish, poultry, and dairy products combined accounted for almost half of the intake of chlorinated organic pesticides by humans in this country in 1968–70 (22), although few registered uses of pesticide chemicals are known to result in significant residues. No registrations are allowed that will result directly in residues in milk, so that those detected must be contributed by indirect means. Steps taken to minimize these indirect sources have included the severe restrictions on the use of chlorinated organic pesticides, the type found most often in dairy products.

The prevalence of organochlorine residues found in the Market Basket surveys has been declining, while that of organophosphorus residues has been rising (48). The fraction of composite samples containing organochlorines declined from 74.2% in 1969–70 to 48% in 1973–74; the fraction of organophosphorus residues rose from 20.6% in 1969–70 to 28% in 1973–74 (it reached 31% in 1972–73). These trends parallel the trends in overall use of the two classes of pesticides.

Table 6

Soybean contamination from surface or subsurface application of insecticides[a]

	Foliage residue, ppm (mg/kg)			
	DDT	Dieldrin	Endrin	Heptachlor
Vapor (V)	5.4	8.5	6.7	5.5
Root uptake (R)	0.8	10.8	38.1	20.6
V/R	6.8	0.8	0.2	0.3

a. Soil treated at 20 ppm (a relatively heavy dosage).

Source: Nash, R.G., Beall, M.L., "Chlorinated Hydrocarbon Insecticides: Root Uptake versus Vapor Contamination of Soybean Foliage," *Science*, **168**, 1109 (1970). Copyright 1970 by the American Association for the Advancement of Science.

Table 7

Residues found in 360 total-diet food composites,[a] United States, August 1973– July 1974

	Composites with residues	Composites with residues reported as trace[b]	Range[c] ppm (mg/kg)
Metals, PCB			
Zinc	360	0	0.1–35.5
Cadmium	211	0	0.01–0.31
Lead	180	0	0.02–1.30
Selenium	97	34	0.05–0.40
Mercury	34	17	0.01–0.04
PCB	14	13	0.050
Pesticides			
Dieldrin	93	17	0.0006–0.0330
DDE	81	20	0.0006–0.0380
BHC	76	12	0.0004–0.0070
DDT	54	15	0.002–0.020
Malathion	53	6	0.003–0.115
Lindane	52	18	0.0003–0.0120
Diazinon	50	20	0.0007–0.0270
Heptachlor epoxide	46	19	0.0005–0.0040
TDE	3	23	0.001–0.005
Arsenic	18	2	0.03–0.60
Endosulfan	17	10	0.003–0.012
HCB	17	8	0.0003–0.0070
Parathion	17	10	0.003–0.022
CIPC	12	0	0.005–0.467
PCA	10	1	0.004–0.050
PCP	10	0	0.010–0.033
Carbaryl	8	4	0.05–0.50
TCNB	8	2	0.001–0.284
Methoxychlor	7	2	0.004–0.009
Methyl parathion	7	6	0.008
PCNB	7	4	0.002–0.005
Ethion	6	3	0.003–0.012
Orthophenylphenol	5	2	0.05–0.20
Leptophos	5	1	0.013–0.090
Perthane	4	0	0.030–2.28
Botran	3	0	0.006–0.067
Toxaphene	3	2	0.163
DCPA (dacthal)	2	0	0.003–0.013
Dicofol (kelthane)	2	1	0.010
Aldrin	1	0	0.001
Captan	1	0	0.178
Chlordane	1	1	d
Heptachlor	1	0	0.004
Phosalone	1	0	0.171
Ronnel	1	0	0.001
Nitrofen	1	0	0.039

a. Composited from samples from 30 different grocery markets in 30 different cities.
b. Detected, but concentration too low to be quantified by the analytical method used.
c. At and above limit of quantitation.
d. Concentration too low to be quantified.

Source: Manske, D.D., Johnson, R.D., "Residues in Food and Feed," *Pestic. Monit. J.*, **10**, 4, 134 (1977).

Pesticides and Wildlife

Although pesticides have been used commonly and increasingly for the better part of a century, only in the 1960's did scientists begin seriously to study their effects on wildlife and other nontarget organisms (49, 50). The data required to support judgments of the hazard to wildlife, therefore, have been meager until very recent years. However, research now has expanded remarkably on the infinitely varied problems of pesticide-wildlife relationships (51, 52).

Occurrence of residues

The occurrence of pesticide residues in wildlife can be illustrated by contrasting two areas: California, where pesticides are used heavily, and where residues in wildlife have been studied intensively for years; and Antarctica, where neither condition holds. Such a contrast highlights the frequency, types, and amounts of residues. In addition, an area more distant than Antarctica from the points of use of pesticides could not be found; contrasting that continent with California suggests differences in the two ecologies that can become the meat of significant judgments.

Between 1963 and 1965, the California Department of Fish and Game and the federal Bureau of Sports Fisheries and Wildlife collected 2100 samples from 86 species of California wildlife and their immediate environments (53). More than half of these samples were analyzed individually for residues, and many of the rest were included in analyses of composite residues. The animals came from throughout California and undoubtedly differed in the degree to which they had been exposed to pesticides or to PCB's that might be mistaken for pesticides.

The study showed that residues of chlorinated hydrocarbon pesticides were present almost universally. The general conclusions were these:

- Residues were common both in wildlife and in their environment.

Table 8

Total dietary intake of pesticides

Pesticide	WHO-FAO acceptable daily intake (μg/kg; ppb)[a]	U.S. daily intake 1965–70 (μg/kg; ppb)
Aldrin-dieldrin	0.1	0.08
Carbaryl	20	0.5
DDT-DDD-DDE	5	0.7
Lindane	12.5	0.05
Bromide	1000	300
Malathion	20	0.1
Parathion	5	0.01

a. Acceptable Daily Intake (ADI) is "the daily intake which, during an entire lifetime, appears to be without appreciable risk on the basis of all the known facts at the time." WHO Technical Report Series 391, World Health Organization, Geneva, Switzerland, 1968, p. 22.

Source: Duggan, R.E., Corneliussen, P.E., "Dietary Intake of Pesticide Chemicals in the United States (III), June 1968–April 1970," *Pestic. Monit. J.*, **5,** 331 (1972).

- Species that depend on aquatic or wetland habitats appear to be exposed more heavily to insecticide residues than are species that depend on dry habitats.
- The exceptions among the dry-land species are carnivorous forms, such as predatory birds, which depend on foods that may be contaminated.
- The moderate to high levels of residues found in the tissues of some species probably result less from direct exposure than from the biological concentration associated with the food chain.

Most samples of water in the California study contained residues. Particulate matter suspended in water tends to adsorb residues, and such matter often contained higher levels than unfiltered water or bottom sediments. Suspended material filtered from water contained from 10,000 to 100,000 times the level of residue remaining in the water after filtration. Soils from both arid and marsh lands contained low levels of residues.

Residues in the tissues of animals ranging from insects through birds varied according to the likely exposure. However, even animals from untreated areas usually contained residues in their fatty tissue. Tule elk, for example, averaged about 14 ppm of DDT and its metabolic breakdown products in their fat (53).

Some species contained very high residues, either because they were intensely exposed on agricultural land or because of peculiarities in their feeding habits. Ring-necked pheasants from the rice-growing areas of the Sacramento Valley, for example, showed residues of several pesticides. They averaged about 58 ppm of DDT alone, with an upper limit of about 2770 ppm.

Waterfowl and all carnivorous birds showed moderate to high values. Residues of DDT, toxaphene, and dieldrin were common in birds that eat fish. Species of birds that had high residue levels in their fat normally had higher concentrations in their ovaries and eggs. Fish, as expected, contained residues, which were present in all species who swim up rivers from the sea to breed. Predatory fish contained the highest levels.

Wildlife in Antarctica differed from that in California in three major ways (54, 55). Residues, where present, were very low; only certain groups of animals contained them; and no residues could be found in the physical environment, including samples of snow and water from widely separated points. [DDT later was found at 40 ppb in snow melt (56).]

Extensive sampling of invertebrate wildlife detected no measurable residues of DDT and DDE (a conversion product of DDT in the body). About one quarter of the Adelie penguin and Weddell seal samples contained small amounts of residues of DDT and DDE. All samples of adult Skua (a wide-ranging bird of the gull family) contained residues. Overall, the highest residues found were 2.8 ppm of DDE in Skuas, 0.44 ppm of DDT in fish, 0.18 ppm of DDT in penguins, and 0.12 ppm of DDT in seals. All of these species but the Skua are limited to the edge of the Antarctic continent.

These residues in Antarctic wildlife have raised interesting questions. The important points are that the residues are present, and that they occur in a pattern that suggests remote origin and transferral via the food chain. There is no evidence that the residues found are harming the wildlife; unfortunately, as we shall see, this is not always the case.

In the United States, nationwide studies of birds indicate that residues of the chlorinated hydrocarbon pesticides are declining with the steady decline in the use of these materials during the past decade. Residues of DDT, DDE, and dieldrin in the wings of adult mallards and black ducks were found in 1972–73 to be lower than previously (57). Residues of DDE and dieldrin in starlings were found to be lower in 1974 than in 1972 (58).

Residue transfer and response of species

Pesticide residues in animal tissue originate in an environment that also contains traces of these residues. Contaminated food must be essentially the only way in which land-dwelling animals acquire residues in their tissues. These residues are confined to the few pesticides that are concentrated in food chains. Primarily these are the chlorinated hydrocarbons; the residues of most other pesticides are metabolized quickly.

Contaminated foods are important sources of residues in aquatic animals, too, particularly in secondary links in a food chain. Residues may enter the aquatic food chain adsorbed on particles suspended in the water and then may be concentrated by organisms that obtain their food by filtering it from the water. However, there is increasing evidence that aquatic organisms in water that contains sublethal amounts of pesticides can acquire residues directly through the gills and skin (59).

Consequences of residues: measurement and judgment

The biggest single problem with pesticide residues in natural environments is the possible movement of persistent residues along food chains in part via concentration in fatty tissues. The movement of residues along food chains can be traced in two general ways: by inference or by direct measurement.

The inferential method depends on chemical analyses of residues, the biology of the species involved, and the linkage of residue levels with biological events. The biological events include individual death rates, abrupt decline in numbers of species, and inhibition of reproduction in species, population-wide. In three well-documented examples of the inferential method, pronounced mortality or inhibition of reproduction occurred among the species at the end of the food chain: the robin (60), the grebe (61), and the lake trout (62). The three sequences are these:

- DDT — leaf — litter — earthworm — robin.
- DDT — lake — plankton — fish — grebe.
- DDT — lake — plankton — small fish — lake trout.

Of particular concern is the linkage of residues in birds' eggs, inhibition of reproduction, and decline of the species. The prime examples are the eagles, falcons and hawks. These species are difficult to study, at best, and experiments are generally impossible. Nevertheless, the bald eagle and peregrine falcon have both declined rapidly in numbers in recent years; residue analyses of their eggs, counts of birds, and direct observation of nests in Great Britain, the U.S., and Canada suggest that the connection between the decline and DDT residues is real (52, 63, 64). Until recently, however, the evidence was entirely inferential.

It is now clear that population declines in some birds may be linked to aberrant metabolism correlated with known pesticide residues in bird tissues. Derangement of calcium metabolism, for example, could result from the breakdown of steroids by microsomal enzymes in the liver induced by low dietary levels of the chlorinated hydrocarbons (65). There is growing experimental evidence of declines in egg weight and eggshell thickness and increases in the egg-breaking habit among birds who prey on animals and fish (63, 66, 67). However, even closely related species can show large differences in their eggshell-thinning response to chlorinated hydrocarbons, and gallinaceous birds such as chickens appear to suffer little or no effect (52).

But how is one to assess those instances of exposure that do not result in obvious biological effects? For example, the realization is growing that food-chain concentration of persistent pesticides often is not observed, at least in aquatic animals (59, 68, 69). Food-chain concentration (misnamed "biomagnification") depends on rates of metabolic degradation and excretion of the compound. Therefore, the movement of residues along food chains in the field can be followed directly by tracing the radioactivity given off by isotopically labeled pesticides (70). This technique has been incorporated into more convenient studies in "model ecosystems" in the laboratory (71). For the few species studied to date, direct measurements of residues along these greatly simplified food chains generally support the conclusions reached earlier by inference.

The fact that pesticide residues occasionally can be degraded to compounds that are significantly toxic poses two major problems in evaluating biological effects: selection of analytical methods that will detect both the parent residue and its important conversion products; and assessment of the significance of the residues found. The conditions, influences, and interactions in an ecological system, moreover, are much more complex than in the laboratory. Many scientists have been aware of these difficulties, but only recently have data begun to come from an approach to toxicology that is based on chemistry and oriented toward ecology. Such an approach involves three steps:

- Study of persistent residues in soils, plants, and water.
- Study of the influences of life forms and environmental forces on the rate and character of the degradation of the residues.
- Study of the effect of the residues on the biological systems in which they are present.

The first of these steps is exemplified by research that showed the persistence and initial movements of DDT and its conversion products in a farm pond (72). The second is illustrated by research that showed that DDT conversion rates vary with the quality of the water (73). In particular, DDT was converted more quickly to DDD where plants and animals were abundant in water.

At the third level of examination, the effects of residues on biological systems, it is clear that different species respond in different ways to the same level of residue. Thus chemical analyses of residues and their conversion products must be combined with other kinds of information to assess the biological importance of residues.

As mentioned previously, the differing responses of different species to residues is exemplified among birds. In Great Britain, as in the U.S., most birds contain pesticide residues. Certain species of herons seem to carry the heaviest residue loads of all birds, yet their numbers are not declining, nor do they seem to be suffering physiological impairment. On the other hand, several species of carnivorous birds that, like herons, are at the end of the food chain, also carry heavy loads, and their numbers *are* declining, apparently because of the effects of differing metabolism on eggshell thickness. These species include the kingfishers, the golden eagle, the peregrine falcon, the sparrow hawk, and the Bermuda petrel (52).

Other work on correlations between residues and declining species shows that ring-necked pheasants in California carry almost twice the pesticide load of Western grebes (53, 61). In both cases, high residue loads seem to inhibit reproduction, but the effect is scarcely noticeable in the pheasants, while reproduction among Western grebes has been virtually at a standstill at some locations. In further illustration of the differing responses among species, fish taken from Clear Lake, Calif., where Western grebes are noticeably affected by residue loads, carry even higher loads with no apparent ill effects.

The physiological bases of the differing responses of different species to pesticide residues are essentially unknown. The differences appear to be innate in the species. It is known that relative immunity to residues can be acquired by vertebrates that live in heavily contaminated environments and reproduce selectively for resistance, although the only species in nature known to have acquired such immunity are fish and frogs in the Mississippi Delta region (74). The biological processes involved seem not to differ from those by which insects acquire resistance to insecticides. Indeed, the U.S. Fish and Wildlife Service has experimented with selective breeding as a means of producing rainbow trout resistant to DDT.

Yet producing insecticide-resistant vertebrates, whether or not intentionally, does not necessarily safeguard wildlife. The "protection" acquired by one species may well pose a hazard to others. In one study (75), 95% of the vertebrate predators (11 species) offered endrin-resistant mosquito fish died after consuming only one fish. The ecological effects of acquired resistance may be profound.

The apparent decline in the general level of residues of organochlorine insecticides has been correlated with improved reproductive success in birds. The residues have been thought to be a cause of declining populations of the brown pelican via thinning of eggshells. Residues of organochlorines in the eggs of a population of brown pelicans were found to have decreased from 1969 through 1973, and in the latter year, for the first time in many years, the birds reproduced well (76). The improvement was attributed to the lower residues as well as to favorable tides, weather, and food supply.

Pesticides and Human Health

Humans have been poisoned fatally by many of the major pesticides, and there is little doubt that almost any pesticide could be capable of causing acute illness. Whether caused by overexposure at work, accidental ingestion, or attempts at suicide, acute illness is all too well documented for pesticides (77). The problem of acute intoxication has been approached by many means, including warning labels, use restrictions, and education of users and physicians. The success of current control efforts can best be gauged by the fact that despite the vast increase in the availability and use of pesticides, the incidence of fatal pesticide poisoning in the U.S. — which held virtually constant at 1 per 1 million population per year for 25 years (78) — has dropped continually during recent years although both total population and total accidental poisoning deaths have increased steadily (Table 9). It seems probable that even more drastic control would fail to bring the figure much below the present level.

An analysis of pesticide deaths in the U.S. is instructive. In 1971, only 14 deaths were reported for all pesticides; this was 4.9% of all deaths due to solid and liquid poisons (79). However, a total of 911 symptomatic cases was reported, of which 318 (35%) involved children under five years of age. These data illustrate dramatically the need for great care in handling not only pesticides, but all toxic materials, where children may have access to them.

Virtually all known pesticide poisonings have been due to direct ingestion or other massive exposure. Especially as the less toxic chlorinated insecticides

Table 9

Accidental poisoning in the United States

	1956	1961	1968	1970	1975	1976
U.S. population (millions)	168.1	182.9	199.4	203.8	213.0	214.6
Accidental poisoning deaths	2635	2996	4109	5299	6271	5730
Accidental poisoning deaths from pesticides, fertilizers, or plant foods	152	111	72	44	30	31
Accidental poisoning deaths per million population from pesticides, fertilizers, or plant foods	0.90	0.61	0.36	0.22	0.14	0.14

Source: Hayes, W.J., Jr., "Toxicology of Pesticides," Williams & Wilkins Co., Baltimore, Md., 1975, 580 pp.

"Vital Statistics of the United States," Vol. 2, Mortality, Part A, 1976, National Center for Health Statistics, U.S. Department of Health, Education, and Welfare, Hyattsville, Md., to be published.

have been replaced by other, more toxic compounds, there has been growing concern over possible occupational exposure in both farming and manufacturing. Recognized incidents of farm-worker poisoning from organophosphorus compounds (80) and poisoning of industrial workers by the chlorinated insecticide Kepone (chlorodecone) (1, 2) indicate a continued urgent need for strict regulations to minimize acute exposure.

Pesticidal compounds should all be handled with caution, regardless of the degree of toxicity indicated by experiments on animals. It is difficult to evaluate the hazard to humans on the basis of animal data (81), and the possibility always exists that the toxicity of a compound will vary markedly with the species.

Reported episodes of illness caused by ingestion of food contaminated by a pesticide have been more common in foreign countries than in the U.S. (77). In some instances the contamination has been accidental; in others, people have eaten foodstuffs, such as seed grain, that had been deliberately treated with a pesticide. In one episode in Turkey, seed grain treated with hexachlorobenzene, an insecticide, was eaten instead of being planted. The result was more than 3000 cases of illness and 330 deaths (82). In the U.S., there is no record of major food poisoning episodes caused by contamination by pesticides. In two cases in this country, people applied the insecticides toxaphene and nicotine to their gardens, ate the crops within the next few days, and became seriously ill. The toxaphene residue found on the crop was 3000 ppm, and the nicotine residue was similar. These levels are in marked contrast with the established tolerances, which were "zero" in both cases (83, 84). In Tijuana, Mexico, bakery products contaminated with parathion caused many cases of severe intoxication (84a). In Alamagordo, N.M., grain treated with an organomercurial fungicide was fed to hogs; a number of them died, and the farmer's family suffered severe mercury poisoning from eating contaminated pork (84b).

Long-term, low-level exposure

The acute or subacute episodes are relatively clear-cut. It is much more difficult to evaluate the effects of long-term exposure of humans to comparatively low levels of many pesticides in the environment and in the body itself. Results of studies in the U.S. and elsewhere generally agree that in the 1960's humans carried a body burden of 3 to more than 20 ppm of chlorinated hydrocarbon insecticides and their conversion products in fatty tissue (Table 10). Much of the total consisted of DDT and related materials, but it also included dieldrin, BHC, and heptachlor epoxide (85). Only DDT and its metabolites had been measured for a significant length of time in people; it was expected to reach a constant level of concentration in the body fat of the population of the U.S., given a constant level of residues in the environment and thus constant rates of intake and excretion by individuals (86, 87). The other compounds were expected to behave similarly.

Measurements for 1970–74 show that the prediction was correct; in fact they show a general decline in the concentration of organochlorines in body fat, reflecting the decline in the use of the materials (Table 11). (The values in Tables 10 and 11 are not directly comparable because of differences in the population samples used.) Residues of organochlorine insecticides also are found in the milk of nursing mothers in the U.S. Of 1436 samples taken in 1975, the majority contained dieldrin, heptachlor epoxide (a breakdown product of heptachlor), and oxychlordane (a breakdown product of chlordane) in concentrations of parts per billion (88). The samples were taken from nursing mothers in 150 randomly selected hospitals across the country, but, as with the Market Basket surveys mentioned earlier, the chemical identities of the residues generally were not confirmed. There is no evidence at present that any group of pesticides other than the organochlorines is stored in the body or that the body contains metabolites resulting from the degradation of other pesticides to which the population is regularly exposed.

The chlorinated hydrocarbon insecticides, particularly DDT, have been in the human environment for more than 30 years; the older pesticides — lead arsenate, mercurials, copper compounds, and other inorganic materials — have been in the environment throughout this century and much of the 19th century. What effect have they had? Simply put, they have had no currently detectable effect on the population. It is not possible to prove a negative — to prove that no effects are occurring; it is possible, however, to examine the basis for the inability to detect any such effects.

Volunteers have tolerated 35 mg of DDT per man per day for periods of 21 months with no detectable effect except that they stored and excreted more of the pesticide's breakdown products than they would have normally (87). This dosage rate is about 200 times the rate found in restaurant meals in the U.S. in 1953 and 1954 (89) and more than 2000 times the average daily intake in 1970 (22). The results of a 21-month study cannot be extrapolated to a lifetime, but the absence of effects is reassuring. Data on the rates of death and illness in the U.S. for the period 1945–65, when DDT came into wide use, show no sharp deviation from the trends in 1925–45, the period immediately preceding the introduction of DDT (Fig. 3).

It has been suggested that some pesticides may cause leukemia. However, the trend in the occurrence of leukemia has not changed since DDT and the other new organic pesticides were introduced (Fig. 3). Nor has there been a sharp

Table 10

Concentration of organochlorine pesticides in body fat of humans (ppm; mg/kg)

Country	Year	Total DDT/DDE[a]	BHC isomers	Dieldrin	Heptachlor epoxide
England	1965–67	3.0	0.3	0.21	0.04
Denmark	1965	3.3		0.20	—
Canada	1966	4.3	0.1	0.22	0.14
U.S.	1968	4.8[b]	—	0.11[b]	—
France	1961	5.2[c]	1.2[c]	—	—
Italy	1966	15.4	0.1	0.68	0.23
India	1964	26.0[c]	1.4	0.04	—

a. DDE calculated as DDT.
b. Northern white.
c. Colorimetric rather than gas chromatographic analysis (tends to be high).

Source: Hayes, W.J., Jr., "Toxicology of Pesticides," Williams & Wilkins Co., Baltimore, Md., 1975.

change in the incidence of aplastic anemia, agranulocytosis, or other diseases of the blood that have sometimes been attributed to pesticides. If individual cases of these diseases are due to pesticides, the evidence is not available.

Suggestions that pesticides cause illness not known to be related to the types caused by acute poisoning have usually stressed diseases whose causes are unknown. Infectious hepatitis has been mentioned as a disease that may be related to exposure to pesticides, but this disease, the camp jaundice of the Civil War, was present long before the modern pesticides. Some have claimed that certain people are hypersensitive or allergic to pesticides, but the evidence for such claims is too sparse to allow them to be discussed objectively and in detail.

There is continuing fear that long-term exposure to pesticides causes cancer (90, 91). One widely-publicized study (92) reported that a number of common pesticides and industrial chemicals, including DDT, were "tumorigenic"—that is, they produced tissue abnormalities in a specified organ in a mouse rather than carcinogenic effects. However, other studies showed that DDT was not tumorigenic in the mouse (93) or was actually antitumorigenic (94).

What is one to believe? Unfortunately, there seems to be no sure predictive test for carcinogenicity at present. Studies such as those cited above often are conducted with levels of chemicals far exceeding actual environmental exposure, in special strains of laboratory animals, and under highly artificial conditions. They are conducted in this way to compensate for the short lifetimes of the test animals relative to man, and for other sound technical reasons, but large areas of doubt remain. An extensive review of test methods and the carcinogenicity of pesticides (6) failed to produce agreement among experts on what tests to perform or how to interpret the results; an effort by EPA to develop a set of explicit and consistent "cancer principles" for use in regulating pesticides was finally abandoned (95).

The steady increases in rates of death from various forms of cancer, including leukemia, appear to have emerged before the introduction and first heavy use of the major organic pesticides, including DDT (Fig. 3). In fact, some cancer rates

Table 11

Chlorinated hydrocarbon pesticides in human adipose tissue, United States

Pesticide	Concentration in lipid (arithmetic mean) (ppm; mg/kg)				
	1970	1971	1972	1973	1974
Total DDT equivalent	11.65	11.55	9.91	8.91	7.83
BHC (beta isomer)	0.60	0.48	0.40	0.37	0.32
Dieldrin	0.27	0.29	0.24	0.24	0.20
Heptachlor epoxide	0.17	0.12	0.12	0.12	0.10
Oxychlordane[a]	—	—	0.15	0.15	0.15
Sample size	1412	1612	1916	1092	898

a. First full year in which oxychlordane was analyzed was 1972.

Source: Environmental Protection Agency.

FIGURE 3

Lack of effect of advent of DDT on death-rate trends for five diseases

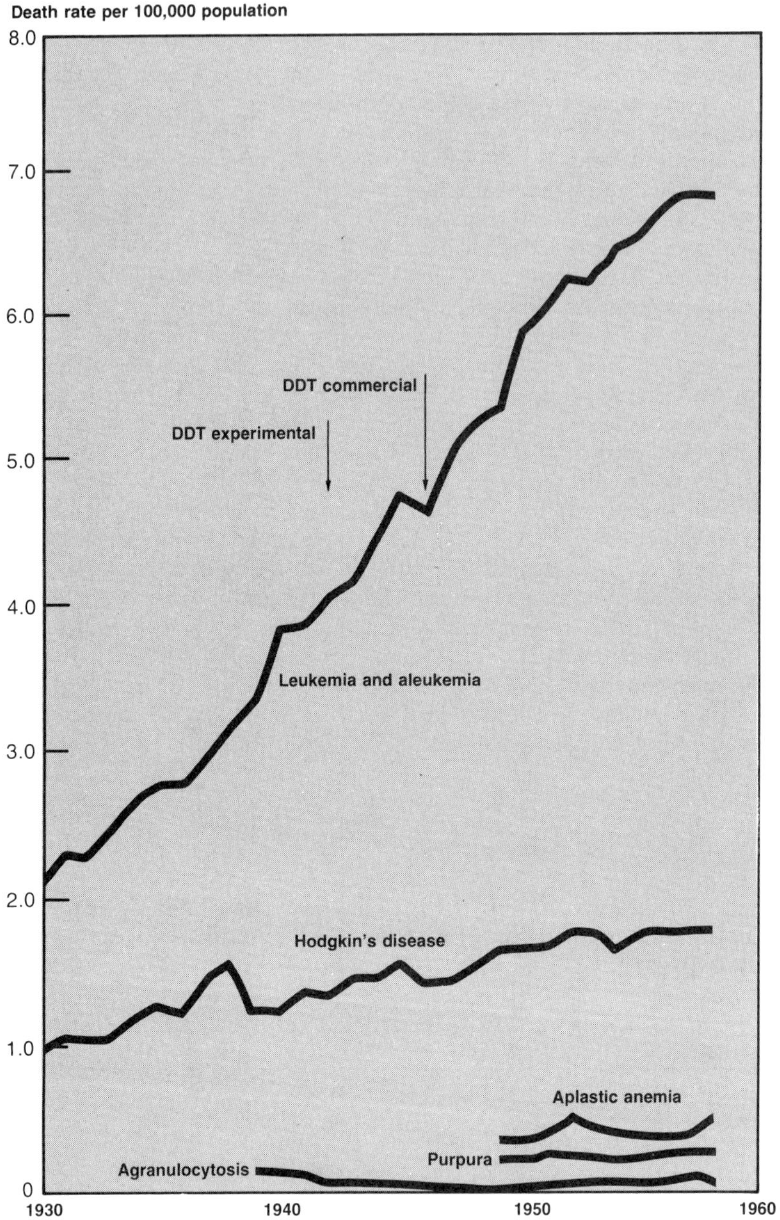

Curves show no marked changes in trends in death rates from leukemia and certain diseases of the blood following the introduction of DDT, which has sometimes been suggested as a cause of all of the diseases shown.

Source: Hayes, W.J., "Proceedings of the Short Course on the Occupational Health Aspects of Pesticides," University of Oklahoma, Norman, Okla., 1964.

started to decline during the following decade (96), with the notable exception of lung cancer attributed to smoking, and the trends overall have shown no clear correlation with the growth in the use of pesticides (3). At the same time, it should be noted that cancers are generally agreed to be caused predominantly by environmental agents, and that the induction period for various forms of the ailment may be in the range of 20 to 30 years.

Certainly, research and testing must continue, and be augmented, to define the nature and causes of cancer and any part that pesticides might play. Carcinogenesis tests are required for registration of any new pesticide, and much emphasis also is being placed on study of existing pesticides. In fact, the cancellations by EPA of most of the uses of aldrin, dieldrin, chlordane, and heptachlor were based mainly on the fact that the compounds had been shown to be carcinogenic in laboratory animals. Carcinogenicity in test animals also was cited in the cancellation of most uses of DDT, but other environmental effects played a major role. In no instance, however, have pesticides tested in animals displayed carcinogenic effects even faintly comparable to those of natural agents to which people have been exposed — aflatoxin, cycasin, pyrrolizidine alkaloids, and many constituents of tobacco smoke, for example (97). Realistic prediction and evaluation of the carcinogenic potential of pesticides must include contributions and interactive effects of simultaneous exposure not only to these agents but also to the host of other chemicals — natural and man-made — to which we are exposed daily (98).

Evaluation of the teratogenic potential of pesticides and related compounds suffers from the same restrictions and uncertainties as the evaluation of carcinogenic effects. There is no doubt that a number of substances consistently produce malformations in the offspring of exposed mothers (3, 75). TCDD is an example of a powerfully toxic and teratogenic compound (99) to which thousands of people have been inadvertently exposed in agriculture, manufacturing, and war. A careful and extensive search has failed so far to provide conclusive evidence of TCDD-caused teratogenic effects in humans or other living things, although part-per-trillion levels have been reported to occur in several samples of fish and shellfish from Southeast Asia (100).

Pesticides by design are biologically active and inherently hazardous. Nevertheless, the number of illnesses and deaths known to have been caused by accidental or deliberate misuse of the materials is far outweighed by the benefits these chemicals have brought in controlling disease-bearing pests and in increasing food and fiber production. In addition, there is no evidence to date that long-term, low-level exposure to pesticides at the concentrations found in our diet or in the environment in the U.S. has had a harmful effect on people, although extensive research is continuing. In other words, the benefits of pesticides still outweigh the risks, and the net effect of the materials on human health in a broad sense is positive.

Recommendation P2: *Efforts should be increased to interpret laboratory data on pesticide toxicity, including carcinogenicity, mutagenicity, and teratogenicity, in terms of the actual existence of hazard under conditions of practical use.*

Recommendation P3: *Research and education emphasizing human safety in the manufacture, transportation, and application of pesticides should receive high priority, especially in relation to nonagricultural uses.*

REDUCING PESTICIDE LEVELS: REGULATORY APPROACHES

Federal and state Laws

Within the United States, both federal and state governments regulate pesticides. In some states, local governments once promulgated regulations, such as the conditions under which pesticides can be applied. The federal laws now preempt local regulations, but often permit state regulations to be the more strict.

Federal regulation of pesticides is authorized under three different acts:

- The Federal Insecticide, Fungicide, and Rodenticide Act authorizes the Environmental Protection Agency (EPA) to register pesticides and prohibits sale or use inconsistent with the instructions on the approved labels. It provides also for registration of pesticide manufacturing and formulating plants and for state certification of applicators who are qualified to use certain restricted pesticides.
- The Federal Food, Drug, and Cosmetic Act authorizes EPA to establish tolerances or exemptions for maximum acceptable levels of residues of pesticides that remain on food crops as a result of field application or on foods as a result of addition during processing. The Food and Drug Administration (FDA) and the Department of Agriculture (USDA) enforce these tolerances.
- The Hazardous Materials Transportation Act of 1958, as amended, authorizes the Department of Transportation to regulate the interstate transport of certain pesticides. Under the Federal Aviation Regulations, the Hazardous Materials Transportation Act authorizes the Department of Transportation to license pilots for aerial application of pesticides.

The Federal Insecticide, Fungicide, and Rodenticide Act (FIFRA) (Public Law 80–104) dates from 1947, but it was amended extensively by the Federal Environmental Pesticide Control Act of 1972 and the FEPCA Amendments of 1975 (Public Law 94–140). FIFRA defines pesticides very broadly to include almost any substance used to control or to mitigate the effects of any unwanted organism. Each pesticide must bear a label that lists all of its ingredients and the concentration of each; instructions for proper, effective, and safe use; and proper precautions and information on antidotes and first aid. The 1972 amendments specified among other requirements that all pesticides registered before 1972 be reregistered. This task had not been completed by the middle of 1978; legislation designed in part to ease the reregistration process was nearing passage. When the requirements of FIFRA, as amended, are fully met, each pesticide also will be categorized either for general use or for restricted use only by certified applicators; in either case, use inconsistent with the instructions on the label will be prohibited. State agencies are expected to certify applicators as properly trained and capable of safely applying restricted pesticides before they are allowed to do so. Early in 1978, EPA categorized 23 pesticides for restricted use and was considering others.

Pesticide registration

A pesticide must be registered with the federal and, where applicable, state governments before it can legally be marketed. The compound must be registered for each separate kind of use; control of the same pest on two different crops, for example, constitutes two different uses. As mentioned before, some 1800 active ingredients are registered for use in pesticides. Each formulation or combination of these pesticide chemicals is considered a separate product and must be registered separately. If two or more companies produce identical products but under separate labels, moreover, each product must be registered. The result is that more than 40,000 pesticides were registered as of late 1977, although the number varies from day to day as new materials are registered and registrations for old ones are cancelled.

An application for registration must provide evidence that the pesticide produces the effects claimed when it is used according to directions and that it will not significantly harm humans, domestic or wild animals, birds, fish, or other elements of the environment if the indicated precautions are observed. If the proposed use involves application to food crops or is apt to result in contamination of food, a tolerance for an acceptable level of residue is required; evidence must be provided that use of the pesticide according to directions will not result in overtolerance residues. The registrant also must provide analytical methods that will permit formulation content and residues to be confirmed routinely.

Several agencies, both governmental and private, pursue monitoring and surveillance programs to assure that the pesticides in actual use comply with the claims on their labels. If variations are found, charges can be brought against the registrant. Penalties for violation may range from simply the correction of the error to cancellation of the registration with or without a fine. Enforcement monitoring is carried out on a planned schedule to assure systematic sampling of all types of pesticides throughout the nation. Imported pesticides, and imported produce bearing pesticidal residues, must comply with the regulations just as if they were coming from a domestic source.

The registration process

Any "interested party" may apply to register a pesticide. Let us assume that we wish to register an existing pesticide — one already registered for one or more other uses — to control an insect pest on a crop grown in almost every state. Performance data, including data on phytotoxicity (toxicity to plants) and crop yields in typical growing areas, have verified that the product is effective under various climatic and field conditions. Residue data for the product, applied to the crop according to label directions under typical field conditions, have verified that the new use does not lead to residues that exceed the previously established tolerances.

When the data have firmly substantiated that, in the new use, the pesticide is effective, safe, and not unduly harmful to the environment, directions and guidelines must be developed and transcribed into an understandable and meaningful label. The label will incorporate directions for the new use with those for previously registered uses. It will explain what pests to use the product against, when to use it, how much of it to use, and how frequently to apply it. The label also must include product name, ingredients and their amounts, manufacturer, use precautions, and the old EPA registration number. With the necessary information compiled, the proposed new label and its volumes of supporting data on performance and safety can be submitted to EPA for an amended registration. Four to six months later, if the agency has no questions on the application, we will receive label approval — a "license to sell" our product for the new use. Should EPA question the data or other aspects of the new use, however, registration may be delayed until we can resolve such questions to the agency's satisfaction.

Suppose, on the other hand, that we have a new chemical with no previously established pesticidal use. We must plan on three to five years to develop performance data representing the typical geographical distribution of the crop, residue data to establish a safe tolerance, and analytical methods for residues and for assay of the technical product and its formulations. In addition, we must compile a variety of other data on the compound: physical and chemical properties; manufacturing and purification processes; compositions of technical and formulated products; safety, including animal and plant metabolism and a substantial number of acute, subchronic, chronic, and other toxicological trials; and environmental impact. Further, if residues are found on crops or crop products that are fed to livestock, the law requires feeding studies in those animals. This product development and analytical work requires considerable expenditure for sophisticated equipment, computers, and instrumentation that further entail the skills of highly trained personnel. When all the needed data have been compiled, we can prepare the petition for a tolerance and a label and submit it to EPA to initiate the registration process. The entire process, up to the point where a new pesticide product is released for sale, requires (in the late-1970's) as much as eight to 10 years and $8 to 12 million in registration costs.

The reregistration requirements of FIFRA, as amended, have led to a third procedure, based on the issuance of a Rebuttable Presumption Against Registration (RPAR). The procedure takes place outside the normal registration/reregistration channels. Where EPA finds upon screening data for a pesticide that it does not meet specified risk criteria, the agency may "presume" that the product poses unreasonable hazard to man or the environment and issue an RPAR. The registrant may rebut the presumption and continue to sell the pesticide during the RPAR proceedings, which originally were intended to last up to six months but in practice have lasted much longer. When the proceedings are completed, EPA must decide whether or not to continue with registration/reregistration in the normal manner or to begin cancellation proceedings. By April 1978, the agency had issued some two dozen RPAR's. The only proceeding completed was for Kepone, whose registration was canceled, although some registrants had elected to withdraw their products voluntarily.

Practical consequences

Although many people have been led to believe that pesticides are subjected to minimal regulation, it should be clear by now that essentially every aspect of a pesticide's life is regulated, from the first days of its initial discovery. At each stage of development, significant regulatory problems challenge both industry and government.

FIFRA, for example, assigns to EPA the responsibility for registering every establishment in which any pesticide is produced — every plant involved in the manufacture, formulation, or packaging of a pesticide product. The owner must report to EPA the types and amounts of pesticides produced; all products in the channels of trade must display both the EPA registration number and a number identifying the plant where the pesticide was manufactured. The manufacturer must maintain accurate records on all pesticides manufactured, shipped, or stored; the records must identify the products by brand names (including copies of labels), amounts produced, quantities shipped, date delivered or received, batch sizes and numbers, amounts designated for disposal, names and addresses of all intermediate and ultimate consignees, including name of carrier, and also must include research data. Records relative to manufacture of the products must be kept for five years, while records relative to research, human safety, disposal, injuries, and adverse environmental effects must be kept for 20 years. Finally, all pesticide plants are subject to federal inspection.

Besides the controls imposed by FIFRA, the pesticide manufacturer must adhere to the rules and regulations imposed by the Fair Packaging and Labeling Act, the Occupational Safety and Health Act, the Hazardous Substances Labeling Act, and various state and local laws. He is regulated from the wording on the label, to the safety of plant and field workers, to the safety of a child in a home where a product may be used, to protection of the environment against any possible adverse effects of his product.

The applicant for registration must provide all chemical, toxicological, and efficacy information required for a new chemical. Under FIFRA, an applicant may not use the information provided by an earlier applicant without the latter's permission. The new applicant may obtain such information by research in his own facilities or he may obtain it from a governmental or private research organization; EPA checks only enough of the research to assure that the information supplied by the applicant is sound and correct.

Most of the industrialized nations of the world have developed some system of regulating pesticide manufacture, distribution, use, and/or residues. Both the systems and the philosophies vary greatly from country to country, as does the effectiveness of the regulations. Considerable effort is made through international bodies to coordinate these regulatory approaches so that differences among nations do not unduly restrict trade in pesticides and in produce treated with pesticides. Both the World Health Organization (WHO) and the Food and Agriculture Organization (FAO) have Expert Committees concerned with various aspects of the problem. These committees in turn advise the *Codex Alimentarius,* an international organization whose Committee on Pesticide Residues attempts to recommend acceptable tolerances or levels of pesticide residues that can be permitted on produce moving in international trade. The Organization for Economic Cooperation and Development also has a committee concerned with methods of regulating pesticides so as to avoid interfering with international commerce.

The complexity of the regulatory system in this country results in enormous expenditures of time and money by both applicant and government. To find an entirely new pesticide good enough to carry through registration and production usually requires the screening of thousands of potentially useful materials. The necessary research and development consume years that effectively reduce the period (normally 17 years) during which the product is protected by patents and thus the time available to the registrant to recover his costs and earn a profit. The costs involved are so high, moreover, that even the expense of registering a new use for an existing product may exceed the potential sales, especially on a minor crop. Society is requiring more extensive safety information and stricter controls, which tend to reduce further the margin between cost of development and return from sales. The substitution of newer, more selective chemicals or microbial pesticides does not ease this economic problem. Permitting more uses to be covered by the same information and more crops to be covered by a single tolerance would help in some degree, as would more specific EPA requirements, described in advance, which would shorten the processing time for applications. On the whole, however, improvements or efficiencies like these can achieve only minor reductions in the costs of pesticide development and marketing.

Improved regulation of pesticides certainly is providing a higher degree of safety to man and his environment. It is doing so, however, at a proportionally higher cost that must be borne in the long run by society in terms of less protection from pests, higher prices, or both. For example, while the realities of present-day analytical sensitivity have made the assignment and enforcement of "zero tolerances" unacceptable, regulatory caution must be exercised in continuing to set tolerances to reflect safety and good agricultural practice rather than simply the current limits of residue detectability.

Another problem area is the accumulation of data required to register pesticides for specialty uses or so-called "minor crops." These are uses which are important to economic crop production, but where the amounts of chemicals required are so small that the cost of registration becomes prohibitive. For example, California produced 38% of the nation's spinach in 1970, requiring for the purpose only 409 pounds of carbaryl insecticide; the state also produced 99% of the domestic olives, requiring only 498 pounds of naphthaleneacetic acid as a growth regulator. While papayas are not common food outside of a few areas, their production remains economically important to several states and depends on small applications of pesticides. If we are not to chance losing from the market a number of food items that add variety and wholesomeness to our diets, more regulatory flexibility will be required.

Finally, if the use of chemicals is to be substantially reduced, consumers must overcome the habits developed over decades and be willing to settle for food of lower quality. USDA grades for raw agricultural commodities and FDA limits for nonhazardous defects (insect fragments, blemishes, mold, etc.) were set during earlier periods of relatively high pesticide use, and producers are legally bound by them to provide quality food. To assure a continued food supply in the face of reduced pest control, government standards on food quality realistically must also be lowered.

Recommendation P4: *Regulatory practices should continue to be based on human and environmental safety, and on good agronomic practice consistent with safety, rather than on the prevailing sensitivity of analytical chemical methods.*

Recommendation P5: *The uses, application rates, and environmental fates of pesticides in nonagricultural applications should be identified and given increased research and regulatory attention.*

Recommendation P6: *Realistic incentives should be provided by law to encourage the development of new, safer, and more effective pesticides and pest-control measures of other types. Such incentives might include extended patent protection, simplification of the registration process, and standardization of test procedures required for registration.*

REDUCING PESTICIDE LEVELS: TECHNICAL APPROACHES

That pesticides have produced great benefits for mankind and can continue to do so is beyond dispute. But those very benefits have led many pesticide users at one time or another, worldwide, to embrace the concept "if some is good, more is better." Hair-raising examples of the misuse of toxic chemicals include multiple overapplications, often by unskilled or illiterate workers; prophylactic and cosmetic uses; and careless handling and disposal of unused chemicals and contaminated containers or equipment. Such incidents, not surprisingly, have achieved a public prominence that often obscured the rewards of the legitimate, controlled uses of pesticides. Nevertheless, even rational application of the compounds undeniably can result in residues in air, water, soil, and living organisms. It is essential, therefore, to recognize the logic in limiting and controlling the prevalence of such residues.

In addition to legal restrictions, a variety of technical methods can help minimize the environmental impact of pesticides without sacrificing the benefits pesticides provide. Of particular current interest are optimizing use, reducing the need for persistent or hazardous pesticides, controlling waste, and finding more effective compounds.

Optimizing the use of pesticides

While pesticides by now have provided at least partial economic control of most of the major agricultural, public health, and forest pests, it has been primarily the threat of adverse environmental effects that has forced a reassessment of the value of heavy or frequent applications of the materials. Such reassessments seek, in general, to reduce to the optimum level the amount of chemical used. A farmer may find, for example, that he has been using more pesticide than really is required to achieve the desired control of a given pest. The amount that is "optimum," moreover, may depend on several variables in addition to simply the cost of chemicals and labor and the degree of control achieved. Depending on the circumstances, less extensive use not only can lower the costs of chemicals and application; it can also increase survival among beneficial predators, parasites, and pollinators; reduce the levels of environmental contamination; reduce the levels of pesticide residues on crops; and decrease the likelihood of pests' developing resistance to the pesticide.

Reducing the amounts of chemicals used, in the absence of other controls, is much more easily said than done. For one thing, the grower must perceive greater overall benefit in not using pesticides than he enjoys currently from using them. While usage might be limited by imposing high taxes on the chemicals or by restricting the number of products that can compete for a given market, these forms of coercion appear counterproductive, could bring undue cost to the public, and perhaps restrict public options in the long run. More positive approaches are suggested in the following sections.

Improved management. Promising results have come recently from grower contracts with professional crop-protection managers who undertake, for a fixed price, to guarantee effective pest control. The pest-control manager increases his profits by using the smallest possible amounts of chemicals in the most efficient way, and the grower receives his profit in any event. Excessive applications of pesticides sometimes are used because the threat of pest invasion causes growers to spray at fixed intervals whether a pest problem exists or not; by the time damage is perceived in a 2000-acre orchard and pest control measures started, the damage can be extensive. Crop management by continuous professional monitoring of developing pest populations and more accurate estimates of "economic threshold" — the amount of damage that can be tolerated — make it possible to reduce greatly the number of essential pesticide applications.

Unnecessary applications can also be reduced by determining some critical period for treatment, which may depend on the life cycle and behavior of both crop and pest. One instance is treatment for sugarcane borer in Louisiana, which was found to be largely wasted after August (101). Another approach is to treat only those areas that are heavily infested by the pest. When the entire watershed of an oak-hickory forest in the Nantahala Mountains in North Carolina was treated for elm spanworm, 0.346 ppb of DDT was found in the mountain water. When DDT was applied only to the 49% of the forest that was infested, damage was controlled but residues could not be detected in the water (102). A study of treatment of 525,000 acres in Salmon National Forest for spruce budworm showed that the target insects were effectively controlled by treating less than 10% of the entire area (103).

Modification of application equipment. Marked reduction in the amount of pesticide used could be achieved in many cases by using improved application equipment. The importance of the quality of application is evident in the results of a study of swath patterns after an aerial application of granules (104). The study was made under conditions as nearly ideal as possible, but still only about 17% of the material was applied at the proper rate (Fig. 4).

The application of dust formulations of pesticides presents exceptional problems. For example, an early study of commercial dusting equipment showed that the pesticide and clay could separate from each other during application, resulting in uneven deposits of the toxicant (105); more uniform application would reduce the total amount of pesticide required. Another study showed that only 10 to 20% of the dust discharged from a conventional duster was deposited on the plant surface at which it was aimed (106). Further, the movement of dust can be difficult to control. For these and many other reasons, the use of dusts is declining, and wettable powders or emulsions are being substituted. Helicopters are being used increasingly for aerial application, although the major proportion of pesticide applications still is made from the ground. The development of better application equipment and more efficient application methods would be of major importance in reducing pesticide losses during spraying. Such losses can amount to as much as 80% (107) and normally are 30 to 60%. If efficiency could only be doubled, the amounts of applied pesticides could often be cut in half.

Modification of formulation. Improved formulations of pesticides could reduce the amount of chemical required to control pests. A small amount of nontoxic Piccopale resin used as an additive in a granular formulation of heptachlor increased the persistence of the insecticide in soil to 250% of its persistence when applied in a standard liquid formulation (Table 12). The additive apparently prevented heptachlor from vaporizing and leaving the soil (108). Less than half the normal amount of insecticide would have to be applied in such a formulation to achieve the same degree of control as with the liquid formulation. A similar effect was found in laboratory work in which two cellulose ether polymers were used as additives with Phosdrin. They more than doubled the residual effectiveness of Phosdrin against Mexican bean beetles (109). Research on additives for liquid formulations that would help to "fix" the pesticide to the plant might be fruitful. Losses caused by washing off by rain could be reduced by such means, allowing use of less chemical and reducing contamination of runoff water.

Additives in the formulation might also improve the performance of systemic insecticides, which are taken up by the plant and then act against pests who feed on it. Three quaternary ammonium salts added to Zectran formulations increased the uptake of the insecticide by cotton plants, thus increasing its effectiveness and reducing the dose required to control pests (110).

In the use of granular heptachlor against the imported fire ant (111), it was possible to reduce the dosage from 2 lb per acre to 1.25 lb by removing 80% by weight of the fine particles from the formulation. The lower dose rate was permitted by the elimination of the loss of active ingredient due to drift of the finer particles away from the target area.

Another technique is to apply ultralow volumes (ULV) of highly concentrated insecticide formulations, or even the technical material, by air to help in reducing the amount of pesticide needed to control some pests (112). This low volume technique has been used successfully to treat the boll weevil with malathion (113). Similarly, application of insecticides as liquefied gas aerosols can reduce the dosage required. Extensive studies with pea aphid showed that only half as much pesticide was required per acre with an aerosol formulation as with a dust (114). Furthermore, heavy rain occurring immediately after application did not

FIGURE 4

Inefficiency of standard aerial application as shown by swath pattern study

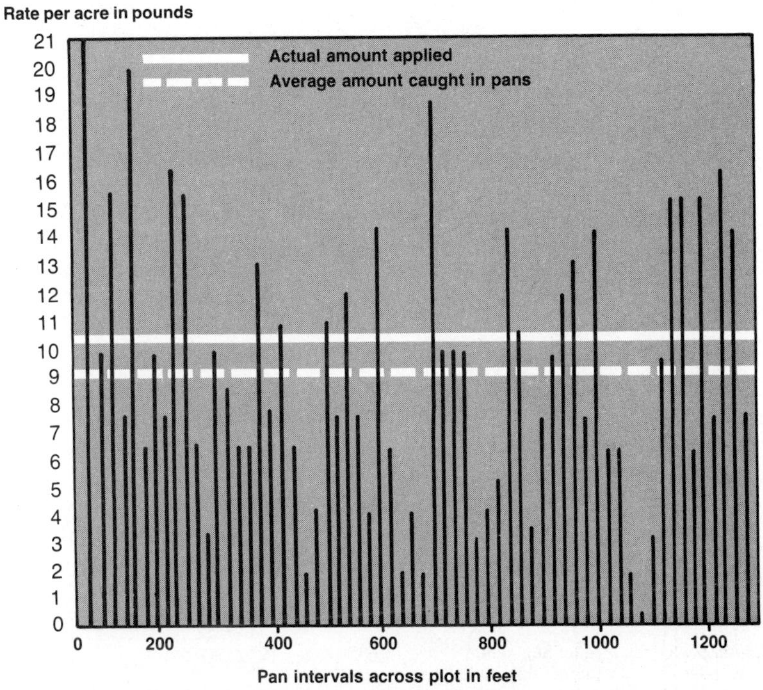

Pans spread across test plot at 20-ft intervals collected granules spread by standard airborne application equipment. Amounts caught in pans show variability of distribution pattern. Only about 17% of material was applied at close to the expected rate (10 lb/acre).

Source: Lofgren, C.S., "Large Scale Tests to Evaluate Multiple Low Dosages of Heptachlor," Report of Methods Improvement Section, Plant Pest Control Division, ARS, U.S. Department of Agriculture, Gulfport, Miss., 1959.

reduce the effectiveness of control. The aerosol method apparently produces particles of concentrated insecticide that stick tenaciously to the plant surfaces, but little recent work has been reported on the use of the technique.

Apparently, appropriate manipulation of pesticide formulations can greatly reduce the environmental mobility of the chemicals as well as reduce the total amount required for effective pest control. Presumably, other formulations could be found that would increase the rate of degradation and so allow any desired persistence to be designed into a pesticide product without having to find and register a totally new active compound.

Recommendation P7: *Major improvements should be sought in methods of pesticide application, including increased application efficiency, control of volatilization and other kinds of movement, and storage stability.*

Reducing the need for pesticides

Although the amounts of pesticides required often could be reduced, chemicals remain man's major weapon against pests for the foreseeable future. Nevertheless, the need for toxic chemicals can be reduced even further by alternative methods of pest control that now are practical as well as beneficial. Major alternative measures can be grouped under broad headings that include biological control, cultural control, genetic resistance in crops, sex attractants (pheromones), and integrated control.

Table 12

The effect of additives on persistence of heptachlor in soil

Additive[a]	Heptachlor applied[b] 0.71 (ppm; mg/kg)	Residue found[b] As heptachlor .09 (ppm; mg/kg)	Residue found[b] As heptachlor epoxide .04 (ppm; mg/kg)	Total (ppm; mg/kg)	Per cent insecticide remaining 18%
Lube oil	0.71	.09	.04	.13	18%
Piccopale resin	0.87	.39	.06	.45	52
Alkylated polystyrene	0.83	.21	.04	.25	30
Aromatic plasticizer	0.80	.19	.04	.23	29
Heavy aromatic naphtha	0.81	.11	.06	.17	21

a. Used at twice the level of heptachlor.
b. Calculated on basis that all insecticide is in top inch of soil.

Addition of Piccopale resin to granular heptachlor formulation increased persistence of insecticide in soil more than did other additives. Increased persistence would allow less pesticide to be used to achieve same effect on pests over same period of time.

Source: Annual Report, Methods Improvement Laboratory, Plant Pest Control Division, Agricultural Research Service, U.S. Department of Agriculture, 1959–1960, p. 13. Plots treated Oct. 1, 1959, and sampled Nov. 12, 1959.

Biological control. Biological control of pests (115) involves the purposeful use of insect parasites, predators, or pathogens — forms of control that, under natural conditions, operate almost continuously (Fig. 5). In several instances, techniques have been developed to produce large numbers of such organisms for release into high-density pest populations by inundation or inoculation; other potentially valuable organisms remain undeveloped (Table 13). The introduction of exotic parasites and predators (classical biological control) has resulted in the complete control of 42 pests and partial or substantial control of another 78 (116).

Biological controls already have achieved spectacular results against several pests. One example is the predacious Vedalia beetle, which was brought from Australia 90 years ago to control the cottony cushion scale on citrus in California. Tiny *Trichogramma* wasp egg-parasites have been used successfully against the cotton bollworm. A formulated suspension of the spores of a bacterial pathogen, *Bacillus thuringiensis*, has been registered under the name Dipel (or Thuricide) as a pesticide to be sprayed on vegetable crops to control caterpillars (117). Formulations of *Bacillus thuringiensis* and nucleopolyhedrosis virus have both been registered for use against the Douglas fir tussock moth.

A revolutionary and important concept of biological control is to eradicate the pest, thus eliminating the need for pesticides (118). In recent years, the Mediter-

FIGURE 5

Average number of natural pest enemies (predators and parasites) per acre in alfalfa fields

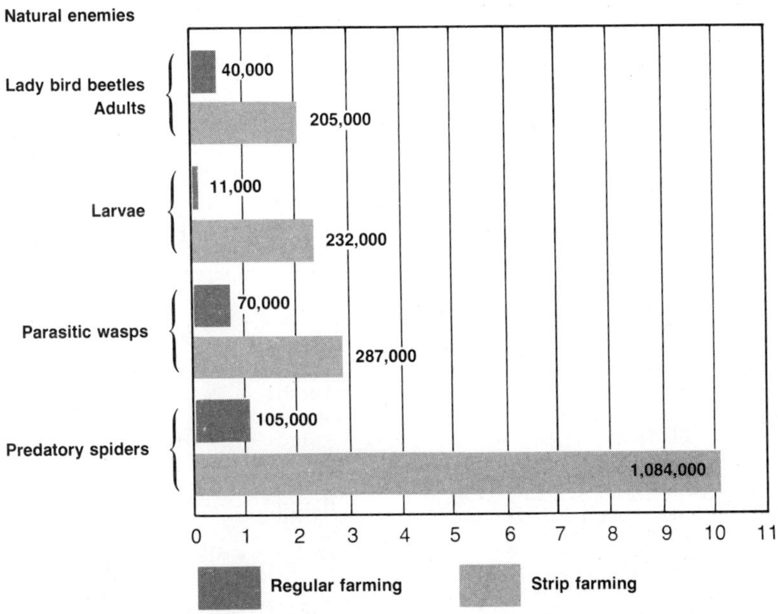

Source: Rudd, R.L., "Pesticides," in "Environment: Resources, Pollution and Society," Murdoch, W.W., ed., Sinauer Associates, Stamford, Conn., pp. 279–301, 1971.

ranean fruit fly has been almost eradicated in Florida and the screw worm fly in the southeastern U.S. and almost throughout the country. Other affected pests include the Khapra beetle, the melon fly, the Parlatoria date scale, and the citrus black fly. Eradication has been tried with the imported fire ant and the cereal leaf beetle. The Mediterranean fruit fly was eradicated from Texas in its most recent incursion into this country. Eradication of just three pests — the boll weevil, the bollworm, and the codling moth — could reduce the amount of insecticide applied annually in the U.S. by an estimated 40% (119).

While the cost of eradicating a pest is high, it may not be excessive in view of the damage the pest does and the cost of successive years of controlling it. The screw worm fly was eradicated from the Southeast at a cost estimated to be equivalent to less than half the damage that the pest could do in a single year if it were uncontrolled. (Recently, several new screw worm infestations have occurred; actual eradication probably is impossible.) The cost of eliminating the Mediterranean fruit fly from Florida was less than the cost of the annual quarantine and regulatory measures involved in living with the pest.

Table 13

Examples of parasites and predators of potential value in pest suppression through inundative releases

Biological control agent	Pest
Nematodes:	
DD–136 (Biotrol NCS)	Codling moth; European corn borer.
Heterotylenchus autumnalis	Faceflies.
Reesimermis nielseni	Mosquitoes.
Parasitic insects:	
Apanteles species	Various caterpillars.
Bracon kirkpatrick	Pink bollworm.
Cuban fly	Sugarcane borer.
Lysiphlebus testaceipes	Aphids.
Macrocentrus ancylivorus	Oriental fruit moth.
Macroterys flavis	Brown soft scale.
Micropletis	Bollworm complex.
Pediobius foviolatus	Mexican bean beetle.
Several tachinid flies	Bollworm complex.
Trichogramma	Various moths and butterflies.
Phytophagous insects:	
Agasicles (beetle)	Alligatorweed.
Bactra veratana (moth)	Nutsedge.
Predacious insects:	
Coccinella (lady bugs)	Aphids.
Cryptolaemus montrouzieri	Mealy bugs and soft scale.
Hippodamia (lady bugs)	Aphids; bollworm complex.
Other:	
Cyprinidon variagatus (saltwater fish)	Mosquitoes.
Dung beetles	Hornflies.
Gambusia (freshwater fish)	Mosquitoes.
Marisa (snail)	Aquatic weeds.
Mollienesia latipinna (saltwater fish)	Mosquitoes.
White amur	Aquatic weeds.

Source: "Integrated Pest Management," Council on Environmental Quality, Washington, D.C., 1972, 41 pp.

Both biological and chemical methods have been used to eradicate pests. With the screw worm, millions of the maggots were raised artificially, the pupae were treated with gamma radiation from cobalt-60, and the resulting sterile male flies were released to compete with native flies. The fly was eradicated from much of the Southeast in a year and a half.

With the Mediterranean fruit fly in Florida, the infested area was bounded by traps baited with a potent synthetic sex attractant (120), a protein hydrolyzate, and malathion. Only about 120 cc of technical malathion per acre was required in this bait. Since the flies are attracted to the bait, it is effective even on those flies that normally would be protected from direct spray by foliage or other cover. The bait apparently injured no wildlife or humans and has not harmed the aquatic environment.

Biological controls have shown none of the common pesticide side effects: toxicity to nontarget organisms, development of resistance among target organisms, and pollution of the environment. They are not hazardous to applicators and, once established, are seemingly quite permanent. On the other hand, they are usually relatively costly and slow to take effect: however, once established, a control species, unlike a chemical agent, becomes *more* concentrated — it reproduces itself, and "cancellation" becomes impossible. Mutations and other unexpected biological properties are an ever-present possibility, so introduction of an exotic species must be accompanied by much research and a certain amount of trepidation.

Cultural control. Cultural control (also termed "environmental manipulation") requires that standard farming practices be changed in ways that either do not favor the pest or do favor its natural enemies. Nearly every agricultural pest is subject to some extent to this form of control (121). With many crops it is now common practice to rotate the crop, plow under plant debris, and time planting or harvesting so as to reduce the numbers of particular pests. Crop rotation, probably one of the oldest and most familiar cultural practices, is used to control soilborne diseases, nematodes, and the root-feeding stages of a number of insects. Wheat has been raised successfully in areas plagued historically by the Hessian fly merely by planting late or harvesting early, thus eliminating the fly's customary host (the wheat) during a critical period in the insect's life cycle.

In Arizona and California, this principle has been applied to the pink bollworm in cotton (122). Crop growth is stopped early (by withholding irrigation water after mid-August), and the stalks are cut and plowed under. The practice deprives the pest of the host, the cotton boll, in which it normally enters larval diapase, the period of dormancy that allows it to overwinter in the field and infest the subsequent crop. In California, strip-cutting of alfalfa — harvesting the crop in alternate strips — leaves enough alfalfa in the field to persuade lygus bugs not to migrate into cotton fields, where they damage young cotton fruit (123); the practice also preserves a stable habitat — the unharvested strips — for the bug's natural enemies.

Genetic resistance in crops. Genetically transmitted resistance to pests can be built into crop plants by selective breeding in the same way as early-maturing and dwarf characteristics have been bred for decades (124). Theoretically, it is possible by selective breeding to create, for any commercial crop, strains that are naturally resistant to insects, diseases, and nematodes. The process is usually very slow, although even now more than 100 plant varieties are resistant to nematodes (125), and a similar number are resistant to plant diseases (126). Even partial resistance to a particular insect can greatly reduce economic damage to a crop as well as the need for pesticides. Among insect pests already partially controlled through plant resistance are the European corn borer, spotted alfalfa aphid, Hessian fly on wheat, and the wheat stem sawfly. Resistant varieties as a rule fit easily into the existing pest control system for any crop and possibly hold our greatest hope for reducing the use of pesticides. It should be observed, however, that natural resistance often is due to natural toxic or repellent chemicals in the plant itself, several of which have been isolated and identified (127).

Sex attractants. Insects often locate mates for breeding by following an airborne trail of chemical sex attractant (a "pheromone") emitted by the other sex. The pheromone is quite species-specific and is effective at extremely low concentrations. A number of these pheromones have been isolated, characterized, and synthesized in the laboratory (128). No insects yet have been controlled completely with natural or synthetic pheromones, but several application techniques have been very promising (129). They include means of attracting mate-seeking insects to poisoned baits, to ultraviolet light traps, to mechanical or sticky traps, or to insecticide-treated areas. A more recent method is designed to spread large amounts of the female gypsy moth pheromone over a wide area to confuse the males and thus prevent mating (130). This principle might be used to control most insects if the necessary pheromone compounds were discovered, registered, and made available in large enough quantities; such natural pest control agents must meet all EPA registration requirements as would any other chemical introduced into the environment.

Other chemical lures include the fragrant food attractant geraniol, which has long been used in the eastern U.S. to trap Japanese beetles. Flytraps have been developed that also use oviposition attractants — ammonia-yielding chemicals or decaying organic matter — to simulate suitable egg-laying sites; to be effective, however, food and oviposition attractants must be applied rather liberally compared to the pheromones.

Integrated control. Integrated control — pest management — combines maximum reliance on natural controls with appropriate use of other techniques: cultural control, pest pathogens, resistant crops, pheromones and other attractants, parasite or predator releases, and chemical pesticides if and when needed (121, 131). Integrated control should not be confused with biological control nor with the use of any single technique; it does not imply nonchemical farming, nor should it be identified with "organic gardening," which uses neither synthetic fertilizers nor pesticides. Integrated control, rather, is a system that embraces the best elements of all methods. For the long run, the integrated approach promises improved pest management with minimum harm to the environment and lower cost to the farmer.

One example of integrated control involves the pink bollworm *(Pectophora gossypiella)* on cotton in Arizona and California. The two practical approaches at present are cultural and chemical. The cultural method is to destroy the plants soon after the cotton has been picked, thus depriving the pest of an essential host. Organophosphorus insecticides are applied during the latter part of the growing season, but only when the bollworm begins to inflict economic damage. Measures to be built into this integrated approach as they are perfected include the use of parasites, sterile-male releases, and the sex pheromone of the parent moth. Two species of parasites have been released in southern Arizona with moderate success; the sterile-male releases are being made in central California to suppress incipient infestations. As mentioned earlier, an extremely promising new technique employs field application of the pink bollworm sex pheromone in quantities that, although minute, are still so much greater than occur in nature that the male is confused and unable to locate a female.

Recommendation P8: *Systematic laboratory and field research on integrated pest management should receive increasing emphasis aimed at the earliest possible practical implementation. Chemists should be strongly encouraged to increase their direct collaboration with scientists of other disciplines in solving the chemical problems of integrated control, including those related to pheromones, natural repellents, and other control substances in addition to conventional pesticides.*

Controlling waste

The production of more than 1.6 billion pounds of pest control chemicals annually in this country is accompanied by the need to dispose of a large amount of chemical by-products as well as by many opportunities for accidental loss of the pesticides themselves. The true magnitude of the problem is difficult to estimate because data on processes and production volumes for individual pesticides usually are proprietary and confidential. In general, however, major producers are well aware of sources of pollution in their manufacturing processes; they monitor emissions of stack gases and discharges of liquid wastes to streams and rivers and already have or are developing treatment facilities and properly designed incinerators to detoxify wastes and reduce their volume (132). Even so, pesticide losses during production could be substantial.

In perspective, some forms of waste disposal, such as incineration, may simply shift the pollution burden. For example, if unrecovered hydrogen sulfide from an organophosphate insecticide plant is burned, the amount of sulfur dioxide that could be released into the air would approach 2 million pounds annually, the undesirable quantity emitted by some electrical generating plants.

Pesticide formulation is a more significant source of pollution than is production of the basic chemicals (132). Small-scale formulators often lack proper waste-treatment facilities and may discharge their wastes to municipal sewers, many times without the help of authoritative information on how to improve their pollution abatement systems. For example (132), the 1970 suspension of several 2,4,5-T registrations resulted in an attempt by formulators to recall some 6 million full containers; one company reportedly accumulated 44,000 gal of dilute 2,4,5-T, but was left with the almost insurmountable problem of safe disposal.

Agricultural applications consume about 60% of the pesticides used in this country; major sources of waste-derived pollution include spillage, disposal of containers, and cleaning of equipment. Of the 40% of the chemicals applied outside of agriculture, as much as 20% of the total is used by homeowners. All of these many seemingly minor sources of pollution, when combined, present a potential threat that requires careful assessment.

Decontamination and disposal. Disposal of surplus pesticides and their containers is a complex problem; procedures have been recommended by EPA (133). Each year, all along the distribution chain, the problem of surplus pesticides is growing as patterns of agricultural application change and legal restrictions are placed on certain chemicals. When a pesticide cannot legally be used for its intended purpose, there are few options: return to the manufacturer (who often must seek a means of disposal also); export for legal use elsewhere; proper incineration; or disposal in specially designated landfills. (Pesticides containing heavy metals, such as mercury, lead, cadmium, or arsenic, are not suitable for incineration or landfilling.) The actual destruction of organic pesticides by soil injection, chemical detoxication, or injection into deep wells is in its infancy.

Disposal of pesticide containers has become a major problem (134). The 116 million pounds of pesticides used in California alone in 1971, for example, were accompanied by more than 17 million containers — from paper bags to 55-gal steel drums — including about 10 million that entered the home. The situation is unusual in that it involves poisons; disposal is a matter of both toxicology and conventional solid-waste handling. The toxicological aspects are the most difficult; if each "empty" container were to retain only 1% of its original contents, California, for example, would be dealing with more than 1 million pounds of "waste" pesticides annually from this source alone.

Pesticide containers are categorized in the same way as their contents; residue and rinse liquids must be disposed of in the manner specified for the particular type of pesticide, most often by incorporation into soil in specified "dumps."

Soil can provide an ideal medium for assimilating and dissipating many organic compounds. Synthetic pesticides are converted to environmentally acceptable products by microbial action, adsorption on soil particles, and chemical decomposition. However, the processes involve conversion of the pesticide to a series of breakdown products whose identity, toxicology, mobility, and persistence should be ascertained before extensive disposal in soil is undertaken (the pertinent information is a routine requirement for registration). Volatilization into air and leaching into water must be avoided.

Selection of the site for soil disposal is of primary importance. Essential considerations include climate, ecology, use, ownership, hydrology and geology, remoteness from centers of population and agronomic production, and low potential for future agricultural, industrial, or recreational use. The disposal site also should be flat; there should be no possibility that mobile pesticides could contaminate surface waters or groundwater by runoff or leaching. The soil, in addition, should be fairly high in organic matter.

High-temperature incineration is generally acceptable and safe for disposing of most flammable organic pesticide wastes (135). Most commercial pesticide formulations degrade below 1000°C; volatile organic materials could be incinerated at as low as 250°C, while more resistant wastes require temperatures as high as 900°C. The design and construction of an incinerator are based on a number of factors: the flash point of the material to be incinerated, its fire point, combustible content, water content, chlorine and sulfur content, inorganic and metal content, etc. The design of feed systems, incineration chambers, scrubber systems, and pollution monitoring devices offers a great challenge for the chemical engineer. Unfortunately, the capital investment and maintenance costs for a large incinerator of sophisticated design are beyond the capability of all but major industrial concerns and governmental agencies, and the problems of transporting toxic chemicals over long distances to a central location to use such an incinerator are formidable.

It is increasingly desirable to reduce the toxicity of surplus or waste pesticides as a preliminary stage in disposal. Chemical treatment can be useful for this purpose as well as for reducing persistence or the resistance to microbiological breakdown (135). Typical treatments include hydrolysis, oxidation, and chlorinolysis. For example, hydrolysis may be used to detoxify common organophosphorus and carbamate pesticides; the half-lives of many of these compounds change rapidly with pH. However, the wastes from such hydrolyses sometimes retain considerable toxicity (i.e., highly toxic parathion is hydrolyzed to moderately toxic p-nitrophenol). Chemical oxidation also is an alternative to incineration. Specific pesticides can be destroyed by oxidants; amitrole (3-aminotriazole), for example, is destroyed by a mixture of ferrous sulfate and hydrogen peroxide (Fenton's reagent) (136). Oxidative reactions initiated by light also can be very effective in destroying pesticides (137).

Recommendation P9: *More attention should be devoted to the role of containers in environmental pollution by pesticides, including container decontamination, reusable containers, possible limitations on package size, and the safety aspects of containers in home use.*

Recommendation P10: *Methods should be developed for the practical chemical or biological disposal of every registered pesticide; this might reasonably be required of registrants before their pesticides are registered.*

More effective chemicals

Another method that often has been proposed for reducing the use of pesticides is to develop more effective and pest-specific substances. Actually, most modern pesticides are remarkably effective and often quite specific when used at the proper (not excessive) level. This is especially so in comparison with pre-World War II chemicals. In those days, carbon disulfide was commonly applied at levels of up to 1000 lb per acre, and the insecticidal lead arsenate was used at 1500 lb per acre. Today, an insecticide application of 1 lb per acre is common, and ultralow volume application involves only a few ounces per acre. Mirex bait has been shown to provide effective ant control at 2 to 4 g per acre (138). (By 1977, registrations of Mirex, a chlorinated hydrocarbon, were being phased out because of the compound's persistence and environmental hazard, including possible carcinogenicity.) It is not so much the pesticides but the delivery systems that must be made more effective.

Several serious problems stand in the way of more species-specific pesticides. One is our lack of knowledge of the factors that influence differential toxicity, especially among closely related species. Another is the very large number of organisms that would have to be considered and, presumably, included in the chemical screening programs by which new pesticides are discovered. Despite such difficulties, great advances in specific control have been made in recent years. For example, Thuricide affects only one class of insect (the Lepidopterae) (117), and the herbicide barban controls wild oats (*Avena fatua*) in spring wheat (139).

However, the major bar to greater specificity may be economic. The more specific the pesticide, the less can be sold. Considering the rapidly increasing cost of pesticide development, the price of a truly specific pesticide — one effective against only a single species — generally would be prohibitive. Perhaps a satisfactory compromise lies in more effective formulations, concomitant use of attractants, and in integrated pest management methods.

WHAT NOW?

The previous edition of this book was published in 1969 at a time of record world production of persistent pesticides, a time of regulatory upheaval, dying birds, shouting critics, and an unclear economic future for new pesticides.

Despite the gaps in knowledge and technology made apparent in this chapter, the outlook for people to benefit from these chemical tools without undue environmental harm now seems more optimistic.

For one thing, the present ability to detect, measure, and monitor pesticide residues in the environment should minimize the frequency of unpleasant surprises from unexpected and unwanted residues. In fact, our analytical capabilities far outstrip the understanding of and ability to measure the parameters of chronic toxicity; we routinely measure residues at levels far below those to which we can yet ascribe with assurance any biological effect. This situation represents one of the principal problems in establishing national or international monitoring systems: what should they monitor? For example, mercury has been a favorite subject for monitoring for many years, yet this element has been found to be almost ubiquitous during human history; it turns out that previously unsuspected methylated mercury compounds, rather than total mercury, are the specific forms most likely to be a cause for concern (140).

The outlook for more rapid and sensitive screening tests to detect potential carcinogens and mutagens now appears promising (141, 142). In addition, several systems have been proposed (59, 143) for setting priorities for the development of new pesticides, and for environmental research on older materials, on the basis of relatively simple measures of bioconcentration, environmental movement and degradability, and toxicity. With these has come extensive research on so-called "model ecosystems" (71) in which the distribution and breakdown of candidate compounds can be measured in greatly simplified living communities under laboratory conditions.

New pesticides now are intentionally selected for biodegradability, and considerable interest has been revived in using natural pest control agents as models for improved and even revolutionary synthetic agents. For example, Japanese research on an insecticidal substance isolated from worms of the genus *Nereis* has led to a synthetic commercial product known as cartap (144); the impractically unstable pyrethrum has provided leads to more stable artificial analogs such as resmethrin, now in widespread home use (145); and research on the juvenile hormones of insects, which regulate their development, has resulted in the pest-controlling synthetic analog, methoprene (146). However, although the use of persistent chlorinated insecticides is being sharply curtailed, there remain a number of problem compounds, including organometallics such as arsenicals and mercurials (whose uses already are severely limited); only further research will determine whether we can continue to tolerate them.

Almost all present-day pesticides have been discovered by screening chemicals empirically for toxicity to insects, weeds, fungi, and other pests. The chemicals must actually be tested against pests because in general we are not yet able to correlate molecular structure definitively with biological activity. Enough is known so that screening need not proceed entirely by chance, but the empirical approach nevertheless will remain the basis of most pesticide discovery and development for the foreseeable future.

A chemist can synthesize from 50 to 150 novel compounds in the course of a year, but only one new pesticide will emerge for every 10,000 compounds screened. The management and coordination of the resources and skills required to develop this pesticide are becoming highly sophisticated. For example, in the "critical path" system (Fig. 6), the overall task is broken into a series of manageable stages or formalized activities which are interrelated in that progress in one stage often depends on the results of work in one or more other stages.

FIGURE 6

"Critical Path" development of a new pesticide: each circle represents a decision point

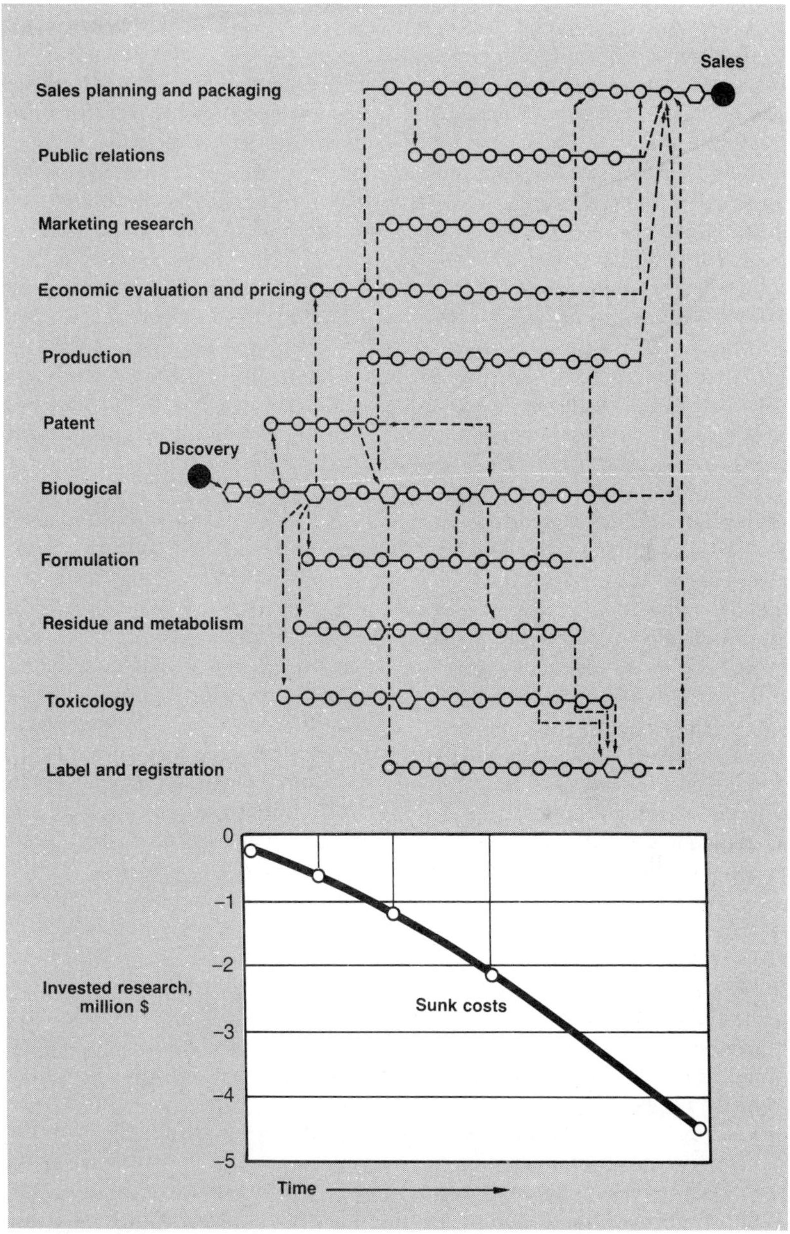

Total time involved in developing a new pesticide typically is eight to 10 years. Total cost is $8 to $10 million when costs of both "winners" and "losers" is averaged over the winners.

Source: Adapted from Johnson, J.E., Blair, E.H., "Cost, Time, and Pesticide Safety," *Chem. Technol.*, **2**, 666 (1972).

The time required to develop a successful pesticide is growing longer and longer, and total costs continually increase as the "winners" must support the "losers." Research costs, meanwhile, also are escalating rapidly. Soon, only a few large companies, operating worldwide, will be able to undertake the discovery and development of new synthetic pesticides. Several important steps could be taken by the Government to stimulate the widest possible participation in the search for safer, cheaper, more effective new pesticides: (1) lengthen the period of effective patent protection in recognition of the fact that often 10 of the 17 years provided currently may be expended in development before any return on investment can be realized; (2) simplify and streamline the registration process, consistent with safety, for example by grouping commodities that require similar pesticide applications; and (3) provide more standardization of test procedures required for registration so that registrants can estimate in advance what kinds of data will be acceptable.

As indicated in previous sections, there is good evidence that most damage done by pesticides has been due to overdose and overexposure, accidental or intentional. There is now a growing awareness of and emphasis on pesticide safety, directed primarily toward user education and understanding rather than more regulation. However, whether we consider a manufacturer faced with a waste disposal problem, a grower or home gardener spraying, or a public official setting regulations, there simply is no substitute for common sense when dealing with pesticides.

The relationship between pesticides and the environment cannot be considered in a vacuum. The two are elements of a complex system in which developments at one point inevitably produce results at another. Those results may be unexpected and their effects may be global. That this is so is clear from a now-classical example — the unanticipated complexities relating to DDT: overuse leading to global distribution in the environment; spectacular gains in public health (and population) by its use, and equally spectacular resurgences in malaria when use of the compound was discontinued (as in Ceylon); declines in bird populations, both real and suspected; unprecedented control of pests like the boll weevil in cotton and the unmistakable signs of growing insect resistance to the insecticide; human deaths in this country from the highly toxic materials that replaced relatively safe DDT when its use was restricted, compared to unknown results of previously increasing levels of DDT in human milk and fat. In this light, it is evident that pesticides in the environment must be approached in relation to a number of benefit-risk considerations. Among these are:

- Maintenance of the capacity to discover and develop safer, more effective pesticides with reduced environmental impact while maintaining a vigorous research effort in all other aspects of pest control including integrated control and its various elements, plant breeding, and minimizing the need for and optimizing the use of chemicals.
- Recognition and evaluation of the importance of monoculture, both in insuring world food supply and in simultaneously encouraging populations and depredations of pests.
- Recognition of the function of pesticides in public health, both in controlling disease vectors and in possibly harmful exposure to residues.
- Maintenance of a strong, global monitoring program on pesticide residues in all phases of the environment, including biota, for those substances *actually known* to be harmful.

- Fostering vigorous research on the environmental impact of pesticides, including the toxicology of low-level exposure of man and animals, analytical methods, and fate in the environment, with a view to developing a better predictive capacity for the future.

LITERATURE CITED

1. "Senate Panel Probes Kepone Disaster," *Chem. Eng. News*, **54**, 6, 17 (1976).
2. "Environmental Quality — 1976," Seventh Annual Report of the Council on Environmental Quality, Washington, D.C., 1976.
3. "Pest Control: An Assessment of Present and Alternative Technologies," Vol. 1 ("Contemporary Pest Control Practices and Prospects"), National Academy of Sciences — National Research Council, Washington, D.C., 1975.
4. "Pesticide Chemistry in the 20th Century," Plimmer, J.R., ed., *ACS Symp. Ser.*, **37** (1977).
5. Fowler, D.L., Mahan, J.N., "The Pesticide Review — 1972," Agricultural Stabilization and Conservation Service, U.S. Department of Agriculture, Washington, D.C., 1973.
6. Mrak, E.M. (chairman), "Report of the Secretary's Commission on Pesticides and Their Relationship to Environmental Health," U.S. Department of Health, Education, and Welfare, Washington, D.C., 1969.
7. "Environmental Quality — 1977," Eighth Annual Report of the Council on Environmental Quality, Washington, D.C., 1977, p. 9.
8. Parke, D.V., "The Biochemistry of Foreign Compounds," Pergamon Press, Oxford, England, 1968, 269 pp.
9. Parka, S.J., Worth, H.M., "Effects of Trifluralin on Fish," *Proc. Southern Weed Conf.*, **18**, 469 (1965).
10. Crosby, D.G., "The Fate of Pesticides in the Environment," *Ann. Rev. Plant Physiol.*, **24**, 467 (1973).
11. Martin, H., "Pesticide Manual," 3rd edition, British Crop Protection Council, Worcester, England, 1972, 535 pp.
12. Stewart, N.E., Millemann, R.E., Breese, W.P., "Acute Toxicity of the Insecticide, Sevin, and Its Hydrolytic Product, 1-Naphthol, to Some Marine Organisms," *Trans. Am. Fish. Soc.*, **96**, 25 (1967).
13. Metcalf, R.L., Fukuto, T.R., March, R.B., "Plant Metabolism of Dithio-systox and Thimet," *J. Econ. Entomol.*, **50**, 338 (1957).
14. Haque, R., Freed, V.H., "Behavior of Pesticides in the Environment: Environmental Chemodynamics," *Residue Rev.*, **52**, 89 (1974).
15. "Azodrin-Wildlife Investigations in California," FWIR Pesticides Investigation Project, California Department of Fish and Game, December 1967.
16. Whitten, J.L., "That We May Live," Van Nostrand, Princeton, N.J., 1966.
17. Gunther, F.A., "Instrumentation in Pesticide Residue Determinations," in "Advances in Pest Control Research," R.L. Metcalf, ed., Interscience, New York, N.Y., 1962, Vol. 5, p. 191.
18. Williams, I.H., "Carbamate Insecticide Residues in Plant Material: Determination by Gas Chromatography," *Residue Rev.*, **38**, 1 (1971).
19. Singmaster, J.A., Crosby, D.G., "Phthalate Esters as Interferences in Chlorinated Hydrocarbon Analysis," *Bull. Environ. Contam. Toxicol.*, **16**, 291 (1976).
20. Reynolds, L.M., "Pesticide Residue Analysis in the Presence of Polychlorobiphenyls (PCB's)," *Residue Rev.*, **34**, 27 (1971).

21. Coon, F.B., Christensen, R., Derse, P.H., "Electron Capture Gas Chromatographic Analysis on Selected Samples of Authentic pre-DDT Origin," Abstracts, 152nd National Meeting, American Chemical Society, New York, N.Y., 1966, A91.

22. Duggan, R.E., Corneliussen, P.E., "Dietary Intake of Pesticide Chemicals in the United States (III), June 1968–April 1970," *Pestic. Monit. J.*, **5,** 331 (1972).

23. "Environmental Pollution by Pesticides," Edwards, C.A., ed., Plenum Press, London, England, 1973, 542 pp.

24. Edwards, C.A., "Persistent Pesticides in the Environment," Chemical Ru. ɔer Co., Cleveland, Ohio, 1970, 71 pp.

25. Edwards, C.A., "Pesticide Residues in Soil and Water," in "Environmental Pollution by Pesticides," Edwards, C.A., ed., Plenum Press, London, England, 1973, p. 409.

26. Kearney, P.C., et al., "Decontamination of Pesticides in Soils," *Residue Rev.*, **29,** 137 (1969).

27. Stevens, L.J., Collier, C.W., Woodham, D.W., "Pesticides in Soil," *Pestic. Monit. J.*, **4,** 145 (1970).

28. Wiersma, G.B., Sand, P.F., Schutzmann, R.L., "National Soils Monitoring Program — Six States, 1967," *Pestic. Monit. J.*, **5,** 223 (1971).

29. Wiersma, G.B., Tai, H., Sand, P.F., "Pesticide Residue Levels in Soils FY 1969, National Soils Monitoring Program," *Pestic. Monit. J.*, **6,** 194 (1972).

30. Wheatley, G.A., "Pesticides in the Atmosphere," in "Environmental Pollution by Pesticides," C.A. Edwards, ed., Plenum Press, London, England, 1973, p. 365.

31. Seiber, J.N., et al., "Determination of Pesticides and Their Transformation Products in Air," in "Environmental Dynamics of Pesticides," R. Haque, V.H. Freed, eds., Plenum Publishing Corp., New York, N.Y., 1975, p. 17.

32. Goldberg, E.D., et al., "Chlorinated Hydrocarbons in the Marine Environment," National Academy of Sciences, Washington, D.C., 1971, 42 pp.

33. Akesson, N.B., Yates, W.E., "Problems Relating to Application of Agricultural Chemicals and Resulting Drift Residues," *Ann. Rev. Entomol.*, **9,** 285 (1964).

34. Cohen, J.M., Pinkerton, C., "Widespread Translocation of Pesticides by Air Transport and Rain-Out," *Adv. Chem. Ser.*, **60,** 163 (1966).

35. Stanley, C.W., et al., "Measurement of Atmospheric Levels of Pesticides," *Environ. Sci. Technol.*, **5,** 430 (1971).

36. Caro, J.H., Taylor, A.W., "Pathways of Loss of Dieldrin from Soils Under Field Conditions," *J. Agric. Food Chem.*, **19,** 379 (1971).

37. Jegier, Z., "Pesticide Residues in the Atmosphere," *Ann. N.Y. Acad. Sci.*, **160,** 143 (1969).

38. Middleton, J.T., "The Presence, Persistence, and Removal of Pesticides in Air," in "Research in Pesticides," Chichester, C.O., ed., Academic Press, New York, N.Y., 1965, p. 191.

39. Brown, E., Nishioka, Y.A., "Pesticides in Water," *Pestic. Monit. J.*, **1,** 38 (1967).

40. Lichtenberg, J.J., et al., "Pesticides in Surface Waters of the United States — A 5-Year Summary, 1964–68," *Pestic. Monit. J.*, **4,** 71 (1970).

41. Butler, P.A., "Organochlorine Residues in Estuarine Mollusks, 1965–72 — National Pesticide Monitoring Program," *Pestic. Monit. J.*, **2,** 238 (1973).

42. Cox, J.L., "Uptake, Assimilation, and Loss of DDT Residues by Euphausia pacifica, a Euphausiid Shrimp," *Fish. Bull.*, **69,** 627 (1971).

43. Jonas, R.B., Pfaender, F.K., "Chlorinated Hydrocarbon Pesticides in Western North Atlantic Ocean," *Environ. Sci. Technol.*, **10,** 770 (1976).

44. Risebrough, R.W., et al., "Pesticides: Transatlantic Movements in the Northeast Trades," *Science*, **159,** 1233 (1968).

45. Nash, R.G., Beall, M.L., "Chlorinated Hydrocarbon Insecticides: Root Uptake Versus Vapor Contamination of Soybean Foliage," *Science,* **168,** 1109 (1970).

46. Lichtenstein, E.P., Schulz, K.R., "Residues of Aldrin and Heptachlor in Soils and Their Translocation into Various Crops," *J. Agric. Food Chem.,* **13,** 57 (1965).

47. Spencer, D.A., "The National Pesticide Monitoring Program," National Agricultural Chemicals Association, Washington, D.C., 1974.

48. Manske, D.D., Johnson, R.D., "Residues in Food and Feed," *Pestic. Monit. J.,* **10,** 4, 134 (1977); *ibid.* **9,** 4, 157 (1976).

49. Dustman, E.H., Stickel, L.F., "The Occurrence and Significance of Pesticide Residues in Wild Animals," *Ann. N.Y. Acad. Sci.,* **160,** 162 (1969).

50. Rudd, R.L., "Pesticides and the Living Landscape," Faber and Faber, London, England, 1964, 320 pp.

51. Pimentel, D., "Ecological Effects of Pesticides on Non-Target Species," Executive Office of the President, Office of Science and Technology, Washington, D.C., 1971, 220 pp.

52. Stickel, L.F., "Pesticide Residues in Birds and Mammals," in "Environmental Pollution by Pesticides," Edwards, C.A., ed., Plenum Press, London, England, 1973, p. 254.

53. Keith, J.O., Hunt, E.G., "Levels of Insecticide Residues in Fish and Wildlife in California," Trans. 31st N. Amer. Wildlife and Natural Resources Conference, p. 150, 1966.

54. George, J.L., Frear, D.E.H., "Pesticides in the Antarctic," *J. Appl. Ecol.,* **3** (Suppl.), 155 (1966).

55. Sladen, W.J.L., Menzie, C.M., Reichel, W.L., "DDT Residues in Adelie Penguins and a Crabeater Seal from Antarctica," *Nature,* **210,** 670 (1966).

56. Peterle, T.J., "DDT in Antarctic Snow," *Nature,* **224,** 620 (1969).

57. White, D.H., Heath, R.G., "Nationwide Residues of Organochlorines in Wings of Adult Mallards and Black Ducks, 1972–73," *Pestic. Monit. J.,* **9,** 196 (1976).

58. White, D.H., "Nationwide Residues of Organochlorines in Starlings, 1974," *Pestic. Monit. J.,* **10,** 10 (1976).

59. Crosby, D.G., "The Toxicant-Wildlife Complex," *Pure Appl. Chem.,* **42,** 233 (1975).

60. Barker, R.J., "Notes on Some Ecological Effects of DDT Sprayed on Elms," *J. Wild. Manage.,* **22,** 269 (1958).

61. Hunt, E.G., Bischoff, A.I., "Inimical Effects on Wildlife of Periodic DDD Applications to Clear Lake," *Calif. Fish Game,* **46,** 91 (1960).

62. Burdick, G.E., et al., "Accumulation of DDT in Lake Trout and the Effect on Reproduction," *Trans. Am. Fish. Soc.,* **93,** 127 (1964).

63. Ratcliffe, D.A., "Decrease in Eggshell Weight in Certain Birds of Prey," *Nature,* **215,** 208 (1967).

64. Risebrough, R.W., et al., "Polychlorinated Biphenyls in the Global Ecosystem," *Nature,* **220,** 1098 (1968).

65. Peakall, D.B., "Pesticide-Induced Enzyme Breakdown of Steroids in Birds," *Nature,* **216,** 505 (1967).

66. Cade, T.J., et al., "DDE Residues and Eggshell Changes in Alaskan Falcons and Hawks," *Science,* **172,** 955 (1971).

67. Hickey, J.J., Anderson, D.W., "Chlorinated Hydrocarbons and Eggshell Changes in Raptorial and Fish-Eating Birds," *Science,* **162,** 271 (1968).

68. Hamelink, J.L., Waybrant, R.C., Ball, R.C., "A Proposal: Exchange Equilibria Control the Degree Chlorinated Hydrocarbons Are Biologically Magnified in Lentic Environments," *Trans. Am. Fish. Soc.,* **100,** 207 (1971).

69. Robinson, J., et al., "Organochlorine Residues in Marine Organisms," *Nature,* **214,** 1307 (1967).

70. Peterle, T.J., "The Use of Isotopes to Study Pesticide Translocation in Natural Environments," *J. Appl. Ecol.*, **3** (suppl.), 181 (1966).

71. Metcalf, R.L., Sangha, G.K., Kapoor I.P., "Model Ecosystem for the Evaluation of Pesticide Biodegradability and Ecological Magnification," *Environ. Sci. Technol.*, **5**, 709 (1971).

72. Bridges, W.R., Kallman, B.J., Andrews, A.K., "Persistence of DDT and Its Metabolites in a Farm Pond," *Trans. Am. Fish. Soc.*, **92**, 421 (1963).

73. Miskus, R.P., Blair, D.P., Casida, J.E., "Conversion of DDT to DDD by Bovine Rumen Fluid, Lake Water, and Reduced Porphyrins," *J. Agric. Food Chem.*, **13**, 481 (1965).

74. Vinson, S.B., Boyd, C.E., Ferguson, D.E., "Resistance to DDT in the Mosquito Fish, *Gambusia affinis*," *Science*, **139**, 217 (1963).

75. Rosato, P., Ferguson, D.H., "The Toxicity of Endrin-Resistant Mosquito Fish to Eleven Species of Vertebrates," *BioScience*, **18**, 783 (1968).

76. Blus, L.J., et al., "Residues of Organochlorines and Heavy Metals in Tissues and Eggs of Brown Pelicans, 1969–73," *Pestic. Monit. J.*, **11**, 40 (1977).

77. Hayes, W.J., Jr., "Toxicology of Pesticides," Williams and Wilkins Co., Baltimore, Md., 1975.

78. Hayes, W.J., Jr., "Occurrences of Poisoning by Pesticides," *Arch. Environ. Health*, **9**, 621 (1964).

79. "Cases of Ingestion During 1971 Reported to the National Clearinghouse for Poison Control Centers," *National Clearinghouse for Poison Control Centers Bulletin*, September-October 1972.

80. Milby, T.H., Ottoboni, F., Mitchell, H.W., "Parathion Residue Poisoning Among Orchard Workers," *J. Am. Med. Assoc.*, **189**, 351 (1964).

81. Hayes, W.J., Jr., "Toxicity of Pesticides to Man: Risks of Present Levels," *Proc. Royal Soc. B.*, **167**, 101 (1967).

82. Cam, C., Nigogosyan, G., "Acquired Toxic Porphyria Cutanea Tarda Due to Hexachlorobenzene," *J. Am. Med. Assoc.*, **183**, 88 (1963).

83. Lemmon, A.B., Bureau of Chemistry, Annual Report for the Calendar Year 1955, Annual Report Department of Agriculture of California, **45**, 128 (1956).

84. McGee, L.C., Reed, H.L., Fleming, J.P., "Accidental Poisoning by Toxaphene," *J. Am. Med. Assoc.*, **149**, 1124 (1952).

84a. Marquez Mayaudon, E., et al, "Problemas de Contaminacion de Alimentos con Pesticidas. Casa Tijuana (1967)," *Salud Public Mex.*, **10**, 293 (1968).

84b. Curley, A., et al, "Organic Mercury Identified as the Cause of Poisoning in Humans and Hogs," *Science*, **172**, 65 (1971).

85. Hoffman, W.S., Fishbein, W.I., Andelman, M.B., "The Pesticide Content of Human Fat Tissue," *Arch. Environ. Health*, **9**, 387 (1964).

86. Durham, W.F., Armstrong, J.F., Quinby, G.E., "DDT and DDE Content of Complete Prepared Meals," *Arch. Environ. Health*, **11**, 641 (1965).

87. Hayes, W.J., Jr., Durham, W.F., Cueto, C., Jr., "The Effect of Known Repeated Doses of Chlorophenothane (DDT) in Man," *J. Am. Med. Assoc.*, **162**, 890 (1956).

88. "National Study to Determine Levels of Chlorinated Hydrocarbon Insecticides in Human Milk, 1975–76," Environmental Protection Agency, Washington, D.C., 1977.

89. Walker, K.C., Goette, M.B., Batchelor, G.S., "Dichlorodiphenyltrichloroethane and Dichlorodiphenyldichloroethylene Content of Prepared Meals," *J. Agric. Food Chem.*, **2**, 1034 (1954).

90. Clayson, D.B., "Chemical Carcinogenesis," Little, Brown & Co., Boston, Mass., 1962.

91. Longgood, W., "The Poisons in Your Food," Simon and Schuster, New York, N.Y., 1960.

92. Innes, J.R.M., et al., "Bioassay of Pesticides and Industrial Chemicals for Tumorigenicity in Mice: A Preliminary Note," *J. Natl. Cancer Inst.*, **42,** 1101 (1969).

93. Gargus, J.L., Paynter, O.E., Reese, W.H., Jr., "Utilization of Newborne Mice in the Bioassay of Chemical Carcinogens," *Toxicol. Appl. Pharmacol.*, **15,** 552 (1969).

94. Laws, E.R., Jr., "Evidence of Antitumorigenic Effects of DDT," *Arch. Environ. Health*, **23,** 181 (1971).

95. Karch, N.J., "Explicit Criteria and Principles for Identifying Carcinogens: A Focus of the Controversy at the Environmental Protection Agency," in "Decision Making in the Environmental Protection Agency," Vol. IIa, Case Studies, National Academy of Sciences — National Research Council, Washington, D.C., 1977.

96. Levin, D.L., "Cancer Rates and Risks," Sixth Annual Report, Department of Health, Education, and Welfare Publications, U.S. Government Printing Office, Washington, D.C., 1974.

97. Barnes, J.M., "Carcinogenic Hazard from Pesticide Residues," *Residue Rev.*, **13,** 69 (1966).

98. "Natural Food Toxicants," Crosby, D.G., ed., reprinted from *J. Agric. Food Chem.*, **17,** 413–538 (May-June 1969), American Chemical Society, Washington, D.C., 1969.

99. Schwetz, B.A., et al., "Toxicology of Chlorinated Dibenzo-*p*-dioxins," *Adv. Chem. Ser.*, **120,** 55 (1973).

100. Baughman, R., Meselson, M., "An Analytical Method for Detecting TCDD (Dioxin): Levels of TCDD in Samples from Vietnam," *Environ. Health Perspect.*, **5,** 27 (1973).

101. Long, W.H., Concienne, E.J., "Critical Period for Controlling the Sugarcane Borer in Sugarcane in Louisiana," *J. Econ. Entomol.*, **57,** 350 (1964).

102. Grzenda, A.R., et al., "DDT Residues in Mountain Stream Water as Influenced by Treatment Practices," *J. Econ. Entomol.*, **57,** 615 (1964).

103. Casebeer, R.L., "Monitoring the 1964 Spruce Budworm Aerial Spray Project," Report of the Forest Service, U.S. Department of Agriculture, 1964.

104. Lofgren, C.S., "Large-Scale Tests to Evaluate Multiple Low Dosages of Heptachlor," Report of the Plant Pest Control Division, Agricultural Research Service, U.S. Department of Agriculture, Gulfport, Miss., 1959.

105. Burkhardt, G., "Observations on Study of Commercial Dusting Equipment," Department of Agricultural Engineering, University of Maryland, College Park, Md., 1946, unpublished.

106. Brittain, R.W., Carleton, W.M., "How Surfaces Affect Pesticidal Dust Deposition," *Agric. Eng.*, **38,** 22 (1957).

107. Hedin, E., May, D.S., Dunstan, G.H., "Distribution of Insecticides Sprayed by Airplane on an Irrigated Corn Plot," *Adv. Chem. Ser.*, **60,** 132 (1966).

108. Barthel, W.P., Annual Report, Plant Pest Control Division, Agricultural Research Service, U.S. Department of Agriculture, Gulfport, Miss., 1959–60.

109. Aller, H.E., Dewey, J.E., "Adjuvants Increasing the Residual Activity of Phosdrin," *J. Econ. Entomol.*, **54,** 508 (1961).

110. Matteson, J.W., Taft, H.M., "The Effect of Various Adjuvants on the Systemic Insecticidal Activity of Phorate and Zectran," *J. Econ. Entomol.*, **57,** 325 (1964).

111. "Interim Federal Specification. Insecticide, Heptachlor, Granulated," USDA AGR-ARS 0-I-00528, March 7, 1961, U.S. Department of Agriculture, Washington, D.C.

112. Messenger, K., "Low Volume Aerial Spraying Will Be Boon to Applicators," *Agric. Chem.*, **18** (Dec.), 63 (1963).

113. Burges, E.D., "Control of the Boll Weevil with Technical Malathion Applied by Aircraft," *J. Econ. Entomol.*, **58,** 414 (1965).

114. Ditman, L.P., et al., "Insecticidal Aerosols for Pea-Aphid Control," *J. Econ. Entomol.*, **39,** 199 (1946).

115. "Theory and Practice of Biological Control," Huffaker, C.B., Messenger, P.S., eds., Academic Press, New York, N.Y., 1976.

116. DeBach, P., "The Use of Imported Natural Enemies in Insect Pest Management Ecology," in "Proceedings: Tall Timbers Conference on Ecological Animal Control by Habitat Management," No. 3, 211–33 (1972).

117. Angus, T.A., "Bacillus Thuringiensis as a Microbial Insecticide," in "Naturally Occurring Insecticides," M. Jacobson and D.G. Crosby, eds., Marcel Dekker, New York, N.Y., 1971, p. 463.

118. "Pest Control: Biological, Physical, and Selected Chemical Methods," Kilgore, W.W., Doutt, R.L., eds., Academic Press, New York, N.Y., 1967, 477 pp.

119. Hall, D.G., "Use of Insecticides in the United States," *Bull. Entomol. Soc. Am.*, **8**, 90 (1962).

120. Beroza, M., Green, N. Gertler, S.I., "New Attractants for the Mediterranean Fruit Fly," *J. Agric. Food Chem.*, **9**, 361 (1961).

121. "Integrated Pest Management," Council on Environmental Quality, Washington, D.C., 1972, 41 pp.

122. Watson, T.F., et al., "Influence of Plowdown Dates and Cultural Practices on Spring Moth Emergence of the Pink Bollworm," *J. Econ. Entomol.*, **67**, 207 (1974).

123. Van den Bosch, R., Stern, V.M., "The Effects of Harvesting Practices on Insect Populations in Alfalfa," in "Proceedings: Tall Timbers Conference on Ecological Animal Control by Habitat Management," No. 1, 47–54 (1969).

124. "Host Plant Resistance to Pests," Hedin, P.A., ed., *ACS Symp. Ser.*, **62** (1977).

125. Good, J.M., "Management of Plant Parasitic Nematode Populations," in "Proceedings: Tall Timbers Conference on Ecological Animal Control by Habitat Management," No. 4, (1972).

126. Johnson, H.W., "Development of Crop Resistance to Diseases and Nematodes," *J. Environ. Qual.*, **1**, 23 (1972).

127. "Natural Pest Control Agents," Crosby D.G., ed., *Adv. Chem. Ser.*, **53** (1966).

128. "Pheromones," Birch, M., ed., North Holland Publishing Co., Amsterdam, Netherlands, 1974.

129. "Pest Management with Insect Sex Attractants," Beroza, M., ed., *ACS Symp. Ser.*, **23** (1976).

130. Marx, J.L., "Insect Control (I): Use of Pheromones," *Science*, **181**, 736 (1973).

131. Marx, J.L., "Applied Ecology: Showing the Way to Better Insect Control," *Science*, **195**, 860 (1977).

132. Lawless, E.W., von Rumker, R., Ferguson, T.L., "The Pollution Potential in Pesticide Manufacturing," Technical Studies Report TS-00-72-04, Office of Water Programs, Environmental Protection Agency, Washington, D.C., 1972, 249 p.

133. *Fed. Regist.*, **39**, 15236, May 1, 1974.

134. Wolfe, H.R., et al., "Health Hazards of Discarded Pesticide Containers," *Arch. Environ. Health*, **3**, 531 (1961).

135. Kennedy, M.V., Stojanovic, B.J., Shuman, F.L., Jr., "Chemical and Thermal Aspects of Pesticide Disposal," *J. Environ. Qual.*, **1**, 63 (1972).

136. Plimmer, J.R., et al., "Amitrol Decomposition by Free Radical Generating Systems and by Soils," *J. Agric. Food Chem.*, **15**, 996 (1967).

137. Crosby, D.G., "Environmental Photooxidation of Pesticides," in "The Degradation of Synthetic Organic Molecules in the Biosphere," National Academy of Sciences, Washington, D.C., 1972, p. 260.

138. Lofgren, C.S., Stringer, C.E., Bartlett, F.J., "Imported Fire Ant Toxic Bait Studies: GC-1283, A Promising Toxicant," *J. Econ. Entomol.*, **55**, 405 (1962).

139. Klingman, G.C., Ashton, F.M., "Weed Science: Principles and Practice," John Wiley & Sons, New York, N.Y., 1975.

140. Wood, J.M., "Environmental Pollution by Mercury," *Adv. Environ. Sci. Technol.*, **2,** 39 (1971).

141. Ames, B.N., et al., "Carcinogens are Mutagens: A Simple Test System Combining Liver Homogenates for Activation and Bacteria for Detection," *Proc. Nat. Acad. Sci. USA*, **70,** 2281 (1973).

142. Bridges, B.A., "Short Term Screening Tests for Carcinogens," *Nature*, **261,** 195 (1976).

143. Goring, C.A.I., "Agricultural Chemicals in the Environment: A Quantitative Viewpoint," in "Organic Chemicals in the Soil Environment," C.A.I. Goring, J.W. Hansker, eds., Vol. 2, p. 793, Marcel Dekker, New York, N.Y., 1972.

144. Konishi, K., "Studies on Organic Insecticides. Part XII. Synthesis of Nereiotoxin and Related Compounds V.," *Agric. Biol. Chem.*, **34,** 935 (1970).

145. "Pyrethrum, the Natural Insecticide," Casida, J.E., ed., Academic Press, New York, N.Y., 1973, 329 p.

146. Henrick, C.A., Staal, G.B., Siddall, J.B., "Alkyl 3, 7, 11-trimethyl-2, 4-decadienoates, New Class of Potent Insect Growth Regulators with Juvenile Hormone Activity," *J. Agric. Food Chem.*, **21,** 354 (1973).

Chapter 8

Radiation in the Environment

8 / Radiation in the Environment

INTRODUCTION

Humans and other organisms have lived from the beginning in an environment containing low levels of radiation from naturally occurring radioactive sources. The fact went undetected until less than a century ago and for some years interested no one but a few specialists. The discovery of nuclear fission and its use in military weapons and in the reactors of a burgeoning electrical power industry have changed this picture completely. The problem of realizing the great benefits to be gained from radioactivity while at the same time controlling its hazard has become the concern of everyone.

"Radiation" in this chapter refers only to emissions from radioactive materials, man-made or natural, and to x-rays from man-made devices. These latter are not, strictly speaking, an environmental problem. For most people, however, they are one of the major sources of exposure to radiation. [The hazard related to theft and sabotage at nuclear facilities is beyond our scope here, but has led the Nuclear Regulatory Commission to strengthen the measures required for physical protection of nuclear plants and materials (1).]

In the past 30 years a literature of amazing proportions has sprung up on radioactivity — its nature, uses, dangers, and control. Much of this material is intelligible only to the specialist, but more general information is available as well (2–16). A discussion of the physical nature of radioactivity is beyond the scope of this chapter, but the topic is covered in most texts in general physics or physical chemistry and in a number of specialized books (17, 18). The nuclear field, including radiation, encompasses major segments of physics, chemistry, biology, metallurgy, and other scientific disciplines and substantial areas in engineering, economics, and politics. For this reason, no doubt, practitioners in the field have become particularly adept at coining specialized words and definitions. Some of these appear in simplified form in Table 1; more detailed and precise information is available from a number of sources (19–22).

Background

The standards of safety and measurement relating to radioactivity are a major concern of a number of groups whose recommendations are in a continuing state of evolution (23, 24). The most influential groups internationally are two committees established by the International Society of Radiology to consider the problems posed by the use of x-rays and radium in medicine. The first group, the International Commission on Radiation Units and Measurements (ICRU), was established in 1925. The U.S. National Bureau of Standards (NBS) published many of ICRU's earlier reports as handbooks, but now the commission issues its reports directly (19, 20). The second group, the International Commission on Radiological Protection (ICRP), was established in 1928 as the International X-Ray and Radiation Commission and was renamed and reorganized in 1950. It operates now through several subcommittees; the one quoted mainly here is Committee II, which deals with permissible doses for internal radiation. ICRP reports have appeared in various journals or as single publications (25–30). Both ICRU and ICRP formulate recommendations, and these eventually become the bases of legal regulations around the world.

Table 1

Glossary of special terms[a]

Actinide elements	Elements 89 through 103 in the Periodic Table. All are classed as metals, and all are radioactive.
Alpha particle (α)	The doubly-charged helium ion, $^{4}He^{++}$.
Beta particle (β)	An electron, e^-, not bound to an atom.
Curie (Ci)	The quantity of a radioactive nuclide undergoing 3.7×10^{10} disintegrations per second. It is approximately equal to the decay rate of 1 gram of radium.
Dose	The amount of radiation absorbed in a unit mass of material.
Electron volt (eV)	The energy necessary to raise one electron through a potential energy of one volt. The energies of the various types of radiation are usually expressed in terms of eV, keV or MeV.
Gamma rays (γ)	Similar to x-rays in being high-energy, short-wavelength, penetrating electromagnetic radiation, but originating in the atomic nucleus itself.
Isotope	One of two or more atomic forms of the same chemical element. Isotopes have the same atomic number, but different atomic weights; that is, their nuclei contain the same number of protons, but different numbers of neutrons. Thus, $^{12}_{6}C$, $^{13}_{6}C$, and $^{14}_{6}C$ are isotopes of carbon; the subscripts denote their common atomic numbers, and the superscripts denote their different mass numbers or approximate atomic weights. Isotopes of an element usually have very nearly the same chemical properties, but somewhat different physical properties.
Neutron (n)	An uncharged elementary particle with a mass slightly greater than the proton. Neutrons are found in the nucleus of every atom heavier than hydrogen, and their release during fission sustains the chain reaction in reactors.
Nuclide	Any atomic form of any element. A specific nuclide is distinguished by its atomic number, atomic mass, and energy state. Nuclide is a more general term than isotope. That is, each isotope of a given element is a nuclide, but it cannot properly be referred to as an isotope except in relation to that element. Thus ^{12}C is a nuclide and is also an isotope of carbon.

(Continued on page 382)

Table 1 (Continued)

Proton particle (p)	The singly-charged hydrogen ion, $^{1}H^{+}$.
Quality factor	For most purposes, this can be taken as equal to one for x- or γ-rays and for β particles; five for protons or slow neutrons; and 10 for α particles. See "rem" below.
Rad	A dose of one rad deposits 100 ergs of energy per gram of receiving material.
Rem	Again a measure of absorbed dose, but takes account of the fact that different types of radiation cause varying amounts of biological damage. The dose in rads times the quality factor and certain other modifying factors gives the dose in rems.
Roentgen (r)	That quantity of x- or γ-radiation such that the associated corpuscular emission per 0.001293 gram (1 cubic centimeter) of air produces, in air, ions carrying one electrostatic unit (esu) of charge of either sign. More simply, 1 r produces 1.61×10^{12} ion pairs per gram of air, representing the absorption of 84 ergs of energy.
Transuranium elements	Elements of higher atomic number than uranium, whose number is 92. These are the man-made elements; as of December 1977, those of atomic number 93 through 106 had been produced. All are classed as metals, and all are radioactive. Note that elements 93 through 103 are also actinide elements.
X-rays	Electromagnetic radiation arising from changes in the electron shells surrounding an atomic nucleus.

Numerical Abbreviations

Prefix	Abbreviation	Relative value
mega	M	one million or 10^6
kilo	k	one thousand or 10^3
—	—	unity or 1
milli	m	one thousandth or 10^{-3}
micro	μ	one millionth or 10^{-6}
nano	n	one billionth or 10^{-9}
pico	p	one trillionth or 10^{-12}

Thus, 1 mCi is 0.001 curie, 1 MeV is a million electron volts, etc.
a. For more detailed information, see citations 19–22.

In the United States, the National Council on Radiation Protection and Measurements (NCRP) was established by the National Bureau of Standards in 1929, originally as the Advisory Committee on X-Ray and Radium Protection. Its reports have been published in various ways, including NBS handbooks. With the growing awareness of the possible adverse effects of radiation on the human "genetic pool," the permissible exposure doses recommended by NCRP and ICRP have been set at conservative levels, particularly for the population at large. The limits recommended by NCRP in 1971 are given in Table 2 (31).

A number of federal agencies, but primarily the Nuclear Regulatory Commission (NRC), become involved in establishing and enforcing the legal standards necessary to insure that the exposure levels recommended by ICRP and NCRP are not exceeded (the tendency is to set the standards even more conservatively than do the international commissions).* The NRC rules are published in the *Code of Federal Regulations, Title 10 (Atomic Energy)*. Part 20 (32) sets standards for protection against radiation; Part 71 (33) deals with the packaging of radioactive materials for shipment; Part 50 (34) governs licensing of facilities, etc. These regulations are amended from time to time, and changes are published in the *Federal Register*.

In 1959 the Federal Radiation Council, composed of cabinet-level officials, was formed to advise the President "with respect to radiation matters." The council's functions were transferred to the Environmental Protection Agency (EPA) in 1970 when the agency was created. The Division of Environmental Radiation, Bureau of Radiological Health, Public Health Service, Department of Health, Education, and Welfare (HEW), was also transferred to the Office of Radiation Programs of EPA. The Public Health Service of HEW continues to review radiological health services throughout the nation and periodically publishes surveys of state and local programs (35). The Department of Transportation regulates movement of radioactive materials (36) and the National Academy of Sciences is asked frequently to form special study projects to consider particularly knotty problems (37–39).

This rather confusing system seems to work, although questions of responsibility arise from time to time. A decision in 1974, for example, makes EPA responsible for establishing standards for radioactivity in the environment as a whole, but leaves NRC in charge of setting concentration limits for activity in effluents from the reactors and plants of the nuclear fuel cycle (40). This decision was challenged by a citizens' group, but upheld by the Supreme Court in 1976 (41).

NUCLIDES OF SPECIAL CONCERN

The nuclear fission process in a reactor produces nuclides of atomic weights ranging roughly from 72 to 161 as well as tritium, the hydrogen isotope of mass three. Practically all of these nuclides are radioactive, and most of them act as parents of chains of active isotopes or daughters that decay through at least several members before becoming stable. In addition, actinide elements such as plutonium are formed by successive neutron additions to the uranium-238 (^{238}U) of the fuel. [When the build-up process on a nucleus forms a beta-emitting nuclide, the decay product is a new element one step higher in the Periodic Table. Thus, ^{238}U captures a neutron to form ^{239}U, which decays to

*This function previously was discharged by the Atomic Energy Commission (AEC). As of Jan 19, 1975, the AEC was eliminated and its functions largely absorbed by the newly established Energy Research and Development Administration (ERDA) and Nuclear Regulatory Commission. References to AEC in the text and bibliography have been left unchanged where that agency was the one involved and where the documents cited were produced under its auspices and are so identified in the originals. As of Oct. 1, 1977, ERDA in turn was largely absorbed into the newly created Department of Energy (DOE).

neptunium-239 (^{239}Np). The latter isotope is also a beta-emitter and decays to plutonium-239 (^{239}Pu). In alpha emission, the mass changes, and the atomic number drops by two, to form a lighter element.] Other radioactive species are produced in the reactor coolant and hardware by neutron capture. As a result of these various processes, several hundred individual radioactive nuclides are present in an operating reactor at any given time.

None of these radioactive species is innocuous, but some are much less a control problem than others. Some nuclides have such short half-lives that they disappear almost instantaneously. Others are formed in very low yield. Still others emit weak beta or gamma radiations that penetrate only very short distances and, accordingly, have relatively low tissue-damage potential. These groups of nuclides can be considered to pose little hazard. A sizable number of others, however, do warrant special concern because of their long half-lives, their biochemistry in living organisms, the difficulty of containing them adequately, or, particularly for the alpha-emitters, the high level of tissue damage

Table 2

NCRP dose-limitation recommendations

Occupational exposure limits	
Whole body (prospective)	5 rems in any one year
Whole body (accumulation to age N years)	(N – 18) X 5 rems
Skin	15 rems in any one year
Hands	75 rems in any one year (25/quarter)
Forearms	30 rems in any one year (10/quarter)
Other organs, tissues, and organ systems	15 rems in any one year (5/quarter)
Pregnant women	0.5 rem in gestation period
Dose limits for the public, or occasionally exposed individuals	
Individual or occasional	0.5 rem in any one year
Students	0.1 rem in any one year
Population dose limits	
Genetic	0.17 rem average per year
Somatic	0.17 rem average per year
Emergency dose limits — lifesaving	
Individual (older than 45 years if possible)	100 rems
Hands and forearms	200 rems, additional (300 rems total)
Emergency dose limits — less urgent	
Individual	25 rems
Hands and forearms	100 rems, total
Family of patient undergoing radiotherapy	
Individual (under 45 years)	0.5 rem in any one year
Individual (over 45 years)	5 rems in any one year

Source: "Basic Radiation Protection Criteria," National Council on Radiation Protection and Measurements, Report No. 39, U.S. Government Printing Office, Washington, D.C., 1971.

they cause if taken into the body. We will include radium-226 (^{226}Ra) among these nuclides despite the fact that it occurs naturally and has nothing to do directly with reactors. The reasons are that ^{226}Ra and its daughters are particularly important in establishing natural background levels of radiation and that the nuclide itself is still used in medicine and industry.

Physical and biological data for selected fission-produced nuclides of special concern appear in Table 3. The "fission yield" columns indicate the number of atoms of the nuclide produced per 100 atoms of ^{235}U or ^{239}Pu undergoing fission in a thermal reactor, where the fission-causing neutrons are slowed down by collision with the light nuclei in the moderator present. The values would differ somewhat in a fast reactor, where there is no moderator. The "biological half-life" in Table 3 is the time required by the body's normal biochemical excretory processes to eliminate half of the material from the critical organ. [The times given, in effect, assume the nuclides to be stable. For shorter-lived species, the fact that they are decaying by physical processes must also be considered. Thus the effective half-life of strontium-89 (^{89}Sr) in bone is actually 50.4 days and that of iodine-131 (^{131}I) in thyroid is 7.6 days.] The maximum permissible concentration (MPC) is the concentration that should not be exceeded in the air or water of the environment of the general populace on a 168 hour/week basis if the maximum permissible body burden (MPBB) is to remain below its stated value.

The picture given by Table 3 is highly simplified. Each of several organs may have separate MPBB's and corresponding MPC's. Also, two sets of numbers are recommended for the MPC's in most cases, depending on whether the chemical form of the nuclide in the environment makes it soluble or insoluble in the body. The values for krypton-85 (^{85}Kr) are on the basis of "submersion" of the body in an infinite cloud of contaminated air. Table 3 contains several other complications as well. The MPBB values are those set by ICRP for occupational exposure. The MPC numbers are those used by the NRC for control purposes; generally they are only 1/10th of the corresponding ICRP values and thus are 10 times more stringent. As a further conservative measure, NRC applies the 10% factor to the lowest value recommended by ICRP, so that while MPBB-MPC numbers in Table 3 usually apply to the same critical organ, this may not always be the case.

Tritium

Tritium (^{3}H) is produced naturally as a result of cosmic ray bombardment of oxygen and nitrogen in the upper atmosphere and also intrudes directly from outer space. It is produced artificially in nuclear detonations and the nuclear fuel cycle. The natural level in the environment is of the order of 10^{-17} to 10^{-18} g per gram of normal hydrogen (^{1}H). Tritium reacts chemically very much like ordinary hydrogen, excepting isotopic effects that are caused by the three-to-one difference in the two masses and are easily discernible (42). Tritium takes part in an isotopic exchange reaction in which it replaces ordinary hydrogen in water up to an equilibrium value; as a result, tritium in the environment is combined almost entirely with oxygen in water (HTO or T_2O). Because water is so prevalent and so important in life processes, its isotopic exchange with tritium holds special interest. The exchange is seen most readily when an organism is exposed initially to tritium; it becomes less apparent as the concentration of tritium in the various body compartments approaches equilibrium (43).

Since tritium generally occurs in the form of water or water vapor, it can be ingested in water (alone or in foods) or absorbed through the skin. The nonwater

Table 3

Nuclear and health physics data for selected fission products

	Nuclear properties				ICRP Recommendations (25)			Maximum permissible conc. (192)[c]	
			Fission yield[a]						
Nuclide	Half-life (187, 188)	Major radiations (187)	^{235}U	^{239}Pu	Critical organ	Biological half-life	MPBB[b] μCi	Air $\mu Ci/cm^3$	Water $\mu Ci/cm^3$
^{3}H	12.26 years	0.0186 MeV β No γ	9×10^{-5}	—	Total body	12 days	2000	2×10^{-7}	3×10^{-3}
^{85}Kr	10.76 years	0.67 MeV β Low γ	1.30	0.56	Skin	—	—	3×10^{-7}	—
^{89}Sr	52 days	1.463 MeV β Low γ	4.76	1.73	Bone	50 years	4	3×10^{-10}	3×10^{-6}
^{90}Sr	28.1 years	0.546 MeV β No γ	5.83	2.24	Bone	50 years	2	3×10^{-11}	3×10^{-7}
^{129}I	1.7×10^{7} Years	0.15 MeV β Low γ	1.0	1.8	Thyroid	138 days	3	2×10^{-11}	6×10^{-8}
^{131}I	8.07 days	0.606 MeV β High γ	2.93	3.69	Thyroid	138 days	0.7	1×10^{-10}	3×10^{-7}
^{137}Cs	30.23 years	0.514 MeV β High γ	6.18	6.56	Total body	70 days	30	4×10^{-7}	2×10^{-3}

a. ^{3}H yield from ^{235}U from refs. (189, 190). Other yields from ref. (191). The values are for fission in thermal reactors. (Units are atoms per 100 atoms of fissionable isotope fissioned.)
b. MPBB = maximum permissible body burden for persons working with radioactivity.
c. MPC values are for air or water consumed by members of the general population.

molecules in foods also may contain small amounts of tritium that has replaced ordinary hydrogen by isotopic exchange. Tritium thus becomes distributed throughout the body to the same degree as water — essentially uniformly. However, because of its relatively short biological half-life, its low average disintegration energy, and its inability to concentrate in the body, tritium is one of the least hazardous of the nuclides. It has the largest maximum permissible body burden for an occupationally exposed "standard man" who is receiving the maximum permissible dose of 5 rem per year (25). In man, the biological half-life of water in the body varies from about seven to 12 days, depending on water and electrolyte intake and hormonal influences. Long-term studies of tritium indicate that while in the body it is not all in the form of water, since the pattern of excretion is the sum of exponentials with half-times of approximately six, 23, and 345 days (44).

Tritium decays in a single step to helium and in the process emits a very low-energy beta ray (maximum energy 18.6 keV; average energy 4.7 keV) with a maximum range of about 6 μm in water; the average range is less than 1 μm in water. The beta emission causes ionization and excitation. Certain additional events, strictly localized to the site of the decay, are initiated primarily by the chemical transmutation of tritium to helium as the former decays by beta emission. There appears, in fact, to be some justification for the belief that the biological effects of tritium transmutation result mainly from the production of helium as opposed to the beta emission (45). Because of the short range of the low-energy beta particles, however, the transmutation effects of tritium are not easily recognized in complex molecular systems within biological structures.

Krypton-85

Krypton-85 (^{85}Kr) in the atmosphere is entirely man-made; the current atmospheric concentration of the nuclide is approximately 18 pCi/m^3 (46). Krypton gas is biochemically unreactive; it is not involved in metabolism and is not incorporated into biological molecules. It is slightly soluble in water, the partition coefficient (the concentration in the fluid divided by the concentration in air) being about 0.05 for water and 0.07 for blood (47, 48). The gas is much more soluble in fat or oil, where the partition coefficient is about 0.5 (49). Little is known of its solubility in tissues other than blood or fat; the usual practice has been to use the blood value for all tissues other than fat. The consensus (5r, 49, 50) is that the dose to the skin establishes the maximum permissible concentration in air.

Measurement of airborne ^{85}Kr at the low concentration now existing or anticipated in the environment is difficult, and no commercial detection system is available. No field monitoring instrument is capable of measuring present levels in the atmosphere. Several laboratory measurement techniques have been developed, however. All involve preliminary concentration of the krypton fraction of the air by cryogenic techniques. The two most common means of counting the ^{85}Kr in this separated fraction use either plastic-scintillation shavings contained in a gas-tight vial or a liquid scintillator in which the krypton is dissolved. In the first method, a 4-ml vial containing fine shavings of a plastic scintillator is filled with pure krypton gas at a known pressure. A scintillator counter then measures the beta pulse-height spectrum. The second technique relies on the high solubility of krypton in many of the commonly used liquid scintillators. The minimum detectable concentration is of the order of a few pCi/m^3 in the original air (51).

Strontium-89 and -90

Strontium-90 (^{90}Sr) and, to a lesser extent, strontium-89 have attracted much public attention because of the hazard they represent in fallout from atmospheric nuclear bomb tests, where the heavier isotope appears in 3.5% and the lighter in 2.6% of weapons fissions. The nuclides' nuclear properties appear in Table 3. Note that ^{90}Sr usually is in radioactive equilibrium with its yttrium-90 daughter, which emits very energetic (2.27 MeV) beta particles.

In the body, both of the strontium radioisotopes seek bone, where they substitute for calcium. The bones of infants and young children take up a large fraction of ingested radiostrontium, while those of mature individuals take up only about one fourth of the amount ingested. The long biological half-lives of the isotopes mean that significant amounts of them do not leave the body in a lifetime. The beta dose to the bone has caused bone tumors in experimental animals, and irradiation of the marrow has produced leukemia. ICRP has reported on the metabolism of strontium in the body (30).

The measurement of radiostrontium in the environment is relatively straightforward, since strontium displays a generally simple chemistry and very little tendency to form colloids or to be adsorbed strongly onto surfaces unless anions (negatively charged ions) are present with which it can form insoluble compounds. Strontium-90 has the additional advantage that the final analysis can be made by separating and measuring the amount of its yttrium-90 daughter present. Strontium-89 is more of a problem than ^{90}Sr, since it is measured as the difference between total radiostrontium and the strontium-90 estimated from the yttrium-90 daughter. This tends to decrease the accuracy of analysis, but does not interfere in the detection of significant amounts of strontium-89. Good separation must be obtained from the barium-140/lanthanum-140 pair before extracting the yttrium 90, since barium and strontium are very similar chemically and yttrium and lanthanum are even more alike.

Strontium is an alkaline earth metal and in general behaves chemically like calcium. The processes by which the two elements are taken up from the soil and transferred from plant to animal to man, however, all discriminate against strontium. Thus its concentration relative to calcium decreases as the two move up the food chain.

Since radioactive species occur in the environment in unweighable amounts, the usual practice is to add to the original sample enough of the inert salts ("carriers") of the element being sought so that the chemical yield through the separations process can be estimated by ordinary methods of chemical analysis. In the case of strontium, the element is collected as a phosphate or carbonate, separated from most other elements by precipitation in strong nitric acid solution, and finally collected as the carbonate for weighing. The yttrium daughter may be isolated as the hydroxide, then converted and weighed as the oxalate.

Strontium from weapons testing probably changes slowly from the oxide to the carbonate in nonurban air and to the sulfate in city atmospheres. Fallout is readily leached from soil with acids and apparently is completely available to vegetation. It moves downward in the soil relatively rapidly; some weapons debris is now more than 30 cm below the surface.

Nuclear power production would yield about 25 MCi of ^{89}Sr and 2.5 MCi of ^{90}Sr per 1000 megawatt electrical years. Strontium-90 is so long-lived that a particular effort is being made to prevent its release into the environment. The effort has been enhanced by the attention shown this radionuclide in the fallout from nuclear weapons tests.

Iodine-129 and -131

The fission process yields significant amounts of six radioactive species of iodine: ^{129}I and ^{131}I through ^{135}I. The half-lives of ^{132}I through ^{135}I, of which the longest is 21 hours for ^{133}I, are so short as to preclude their accumulation in the environment, so that only ^{129}I and ^{131}I will be considered here. The chemical forms of radioiodine released from nuclear tests or reactors are elemental (I_2), organic (most probably methyl iodide), elemental iodine adsorbed on particles, and hypoiodous acid (HOI) (52). In the environment these mix with the natural species of stable iodine (probably organic) and are thought to be converted readily to the same chemical forms, although this has not been proved.

Ingested radioiodine is absorbed rapidly from the alimentary canal in normal persons, especially if they are fasting; some is absorbed in the stomach, but the greater part in the small intestine. The element is then taken into the blood in the form of iodide ion (I^-). The thyroid, salivary, gastric, and mammary glands concentrate the iodide; the thyroid holds about one third of the iodine in the body (53). Numerous factors influence the uptake of radioiodine by the thyroid. Uptake may be decreased by excess iodine or iodide in foods, by iodized salt, and by certain hormones; it may be increased by dietary iodine deficiency, by chronic liver disease, or by any factor that tends to increase the rate of synthesis of the hormone thyroxin by the thyroid gland (54).

When elemental iodine vapor passes through the nose and mouth, the body takes up approximately 30% of it at normal breathing rates. Part of the uptake, apparently, is transported to the pharynx, swallowed, and absorbed in the small intestine. Some may be absorbed directly into the bloodstream through the surfaces of the nasal cavity (55). Retention of inhaled methyl iodide in humans depends also on respiratory parameters; low flow rates are associated with high retention and vice versa (56). At similar breathing rates, methyl iodide evidently is retained about twice as well as elemental iodine (55).

Iodine is lost from the body through the kidneys and, to a lesser extent, in the feces, saliva, and sweat. Urinary iodine is derived from inorganic iodine in the blood plasma, whereas fecal excretion results mainly from incomplete reabsorption of thyroxine or its conjugates (larger molecules that contain thyroxine) that have passed into the alimentary tract with the bile. Part of the fecal excretion may arise from incomplete absorption of iodine in organic form in foods.

The air-grass-cow-milk food chain is the "critical" (most important) pathway to be considered for radioiodine released to the atmosphere. Among fresh fission products, the iodines, and especially ^{131}I, are the nuclides most likely to reach dangerous concentrations in foods. In milk, these concentrations reach a maximum as early as two to four days after deposition. Children and adults take up about the same fraction of ingested ^{131}I, but the effects may differ markedly. The thyroid of a one-year-old child, for example, weighs 2 g and that of an adult 20 g; for a given intake, therefore, the radiation dose to the thyroid would be 10 times greater in the child than in the adult (57).

Iodine-129 is released from normally operating power reactors at a rate of tens of picocuries per year, but the rate is appreciably higher for fuel reprocessing plants (58). AEC estimated that a modern reprocessing plant releases 0.1% of the ^{129}I in the fuel being reprocessed (59). The dosimetric considerations resulting from a release of this magnitude have not been resolved; the inhalation dose is low, but uncertainties exist over the extent to which ^{129}I mixes with stable environmental iodine.

The National Academy of Sciences has reported the most recent estimates of risk from radioactive iodine in the body (14a). The responsible committee concluded that "on the assumption of linearity in dose response, even in the low dose range, the risk of thyroid cancer appearing in adolescence or young adulthood (from birth to 25 to 30 years) after irradiation in childhood may be estimated to be of the order of 1.6 to 9.3 cases per year per million children exposed per rem . . . The actual risk of tumor induction during childhood is lower than this and during adolescence it is higher . . . . "

Analytically, ^{131}I is usually determined by gamma-ray spectrometry. This may be done directly on solid (soil) or semisolid (thyroid) samples, or on ion exchange columns used to concentrate ionic iodine from large volumes of water or milk. Gaseous iodine in stack gases can be collected on charcoal beds for analysis, and methyl iodide and other organic iodine compounds can be collected on similar beds saturated with potassium iodide. The ^{131}I content, again, is determined by gamma-ray spectrometry (8). For ^{129}I, collection and analysis are more complicated because of the isotope's very low concentrations in environmental samples and its low specific activity. Sample preparation and subsequent iodine extraction methods have been described (10). The separated ^{129}I is determined by liquid scintillation counting or neutron activation.

Cesium-137

Like strontium-90 and strontium-89, cesium-137 (^{137}Cs) has caused great public concern because of its injection into the atmosphere by nuclear weapons testing. The nuclide is a high-yield fission product in bombs, appearing in 5.6% of weapons fissions; upon decay, 89% of it goes to barium-137m (m means metastable), which immediately gives off 0.66 MeV gamma rays. For these reasons, ^{137}Cs in debris from past weapons testing probably is the chief source of present external irradiation from fallout. Although fallout has been the source of most of the attention paid to ^{137}Cs, the nuclide also is a high-yield fission product in light-water (thermal) reactors. Such reactors produce it at an estimated rate of about 3.5 MCi per 1000 megawatt electrical years.

Cesium is an alkali metal and, like sodium, distributes itself rather uniformly in the soft tissues of the body. Thus the whole body is considered the critical organ for ^{137}Cs. Its biological half-life, 70 days, is long enough to make it a genetic hazard.

The most sensitive method for detecting cesium is to measure its beta activity in a chemically separated fraction, but a few disintegrations per minute can be detected by modern gamma spectrometry. It is possible to measure cesium in the living human with a sodium iodide crystal and large shield. Such whole body counters have been used to follow the changes in the content of ^{137}Cs in humans following weapons tests.

Cesium is a naturally occurring trace metal in the body and in general behaves chemically like potassium. While cesium compounds tend to be highly soluble, the ion is readily adsorbed on surfaces and on materials such as clay particles in the soil. Such adsorption can be prevented or reversed by acidification, but this process apparently does not affect the behavior of cesium in the environment. Cesium-137 from fallout, for example, has not moved downward in the soil as rapidly as strontium-90, even though cesium compounds generally are more water-soluble than those of strontium. The analytical chemistry of cesium is based almost entirely on formation of complexes such as the one with ammonium phosphomolybdate. The sample can be purified to some extent with ion exchange resins; as a rule the nuclide is finally collected for analysis as the chloroplatinate.

Radium-226, radon-222

The isotopes of radium and radon are products of the three major radioactive decay chains that occur in nature: the "uranium series," derived from ^{238}U; the "thorium series," headed by ^{232}Th; and the "actinium series," coming from ^{235}U. Basically, each parent decays by a series of alpha particle emissions; each emission reduces the atomic number by two and the mass by four until the decay culminates in one of the stable lead isotopes: ^{206}Pb, ^{208}Pb, ^{204}Pb, ^{207}Pb. (This is a highly oversimplified picture since beta-decay and complicated branching occur in each chain.) Along the way, the uranium and actinium series produce one radium isotope each, and the thorium series produces two (because of complex branching); each of the three series produces one radon isotope. Thus four radium isotopes (^{223}Ra, ^{224}Ra, ^{226}Ra, ^{228}Ra) and three radon isotopes (^{219}Rn, ^{220}Rn, ^{222}Rn) are found in nature. Of these, however, the two normally referred to as "radium" and "radon" are ^{226}Ra and ^{222}Rn, primarily because each has by far the longest half-life in its group (2q, 17) and therefore is the most abundant. These two isotopes occur in the uranium-238 series:

$$^{238}U \;\alpha\!\rightarrow\; ^{234}Th \;\beta\!\rightarrow\; ^{234}Pa \;\beta\!\rightarrow\; ^{234}U \;\alpha\!\rightarrow\; ^{230}Th \;\alpha\!\rightarrow\; ^{226}Ra$$

$$^{222}Rn \rightarrow \text{(complex branching)} \rightarrow \text{stable } ^{206}Pb.$$

If the entire uranium series is left chemically undisturbed for sufficient time, it reaches equilibrium — that is, at each step a transmutation of an atom of a particular isotope to a new element is balanced by the formation of a new atom by decay of the immediate parent. Under this condition the number of atoms of each isotope (or, roughly, the amount present) is proportional to its half-life. For this reason, a uranium ore will contain about 2.5 mg of radium per 1000 kg. The isotope ^{226}Ra is essentially a pure alpha-emitter and by itself contributes very little hazard directly to the environment. Its daughter, ^{222}Rn, however, is an inert gas; it can diffuse away from its parent and, as it decays, produces a series of gamma-emitting daughters that deposit out as they are formed, thus adding to the natural radioactive background.

Radium, of course, is the best known discovery of Pierre and Marie Curie at the turn of the century, and for a considerable time the extreme hazard of even microscopic amounts of the material was not widely recognized. "Radium waters," "radium belts," and similar patented nostrums enjoyed popularity in the 1920's, and it was fashionable to "take the waters" at resorts built around springs containing unusually high amounts of radioactivity or to vacation in other areas with naturally high radiation backgrounds (60). The latter activity probably did relatively little harm, but some of the nostrum-takers had reason later to regret their actions. Radium is an alkaline earth element, and thus its chemistry is very similar to that of calcium. Accordingly it can be metabolized into bone from which it is released only very slowly. The constant irradiation of the bone with energetic alpha particles of the radium and its daughters not only weakens the bone structure physically, but in many cases induces cancers and various blood diseases. The situation that eventually dramatized the hazard of ^{226}Ra to the general public arose from the "radium dial-painter" episode.

Around the World War I period, a number of women were employed in both the United States and Switzerland to paint the numbers on timepieces with luminescent paint containing ^{226}Ra. To do this, they would "point-up" the brushes with their lips as they worked. Many of these unfortunate individuals later developed bone cancer and aplastic anemias (60–62).

Once the hazard of radium was recognized, research was undertaken to establish the maximum amount that could be tolerated in the body (the maximum permissible body burden). The concepts developed in devising protective measures against the radium hazard became the basis of the various control structures that emerged when the large-scale production of radioactivity came about as a result of the discovery of fission in the early World War II period.

A considerable research program has been undertaken over the years based on collection of data on the life histories, causes of death, and posthumous examinations of the bodies of the"radium dial-painters" and individuals who were injected with radium for medical reasons or who used some of the radioactive nostrums popular in earlier years. Much of this work was formerly centered at the Massachusetts Institute of Technology. MIT still is involved, but the major effort is centered now at the Center for Human Radiobiology of Argonne National Laboratory (62). Since direct experimentation on the effects of ingested alpha-emitters on the human body is impossible, advantage is being taken of the existence of an inadvertently exposed population to learn as much as possible in order to avoid similar tragic situations in the future.

As mentioned, radium is an alkaline earth element and thus exists in solution only as the +2 cation (positively charged ion) with relatively little tendency to form complex ions. Its chemistry is particularly similar to that of barium, the next lightest element in the series, and clean separation of the two was a very difficult chemical problem until the advent of some of the newer separation methods that have been developed in recent years. The radiochemistry of radium has been summarized (63).

The plutonium isotopes

Plutonium is a transuranic element, atomic number 94, produced by neutron reactions in uranium; ^{238}U is the natural precursor of the isotopes of plutonium that are important biologically and radiologically. Minute amounts of naturally occurring ^{239}Pu have been isolated from pitchblende, but for all practical purposes the natural level of plutonium is zero. However, some plutonium, primarily ^{239}Pu, is spread over the globe at a very low level from weapons tests. The plutonium of interest in a reactor context is all man-made and consists of the isotopes ^{238}Pu through ^{242}Pu. Of these, ^{239}Pu is the most significant both physically and biologically, since it is produced in the greatest abundance. The nuclide is also, perhaps, the ultimate example of the good and the bad of artificial radioactivity. Because it is fissionable, it is the key to the breeder reactor concept that offers one of the possible solutions to energy-shortage problems over the next century; yet, on a weight basis, it is one of the most biologically hazardous substances known.

The isotopes ^{238}Pu, ^{239}Pu, ^{240}Pu, and ^{242}Pu decay by alpha-emission and ^{241}Pu by beta-emission (Table 4). The decay product of ^{241}Pu is americium-241, which is an alpha-emitter. The relative proportions of these isotopes in a given sample of reactor fuel depend on the fuel's history: the time it has been in a reactor, the neutron flux, the neutron spectrum, and the number of times that it has been

Table 4

Nuclear and health physics data for nonfission product nuclides

	Nuclear properties (187, 188)		ICRP recomendations (25, 26)			USAEC, MPC[a] (192)	
Nuclide	Half-life	Major radiations	Critical organ	Biological half-life	MPBB[b] μCi	Air μCi/Cm^3	Water μCi/Cm^3
^{226}Ra	1600 years	4.78 MeV α Low γ	Bone	45 years	0.1	3 x 10^{-12}	3 x 10^{-8}
^{237}Np	2.14 x 10^6 years	4.78 MeV α	Bone	200 years	0.06	1 x 10^{-13}	3 x 10^{-6}
^{238}Pu	86 years	5.5 MeV α	Bone	200 years	0.04	7 x 10^{-14}	5 x 10^{-6}
^{239}Pu	24,400 years	5.16 MeV α	Bone	200 years	0.04	6 x 10^{-14}	5 x 10^{-6}
^{240}Pu	6580 years	5.17 MeV α	Bone	200 years	0.04	6 x 10^{-14}	5 x 10^{-6}
^{241}Pu	13.2 years	0.021 MeV β	Bone	200 years	0.9	3 x 10^{-12}	2 x 10^{-4}
^{242}Pu	379,000 years	4.90 MeV α	Bone	200 years	0.05	6 x 10^{-14}	5 x 10^{-6}
^{243}Pu	4.98 hours	0.58 MeV β	Bone	200 years	7.0	6 x 10^{-8}	3 x 10^{-4}
^{241}Am	458 years	5.49 MeV α 0.06 MeV γ	Bone	200 years	0.05	2 x 10^{-13}	4 x 10^{-6}
^{243}Am	7370 years	5.28 MeV α 0.075 MeV γ	Bone	200 years	0.05	2 x 10^{-13}	4 x 10^{-6}
^{242}Cm	163 days	6.12 MeV α	Liver	200 years	0.05	4 x 10^{-12}	2 x 10^{-5}
^{244}Cm	17.6 years	5.81 MeV α	Bone	200 years	0.1	3 x 10^{-13}	7 x 10^{-6}

a. MPC = Maximum Permissible Concentration for exposure of the general population.
b. MPBB = Maximum Permissible Body Burden for occupational exposure.

reprocessed (recycled). The major constituent, by weight, however, is always ^{239}Pu. The fraction of the total mass represented by each isotope in a representative mid-cycle sample of plutonium, and the contribution of each to the total radioactivity, are compared below:

Isotope	% of total mass	% contribution to total radioactivity
^{238}Pu	<1	2.0
^{239}Pu	62	0.5
^{240}Pu	33	1.5
^{241}Pu	4	96.0
^{242}Pu	1	–

The mass percentage of ^{239}Pu will vary from about 55% to a high of 78%, depending on the history of the fuel and its position in the fuel cycle (64). Although the radioactivity of ^{241}Pu is high, the radiation emitted is low-energy beta. In older fuels, however, the alpha activity of the nuclide's americium-241 daughter can actually equal or even exceed that of the ^{239}Pu present because of the former's much shorter half-life and correspondingly higher radioactivity per unit of weight.

The radiotoxicity of plutonium stems from two facts: the more abundant isotopes are alpha-emitters, with physical and biological half-lives that are long relative to the lifespan of man; and a major fraction of that taken up by the body deposits in the skeleton and liver. The biological hazard of plutonium arises from its long retention time in the body and the consequent possibility of damage, particularly to bone, from the highly energetic alpha particles. The nuclide is an internal emitter in the true sense of the term; the low-energy, low-intensity x- and gamma-radiations from plutonium isotopes contribute negligibly to the external exposure of man.

Plutonium enters the human body mainly by inhalation and ingestion. Except in rare and unusual circumstances, absorption through the skin is a minor pathway, although it may be an important one for workers in plutonium-handling facilities, such as reactor fuel fabrication plants, since the element can enter the body through breaks in the skin. For the general population, then, inhalation and ingestion would be the likely routes of entry for plutonium in an economy powered by liquid metal fast breeder reactors; this would be true in both normal and abnormal (accidental) operating conditions. (The LMFBR is the version of the breeder reactor that to date has received most of the development emphasis in this country.) Of the two routes of entry into the body, inhalation is by far the more important: uptake of plutonium by the pulmonary route is about 1000 times greater than absorption through the gastrointestinal tract (65). This is true particularly for plutonium in particles of 0.1 to 1.0 μm diameter, the optimum size range for retention in the lungs and solubilization into the bloodstream. Particles released from reprocessing or fuel-fabrication plants will be in this size range. The reason is that the High Efficiency Particulate Air (HEPA) filters used to remove particles from the air exhausted by such facilities have an efficiency minimum at about 0.1 μm, although they still remove 99+% of the particles of this size. The removal efficiency is higher for both larger and smaller particles. Once across the lung-bloodstream barrier, a given amount of soluble plutonium will deposit in about equal fractions in bone and liver. Experience has shown that the HEPA filters now available are generally adequate in use and can be made more so by placing them in series. Where problems have occurred, they resulted almost invariably from poor installation or maintenance. There is also a sufficiency of regulation in relation to emissions treatment — here the difficulties are those of insuring compliance and enforcement. While additional chemical research can contribute to a certain extent, the greater need perhaps is for a systems approach that would guarantee that existing regulations are followed and that the available mechanisms for particulate removal are utilized intelligently.

Much of our present knowledge of the behavior of plutonium in the environment stems from studies of fallout. The present concentrations of plutonium deposited on the ground in the U.S. average about 2 mCi/km^2; the concentrations of plutonium in air average about 0.1×10^{-15} Ci/SCM (standard cubic meter) at ground level (66). What data are available indicate that plutonium dioxide (PuO_2) is not taken up easily by plants from soil. The uptake coefficients range from 10^{-6} to 10^{-4} — that is, the concentration of plutonium in plants is 10^{-6} to 10^{-4} times the concentration in the soil on which they were grown (65). The figures vary with plant type, and the less acidic the soil (the higher its pH) the greater the percentage uptake of plutonium. Plutonium is fixed readily in soil and migrates downward very slowly, but the effects of long-term weathering on the biological availability of the element are not well known. The transfer of plutonium from ingested food to body tissue is also quite low in various animals, including man; gastrointestinal uptake coefficients range from 10^{-5} to 10^{-4} (67, 68). Because the uptake coefficients are relatively small, the concentrations of plutonium in tissue (including bone) decrease as the metal moves up the food chain in both terrestrial and aquatic systems. In other words, the data at hand indicate that plutonium accumulates only minimally in food chains that lead ultimately to man (69).

Intake of plutonium in a more-or-less standard U.S. diet amounted to 2.5 pCi/year in the mid-1960's and has diminished somewhat since then as fallout has decreased (66). The total body burden of plutonium in individuals in the general population is of the order of 4 to 5 pCi at present (66). This body burden is equivalent to essentially two years of inhalation at a concentration of 0.1×10^{-15} Ci/SCM in ambient air. In addition, it is consistent with the presumed mechanism of pulmonary uptake. The body burden of plutonium in the general population is distributed as follows (66):

Organ	Activity in total organ, pCi
Lung	0.3
Lymph nodes	0.03
Liver	1.4
Kidney	0.1
Bone	1.4

The rest of the body burden is distributed in low concentrations throughout the system. (Note that roughly 10% of the plutonium remains in the lungs; the balance is solubilized into the bloodstream and distributed to the rest of the body.) The amount in lymph nodes may be important from a carcinogenic standpoint because of the small body mass, 15 g, in which the 0.03 pCi is contained. It is apparent also that the partition between bone and liver is 50/50, as found in inhalation studies with laboratory animals.

Other transuranium nuclides

Because of the complexities of differing half-lives, degree of burnout by neutron capture or fission, formation routes, and the like, nuclear power operations will produce relatively large amounts of only five nonplutonium transuranic nuclides: neptunium-237, americium-241 and -243, and curium-242 and -244. Other heavy isotopes will always be present, but in much smaller quantities than these five. All of the five are alpha-emitters (Table 4). Because they differ so widely in half-life, however, their radiation hazard on a weight basis ranges from the relatively innocuous neptunium-237 (^{237}Np) to the extremely toxic curium-242 (^{242}Cm). All are bone seekers. All have estimated biological half-lives of 200 years, so that the body will excrete very little of the longer-lived species over a normal lifetime. These properties together account for the extremely low permissible body burdens and concentrations assigned to the five nuclides. The tolerances, in fact, are comparable to those for ^{239}Pu itself.

The chemistry of neptunium in solution is complicated (70–74), so it is difficult to predict what its chemical form would be if it entered the environment. The element probably would occur in either the (IV) or (V) oxidation state. The former tends strongly to form complex ions, so neptunium in that form would be expected to be highly mobile in nature.

Americium can be oxidized to higher states in the laboratory, but for all practical purposes americium and curium exist only as +3 ions in solution, behave like the rare earth elements, and can be separated from that group or from each other only with some difficulty (72–75). Because of their high ionic charge, the two nuclides are adsorbed readily on natural materials having ion exchange properties if no complexing ions are present. If such ions are present, both americium and curium could be fairly mobile in water.

Standard methods do not exist for detecting neptunium, americium, and curium at the extremely low levels at which they occur in the environment. Analytical techniques have been developed, however, for less than picocurie amounts in urine, blood, and other biological materials. With these methods, the nuclide usually is separated in carrier-free form, electrodeposited on a counting plate, and determined by measuring its alpha emission in a spectrometer. Americium-241 can be determined sometimes by measuring its 60 keV gamma ray; ^{243}Am can be determined by measuring the gamma emission of its ^{239}Np daughter (half-life 2.35 days) if the two are known to be in equilibrium. The nuclides usually are separated by solvent extraction; thenoyltrifluoroacetone (TTA) is used frequently for neptunium and bis(2-ethylhexyl)phosphate (DEHPA) for americium and curium. The latter two elements also can be isolated by ion exchange.

Recommendation R1. *In cases of significant accidental exposure, excretion of the bone-seeking strontium radioisotopes can be accelerated by prompt medical treatment. While some progress has been made, similar treatments are much less available for the transuranics and rare earths. Research in search of suitable techniques should continue for all of the "nuclides of special concern."*

Carbon-14

At the time this chapter was organized it was not recognized that carbon-14 (^{14}C) could be a "nuclide of concern" in the nuclear power cycle, simply because the mechanisms whereby it could be produced are not obvious. Upon examination, however, it is evident that there are such mechanisms (neutron reactions of various types with ^{14}N, ^{17}O, and ^{13}C), and that in the aggregate they do produce significant quantities of ^{14}C, particularly in high temperature gas-cooled reactors (76). Very few direct measurements of ^{14}C production have been made, but it is estimated that perhaps one half to two thirds of the amount formed is released during reactor operation and the remainder when the fuel is dissolved for reprocessing. Presumably it is emitted as the dioxide ($^{14}CO_2$).

Carbon-14 emits a very low-energy beta particle and requires special care in measurement. It is produced continuously in nature by cosmic ray reactions, and the present estimate is that the material put into the atmosphere from the nuclear energy cycle (assuming no efforts are made to trap it in the meantime) will not equal the natural tropospheric inventory until the year 2020, when the ^{14}C is projected to deliver an average whole-body dose of 0.8 mrad/year. However, because of its long half-life, 5730 years, the nuclide would continue to build up indefinitely. If the "dose commitment" to the world population over the next 100 years is calculated, a value some 50 times that from the krypton-85 from the nuclear cycle is obtained (again assuming that no intervening efforts are made to trap and remove either gas).

Carbon dioxide, of course, is a reactive molecule in contrast to krypton. The elaborate systems being developed to remove krypton from nuclear off-gas streams can thus be relatively easily modified to also take out the ^{14}C for long-term storage as a waste.

SOURCES AND CONTAINMENT

Calculation of the average dose from environmental radioactivity for members of a large population is a difficult problem, complicated by the fact that many of the data required are not available. Estimates of past and present exposures are thought to range from about half to about twice the true values; estimates of future exposures probably are within 1/10th to 10 times the true values (14b). It is not surprising, therefore, that different groups of estimators derive different values, although all agree in general on the relative importance of the exposures generated by individual sources of radiation.

The United Nations Scientific Committee on the Effects of Atomic Radiation (UNSCEAR) has prepared six reports on population exposures for the UN General Assembly; the most recent one appeared in 1972 (13). A similar analysis (5) was made for the U.S. by an interagency group under the auspices of the Environmental Protection Agency when EPA was assigned the responsibilities held formerly by the Federal Radiation Council. Table 5 is based largely on these two reports. The natural background figures are for exposure to the gonads and are from the 1972 UNSCEAR report (the EPA figures for natural background in the U.S. are about 50% higher, but are for the whole body, as are other EPA figures quoted in the table). The diagnostic radiology values (chiefly exposure from x-ray examinations, including dental) are in terms of the "genetically significant dose."

The exposures shown in Table 5 are averages taken over the entire population of the United States. This approach was taken presumably as a means of obtaining a common base in order to evaluate the relative importance of the different sources of radiation. Exposures obviously will deviate widely from these averages in individual cases, as, for example, for a patient undergoing cobalt-60 therapy. [A recent study by the National Council on Radiation Protection and Measurements includes, where possible, the variability of exposure with cir-

Table 5

Estimates of radiation doses to the United States population

Source	Index used	Ref.	Annual dose per person (mrem) for years:				
			1960	1970	1980	1990	2000
Natural							
Cosmic rays	Gonadal	(13n)	28.3	28.3	28.3	28.3	28.3
External gamma	Gonadal	(13n)	44.0	44.0	44.0	44.0	44.0
Internal	Gonadal	(13n)	20.7	20.7	20.7	20.7	20.7
Subtotals[a]			**93.0**	**93.0**	**93.0**	**93.0**	**93.0**
Medical							
Radiographic examinations[b]	GSD	(84)	16.0	20.0	20.0	20.0	20.0
Radiopharmaceuticals	Whole-body	(5s)	0.4	2.0	14.0	14.0	16.0
Subtotals			**16.4**	**22.0**	**34.0**	**34.0**	**36.0**
Nuclear fuel cycle							
Reactors	Whole-body	(5t)	0.0001	0.0091	0.026	0.082	0.17
Fuel reprocessing	Whole-body	(5u)	—	0.008	0.02	0.09	0.2
Worldwide ^{3}H	Whole-body	(5v)	0.02	0.04	0.03	0.02	0.03
Worldwide ^{85}Kr	Whole-body	(5w)	0.0001	0.0004	0.003	0.01	0.04
Subtotals			**0.029**	**0.058**	**0.076**	**0.20**	**0.44**
Misc. DOE[c]	Whole-body	(5x)	0.04	0.01	0.01	0.01	0.01
Grand totals			**117**	**122**	**134**	**133**	**136**

a. EPA (5l) gives 130 mrem, whole-body dose.
b. In millirads. 1964 estimate used for 1960. 1970 estimate extended across table lacking other information. See text.
c. Includes peaceful nuclear explosives, Nevada Test Site, other Department of Energy installations.

Notes: Other medical-type exposures are discussed in the text.

Subtotals and totals were computed in the table for comparison purposes, but are not strictly legitimate since different exposure indices are used.

cumstance, which should make it possible to estimate exposures for various population groups within the country (77)].

The figures in Table 5, present and projected, are dominated by the natural and diagnostic radiology exposures; fallout is a smaller but significant third source. While the dosage associated with nuclear power is only a small fraction of the total, the estimates for that source are based on accident-free operations. It is true that over the past 30 years, atomic energy facilities have had a remarkably good safety record — far better than in coal mining, for example. It is also true that strong arguments can be made that reactors are inherently incapable of behaving like an atomic bomb (78, 79). Nevertheless, the large inventory of lethal materials in the core of an operating reactor is a potential hazard that cannot be ignored.

The UNSCEAR report (13) considers natural radioactivity in detail, as does a second EPA publication (6). The doses in Table 5 are averages, and specific doses will vary from person to person, depending largely on geographical location and way of life. Jet-setters and airplane pilots, for example, receive more then average exposure from cosmic radiation, since they spend much of their time at high altitudes.

Cosmic ray dose rates vary with altitude and with latitude up to about 50° north or south. At 45°N, the values range from 40 mrem/year at sea level to 200 mrem/year at 8000 ft (5a) (these are EPA figures; UNSCEAR values are lower, but use a different index of exposure). Using such relationships, the EPA-sponsored group estimated doses for each county in the United States and used these data to calculate averages for each major political unit (states, etc.) of the nation. These values range from about 30 mrem/year in Puerto Rico to 130 mrem/year in Wyoming.

The significant external exposures arise primarily from naturally occurring potassium-40 (^{40}K) and the decay products of the uranium and thorium series of elements. Exposures from radon and radon daughters vary significantly with atmospheric conditions, and they vary also with the condition of the surface (soil moisture, porosity, cultivation, pavement, etc.). In addition to variation with geological and geographical factors, therefore, differences occur with time and specific location. Different investigators have reported several hundred measurements with scintillation counters, and using these data the EPA group again estimated average exposures from external gammas for each state. Ninety percent of all areas fall in the range of 15 to 130 mrem/year, while 90% of the population falls in the range of 30 to 95 mrem/year (5b).

The natural radionuclides that contribute to internal radiation in any significant way include ^{3}H, ^{14}C, ^{40}K, and ^{226}Ra and its decay products, including ^{222}Rn. This last is the only significant radionuclide leading to widespread exposure through inhalation; the others enter the body primarily in food. Radon is released from soil, rock, and building materials and is contained in natural gas and other fossil fuels. Various factors tend to equalize the concentrations of the other nuclides in foods, so the EPA group did not assess doses from them by geographical or political units; instead, it calculated an average value (25 mrem per person, whole-body basis) for the U.S. as a whole. Well over 90% of the internal dose comes from ^{40}K, the naturally occurring radioactive isotope of potassium.

Exposures from Medical Radiation

Exposure to radiation in medicine and dentistry stems from several sources. The largest and most widespread source for the population as a whole is the use of x-rays for diagnostic purposes. The rapidly increasing use of radiopharmaceuticals for diagnosis and therapy and the use of x- and gamma-rays for therapy also contribute, however, as do the higher-than-normal general exposures received by medical and dental personnel who administer the irradiations. Extracting a meaningful "average" exposure for a large population is particularly difficult. The data are limited and often not easy to interpret, and some groups in the society are exposed much more than others. Also, in contrast to exposure to natural radiation, where everyone receives roughly an equivalent whole-body dose at a low but continuous rate, medical and dental work generally involves irradiation of a limited portion of the body intermittently at high dose rates. Nevertheless, investigators agree in general that medical-dental exposure is currently by far the largest man-made source.

Since the major exposure in medical use of radiation is to a limited part of the body, the choice of a suitable index for comparison purposes is particularly important. One that is used frequently is the "genetically significant dose" (GSD). Radiation that reaches the gonads can cause mutations in the genes carried by the sperm or ova. These mutations do not affect the individual, but are passed from generation to generation. If persons who carry similar mutations mate, serious effects can result in the offspring. The genetic injury to a population thus depends on the number of mutant genes introduced. One measure of potential genetic damage from radiation exposure is the per-capita gonadal dose. When this index is corrected by allowing for age distribution and for the probability that various members of a given population will have additional children, the per-capita gonadal dose becomes the "genetically significant dose." (For simplicity, some GSD calculations consider only the under-30 age group. The gonad-exposure and abdominal exposure indices will naturally be higher than the GSD, since in those cases all members of the population are considered regardless of age.) Indices are used also for the thyroid and bone marrow, but breakdown beyond that point generally is not attempted because of lack of survey data.

The United Nations' SCEAR group reviewed medical radiation in 1962 (80) and again in 1972 (13) for all countries from which data were available. The EPA study (5c) confined itself to the U.S.; most of its conclusions were based on two surveys made by the U.S. Public Health Service, in 1964 (81–83) and 1970 (84). The 1962 UNSCEAR report showed variations in GSD from diagnostic examinations of from 6.8 mrem in Leiden, Holland, to 58.2 in France. The range for the 1972 summary was 8.6 (Sheffield, U.K.) to 75.3 (New Orleans). Comparable figures for exposures from mass x-ray surveys of the type conducted for tuberculosis were 0.02 to 1.90 mrem in the 1962 survey and 0.05 to 0.44 mrem in 1972 (13a).

The difficulties encountered in making such analyses are indicated by the published results of the U.S. Public Health Service surveys made in 1964 and 1970. Using one model for the dose received by the gonads during x-ray irradiations of various parts of the body gave GSD's originally published as 55 mrads for 1964 and 36 for 1970. Reexamination and refinement of the model modified these figures, respectively, to 16 and 20 (84), the numbers used in Table 5. The EPA report assumed a value of 72 mrem (abdominal dose), but emphasizes that the present uncertainties make it impossible to draw firm conclusions as to possible changes with time, so the authors made no attempt to project future levels.

The GSD from dental examinations is very low; UNSCEAR puts it at less than

0.01 mrad/person/year (13b). The thyroid dose index, of course, is affected much more; for the annual per capita dose in the U.S., EPA estimates (5c):

Head and neck examinations	7 mrem
Chest and thorax examinations	19 mrem
Dental examinations	14 mrem
Total thyroid dose index	**40 mrem/person/year average**

Although diagnostic radiology dominates the exposure picture in medicine, other sources exist. About 4 million persons in the U.S. were diagnosed or treated with radioisotopes (including radiopharmaceuticals) in 1970, compared with half that number in 1965, and such usage appears to be accelerating (85, 86). The two UNSCEAR reports show that this caused the GSD to vary from 0.01 to 0.40 mrem average/person/year for the countries where data were available (13c), while EPA estimates a value of 0.26 for the U.S. (5d). EPA also estimated indices for other body organs, the total being approximately 1 mrem/person/year (5e). A sizable fraction of the total in the past has been due to the use of ^{131}I for determining thyroid function. A new technique (87) using the soft gamma rays from ^{241}Am to excite the characteristic x-rays of iodine for measurement shows promise of reducing thyroid exposures drastically. The ^{241}Am source is placed over the thyroid externally to cause the excitation, so that no radioactivity is introduced into the body.

The primary therapeutic uses of radioisotopes are ^{131}I for thyroid treatment and ^{32}P for polycythemia. The 1962 UNSCEAR report showed GSD values ranging from 0.15 mrem (United Kingdom) to 0.40 (Canada). The U.S. average was 0.24 mrem/person/year across the entire population (13d). The Public Health Service has prepared a survey report on the use of radionuclides in medicine (88).

The development of nuclear medicine techniques requires close cooperation between radiochemists and medical personnel. The chemists devise methods for preparing a desired isotope in high radiochemical (free of other isotopes) and chemical purity and for preparing it as a compound suitable for administration. They then work with the medical people through the various stages of animal and clinical testing that are the necessary prelude to formal acceptance of the isotope as a routine diagnostic or therapeutic hospital tool.

Recommendation R2: *Radioisotopes unquestionably have furnished a powerful new diagnostic tool to medical science. Efforts should continue to develop techniques for minimizing patient exposure by optimizing the chemical forms used in introducing the material into the body and by working out methods utilizing short-lived isotopes to minimize overall retention times. The problem of adequate regulatory control of the use and disposal of radioisotopes used in medicine should receive constant attention in view of the rapid expansion in the use of such materials.*

It will be noted that some of the last sources of exposure discussed above would appear to contribute relatively large doses to the U.S. average. They were not included in Table 5, primarily because the available data are so limited and uncertain. They should still be kept in mind as contributors to the national average. They are obviously of most vital importance to the relatively small fraction (at least for radiation workers and radiotherapy patients) of the general population involved.

About half of all cancer patients receive radiotherapy during treatment. A large fraction of other uses of radiation in treating disease is for dermatological purposes, with the head area being mostly affected. UNSCEAR gives (13e):

	GSD's (mrads)		
	Neoplastic	Nonneoplastic	Total
West Germany	0	2.2	2.2
France	2.5	3.1	5.6
United Kingdom	0.52	4.47	4.99
Netherlands	1.0	3.1–12.1	4.1–13.1
Japan	–	–	0.98

For the U.S. population, EPA gives a GSD of 5 mrems from cancer treatments (5f). No value was estimated for radiotherapy for nonmalignant diseases.

Medical and paramedical personnel who administer radiation for diagnosis or therapy are subject to above-normal exposure. The population at risk — medical and dental x-ray workers, radioisotope workers, and radium workers — in 1968 received doses in the range of 125 to 540 mrem. A population of 454,000 such individuals was surveyed. Averaging this dose over the population of the U.S. gave an annual whole-body per capita dose of 0.56 mrem (5g).

Miscellaneous Sources of Ionizing Radiation

UNSCEAR lists 56 consumer products containing radioactivity and says the list is not complete (13f). The general categories include radioluminous products; electronic and electrical devices; antistatic devices; smoke detectors; uranium- or thorium-containing ceramics, glassware, alloys, etc.; scientific instruments; and certain building materials. EPA considered a more restricted group of miscellaneous sources, but included high-altitude air travel (5h).

The largest and most ubiquitous source of miscellaneous radiation probably is self-luminous wristwatches and clocks. The radium-226 that has been used to activate such devices led to a GSD of about 2 mrad/person/year in 1960 (13g). ^{226}Ra in luminous paints is gradually being replaced, however, generally by tritium or promethium-147 (^{147}Pm) (5i). Both emit soft beta but no gamma radiation. ^{147}Pm gives doses in the range of microrems per year (45); its shorter half-life means a substantial loss of activity over a few years, but this may be tolerable in certain applications. Tritium paints are formed by replacing some of the hydrogen in a polymer with ^{3}H. Self-radiation damage causes a very slow release of radioactivity, but at a rate that tests have shown to present no hazard.

The cathode-ray tubes and some other components of television receivers produce x-rays. This was not a particular problem with black-and-white sets. The higher voltages and beam currents of color receivers, however, became a matter of concern, particularly where faulty high-voltage control of shunt-regulator tubes occurred. A survey (89) of 1124 color television sets was made in Washington, D.C., in 1968, concurrently with a study of family viewing habits. The great majority of the sets did not give off x-rays at a measurable level, but some were strong enough emitters to produce an appreciable dose, particularly for the under-15 age group. Performance standards have since been established

(90), and under typical operating conditions the x-ray emission from recently-built color television sets is indistinguishable from natural background. Whereas the GSD from this source was estimated to be 0.5 mrad/year in 1967 (13h), it is expected to become very small as older sets are replaced.

EPA estimates (5j) that the whole-body radiation dose from miscellaneous sources for the U.S. population in 1970 was 2.6 mrem. Because of factors such as lower usage of ^{226}Ra in luminous paint products, and better-engineered television sets, the agency projects a steady decrease to 1.1 mrem by 1990.

Occupational Exposure

People who are occupationally exposed to radioactive materials include medical-dental workers, uranium miners and others associated with the nuclear fuel cycle, employees of certain Department of Energy and private installations that produce radiopharmaceuticals, persons handling radioactive wastes at commercial burial sites, industrial radiographers, and x-ray crystallographers. The total appears to be in the range of 1 to 2 per 1000 population in most industrialized countries. Data on occupational exposure are both meagre and uncertain. The UNSCEAR report for 1972, however, concluded that the mean annual recorded dose for most radiation workers lies in the range of 0.2 to 0.6 rad/year. The group's 1962 report had concluded that the likely upper limit for GSD as an average for the entire population was 0.5 mrem/person/year; this conclusion appeared still to be valid in 1972 (13i). The International Commission on Radiological Protection has established recommended maximum permissible doses for radiation workers (27).

Projected Production of Radioactive Elements

The remaining sources of environmental exposure to radiation shown in Table 5 are associated primarily with the discovery of nuclear fission. The estimate that natural exposure will still be dominant in the year 2000 assumes tacitly that man will be able rigorously to contain all radioactivity arising from his use of the fission process. The amounts of radioactive materials to be produced are sobering. Nevertheless, there is no apparent technological reason why the activity cannot be contained, although the problem requires maximum attention. And in case of a nuclear war, all bets are off. (Some of the estimates that follow are based on reports prepared before it became obvious that the development of commercial fast breeder reactors would be much slower than anticipated. At the very least, some displacement of the quoted time scales should be applied.)

The primary sources of man-made radionuclides are nuclear reactors, nuclear detonations, and, to a much lesser extent, charged ion accelerators. Secondary sources of radioactivity in the environment are more numerous. They include reactor fuel fabrication and processing plants, research laboratories, transportation and waste storage facilities, hospitals, industrial users, and the release of natural radioactivity through mining and ore refining.

The radioactive products formed in fission power reactors will be the major problem for some decades. Projections of the amounts to be produced will vary with the estimator's assumptions of the growth rate of nuclear power. Once the assumption is made, however, reasonably good values can be calculated for individual fission products and their totals, since fission yields are well-known for the three power-producing isotopes, ^{235}U, ^{239}Pu, and ^{233}U, and vary little relative to the other uncertainties. The case of ^{239}Pu and the other long-lived actinides is somewhat more complicated, since assumptions also must be made of the extent to which ^{239}Pu produced in uranium-based light-water reactors will itself be used as a fuel and the degree to which isotopes such as ^{238}Pu, ^{244}Cm, and ^{242}Cm will be produced deliberately because of their value as heat or power sources. If no ^{239}Pu is recycled, and light-water reactors are the only type used, one projection (91) indicates an accumulation of 907 t (metric tons) of plutonium in this country by 1990. If plutonium is recycled as a fuel, and fast breeder reactors play an increasing role, the stockpile in the same year would be 1 t, although substantial amounts would be in the reactors themselves. The stockpile would be made up essentially of plutonium in the process of being fabricated or awaiting processing recovery. (In April 1977, President Carter said that his Administration's policy in part would be to "defer indefinitely the commercial reprocessing and recycling of the plutonium produced in the U.S. nuclear power programs.")

Table 6 gives one prediction of the annual production rates for certain selected fission products, again based on the assumption that the bulk of the plutonium will be recycled and that breeder reactors will play an increasingly important role during the 1980's. [Some of the fission products are stable when formed; other, short-lived species decay rapidly to stable daughters. Fission product mixtures in reactor wastes accordingly are being considered seriously as materials that could be "mined" for scarce elements like rhodium, palladium, and xenon (92, 93)]

As indicated earlier, predicting the production and accumulation of transuranium elements is complicated by the fact that they are created not only in normal reactor operation, but also may be manufactured deliberately. One projection, assuming no plutonium recycle before 1980 (94), is shown in Figure 1. The ^{237}Np curve represents the amount of the isotope available annually by recovery from commercial reactor fuels. Part or all of this material, however, could be put back into a reactor to produce ^{238}Pu for use as a concentrated power or heat source, and the curve for that isotope on the figure assumes that all of the ^{237}Np produced is thus utilized. (Neptunium-237 captures a neutron to form ^{238}Np which decays by beta emission to ^{238}Pu. The latter can be formed also by

Table 6

Projected annual production of selected fission products (kilograms per year)

Year	Isotopes and half-lives[a]						
	^{85}Kr (10.76 years)	^{90}Sr (28.1 years)	^{106}Ru (367 days)	^{137}Cs (30.23 years)	^{144}Ce (284.9 days)	^{147}Pm (2.5 years)	^{3}H (12.26 years)
1970	1	19	3	42	7	2	0.001
1975	23	532	69	1220	172	36	0.033
1980	58	1320	202	3610	432	180	0.086
1985	130	2940	488	8090	1080	457	0.188
1990	223	4930	1150	14100	2050	900	0.324

a. Half-lives from refs. 187, 188.

Source: Deonigi, D.E., McKee, R.W., Haffner, D.R., "Isotope Production and Availability from Power Reactors," U.S. Atomic Energy Commission, Report BNWL-716, Washington, D.C., July 1968.

irradiating ^{241}Am to form ^{242}Am, which decays to ^{242}Cm, which in turn decays by alpha emission to ^{238}Pu. This last approach would affect the ^{241}Am and ^{242}Cm curves. The details of Figure 1 obviously can vary drastically, depending on the options chosen.)

Fallout

The character of the radioactive fallout from a nuclear detonation in the atmosphere is affected by the size of the detonation and its height above the surface. If the burst is low enough to suck up soil or water, elements in those materials become radioactive by neutron-capture processes in the fireball. Most of this radioactivity will settle within 24 hours as local fallout. If the blast is high enough so that it exerts no suction at the surface, very small particles are formed as melted bomb components and fission products cool and condense. The

FIGURE 1

Nuclide production from domestic power reactors

(^{237}Np and ^{238}Pu, tens of kilograms)

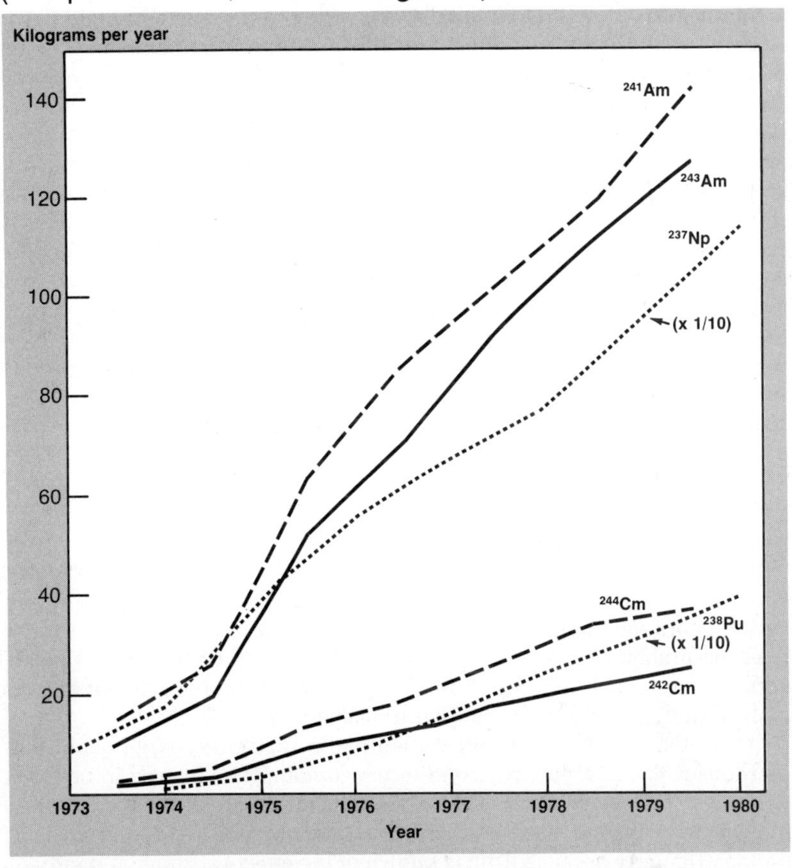

Source: Wood, P.M., "Isotope Production in Commercial Power Reactors," *Iso. Rad. Tech.*, **9**, 129 (1971–72).

particles form small agglomerates that eventually settle back to earth as worldwide or delayed fallout. If a burst of this last type is relatively small, the fission products are confined to the troposphere and are washed down by rain or snow in one day to four weeks. Since winds are mainly easterly or westerly, tropospheric fallout reaches the earth in an irregular band centered roughly at about the latitude of the detonation. High-yield (megaton) explosions propel the fission products into the stratosphere. Mixing is minimal in the stratosphere, and precipitation is nil, so material can remain there months or years before it returns to earth. The mechanism by which the fallout finally reaches the surface is involved and still debated (95, 96).

The 1961–62 series of large-scale, high-yield atmospheric tests of nuclear weapons by the United States and the Soviet Union introduced radioactive material into the stratosphere. The last of these U.S.-Soviet detonations was in 1962, and fallout peaked in 1963, reaching a total exposure of about 13 mrem/person/year (5k) before it began to decline. During the past few years, the relatively few atmospheric tests by the French and the Chinese have been enough to maintain a constant or slightly increasing annual deposition of fallout (Table 5 — the Table assumes that future testing patterns will be similar to those of 1965–70).

The EPA-sponsored group has calculated external gamma radiation doses and the internal exposure through inhalation and ingestion from fallout (97). About half of the total external component results from the accumulation of ^{137}Cs from past nuclear tests; shorter-lived species are important for the first few years after deposition. The external dose per person in the United States in 1969 was an estimated 0.9 mrem/year and was assumed to remain relatively constant to the year 2000. This contrasts with an external dose of 5.9 mrem/person/year in 1963. The calculation of internal dose is more complicated, since the effect of each nuclide on the various organs must be considered, and long-lived nuclides like ^{90}Sr and ^{239}Pu will deliver doses over most of a lifetime. EPA calculated total internal exposures of 6.9 mrem/year for 1963 and 3.1 mrem/year for 1970. The UNSCEAR summary (13j) concludes that the dose commitment, the cumulative dose to the year 2000 due to nuclear tests performed before 1971, is 180 mrads/person to the bone cells of the world population. The ^{137}Cs component appears to be the main contributor to the total dose commitment (13k). A compilation of current data on fallout became available in 1974 (98).

The Nuclear Fuel Cycle

The nuclear fuel cycle comprises the various steps involved in using uranium as the basic fuel for producing electrical power (Fig. 2). Each step involves the possibility of producing radioactive waste in gaseous, liquid, or solid form, but the wastes problem is relatively less serious until after the enrichment stage. The width of the arrows in Figure 2 is a rough indication of the potential hazard involved in transporting the material from step to step.

A point that has not had much attention is that rather large quantities of hazardous chemicals, particularly fluorine compounds, are used in portions of the nuclear fuel cycle. These materials apparently have not been a serious problem. Nevertheless, hazards of ordinary chemical pollutants must be kept in mind, particularly because little is known of the effects of combined radiation-chemical exposures.

Recommendation R3: *Relatively little is known of the possible synergistic or antagonistic effects of radiation combined with chemical pollutants on humans and ecological systems. This area of research should be vigorously pursued.*

Uranium mining, milling, and refining

Uranium is mined either in open pits (about 40% of the total) or underground. The process apparently does not endanger the general public, since measured radiation in mining communities and areas is in the same general range as in nonmining areas (99, 100). A small amount of radon must enter the atmosphere, but it cannot be distinguished from natural background. In confined underground workings, however, the situation can be different. Governmental control of uranium mine conditions was not enforced strictly until the late 1960's, and the incidence of lung cancer among miners appears to be higher than normal, presumably because of radon daughter exposure (101). Regulation currently is much better; the U.S. Bureau of Mines is responsible for enforcement. A new permissible exposure limit of four "working level months" (WLM) per year was imposed on July 1, 1971 (102). The new limit was three times more stringent than formerly. (A WLM is exposure to the radiations of radon daughters in equilibrium with 100 pCi per liter of radon for one month.) Improved ventilation is the chief technique for reducing radon concentrations. The Bureau of Mines also is sponsoring research on other methods, including the development of solid absorbers (103).

The uranium ores mined today vary from about 0.1 to 1.0% uranium oxide (U_3O_8) and are concentrated at mills near major mining sites to an oxide or salt assaying 70 to 90% U_3O_8. The daughters of the uranium series, including the ^{226}Ra, go with the tailings; the tailings accumulated by the end of 1970 contained an estimated 5300 Ci of radium (2a). The tailings normally are piled near the mill

FIGURE 2
The nuclear fuel cycle

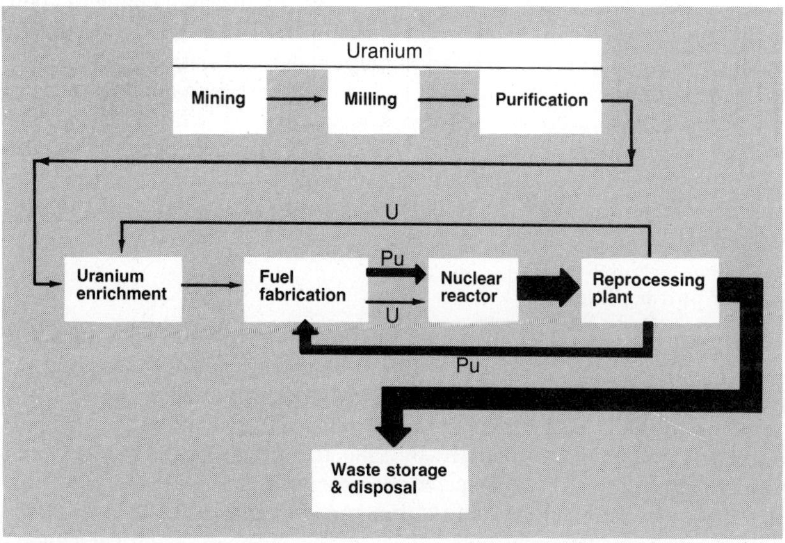

or collected in drying ponds. Early studies (104) around mill sites showed only normal background for radon concentrations and external gamma radiation unless the readings were taken directly over or immediately downwind of the tailings piles. An environmental analysis by EPA (105), however, would indicate that some individuals downwind receive substantial radiation exposure and that, while the 100-year dose commitment for the country at large is small, it exists. This problem accordingly is receiving new attention by the Nuclear Regulatory Commission.

Water supplies of certain western towns have become contaminated by seepage from tailings (2a, 106). At Grand Junction, Colo., local contractors used tailing material as backfill or in concrete for foundations (107, 108). Radon concentrations in a number of buildings were above background and, in some cases, were high enough to call for remedial action (106).

Concentrates from the mills are sent to centralized plants for purification and conversion to the metal or to an intermediate compound (UO_3, "orange oxide," or UF_4, "green salt"). Older plants of this type undoubtedly released uranium dusts to the atmosphere, but the air control and filtration systems of modern installations have largely eliminated this hazard. The monetary value of the uranium adds to the inducement to avoid such losses (2b).

Isotope enrichment and fuel fabrication

Salts from the refineries are sent to one of the gaseous diffusion plants at Oak Ridge, Tenn.; Paducah, Ky.; or Portsmouth, Ohio. In these facilities the uranium is converted to uranium hexafluoride (UF_6), which is passed through a cascade of porous barriers to raise the concentration of ^{235}U above its natural level of 0.72%. From the gaseous diffusion plants the enriched material goes to any of a number of private or governmental facilities for fabrication into reactor fuel elements.

Under present circumstances, the steps in the nuclear fuel cycle up to this point have involved materials of relatively low radiation intensity, such as natural or slightly enriched uranium; they have not involved potential environmental impacts of widespread significance. This picture will change somewhat for the fabrication plants if plutonium assumes a larger role in reactor fuels. Elaborate arrays of glove boxes and complicated ventilation systems will be required, particularly since the material is a powder during much of the fabrication process. [Glove boxes are large sealed boxes in which the material is handled by operators working from the outside through long rubber gloves. Controlled atmospheres, usually nitrogen from which the oxygen and water vapor have been removed, are frequently necessary. Ventilating gas leaves the box through High Efficiency Particulate Air (HEPA) filters (109).]

Nuclear power plants

About 194 nuclear power plants were operating worldwide by the end of 1976; 65 of them were in the U.S. and had a total generating capacity exceeding 47,000 megawatts (Mw) (110). The estimated growth rate of nuclear power has been shrinking in the face of events of the 1970's. AEC estimated in 1973 that U.S. capacity would reach 132,000 Mw by 1980 and 1.2 million Mw, one third of it from breeder reactors, by the end of the year 2000 (111). ERDA projections in 1976 showed only 69,000 Mw by 1980 and no more than 500,000 Mw by 2000. In 1970, nuclear plants contributed 2.2% of the nation's electrical generating capacity, a proportion that has been expected to reach 50% by 2000 (112). In 1977, nuclear power contributed more than 11% of the nation's electricity output (113) and was expected to reach about 28% by 1985 (110). AEC reported also that 72 research reactors were operating in the U.S. at the end of 1970; in the same year the International Atomic Energy Agency listed 367 such reactors in 47 countries (2c). Besides research installations, which generally are much smaller than power reactors, this and a few other countries have a limited number of reactors for producing plutonium and other transuranic nuclides. The power reactor area, however, is the one that is projected to expand markedly in the next few decades, if less rapidly than predicted at first.

Nuclear power plants in the U.S. today are predominantly light-water reactors (LWR's) of two types: the pressurized water reactor (PWR), which accounts for about 60% of the total, and the boiling water reactor (BWR). Concise descriptions of these and the next generation of reactors, the breeders, are available (2h, 12b, 114, 115a, 116, 117a). (The breeders use excess neutrons from the active core to produce fissionable ^{239}Pu or ^{233}U in uranium or thorium blankets; they produce more fuel than they consume.) Two high-temperature, gas-cooled reactors (HTGR's), a type that is much more prevalent in the United Kingdom, have also been built in the U.S.

In a typical PWR (Fig. 3), the fuel is UO_2, enriched to 2.5 to 3.2% ^{235}U, and sealed hermetically in tubes, usually made of a zirconium alloy (112, 114). This tubing or cladding constitutes the first barrier against escape of highly radioactive fission products. The heavy steel reactor vessel (walls more than 8 in thick) surrounding the assemblies of fuel rods serves as a second barrier, and many feet

FIGURE 3

Pressurized water reactor (PWR)

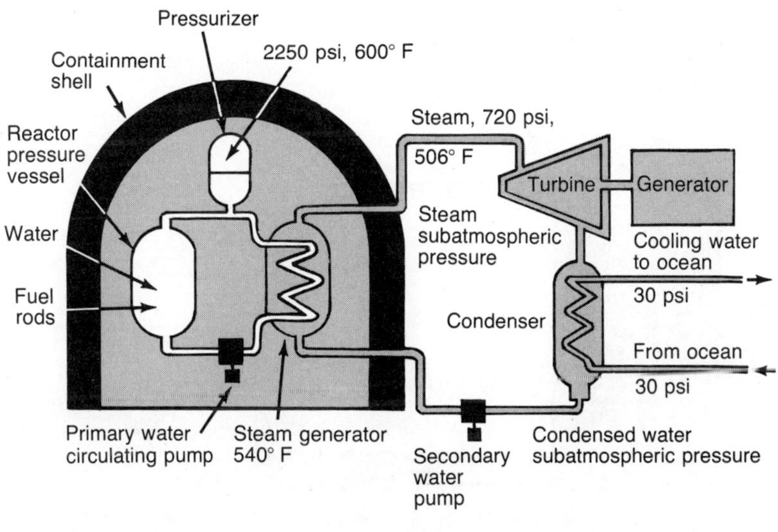

Source: Hottel, H.C., Howard, J.B., "New Energy Technology, Some Facts and Assessments," MIT Press, Cambridge, Mass., 1971, p. 232.
Benedict, M., "Electric Power from Nuclear Fission," Proc. Nat. Acad. Sci. USA, 68, 1923 (1971).

of concrete shielding around the reactor vessel serve as a third. The entire assembly is enclosed in a massive, reinforced, steel-lined, concrete containment shell, sized and designed to withstand the pressure and temperature that would be generated if all of the coolant in the primary system were converted suddenly to steam (117b). In a PWR, the steam driving the turbine is produced in a secondary water loop and does not contact the fuel rods themselves. [The liquid metal fast breeder reactor will carry this one step further. A primary liquid-sodium loop extracts heat from the fuel and transfers it to a secondary sodium loop in an intermediate heat exchanger. From the secondary sodium loop, another exchanger transfers heat to a water loop where steam for driving the turbine is formed (114).]

The typical BWR (Fig. 4) has only a single heat-exchange loop, and steam is formed by direct contact of the cooling water with the clad fuel assemblies (115b). The steel reactor vessel is 6 to 7 in thick and surrounded by reinforced concrete. Newer BWR's have a separate, free-standing containment shell within the reactor building (117b). The containment-shell concept is in contrast to Soviet practice, where reactors are housed in buildings similar to conventional fossil-fuel electricity generating plants.

Gas-cooled reactors in this country use helium under pressure to cool the fuel assemblies (the English use carbon dioxide). The hot gas is passed through a steam generator to produce steam for driving the turbogenerator and then is returned to the reactor in a closed loop. In fast breeder development, emphasis to date has been on the LMFBR design, but the gas-cooled fast breeder reactor (GCFBR) also has received attention. A third type, the molten salt breeder reactor (MSBR), would use a fused mixture of uranium and thorium fluorides dissolved in beryllium and lithium fluorides for both fuel and coolant.

Reactor effluents. Radioactive effluents from a reactor can be either gaseous or liquid. Small quantities of radioactive and nonradioactive gases as well as

FIGURE 4

Boiling water reactor (BWR)

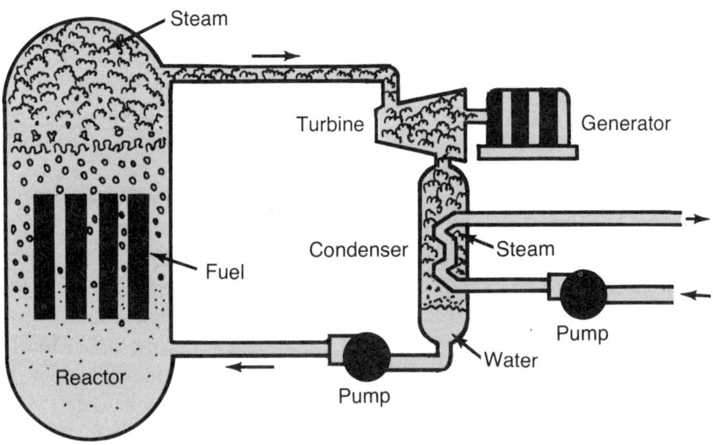

Source: "Nuclear Power and the Environment," International Atomic Energy Agency, Vienna, Austria, 1973, p. 52.

soluble and insoluble solids are formed in the primary coolant by neutron activation of corrosion products, water, and traces of dissolved air; fission products also may enter the primary system from leaks in the fuel cladding. Leakage from fuels with failed cladding also can occur during refueling or during storage of spent fuels under water in canals. The mixtures of nuclides in liquid or gaseous effluents will vary with the reactor type, and the basic cleanup techniques will vary correspondingly.

Most of the gases produced in a reactor by activation are isotopes of oxygen and nitrogen with half-lives of less than 30 seconds; they are a problem only in terms of shielding within the plant. Gases from failed fuel elements are primarily isotopes of krypton and xenon. Many of these have relatively short half-lives, so a "delay and decay" step is a feature of all waste treatment schemes. In a PWR, gases are stripped continuously from the primary coolant in a small side circuit and transferred to gas storage tanks. There they are held until the plant is shut down and the primary loop depressurized. The tanks are then bled to the atmosphere at controlled release rates after having been held for a minimum of 30 to 60 days. During this period, most of the nuclides have decayed to stable daughters, so krypton-85 and small amounts of xenon-131m are substantially the only radioactive gases discharged to the environment (117c).

In BWR's, the noncondensable gases carried by the steam (primarily oxygen, hydrogen, nitrogen, krypton, and xenon) collect in the turbine condenser where they are removed continuously by an air ejector and discharged to the off-gas system. Most BWR plants now operating provide a minimum holdup time of about 30 minutes. In all plants now being designed or built, and in some that are operating, provisions are being made to increase this time substantially, chiefly by incorporating charcoal beds into the gas treatment system to adsorb krypton, xenon, iodine, and particulate material selectively. Delay times of 20 hours for kryptons and 15 days for xenons are obtainable in systems operated at ambient temperatures. If operation is at 0°F (–17.8°C), delay times of two days for kryptons and about six weeks for xenons can be achieved. For BWR's with a 30-minute delay time, the rates of emission of gaseous radioactivity are substantially higher than for PWR's, but are well within currently specified limits. The addition of charcoal beds is expected to reduce these emission rates to perhaps one 1000th of current levels (117d).

[Perspective can be found in comparisons of radioactivity in gaseous emissions from nuclear power plants and from coal-fired plants of similar generating capacity. Coal contains traces of the naturally occurring, radioactive uranium and thorium series of elements. One study has concluded that if the physical and biological behavior of the pertinent nuclides is taken into account, the coal plants add more radioactivity to the environment than the nuclear plants (118). A second study concluded that this is true for an older coal-fired station as compared to a PWR, but not when compared to a BWR (119). Newer coal-burning facilities with better control of particulate emissions presumably release very little radioactivity to the environment. A third study assessed not just generating plants, but total fuel cycles (120). The study concluded that, on the basis of radiological impact alone, the long-term population impact of the nuclear fuel cycle appears to exceed that of the coal fuel cycle by a significant margin.]

BWR gas treatment systems also incorporate a "recombiner" (117e). The intense radiation in the reactor core decomposes some of the water in the cooling system to hydrogen and oxygen, and it is desirable to recombine these gases to reduce the volume and to eliminate a possibly explosive mixture. A typical recombiner consists of a steel tank containing a replaceable catalyst-carrying cartridge or bed. In a PWR, the high pressure causes the hydrogen and oxygen to recombine without special treatment. Both PWR and BWR off-gas systems also contain High Efficiency Particulate Air filters. These remove radioactive solid particles formed when a gaseous parent nuclide decays to a radioactive particulate daughter or is itself adsorbed on a solid particle.

Primary coolant water discharged from both PWR and BWR plants is treated by more-or-less conventional means, and potentially radioactive liquids from vents, drains, valve leaks, etc., are collected for similar processing. The "delay and decay" tactic can be used with liquid wastes, although as a rule these move through the plant slowly enough to eliminate many short-lived species. Beds of cation and anion exchange resins remove ions in solution, and filtration or centrifugation takes out suspended particulate matter. Evaporators concentrate aqueous wastes, reducing their volume by 10 to 50 times. The residues are incorporated with a binder, such as cement, in 55-gal drums for off-site disposal. Miscellaneous solid contaminated waste, such as rags or paper, is compacted into similar drums (117f).

Tritium. One radioactive component of liquid wastes that cannot be removed by conventional water treatment is tritium, since it exchanges rapidly with the hydrogen in ordinary water to form a heavier form of water, usually HTO. In BWR's, tritium is formed mainly in the fuel rods by ternary fission or in the cooling water by neutron capture by the very small amount of naturally occurring deuterium (^2H or D) present. (Much more will be formed by the latter mechanism in heavy water reactors of the type used by the Canadians.) Very little of the tritium produced in the fuel rods escapes in the reactor; instead, the nuclide is released in the fuel reprocessing plant. Ternary fission or neutron capture by deuterium are also sources of tritium in PWR's. In these reactors, however, boron is dissolved in the primary cooling water as a reactor control device, and some lithium may get into the system from chemicals used to control acidity. Neutron reactions with these two light elements yield tritium, so the production of the nuclide is substantially higher in PWR's than in BWR's (117g).

Tritium is produced naturally in the environment, as mentioned earlier, by cosmic ray bombardment of oxygen and nitrogen in the upper atmosphere and by direct intrusion from outer space. The amount of tritium from natural sources far exceeds that produced by current nuclear power activities, but is substantially less than that injected into the atmosphere by testing of nuclear explosives. Fusion-type bombs produce hundreds of times more tritium than the fission type. Thus most of the residual tritium now in the atmosphere stems from pre-1963 U.S., Soviet, and British tests; small amounts have been added by more recent French and Chinese detonations.

Figure 5 shows the estimated world inventory of tritium to the year 2000. Lower estimates have been made (117h, 121), but all sources agree that the tritium produced in reactors will not reach the natural production rate until 1985–90. The amount of natural material in the environment cannot be reduced, but tritium produced in reactors presumably can be separated and contained at the fuel reprocessing stage. Research on methods of doing so is under way; if it succeeds, reactor contribution of tritium to the environment can be maintained indefinitely at its present level — well below natural background. The projections of Figure 5 were used by the EPA group to calculate the estimated doses from tritium shown in Table 5.

Allowable levels. Standards for radioactivity levels are established by the

Nuclear Regulatory Commission and are published periodically (32). To obtain a license to operate a reactor, an applicant must demonstrate his ability to establish an adequate monitoring and surveillance system for radioactivity releases. Operators are required also to report semiannually the quantities of discharged nuclides and the environmental levels of radiation and radioactivity that result from plant operation (34). AEC used these data, and similar information from nuclear facilities other than reactors (122), in a detailed two-year study of the impact of nuclear power to the year 2000 in 303,320 square miles in the central United States (123). This "Year 2000 Study" may be extended until the entire country has been considered essentially county by county. The first report concludes that, on the average, nuclear facilities will increase the dose rate per individual in the year 2000 by about 0.2 mrem/year over natural background.

EPA has studied intensively the effluents and their effect on the surrounding environment at the Yankee Nuclear Power Station (Haddam Neck, Conn.), which is an older PWR (7), and at the Dresden Nuclear Power Station (Morris, Ill.), an older BWR (8). Similar studies are under way or planned at newer reactors of both types (11). The Yankee and Dresden studies both concluded that the impact of the plants on the surrounding environment was minimal; radioactivity above the natural background was detected only with great difficulty. While there are some dissenters, such surveys have led most observers to decide that LWR's in normal operation do not greatly threaten the environment or public safety. The main concern is the possibility of accidents.

FIGURE 5

Projected world inventory of tritium in atmosphere and surface waters

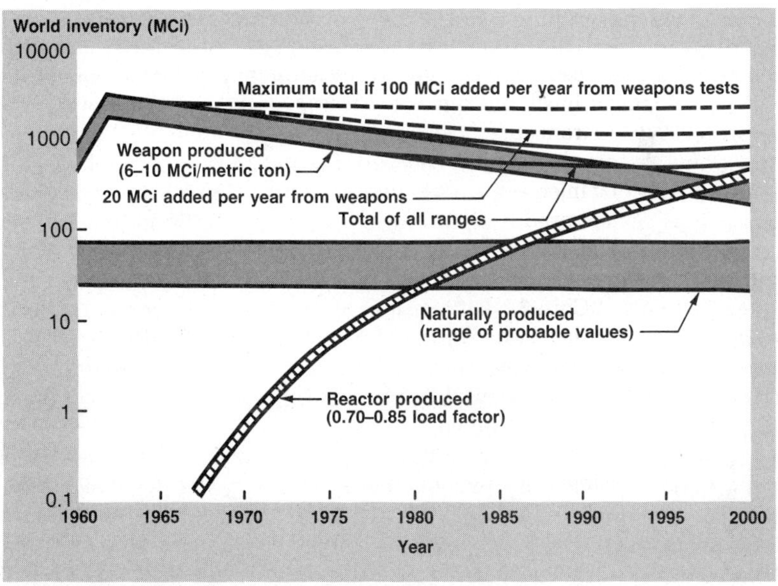

Source: Klement, A.W., Jr., et al, "Estimates of Ionizing Radiation Doses in the United States, 1960–2000," U.S. Environmental Protection Agency, ORP/CSD 72-1, Washington, D.C., 1972, p. 44.

Recommendation R4: *One of the important sources of occupational radiation doses within nuclear power stations results from the chronic buildup of radioactive materials, especially cobalt-58 and -60, which accumulate in the primary system. Reduction of this source of radioactivity by off-line or preferably on-line chemical cleaning of the primary system should be given greater attention.*

Reactor mishaps. Reactors, like any complex facility, will have occasional maintenance problems. Pumps malfunction and must be repaired or replaced; steam lines, water lines, and heat exchangers leak; bearings wear out; a leaky fuel pin may require that part of the cooling water be given special treatment. Such problems are expected, accepted, and corrected as they occur. The possibilities for more serious accidents lie in several directions. Excess reactivity may be added to the core, causing a nuclear excursion, or coolant may be lost partially or totally. The effect in either case will be an abrupt rise in temperature in part or all of the core. There is no chance in an LWR that this will create an "atom bomb" (14, 79, 117i). Depending on the rate of temperature rise and the time required for backup safety controls to operate, however, core components can melt and rapid, energy-releasing chemical reactions can take place in the molten material, possibly causing an explosion. If melt-down or an explosion occur, fission products will be released from the fuel.

Relatively few reactor accidents are known to have happened worldwide in the past 20 years, but there are examples of the effects of excess reactivity and of partial loss of coolant. The most serious accident occurred in 1961 at an experimental facility, the Army Stationary Low-Power Reactor (SL-1), at the National Reactor Testing Station in Idaho. The mishap killed the three operators in the building, so there were no survivors to testify to the cause. The reactor had been shut down for maintenance, but the evidence indicated that a control rod had been removed for unknown reasons, allowing the reactor promptly to go critical. The resulting steam explosion did extensive damage and caused high radiation levels within the building. Most of the external contamination was limited to about 3 acres around the plant; the only exception was gaseous ^{131}I, which could be detected on vegetation many miles downwind (2d). (It should be noted that this reactor had no containment vessel of the kind used with normal power reactors.)

In 1966 a partial blockage of coolant occurred in the sodium-cooled, fast-breeder Enrico Fermi reactor near Monroe, Mich. Local heating in the core caused the partial melt-down of two of the 105 fuel subassemblies, which released about 10,000 curies of fission products to the primary sodium loop. The radioactivity was contained completely, no contamination was detected outside the containment shell, and no exposure of employees or the public resulted. Cleanup, repairs, and the resolution of various legal difficulties imposed tremendous expense on the reactor operators, and four years elapsed before the facility could be brought back to full power (2f). After operating for an equal period the reactor was shut down.

Sudden and complete loss of coolant presumably would be the worst situation of all, but since such an event has never occurred, the possible results are speculative and subject to much debate (79, 117j, 124–126). About the only conceivable way that coolant could be lost completely in an LWR is by rupture of the pressure vessel or breakage of a main water pipe. These are massive structures, and defenders of nuclear power argue that experience with comparable, nonnuclear facilities, operating at substantially higher pressure than reactors, justifies estimates that a major pipe break would only occur between once in 1 million and once in 10 million operating years (80). All LWR's have elaborate backup safety systems, including emergency core cooling systems (ECCS) that automatically flood the reactor core with water to prevent heatup in case of a coolant loss (117k). The infallibility or fallibility of the ECCS is, again, a subject of much current disagreement. Relatively small prototypes have not always worked perfectly in tests (127). The disagreement occurs over the extent to which the results of the small-scale tests can be extrapolated to full-scale ECCS's, which have not been tested because of the great expense involved. In an inadvertent test at the San Onofre-1 reactor in California, the ECCS worked perfectly (128). The effects of a broken blade in the turbine tripped the ECCS, even though the core never lost its cooling water.

The most recent serious reactor incident in the United States was not a result of a failure in the reactor itself, but of a fire that incapacitated a disturbing number of the in-depth array of safety devices before the blaze was brought under control (129). The fire occurred at the Brown's Ferry Reactors 1-2 in Alabama in March of 1975 and has had widespread repercussions in causing reexaminations of similar systems in other existing installations.

A very comprehensive analysis of the reactor safety problem (the "Rassmussen report"), prepared under the aegis of the Massachusetts Institute of Technology, has been published (130). EPA also included a review of nuclear reactors (105b) and fuel processing plants (105c) in its 1973 analysis of the nuclear fuel cycle.

Fuel reprocessing plants

The buildup of fission products and loss of reactivity permit the "burning" of only about 50% of the ^{235}U in a typical LWR fuel before it must be replaced. The great economic value of the remaining ^{235}U and of the fissionable ^{239}Pu produced during the time in the reactor is the chief argument for reprocessing the fuel to recover the two isotopes.

The Administration's position (in 1978) is that the value of plutonium recycle does not warrant the risks of nuclear proliferation arising from the weapons potential of separated plutonium — hence the decision to delay commercial reprocessing indefinitely.

By early 1978, very little spent fuel from commercial power reactors in the U.S. had ever been reprocessed; virtually all spent fuel assemblies from such reactors remained in interim storage. The Nuclear Regulatory Commission (NRC), which had been studying the regulatory aspects of reprocessing and related activities, had terminated the work. NRC also had stopped processing applications to build, operate, or modify reprocessing facilities.

Government-owned fuel reprocessing plants have been built at Hanford, Wash., Oak Ridge, Tenn., Savannah River, S.C., and the National Reactor Testing Station in Idaho (117t). Nuclear Fuel Services (NFS) operated the first commercial reprocessing plant under an AEC license from 1966 to 1972, when it closed down to expand capacity. NFS never reopened and in September 1976 withdrew from reprocessing because it could not economically comply with the by-then much stiffer licensing requirements. General Electric completed a second commercial plant, at Morris, Ill., in 1974, but later suspended its reprocessing plans indefinitely because it could not economically correct severe operating problems encountered in the separations scheme. Allied–General Nuclear Services completed a separations plant at Barnwell, S.C., by early 1977, but did not start to build its planned plutonium conversion and waste solidification facilities because of the Administration's policy. Thus by mid-1978 no commercial reprocessing plant was operating in the U.S. The plants mentioned and more would be required if the nuclear power industry expands as projected and if the Government ceased to oppose plutonium recycle. Each plant would serve a number of power reactors.

Processing steps. Reference 117l gives a general description of the equipment needed for the various steps in fuel reprocessing. (All manipulations must be done remotely behind heavy shielding because of the intense gamma activity of the fission products.) Reference 2e summarizes the separations chemistry used. The clad fuel rods are first cut into small pieces, which are transferred to a dissolver. With the usual zircalloy-clad UO_2 fuels, nitric acid is used to dissolve the uranium and its reaction products, leaving the cladding hulls as residue. (For uranium metal fuels, the cladding may be dissolved in a separate step or removed mechanically.) The hulls are removed as solid waste. The subsequent separation chemistry varies in detail at different sites, but all use some version of the Purex process, a solvent extraction system using tributyl phosphate dissolved in kerosene as the extraction agent. The uranium, plutonium, and fission products are separated into individual fractions in a series of extraction and stripping steps under controlled chemical conditions. The transplutonium elements remain with the fission products in the aqueous phase, as does about 0.5% of the plutonium (131). The separated uranium is returned to the enrichment plants, and the plutonium at present is converted to a solid and stored. The fission-product fraction is the "high-level waste" that is of major concern in terms of disposal or long-term storage.

Nuclide releases. Since the fuel is dissolved completely, all of the radioactive fission-product gases are released. Current practice, however, is to store the used fuel elements in fuel pool storage facilities at the reactor for "cooling," so that all of the short-lived species have decayed to stable daughters by the time fuel is shipped to the reprocessing plant. The most important gaseous radioactive nuclides remaining are tritium (half-life 12.33 years), ^{129}I (half-life 1.5 x 10^7 years), ^{85}Kr (half-life 10.73 years), and perhaps traces of ^{131}I (half-life 8.01 days). The cooling period nominally is 90 to 150 days, but in practice has been much longer in most cases to date owing to the lack of operating reprocessing plants in this country. The lack of reprocessing facilities, in fact, is creating an increasingly serious storage problem for privately-owned reactors.

Most of the tritium is converted to water in the dissolver solution, but essentially all of it leaves the plant as water vapor in the gaseous effluents. An intensive EPA survey (9) of the Nuclear Fuel Services (NSF) separations plant and its surroundings concluded that the stack contribution of tritium was only in the range of 0.5–3.0% of the natural background (including weapons debris). Nevertheless, studies are being made on methods of eliminating tritium completely. One technique would be to remove the tritium from the fuel as a gas before the dissolving step (132). In a second scheme, all process liquids would be recycled continuously until the tritium reached a predetermined level, at which point the liquids would be removed and stored as active waste. (Both the Morris and Barnwell plants are designed so that no liquid effluents would be released; all would be stored.)

The amount of ^{131}I in the fuel will depend on the degree of cooling that occurs before processing begins. This period generally has been of the order of at least a year in the limited commercial experience to date, so none of the isotope has been detected in the local environment. Should breeder reactors enter the picture, however, ^{131}I would require attention, since breeder fuels would be of higher initial enrichment, would be carried to greater burnup, and would be given less cooling time before reprocessing. (These factors also would complicate the transportation problem.)

Originally, ^{129}I was not considered a potential hazard because of its long half-life and correspondingly low specific activity (disintegrations per second per gram) and low fission yield. The long half-life, however, means that any ^{129}I that enters the environment will be essentially a permanent contaminant and conceivably could build up to a dangerous degree in the grass-cow-milk-human thyroid chain. An EPA study (10) around the NFS plant did indeed find ^{129}I at levels substantially above background. Its concentration was equivalent to 7% of the specific activity that, if present in the human thyroid, would result in the individual dose limit specified by the Federal Radiation Council, but even that level is a matter for concern in view of the relatively short time the plant operated. Two units in the plant were designed specifically to remove iodine: the off-gas system included a chemical scrubber that washed the gaseous effluent with a solution of mercurous and mercuric nitrate; and the gases were passed over a solid adsorption medium (ceramic berl saddles coated with silver nitrate). Improvements in the system were planned for installation when it was thought that the plant would reopen, and work on the problem is continuing. The silver form of a synthetic zeolite shows promise for iodine adsorption (133), and Oak Ridge National Laboratory is working on a scrubbing system employing concentrated nitric acid (134).

All of the ^{85}Kr in the fuel is released at present to the atmosphere. It accumulates there because of its relatively long half-life and chemical inertness, but is not considered a problem at this time. The 0.037 mrem/year dose per person for ^{85}Kr in the year 2000 (Table 5) is based on the assumption that current practice will not change; the estimate reflects a steady increase to about 100 times the value for 1970. Research is being actively pursued on methods for trapping both the xenons and kryptons in off-gases (97). In one technique, the noble gases are distilled from a column maintained at the temperature of liquid nitrogen (−196°C) and collected in pressurized cylinders for storage (117m, 135). Also being developed is an absorption process in which the dried off-gases are contacted with a liquid fluorocarbon at −32°C and about 27 atmospheres (400 psi). The fluorocarbon selectively absorbs the krypton and xenon, which are stripped out in a second column for bottling and storage (117m, 136, 137). Both processes have been demonstrated on a pilot-plant scale, but not at the commercial level.

Reference 117n summarizes the average radioactive releases from the gaseous and liquid effluents from the Nuclear Fuel Services plant during 1966–71. As a fraction of the limits established for the plant, the maximum annual releases were 7.2% for ^{85}Kr, less than 1% for ^{131}I, and 35% for radioactive particulate material. The maximum annual release of radioactivity in the treated liquid effluents was 22% of the licensed limit. Releases at such levels are not anticipated in the future, however, at any other modern reprocessing plant.

Recommendation R5: *Considerable progress has been made in devising methods for removing radioactive species from gaseous and liquid effluents. Efforts to improve such techniques should continue with particular attention* *paid to the tritium, iodine, and xenon-krypton release problems and with even more emphasis on removal of trace transuranics from emissions and effluents.*

Recommendation R6: *Strong emphasis should be placed on developing more efficient methods for complete removal of actinides and other long-lived species in reprocessing light water and breeder reactor fuels in the event that circumstances call for the country to reconsider its present position on the reprocessing question. Innovative processes having the effect of simplifying the final waste disposal problem should receive particular attention.*

Transportation of Radioactive Materials

Spent fuels must be moved from reactors to reprocessing plants and eventually, as solid waste, from those plants to federal repositories for long-term storage or disposal. ("Storage" and "disposal" often are used interchangeably. More precisely, storage refers to methods of emplacement of wastes in such a manner and with the intent that they can be retrieved; disposal applies to techniques of emplacement that are considered final and that for all practical purposes make the materials irretrievable.) Storage and disposal will not occur to a significant degree until about 1985 or later. Reprocessing plants are required to solidify liquid high-level wastes within five years after they are generated and are allowed an additional five years before the wastes must be shipped to the repository.

The Department of Transportation (DOT) has primary responsibility for regulating (33) the shipment of radioactive materials in the United States; NRC specifies the criteria for packaging and shipping fissile and large amounts of radioactive substances (36). The International Atomic Energy Agency has developed regulations for movements between countries (138). The rules and regulations are complex, since they apply to a variety of types of shipments. Isotopes are classified in seven "transport groups." The most hazardous species, such as ^{239}Pu, ^{241}Am, and ^{226}Ra, are in group I; the least hazardous, such as certain pharmaceuticals labeled with tritium, are in group VII. Each of the two classes of shipments, A and B, has its own specifications for type of package, permissible types of transport, etc., and for the quantity in curies of an isotope in a particular transport group that can be shipped. Shipment of a transport group I isotope, for example, is limited to 0.001 Ci under the A regulations, but 20 Ci may be shipped under the B regulations, which are much more stringent.

Prototypes of all packages used to ship radioactive material must pass a series of tests to demonstrate their integrity under conditions of heat and cold, exposure to water sprays, vibration, free fall on a hard surface, etc. These tests and the shipping regulations are particularly detailed for the class B shipments. Where the large amounts of radioactivity represented by spent fuel assemblies or solidified high-level waste are being moved, the standards require that the packaging shall prevent the uncontrolled loss or disposal of the contents, retain shielding efficiency, assure nuclear criticality safety, and provide adequate heat dissipation, not only in normal conditions of transport, but also in severe accidents. The packages must withstand the following tests in sequence: a 30-ft fall onto an essentially unyielding surface; a 40-in drop onto a 6-in diameter steel pin; 30 minutes in a 1470°F fire; and, for fissile materials, an 8-hour immersion in 3 ft of water. Containers that meet these standards and provide the needed shielding are necessarily massive. Only a few have been constructed to date;

those designed for truck shipments weigh 20 to 25 t, and one unit for rail car use weighs 65 t.

Numerous analyses have been made of the probability that a large shipment of radioactivity will be involved in a serious transportation accident (2g, 117o, 139, 140). The last of these concludes that a shipment of fuel or waste will be involved in a transportation accident about once every 10 years and that one accident out of 100 will be severe. The other analyses reach somewhat similar conclusions.

Radioactive Waste Management

The management of radioactive wastes is perhaps the most vexing problem of nuclear electric power. The following discussion is restricted to activities in the U.S., but the problem is international. A preponderance of the research and development on high-level waste management has been done in this country, but it would appear that certain European nations will take the lead in demonstrating potential application of the known information.

The bulk of the waste material that is either inherently radioactive or is contaminated with radioactivity is generated either in the nuclear power cycle or in government-owned production and research facilities. Smaller amounts, however, can arise in departments of nuclear medicine in hospitals, in university research laboratories, in industrial plants using radioisotopes for various applications, etc. All users of significant amounts of radioactivity must be licensed by NRC or by a "license-agreement state" — a state that has assumed from NRC the responsibility of regulating the distribution, handling, and use of radioactive materials. Establishment of a waste-disposal system is part of the licensing procedure; the ground rules for disposal are established by NRC (141). Wastes are classified as low-, intermediate-, and high-level, depending on their concentration of radioactivity and their potential hazard.

Low-level wastes

Low-level wastes are those that can be discharged to the environment under controlled conditions after reasonable dilution or simple processing. Liquid low-level wastes usually contain less than a microcurie of radioactivity per gallon and can be disposed of to the environment if diluted to less than 10% of the maximum permissible concentration for continuous occupational exposure. Solid low-level wastes can be paper, wood, biological materials, etc. Because of the difficulty of monitoring for low radioactivity in such materials, all solid waste generated in "suspect" areas is often treated as if it were contaminated, which adds considerably to the amount of material that is handled as low-level solid waste. Combustible material may be burned, providing the radioactivity in the gaseous effluents does not exceed NRC limits. Combustion is used infrequently, however, except in hospitals and similar institutions where biological materials must be disposed of for sanitary reasons.

In the past, a certain amount of low-level waste was treated by packaging in 55-gal steel drums, adding concrete and allowing it to harden, and dumping the drums at preselected sites in the ocean. This technique has been largely discontinued in the U.S., but still is used to some degree by other countries. By far the bulk of the low-level waste is disposed of by near-surface burial on land (142). Large metal containers are placed in a trench in carefully chosen and monitored sites and covered with several feet of earth. This was done only at AEC sites until 1963, when commercial organizations were first licensed to establish disposal areas. There are now about half a dozen such areas in the country. Wastes generated by AEC and ERDA have amounted to 1.5 to 2.0 million cubic feet per year since 1963 and have required about 25 acres of land annually for burial. The volume of waste available for commercial burial at non-AEC sites was estimated in 1969 to reach 2 million cubic feet per year by 1975 and 3 million by 1980 (143). Standard baling machines are being used to compress the wastes, reducing their volume some two to 10 times before burial (143). Recent evidence of radioactivity leakage from the disposal trenches used by commercial operators has generated discussion of the advisability of the Federal Government's owning and operating all radioactivity burial grounds.

Intermediate-level wastes

Wastes are classified somewhat arbitrarily, and the definition of the intermediate level will vary from site to site, depending on the particular waste-management scheme. Oak Ridge National Laboratory defined intermediate wastes as those that contain 10^4 to 10^6 times their maximum permissible concentration. On that basis it is assumed that 200 gal of intermediate-level liquid waste will be produced per metric ton of fuel processed. From this assumption, Oak Ridge projected an annual generation of 3.2×10^6 gal of intermediate wastes by the year 2000 and a total accumulation of 4.9×10^7 gal to that date (144a). This estimate, and the one in the preceding paragraph, may be overly high in view of the slippage in planned reactor construction that has occurred since the projections were made.

Materials at the intermediate level of activity require more than simple processing. They may be treated by methods like evaporation, ion exchange resin column, flocculation, etc., to split them into a high-level concentrate and a low-level liquid. As the quantities generated increase, however, it is probable that intermediate-level material will simply be combined and solidified with high-level wastes. Oak Ridge also has been studying hydraulic fracturing, a method borrowed from the oil industry, for direct disposal of intermediate wastes. Wells are drilled into shale beds or other suitable geologic formations, and lateral cuts are made by jets of a sand-water mixture or by shaped explosive charges. A mixture of water, waste, and cement is pumped into the well casing under high pressure and forced into the cracks in the rock, where it hardens to form a closed, water-tight chamber. Test corings around the area of a pilot plant built to test this technique have shown satisfactory containment of the radioactivity (143).

Cladding hulls from fuel reprocessing plants might be considered intermediate-level solid waste. The elements in the alloy used to clad fuel are activated by neutron capture while in the reactor, and some small amount of fission products and actinides diffuses into the metal. Oak Ridge projected an annual production of 87 cu ft of solid waste from LWR and LMFBR fuel processing in the year 2000 and an accumulated volume of 1030 cu ft, requiring 980 acres for burial (144b). The nuclides formed by activation are relatively short-lived compared with some of the fission products, so hulls and similar materials generally are stored for long periods (years) to allow decay to occur. They can then be placed in burial sites if demonstrated to be sufficiently free of actinide contamination.

Another type of solid waste that might be considered in the intermediate class is contaminated obsolete equipment from fuel fabrication plants, reactors, etc.

Normally such items can be decontaminated to the point where they can be buried at licensed sites. Highly alpha-contaminated units of this type that cannot be cleaned sufficiently are a disposal problem that presumably will become more urgent as first-generation facilities in the nuclear fuel cycle, including reactors, eventually are decommissioned and if plutonium is used to a greater extent in fuel fabrication.

Recommendation R7: *The problem of decontamination or disposal of alpha-contaminated equipment should receive a high level of attention.*

High-level wastes

Process solutions from fuel reprocessing plants constitute essentially all of the high-level wastes. They present one of two separate problems, depending on whether or not they contain long-lived alpha emitters. If these are not present, the isotopes of primary concern are ^{90}Sr (half-life 28.1 years) and ^{137}Cs (half-life 30.2 years); methods must be devised to isolate such wastes for 600 to 700 years to allow them to decay to an innocuous level. If substantial amounts of alpha-emitters are present, the time of isolation must be of the order of several hundred thousand years. Guaranteeing this isolation certainly presents a unique technological challenge to the human race, whose written history goes back only a small fraction of that time. The situation has been studied a great deal [one report (145) lists 345 literature references in *Nuclear Abstracts* in the first six months of 1973] and must be studied much more to be certain that future generations are not exposed to a hazard. The problem is difficult, but not necessarily insoluble.

The Purex solvent extraction process used in fuel reprocessing is a nitric acid system. Because of variations used by different processors, however, the first-cycle aqueous raffinate that constitutes the very high-level waste may range all the way from a simple acid solution of fission product nitrates to a solution that also is high in nitrate salts of various inert cations. (The first-cycle raffinate is the stream that has undergone the first cycle of uranium and plutonium extraction but contains transplutonium elements and fission products and small amounts of unextracted plutonium and uranium.) Some systems use a later step involving sulfate to separate uranium and plutonium; if the resulting aqueous wastes are mixed back with the earlier raffinate, the sulfate also will be present. Because the added salts and the sulfate complicate both the interim storage of the material as a liquid and its eventual solidification, the tendency is toward using modifications that yield an unsalted first-cycle raffinate. This solution typically will be rather strongly acidic (6*N* HNO_3) and will contain some inert cations (0.3*N*), and substantially more fission product cations (1–2*N*), depending on the degree of ^{235}U burnup in the reactor (*N* means "normal"; the larger the numerical coefficient of *N*, the higher the concentration of the given substance in the solution).

Past and current practice has been to store the liquid wastes in tanks, which everyone understands to be purely an interim measure. Evaporation can be used to reduce the volume of the wastes and to remove part of the nitric acid, but the remaining solution cannot be concentrated to much more than 4*N* in total metallic ions without the danger of precipitation. This may lead to problems during extended storage (146), although in some cases at Hanford precipitation has been caused deliberately and the wastes solidified in place in the tanks (117p).

The volumes of condensed raffinates may vary from 40 to several hundred gallons per metric ton of processed fuel. Activity levels may be as high as several thousands of curies per gallon, and heat generation rates may be 20 to 30 Btu per hour per gallon. The liquids are stored as acid solutions in stainless steel tanks or are neutralized for storage in carbon-steel tanks. Neutralized wastes are allowed to self-heat and the resulting vapors condensed and returned to the tank, or they are cooled by water coils submerged in the liquid. In the U.S. the tanks (0.33 to 3.0 million-gallon capacity) are encased in concrete and buried underground. Department of Energy (DOE) tank farms, comprising a total of about 200 tanks, are located at Hanford (perhaps 75% of the total), at the Savannah River plant, and at the Idaho Chemical Processing Plant near Idaho Falls (117q, 144c).

The tank farm system has not been free of problems. During 1958–73, AEC reported 17 leaks involving a total of 447,000 gal of waste (147, 148). The liquid soaks into the ground, but the radioactive species appear to be held sufficiently by the ion exchange action of the soil so that the activity remains within 10 or 15 ft of the tank and well above the water table (144c). The fact that the leaks occur at all, however, is a matter of considerable concern. As a result, each new generation of tanks has incorporated additional safeguards and more sophisticated instrumentation (117r, 144d).

Solidification. Converting the liquid wastes to solids offers many advantages for long-term storage (149) or ultimate disposal. A very considerable amount of research has been undertaken to find methods leading to maximum immobilization of the active species. Four processes have been developed in the U.S. and demonstrated on an engineering scale. These are the pot, spray, phosphate glass, and fluidized bed techniques (117s, 144e, 146). In each of the four, heat is applied to raise the temperature of the waste to 400° to 1200°C. The volatile components (mostly water and nitrates) are driven off, leaving a solid or a melt that will cool to a solid.

In the pot calcination method, the wastes are evaporated to dryness and converted to oxides by controlled heating on a batch basis in the "pot" that serves as the final container. Because the pots are the processing vessels, they are exposed to severely corrosive conditions during calcination and must be made of stainless steel. Pot calcination is a simple process and applicable to a variety of feed concentrations. The product is somewhat more water-leachable than the glassy solids produced by some of the other methods.

In the spray calciner, liquid waste containing melt-making additives is fed through an atomizing nozzle into the top of a heated cylindrical tower. As the liquid falls, it is calcined to a powder that collects in a continuous melter at the bottom of the column. Here it is melted at 800° to 1200°C. The molten calcine flows through an overflow weir or freeze valve into the receiver-storage pot where it cools and solidifies. These pots may be made of mild steel, but platinum is the only metal found to date that reliably withstands the heat and corrosive conditions of the melter. The product of the spray process is a tough, rocklike, microcrystalline material of moderately low solubility in water.

The first step in the phosphate glass process is the evaporation of the waste plus additives (phosphoric acid and alkali metal oxides) to a thick, syrupy, aqueous phosphate slurry. This material is fed to a melter, where the remaining volatile components are driven off, and then it is heated to 1000° to 1200°C to form a molten glass. As in the spray process, this melt flows into a receiver-storage pot for cooling. These pots, again, can be of mild steel, and the melter must be of platinum. The final product is a monolithic, moderately brittle glass with low solubility in water solutions.

The fluidized-bed technique was developed primarily for use with the aluminum nitrate wastes from earlier separation processes, and development for the more complex Purex wastes has been somewhat limited. In this technique the liquid waste is atomized into a bed of heated granular solids that is agitated (fluidized) continuously by sparging air upward through the bed. The waste calcines as a coating on the particles of the bed, which are drawn off continuously and transferred to storage bins. The product is granular, with an average particle size of about 50 μm. The granules generally are spherical and are moderately soft and friable.

Most waste management plans call for interim storage of solidified materials for as long as 30 years before final disposal. This is primarily because of the heat released by the fission products, which can raise initial temperatures at the centerline of a 6-in-diameter pot filled with solidified waste to almost 1000°C. Concepts for interim storage accordingly have envisaged water-filled canals, air-cooled annular bins, or air-cooled concrete vaults. An essential safeguard is a thoroughly reliable backup method for removing heat should the primary cooling system fail.

The rate of heating by the fission products drops sharply during the first few years after removal from the reactor as the shorter-lived species decay to stable forms. These latter are different elements than those present originally, a fact that produces changes in the properties of the calcined or melted materials. For most of the solidification processes, clear definition of the effect of these changes remains to be determined.

Recommendation R8: *Research on waste solidification methods should be maintained at a high level because of the obvious advantages to be gained in terms of the final disposal problem. Particular attention should be paid to the effects of chemical changes with time upon the water solubility and physical properties of the solid.*

Permanent disposal

The time will arrive eventually when the problem of disposing of the solidified materials permanently must be faced. Various alternatives have been considered (150) (Fig. 6).

The Congressional Joint Committee on Atomic Energy asked the National Academy of Sciences a number of years ago to study the disposal problem and make recommendations. NAS did so and concluded in 1957 that disposal in abandoned salt mines was an attractive possibility (37). The engineering evaluations and development work that have been carried out (primarily by Oak Ridge National Laboratory) accordingly have concentrated on this concept. NAS has reviewed the disposal problem periodically and has a permanent Committee on Radioactive Waste Management. NAS reviewed the salt bed concept again in 1970 (38).

There are several arguments for bedded salt deposits as repositories: salt is easy to mine; in time it will flow to seal the containers in place; and the fact that the salt is there proves that water has not intruded for millions of years. AEC selected a site near Lyons, Kan., and the engineering and construction had proceeded almost to the demonstration point when the discovery of a number of man-made wells near the mine raised doubts of its future vulnerability to groundwater intrusion; the project was abandoned in 1972. Early in 1978, a DOE

FIGURE 6

The "taxonomy" of nuclear waste disposal options

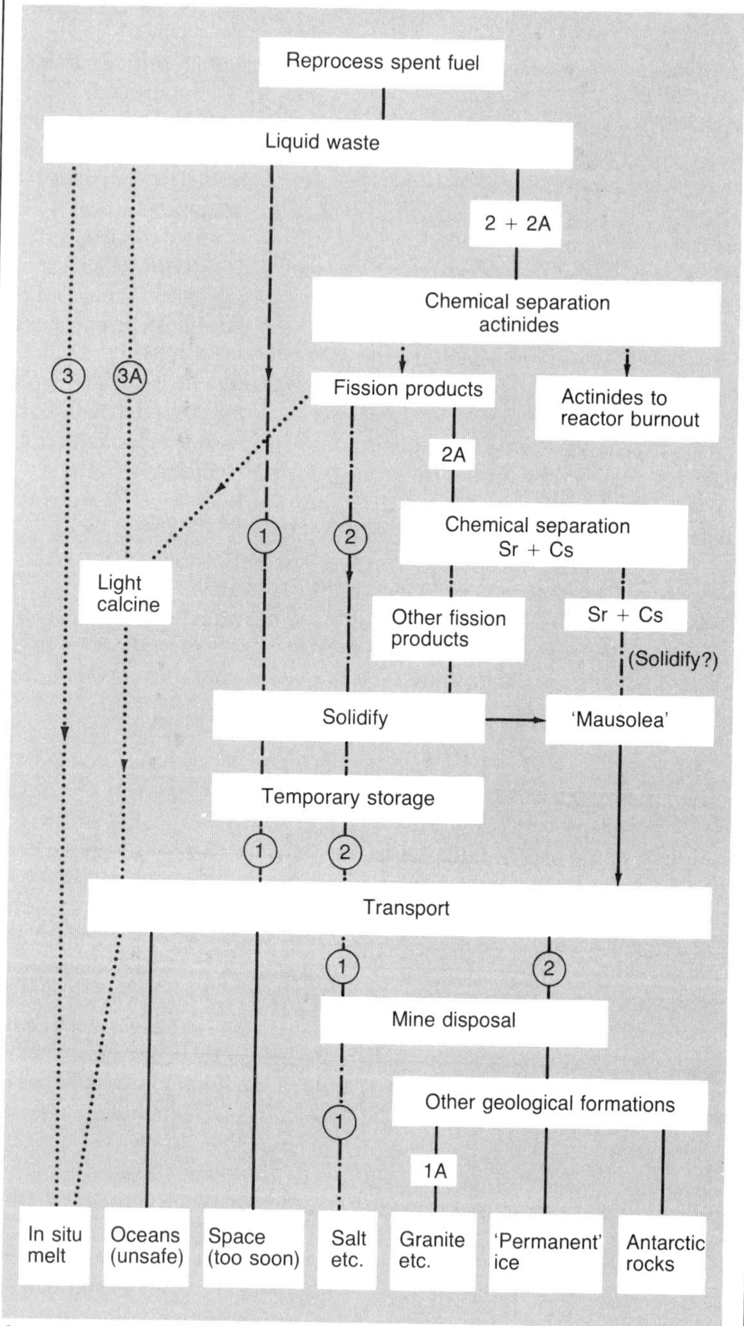

Source: Kubo, A.S., Rose, D.J., "Disposal of Nuclear Wastes," *Science*, 182, 1205 (1973).

task force reported that salt beds in southeastern New Mexico would provide safe repositories.

Other types of geologic structures can be considered for deep burial of active wastes (Fig. 6). An extensive study of the bedrock under the DOE facilities at Savannah River has been undertaken to determine if vaults mined in the rock, well below existing groundwater aquifers, could be used for waste disposal (39). Burial in Antarctic rocks, where all groundwater is frozen to a depth of 1 km, has been suggested, as has disposal in deep holes drilled in one of the long-lasting ice sheets such as those of Greenland or Antarctica. Another possibility is to shoot the material into space (151). This does not seem economically feasible today, but may become so in the future. One of the problems would be designing a container guaranteed to withstand the effects of an aborted flight.

The Lawrence Livermore Laboratory has suggested the "*in situ* melt" concept. A large cavity would be blasted into a deep impervious rock stratum by nuclear or conventional explosives and the liquid wastes simply poured in directly. Here the wastes would boil themselves to dryness and, with some of the surrounding rock, form a melt that would solidify to a permanent, glassy mass.

The "mausolea" of Figure 6 refer to carefully engineered, near-surface underground vaults that would be used to store solidified wastes in retrievable form until a completely satisfactory method of final disposal can be developed. The idea is an extension of the interim storage concept and would entail the costs of very long maintenance and surveillance to be certain the materials were contained safely.

The ultimate disposal of high-level wastes would be simplified considerably if plutonium and other actinide elements could be removed essentially completely and either handled separately or destroyed by long irradiation in a reactor (152). Destruction by irradiation is possible because, as successive neutrons are captured, new isotopes are formed, some of which have very high fission cross sections. The original targets — the plutonium and actinide elements — eventually are "burned out" and converted to fission products. This is a very long process at the level of neutron fluxes available in most present-day reactors, although the seriousness of the actinide disposal problem conceivably could justify the design and construction of special, high-flux reactors dedicated only to this purpose. The advent of fast breeders could make the concept more attractive, particularly since the actinide isotopes are fissionable by fast neutrons and thus would serve as supplementary fuel.

Ultimate disposal of high-level wastes might be eased also if the cesium and strontium could be separated from the other fission products, since these are the two chief heat-producing fractions. Their removal would simplify considerably the engineering problems associated with burying the remainder of the material underground. The cesium and strontium presumably would be stored separately in mausolea, although the problem of their ultimate disposal eventually would have to be faced. It has been suggested also that the actinides could be recovered relatively simply for use as fuel after storing the wastes retrievably for a long enough period to permit the fission products to decay (153).

Recommendation R9: *A major effort should be made to simplify the radioactive waste disposal problem by separating the actinides, and the possibility of destroying the separated actinides in nuclear reactors should be thoroughly investigated.*

Recommendation R10: *Research should continue to find economical means of isolating cesium and strontium from radioactive wastes for separate management, although these elements are not as critical a problem as are the actinides.*

By 1978, DOE was at work on plans to establish full-scale high-level waste repositories in deep geologic formations. The first of these, to be ready for use sometime after 1985, would be built in a large salt bed; other types of geologic formations were being considered for later repositories. At the same time, the Nuclear Regulatory Commission was reviewing the status of waste-related technologies (154) and developing performance criteria for wastes and repositories. The planning for long-term storage was to accommodate both high-level wastes from fuel reprocessing and, in the event that widespread reprocessing did not materialize, the spent fuel assemblies themselves. In fact, the DOE task force mentioned earlier recommended that operations begin with the interment of perhaps 1000 spent reactor fuel assemblies in a specially excavated repository in salt beds in southeastern New Mexico.

Nevada Test Site

Nuclear explosives are tested, for both military and peaceful uses, on a large, government reservation northwest of Las Vegas, Nev. An extensive year-long monitoring program was conducted during 1961–62 to determine the radiation exposure of the population living near the site, and a similar study was conducted in 1969–70. During the first period, when atmospheric testing had been resumed following the moratorium of 1958, the calculated whole-body dose to the exposed population of 18,000 was 47 mrem/person. No exposure from test site activities could be detected above the worldwide fallout background during the second period after atmospheric testing had been banned for some years (5q).

Tests of nuclear reactor engines were conducted at the Nuclear Rocket Development Station adjacent to the Nevada Test Site for 10 years beginning in 1959. Radiological measurements were made during each test. External gamma exposure to the population was too low to be significant, but some ^{131}I and ^{133}I were detected in milk from nearby ranches. The average thyroid dose from the milk to the exposed group over the 10-year period was calculated to be about 3 mrem/person/year (5m).

Two underground nuclear tests have been conducted at the test site on Amchitka Island, Alaska. No radioactivity above natural and fallout background was detected off the site following the tests (5n).

No projections of exposures from nuclear weapons tests are shown in Table 5 because of the uncertainties in predicting their number in the future.

Peaceful Nuclear Explosives

Nuclear explosives represent huge sources of energy, and many practical uses for them have been suggested. The problem is to avoid contaminating the environment with radioactivity. The two general classes of application are "cratering" and "contained" detonations. Cratering explosions, for example, could be used to uncover mineral deposits, create harbors, and build canals, reservoirs, and earthen dams. Contained explosions might be used to increase the permeability of gas- and oil-bearing strata, to stimulate geothermal heat, and to create underground cavities for storing natural gas or disposing of industrial wastes. Six cratering experiments had been conducted in the U.S. and three in the U.S.S.R. by 1973; each nation had conducted five contained explosions. The U.S. tests have been experimental, while the Soviets have actually used their devices to increase production in oil and gas fields, to construct a reservoir, and to eliminate a runaway gas well (13l).

Three of the U.S. contained tests were in natural-gas-containing strata. The first, "Gasbuggy," was conducted near Farmington, N.M., in 1967; the second, "Rulison," near Grand Valley, Colo., in 1969; and the third, "Rio Blanco," also in Colorado, in 1973. No radioactivity was detected off the test sites during the detonation phase, but ^{3}H, ^{14}C, and ^{85}Kr were found in the gas when wells were drilled. The results of Gasbuggy indicated that tritium was the isotope of most concern, so the Rulison device was redesigned to yield less of the nuclide. In the event, tritium production was reduced about 80% from the Gasbuggy level. In both tests, the gas was flared at the well-head, not distributed commercially. Investigators did, however, calculate the whole-body dose that would have been delivered to inhabitants of the Los Angeles basin if the effluent from Gasbuggy had been injected into that area's natural gas distribution system. The maximum exposure would have been slightly more than 0.1 mrem and almost completely from the tritium (13m). In the Rio Blanco experiment, three 30-kiloton devices were detonated simultaneously. As with Rulison they were designed for low tritium yield, and the gas was flared at the well-head (155).

Again, no projections of exposures from peaceful uses of nuclear explosives are included in Table 5 because of their uncertain future.

Other Federal Installations

EPA lists 22 major DOE installations other than the three gaseous diffusion plants (5o). These include facilities like the Argonne, Brookhaven, and Oak Ridge National Laboratories, the National Accelerator Laboratory, the National Reactor Testing Station, and Los Alamos Scientific Laboratory. Most of these facilities are primarily for research and development, although some are active in certain phases of nuclear materials production. Most of the installations handle radioactive materials in varying degree. All have environmental surveillance programs whose extents are determined by the amounts of active isotopes used. Periodic reports from these programs led EPA to conclude that the whole-body dose from the facilities was in the range of 0.011 to 0.014 mrem when averaged over the entire U.S. population, although the 1970 bone dosage was substantially higher — 4.8 mrem — for the "population at risk" (about 100,000 persons working at or living near the facilities).

The great majority of the DOE installations are not a problem to their neighbors. In 1969, however, a major fire at the Rocky Flats facility in Colorado resulted in the discovery that a large area around the plant was contaminated by plutonium at a level that, though low, was significantly above the background level from fallout. The event, and the national attention devoted to it by the press (156), led to a thorough review of safety measures in all facilities handling plutonium.

TRANSPORT AND DISTRIBUTION

As scientists have examined the effects of man's activities on his environment more closely, it has become apparent that very few of the less desirable materials — pesticides, radioactivity, trace metals, other pollutants — that enter natural systems remain at their original points of deposition. Almost without exception they are transported by physical and biological means to varying distances at rates affected by many variables, not the least of which is the chemical nature of the individual pollutant. Of particular concern are the mechanisms by which noxious materials enter the human food chain or otherwise affect man directly. Also highly important are the effects on ecological systems produced by pollutants as they move about in the environment. Ideally, one would like to understand the mechanisms involved well enough to be able to project the movement and fate of contaminants with sufficient confidence to permit steps to be taken to minimize undesirable consequences. In a few favorable cases this can be done, but the problem is extremely complex and many serious gaps remain in the needed information.

The following discussion is necessarily brief. Excellent and somewhat more extended summaries have been prepared of the transport of radioactive species in air (2i), in water systems (2j), and in the food chain from soil to man (2k). An extensive literature has been developed on various phases of the transport of radioactivity in the environment, only a small part of which can be cited here (65, 69, 115, 125, 157–159).

Atmospheric Transport

Our direct experience in the atmospheric transport of radionuclides is limited almost entirely to the distribution of debris from nuclear weapons testing. A few tracer experiments have been based on the deliberate release of short-lived radioiodine, and an unplanned opportunity for study occurred as a result of a major release of radioactivity in the Windscale reactor accident in Great Britain in 1957 (2l). More recently, atmospheric transport has been studied using tracers such as krypton-85, sulfur hexafluoride, and CD_4 [methane (CH_4) with each of the four hydrogen atoms replaced by deuterium (D), the hydrogen isotope of mass 2]. In addition, over the past 50 years much information has been collected on the transport of nonradioactive atmospheric contaminants from surface sources such as smokestacks. The modeling of such emissions is far from perfect under all conditions, but it is reasonably adequate for many predictive purposes and can be applied directly in most cases to radioactive contamination from the nuclear fuel cycle (160). Information on the transport of both radioactive and nonradioactive species in the upper atmosphere is much more sketchy. Its importance is growing, however, not only because of nuclear testing, but also because of the advent of high-altitude rockets, satellites, and aircraft.

The atmosphere can be pictured in simplified form as a series of layers: the troposphere, the stratosphere, the mesosphere, and the ionosphere. The relative thicknesses of these layers vary somewhat with latitude and with the season of the year. In the temperate zones the troposphere extends 10 to 12 km upward from the surface and is separated from the stratosphere by an imaginary boundary called the tropopause. All of our weather occurs in the troposphere. As one rises through the troposphere, the temperature normally drops, then remains constant at –50° to –60°C through the stratosphere, which extends upward another 20 km. The temperature then begins to rise through the mesosphere, which extends into the ionosphere, a densely ionized region that begins at about 60 km above the surface (2m). Occasionally, special conditions produce in the troposphere a layer of warm air over colder air, so that temperature increases rather than decreases with height. This condition, an inversion, can have serious results, since noxious materials released at the earth's surface are confined and can build to dangerous concentrations. Inversions, moreover, can remain stable for considerable times.

Gases or aerosols introduced into the atmosphere are diluted by either molecular or turbulent diffusion. (The diffusivity coefficient will vary from 0.2 cm^2/second for molecular diffusion to as much as 10^{11} cm^2/second for turbulent diffusion under cyclone conditions.) Turbulent diffusion is a very complex phenomenon, but statistical methods have been developed that make it possible to predict with varying degrees of certainty the manner in which a contaminant will diffuse in the atmosphere under a given set of meteorological conditions (2m).

The trace gas ozone in the stratosphere absorbs more than 99% of the ultraviolet light from the sun, and it is probable that life did not develop on land until this ozone shield was in being. Concern has been expressed in recent years over the possible destruction of atmospheric ozone by man's activities, including the supersonic transport and the halomethanes (fluorocarbons) used in aerosol spray cans. Thermonuclear explosions in the stratosphere are another possible destructive agent. Their effect, however, is due not to the radioactivity involved but to the large amount of nitrogen oxides generated by the intense heat of the blasts. These oxides may react with and destroy the ozone (161).

Fallout studies

Studies of radioactive fallout, as indicated earlier, have furnished most of what we know of the transport of radioactive materials in the environment. Low-yield nuclear bomb tests produce particles and gases that remain in the troposphere for some time or even indefinitely, as well as the heavier materials that quickly deposit locally. The particles are deposited on the ground, mostly in precipitation, after a mean airborne life of about one month; gaseous iodine tends to be adsorbed on ambient particles and follow the same pattern; the noble gases — argon, neon, krypton, xenon — remain airborne indefinitely. High-yield tests inject a large fraction of their debris into the stratosphere. The debris is then transferred back into the troposphere during a mean life of about one year. The mean life appears to be longer for equatorial tests and somewhat shorter for tests in the Arctic.

Material in the troposphere travels generally east or west and reaches the surface relatively quickly, so that it deposits as fallout in a relatively narrow band of latitude, about 30°. Debris from individual nuclear tests is still concentrated enough to be detected on its second pass around the earth, and even a third-pass detection has been reported. The longer residence time of debris in the stratosphere allows it to spread farther north and south; in fact, part of the stratospheric radioactivity from a detonation in one hemisphere deposits in the other. The movement of debris from the stratosphere to earth is highly season-dependent. Parcels of stratospheric radioactivity enter the troposphere in the spring, although such transfer has been more marked in the northern than in the southern hemisphere. The rapid deposition of this debris once it enters the troposphere produces the spring peak in fallout.

The northern hemisphere has experienced about three times as much fallout as the southern hemisphere as a result of the relative yields of tests in the two hemispheres and the transfer of debris between them. The maximum fallout appears in the mid-latitudes — 30° to 50° north or south — even though the high-yield tests took place around the equator or in the Arctic. This is because the movement of debris from stratosphere to troposphere occurs preferentially in the mid-latitudes.

Surface sources

Also to be considered is airborne radioactivity from surface sources, such as stacks. Particles from such sources deposit quite locally; the noble gases, again, will remain airborne indefinitely; tritium and iodine will exhibit intermediate behavior. The precise pattern of deposition will depend on the immediate atmospheric conditions. All nuclear facilities in the U.S. are required to calculate, for their surface emissions, the pattern of radionuclide deposition and the consequent dose to man for various conditions. Both routine releases and accidents are covered. The resulting calculated doses depend very much on the population density for external exposure and on local agricultural factors for internal exposure. The transuranic elements are considered hazardous only when inhaled; the doses are controlled largely by the concentrations of the elements in the air during initial deposition, although in arid regions a second exposure can result when air currents resuspend settled debris. Fission products do not appear to present an inhalation problem.

Transport in Water Systems

While diffusion equations have been developed that in favorable conditions predict the air transport of radionuclides reasonably well, such equations do not currently exist for radionuclides dispersed in water. For one thing, the mixing process depends highly on a number of physical factors: depth of water, type of bottom, shoreline configuration, tidal factors, wind, temperature, and the depth at which the pollutant is introduced. Each aquatic system, moreover, has its own, not necessarily constant mixing characteristics, and so each system must be considered individually. The problem is complicated further by the fact that other factors, physical and biological, are involved in addition to the diffusion process. The pollutant may settle to the bottom, be taken up by organisms, be adsorbed on organic or inorganic suspended material, etc. (Fig. 7).

The horizontal and vertical movement of radioactive species in the oceans has been studied primarily in connection with fallout from nuclear bomb tests conducted from islands in the Pacific. The oceans, like the atmosphere, can be viewed as a series of layers. The top layer varies in depth from a few to several hundred meters and is a region of rapid mixing caused by wind. Under this mixing layer is an intermediate zone, the thermocline or pycnocline, which may be as deep as 1000 m. Temperatures gradually decrease and density increases passing down through the thermocline, which thus acts to a degree as a diffusion barrier between the mixing layer on top and the cold (1° to 4°C) waters that form the bottom layer and constitute the bulk of the oceans. The oceans are about 3800 m deep, on the average, but where trenches occur are as deep as 7000 m.

Because of the structure of the seas, debris from nuclear explosions at the surface tends to disperse horizontally more than vertically. Radioactivity from some of the Bikini Atoll tests has been found to spread laterally as much as 193 km in 40 days while diffusing downward only 30 to 60 m (162). The direction of drift depends largely on the currents in the surface layer, and these in turn depend primarily on the direction of the prevailing winds. Radioactivity rising from the ocean floor (as from nuclear waste dumping) would be expected, again, to diffuse more rapidly horizontally than vertically. One investigator, using a

FIGURE 7

Transport pathways in aquatic systems

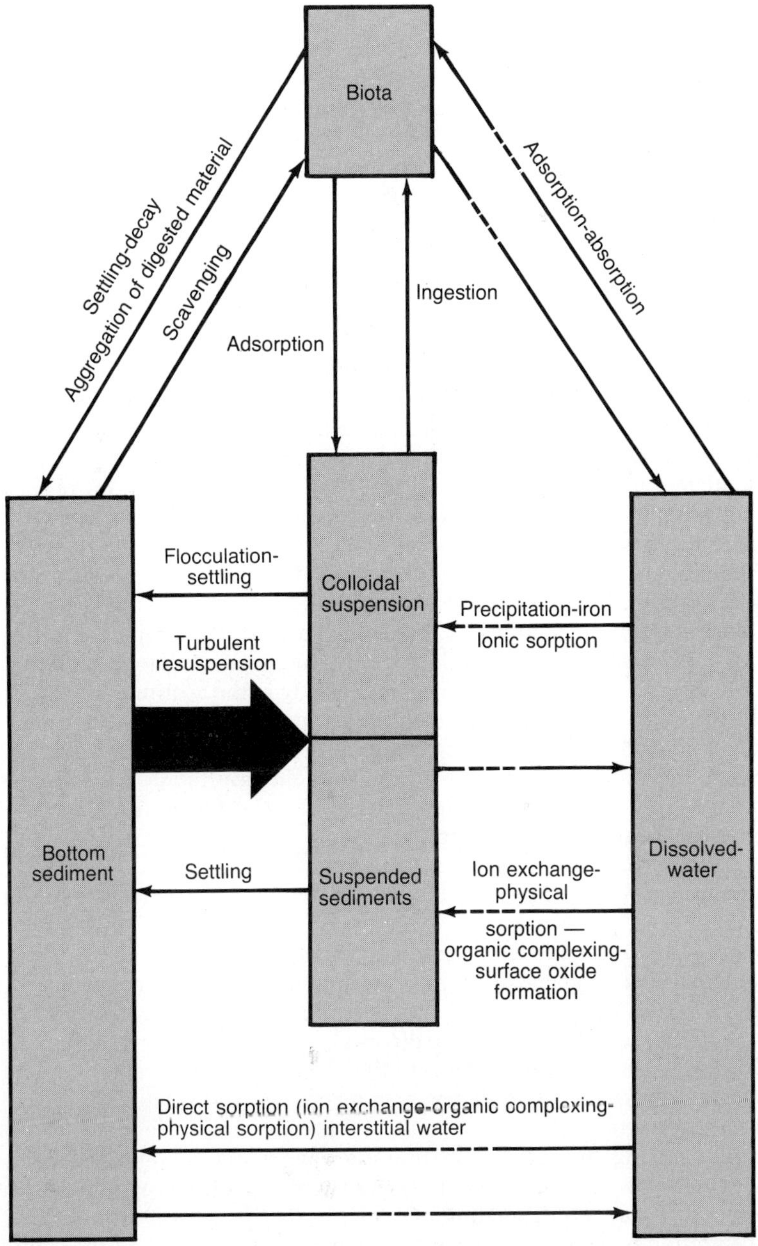

Source: Eisenbud, M., "Environmental Radioactivity," 2nd ed., Academic Press, New York, N.Y., 1973, p. 210.

simplified model, has estimated that radioactive material from a single source would spread horizontally over most of the North Atlantic in from 40 to 400 years; the same material, on the other hand, would require 600 years to diffuse from 4000 m deep to 1000 m (163).

The Columbia study

Coastal waters and estuaries warrant particular concern. They are important commercially, and coastal sites, moreover, are technically desirable for nuclear reactors because of the abundance of cooling water. Coastal waters and estuaries are complex systems, however, and current understanding of them will support only very crude estimates of the probable diffusion patterns of introduced radioactivity. The most highly studied river-estuary-continental shelf system perhaps comprises the Columbia River and the coastal areas near the point where it empties into the Pacific (164). The Hanford plutonium project was established on the Columbia in the southeastern part of the state of Washington during World War II. Three nuclear reactors were built at that time, and more were added later; in the period 1955–64, eight reactors were taking cooling water from the river. Currently, three reactors are operable and sometimes operating. The water for two of them is pretreated, run through the reactor, and returned to the river; the third and newest facility is cooled by demineralized water, which is recirculated instead of being discharged after one pass through the reactor.

The Columbia River study has been under way in various forms since 1943. In spite of the vast amount of information that has been collected, however, it is still not possible to devise any but very crude models for radioactive distributions downstream, in the estuary, and at the point of outflow into the Pacific — the system is simply too complex and variable. Most of the radioactivity that enters the river from the reactors does so not as fission products but as radionuclides formed by activation of trace elements originally in the cooling water or picked up by the water as it passes through the tubes holding the encapsulated fuel. The isotopes zinc-65 and phosphorus-32 have been studied especially well, as much because their nuclear characteristics make them easy to measure at very low concentrations as for any other reason. These two, plus chromium-51, copper-64, arsenic-76, and neptunium-239, are the most abundant in the cooling water. Iodine-131, manganese-54, scandium-46, and strontium-90 are measured routinely also, however, because they may be important in radiation exposure of humans.

Many of the radionuclides become associated very quickly with inorganic and organic particles suspended in the river and end up in sediments when such material settles in areas like the basins behind the dams on the river. A few radionuclides probably precipitate directly, without associating with particles, and a few others remain in true solution. Certain elements that are held on particulate matter by ion-exchange action apparently are displaced by sodium and magnesium and returned to solution when the river encounters the ocean, with its much higher content of sodium and magnesium ions. Some elements are concentrated by specific organisms, although not by all, and by a given species in a variable manner, depending on when in its life cycle it is exposed. This concentration phenomenon may increase for certain elements as one moves up the food chain from one-celled species to fish and shellfish. For 1964, however, when eight reactors were operating at Hanford, the dose that might have been received by an individual eating an extraordinary quantity of fish caught in the vicinity has been estimated to be well below limits recommended by the ICRP. As the number of operating reactors has decreased, this estimate has been lowered substantially because smaller amounts of nuclides are entering the river (164). A brief discussion of the biological uptake of radionuclides and tabular summaries of the concentration factors for various elements in freshwater and marine organisms are available (2n); the same author discussed the function of sediments and suspended solids as reservoirs for radioactive nuclides that can be passed by way of bottom-feeding biota to higher levels in the food chain. Also available is a very thorough review of radioactivity in the marine environment as it existed in 1971 (165).

Radionuclides in Soils

Radionuclides in soils and food chains are closely interdependent (2k): nuclides can move from soils into plants; and the bulk of the human diet in all parts of the world is made up of foods, including meat and milk, derived from agriculture, as opposed to food from aquatic sources (166).

Most of the naturally occurring radionuclides, such as potassium-40, occur in soils and are incorporated into plants by metabolic processes. Plants also can pick up artificial radioactivity from soils, where it arrives by deposition of airborne debris, such as fallout, or by sorption of waterborne material, such as leakage from underground waste storage tanks. Airborne material works its way through the surface of the soil and downward by solution processes or mechanical action. Water carrying radioactivity to a given soil may be under the surface already and, in fact, may never appear aboveground.

The elements that plants take up from soils — whether contaminants or nutrients, whether radioactive or not — exist in soil in three general forms: ions in solution in the water in the soil; ions held by ion-exchange forces on the surfaces of the inorganic and organic constituents of the soil; and elements contained in the soil in insoluble or sparingly soluble form. Of these three forms, only ions in true solution can be absorbed by the plant root. Equilibria exist, however, between these ions and those held to the soil by ion exchange; when the dissolved ions are depleted, those held to the soil are released into solution to replace them. The nutrient ions in solution at any given time, moreover, are far exceeded by those held to the soil, so that the latter serve as a nutrient pool vital to the life of plants. Elements in the third general form — insoluble or only slightly soluble — also may serve as a nutrient pool; they enter solution, however, not in response to equilibrium forces, but to changes in the acidity and other characteristics of the soil.

The ion-exchange properties of soil are thus fundamentally important to the behavior of soil contaminants. Clays have very high ion-exchange capacities. Clay particles are smaller than 2 μm in diameter, so the surface-to-volume ratio is very high and a few pounds of the material provide many thousands of square feet of surface where ions can be held and exchange can take place. In sands, on the other hand, the average particle is relatively large, the available reaction surface is low, and ion-exchange capacity is limited. Most soils are mixtures of clay, sand, and silt.

The depth to which a radioactive species will penetrate thus depends not only on its own chemical nature, but on the nature of the soil involved. Strontium-90, for example, can now be found at depths of 30 cm in some soils, whereas 10 years ago all of the radionuclide could be collected by sampling to 20 cm. Cesium-137 has remained closer to the surface because of its high affinity for clays and organic matter. The most surprising element is plutonium. It has penetrated at least as deeply as the ^{137}Cs in spite of the high positive charge of its ion and the insolubility of its oxide. The movement of plutonium may be due to the increased solubility of very small particles or to the formation of soluble organic complexes. Work at Argonne National Laboratory in connection with the waste storage problem suggests in addition that hydrolysis products of Pu^{+4} may be involved in the comparatively rapid movement of the element into soils. At any rate, plutonium is taken up to a very small degree by plants and, if ingested by animals, is absorbed through the gut to an even smaller degree (166a).

The penetration of radionuclides into the ground can be affected by fertilizers, by neutralizers (lime) for acidic soils, or by irrigation practices. Radionuclides in soil will move to a considerable degree with surface runoff, either because they dissolve in it or are moved physically by surface erosion. The extent of erosion depends on the vegetative cover; very little transfer of radionuclides occurs from areas with well-developed root mats. This is why flat pastureland is preferred when sampling soil to evaluate total deposition of fallout or other radionuclides.

In arid regions, some radioactivity is transferred by wind action, and resuspended radionuclides can be measured in air many years after they are deposited. As with waterborne transfer, radionuclides can build up in low spots and other collecting areas to many times the average concentration for the region. Resuspension by wind could be a particular problem in a plutonium-contaminated region since, while the element is somewhat innocuous when ingested, it is extremely hazardous when inhaled.

Recommendation R11: *Further research on the geochemical behavior of plutonium and the other actinides in soils and rocks, as well as on their chemistry in the other portions of the environment, is required with high priority because of the toxicity of this man-made group of very heavy elements.*

Radionuclides in the Food Chain

The pattern of transfer of radionuclides from soils into food chains is somewhat less complex than the pattern for aquatic systems. Still, the picture for soil is affected by many variables. They include the chemistry of the element involved; the characteristics of soils; the types of plants or animals affected; and, with plants, whether the edible parts are above or below the ground. The most important difference between the terrestrial and aquatic cases is that material deposited on land can lodge directly on the food of animals or man. This does not happen in aquatic systems, where the nuclide is mixed and diluted before being absorbed by any components of the chain, thus providing time for short half-life species to decay and for the level of contamination of the entire food web to decline. Aquatic food chains are thus but a minor source of exposure compared with terrestrial food chains (166b).

Radioactive contamination, as mentioned earlier, can be deposited on soils directly from the atmosphere, can be absorbed from water, or can remain as residue after contaminated water has evaporated. Airborne material can settle either on the soil or on parts of plants. Man may eat the plants directly as cereals or vegetables or indirectly as milk or meat from animals that have eaten the plants or products prepared from them. Radionuclides in the soil from either air or water can be transferred to vegetation by absorption and thence to man either directly or indirectly. Transfer pathways have been difficult to study in detail in the field. The reason is that root uptake is obscured by direct fallout of radioactive material on plants' leaves, where it may be absorbed metabolically or eaten by animals. For relatively short-lived species like ^{131}I such foliar contamination is the major route of entry of radionuclides into the food chain. Root absorption is a much more leisurely process that allows time for most of the short-lived material to decay before it is absorbed and metabolized by the plant. Among the longer-lived radionuclides, ^{90}Sr shows the greatest uptake by root absorption, ^{137}Cs is next, and the plutonium isotopes a very distant third. The chemical availability of these long-lived species also may vary with time, a problem that is currently being studied.

Radioactivity may deposit on a plant as particulate matter, as a condensed vapor, or in solution in rain. The length of time the activity remains is affected by the mode of deposition and by the physical structure of the plant; the latter characteristic also determines whether the contamination will tend to concentrate in the leaves, the flowers, or the basal structures. With normal rainfall, 80 to 90% of the deposit apparently will be washed to the ground in a month (166c).

Plants acquire relatively little radioactivity by foliar absorption, and the mechanisms at play are not well understood. In root absorption, the ions apparently move first into "free space" in the root structure. This step occurs by passive diffusion, which requires no metabolic energy and favors the movement of multivalent ions over univalent ions. (Multivalent ions lose more than one electron in the ionization process and thus have a higher positive charge than univalent ions, which lose only one electron.) In the second step, which does require metabolic energy, ions move from the free space to the conducting tissues in the center of the root. In this step the ionic preference is reversed, so that univalent ions tend to accumulate in the conducting tissues. In the third and last step, which also requires energy, water and nutrients move from the conducting tissues into the vascular tissues or circulatory system of the shoot, the aboveground part of the plant. Overall, then, relative to the external supply of ions, those of lower valency usually move to the aboveground part of the plant, while those of higher valency tend to remain in the root. Thus the shoot will absorb monovalent ^{137}Cs, for example, far more efficiently than quadrivalent ^{95}Zr (166d). The degree of contamination of the food chain by a particular isotope, therefore, depends importantly on whether the edible part of the plant is above or below the ground.

The artificial radionuclides of greatest importance in the food contamination problem are iodine-131, strontium-90, and cesium-137. Detailed reviews of the distribution and behavior of each are available (166, 167). In addition, the International Atomic Energy Agency has sponsored a number of symposia on soil-plant relationships. These meetings have emphasized the use of radioisotopes in studying plant physiology and the characteristics of soil, but much of the material pertains directly to food contamination by the soil-plant route (168–170). The ultimate case of radioactive contamination — nuclear war — was the subject of a symposium at Brookhaven National Laboratory (171).

There is currently a debate closely related to the question of radioactivity uptake by plants, although the material involved is not part of a food chain and the radioactivity is of natural origin. It has been suggested (172) that the agent in tobacco smoke leading to lung cancer is radioactivity arising from the radon decay chain. According to this theory, the polonium-210/lead-210 daughters in the chain concentrate in the glandular heads of tobacco leaf trichomes (threadlike protuberances on the surfaces), and these heads are rendered insoluble by high temperatures during smoking. These "hot particles" then lodge in the lung where they cause a disproportionate amount of damage because of the concentrated radioactivity. The "hot particle" theory has been challenged (173) on the basis that while the lung cells immediately adjacent to the particle would be destroyed, only a relatively few at the periphery of the damaged area would acquire the mutations leading to cancer. From this view, if radioactivity is the carcinogenic agent in tobacco smoke, its action is more diffuse but affects the lung as a whole.

EFFECTS OF IONIZING RADIATION

Effects on Humans

The biological effects of radiation undoubtedly have been studied in more detail than those of any other agent to which man exposes himself. A tremendous body of literature deals with the basic and practical implications of the interactions of ionizing radiation with the complex biochemical systems that constitute living matter. It is not possible here to do more than outline the present scientific consensus and indicate the uncertainties. More extensive treatment of the basics and of special problems, particularly related to differing dose rates and qualities of radiation or to patterns of distribution of radionuclides in the body, are available (2p, 13, 14, 31, 174).

Radiation produces its effects because it introduces energy into matter — into some chemical system — and that energy somehow must be absorbed or reemitted before the system can return to equilibrium. The primary effect is ionization; that is, electrons are knocked out of or added to atoms or molecules. This produces unusual chemical species that may exist only for a minute fraction of a second but are extremely reactive chemically. The major effects of radiation in living tissue probably are due to such "transient" species formed from the decomposition of water and reacting with the enzymic and other delicately balanced systems that control the metabolism of the cells. Some atoms or molecules may not actually ionize but become "excited" (contain more than their normal amount of energy) and thus very reactive and capable of causing similar damage. Radiation also may break chemical bonds. This can be particularly serious if it occurs in the genes of the sex cells, since the genetic material may reform in ways that produce mutations or aberrations in subsequent offspring.

Generally speaking, the biological effects of ionizing radiation can be correlated with the dosage received. The effects on human individuals and populations can be considered best in terms of two levels of exposure:

- Acute high-level: Whole-body exposures greater than tens of rads in a few hours, as could result from nuclear warfare and from catastrophic nuclear accidents and other localized accidents.
- Chronic low-level: Whole-body exposure, of the order of a few rads per year or less, as could result from normal occupational exposure, fallout to date, a normally operating nuclear power industry, and miscellaneous sources such as television, jet travel, and luminous watch dials.

Acute, high-level exposures can and have produced directly observable manifestations. Chronic, low-level exposures, as defined above, have not produced detectable deleterious effects in human individuals or populations or in other living organisms (with the possible exception of exposure of the foetus). This does not mean that such effects do not occur — it means that if they do occur, they occur so rarely as to be indistinguishable from changes caused by the many other stresses undergone by the organism during its lifetime.

Acute, high-level exposure

High-level doses of ionizing radiation may produce both readily apparent and delayed effects in humans or other organisms. The readily apparent effects of whole-body exposure to acute, high-level doses are summarized in Table 7; "readily apparent" means that the effects appear within a period of weeks. Exposure to high levels of radiation, however, also can produce effects that may not develop until some years after the exposure. These effects can occur either in the exposed individual (somatic effects) or in the offspring of the exposed individual and succeeding generations (genetic effects). (The probabilities that late somatic or genetic effects will occur can be estimated from the relationships discussed in the next section.) The principal somatic effects are the induction of leukemia and other forms of cancer. Cataracts also may develop if the lens of the eye receives a heavy dose of x-rays, gamma rays, beta particles, or neutrons; neutrons are believed to be particularly damaging. In addition, research on experimental animals indicates that high-level, whole-body irradiation shortens life spans, even when other effects do not appear. Such exposure may somehow accelerate the aging mechanism or it may weaken the body's defense mechanisms, increasing susceptibility to the usual causes of death.

Table 7

Representative dose-effect relationships in human beings for whole-body, acute[a] irradiation

Nature of effect	Rads of X or gamma radiation
Minimal dose detectable by chromosome analysis or other specialized analyses, but not by blood changes	5–25
Minimal dose readily detectable in a specific individual (e.g., one who presents himself as a possible exposure case)	50–75
Minimal dose likely to produce vomiting in about 10% of people so exposed	75–125
Dose likely to produce transient disability and clear blood changes in a majority of people so exposed	150–200
Dose likely to cause death in about 50% of people exposed	300

a. Essentially a "one-shot" exposure, i.e., delivered over a period of seconds or minutes. As an informal calibration point, the incident skin-dose rate during nonintensified fluoroscopic examination is in the range of 4 to 10 rads per minute (14c).

Source: "Review of the Current State of Radiation Protection Philosophy," National Council on Radiation Protection and Measurements, Report No. 43, U.S. Government Printing Office, Washington, D.C., 1975.

Chronic, low-level exposure

We stated earlier that effects from chronic, low-level exposures have not been observed. Nevertheless, current understanding of the mechanisms of interaction of radiation with biological systems requires us to assume that any level of radiation may be harmful to some degree — that is, to assume that there is no dosage threshold below which no damage occurs. Many have taken for granted that exposure to radiation in addition to the natural background and of the same or lower magnitude represents a risk so small compared with the other hazards of life that any associated, nontrivial benefit gained by such exposure would far offset whatever harm resulted from it. But there has been public pressure to estimate the probabilities or frequencies of the effects of such exposure. The argument is that, since any level or radiation may cause some harm and since entire populations of nations or of the world could be exposed to additional, low-level radiation if care is not taken, the extent of the absolute harm could increase even though the damage might not be detectable.

The BEIR committee (14) of the National Academy of Sciences — National Research Council has felt that there was an advantage in considering quantitative risk estimates, despite the recognized uncertainties in the data and calculations. The overall numerical values from the BEIR report can be summarized as follows:

- It is estimated that exposure of the parents to 170 mrem per year (or 5 rem over the 30 years of the usual reproduction period) would cause in the first generation between about 200 and 3600 serious genetic disabilities per year in the U.S. population, based on 3.6 million births per year.
- It is estimated that the same exposure as above *could* cause from roughly 3000 to roughly 15,000 cancer deaths annually in the U.S. population, with 6000 being the most likely number. ("Could" is used in the preceding sentence because some scientists feel that because of the efficiency of the body's repair mechanisms at the very low dose rates involved, the true effect might approach zero production of cancer.) However, even if the upper limit risk estimates are used in evaluating the effects of exposures expected from man's future activities, they could be considered qualitatively to support the statement in the quotation above that they ". . . represent a level of risk . . . small compared with the other hazards of life. . ."

There is not complete agreement on the validity of attempting even semiquantitative estimates of this type. The position of the National Council on Radiation Protection and Measurements is (31):

"The NCRP continues to hold the view that risk estimates for radiogenetic cancers at low doses and low dose rates derived on the basis of linear (proportional) extrapolation from the rising portions of the dose-incidence curves at high doses and high dose rates, as described and discussed in subsequent sections of this report, cannot be expected to provide realistic estimates of the actual risks from low-level, low LET [linear energy transfer] radiations, and have such a high probability of overestimating the actual risk as to be of only marginal value, if any, for purposes of *realistic* risk-benefit evaluation." (Emphasis in the original.)

Radiation Effects on Ecological Systems

An ecological system is not simply a collection of plants, animals, and microorganisms that happen to live in a given area, but rather a community of interdependent organisms. This being so, serious disruption of one part of the pattern may upset the balance of the entire system. Such disruptions can be natural, such as drought, floods, or fires, or they can be man-made, such as bulldozers, pollution, and, again, fires. Radioecology is the study of radiation as one such potentially disruptive agent. The radioecologist has two major concerns: one of them is the movement of radioactive species through ecological systems (considered briefly in the sections on Transport); the other is the effects of radiation on individual organisms and the consequences of these effects in the ecological community to which the organism belongs (157).

A severe disruption of an ecosystem triggers a series of events that, in general, lead eventually to the system's full recovery. After a fire, for example — unless it is catastrophic or is followed by flooding and erosion that cause severe losses of soil nutrients — new vegetation emerges from seeds that remained unharmed beneath the ground or are carried in by mechanisms like wind and rain, and the burned-out area begins to recover. In the process it goes through a number of stages; weeds may predominate for a time, then bushes of particular types, and so on. These stages or "successions" are not stable, although some may last for a considerable period. Usually the series of successions reaches the climax, the point where the ecosystem is once again stable. The rate of such recovery can vary widely, and desert systems are particularly slow to reach the climax. The depression left by the first atomic bomb blast in the desert at Alamogordo, N.M., in 1944 is revegetated, but still not to its original state (175). On the other hand, recovery at the tropical Eniwetok Atoll after the Nectar test in 1954 was essentially complete within a few years (176).

Natural or man-made radioactivity in the environment is sufficiently intense in some cases for its movement to be traced with sensitive instruments, but there have been very few cases where effects on organisms could be observed directly. In such cases, furthermore, the radiation effects are usually masked by other events occurring at the same time as the radiation exposure. The fireball and blast wave of atomic detonations at various test sites are examples, although some studies can be made of changes caused by the heavy local fallout (177). Similarly, the level of radionuclides in reactor cooling water is too low to allow their effects to be seen with any certainty, and the situation is further complicated by the temperature increase as the water goes through the reactor (178, 179).

For such reasons, research on the ecological effects of radiation has been done necessarily at much higher levels than now exist in the environment — in fact, under conditions more comparable to those of nuclear war than to those of a peacetime nuclear power economy. Various experimental arrays have been built where large radiation sources can be exposed by remote control in the center of large, fenced-off areas in selected ecosystems. In this country, Brookhaven National Laboratory established a 9500 Ci, ^{137}Cs source in an oak-pine forest in New York in 1961 (180). Emory University in Georgia had both a gamma irradiation facility on its campus and access to a 10,000-acre reservation surrounding a bare (air shielded) reactor operated by the Lockheed Aircraft Corp. (181). A number of other, smaller exposure arrays have been built since, and numerous experiments have been done with simulated fallout particles and with individual organisms under controlled laboratory conditions (157, 171, 182).

The data obtained by these various means are now very numerous, although not always easy to interpret. A few generalizations may be made. The Brookhaven experiments allowed estimates to be made of the acute exposures required to affect dominant species of major North American vegetations, and these estimates have been tabulated (180). Among the more surprising results were the much higher susceptibility of pines as compared to hardwoods and of forest trees as compared to herbaceous annuals. Based on the estimates for the individual species, the Brookhaven group considered effects on major ecosystems (Table 8). The relatively low values shown for the "city" and "agriculture" ecosystems are due to the fact that man is the dominant organism in both situations. The severity of the effects of radiation on these systems will depend on man's ability to survive and to function normally.

The Emory University group reached essentially the same conclusions as the Brookhaven group. They would rank ecosystems in increasing order of resistance to radiation stress as coniferous forests, rain forests, deciduous forests, grasslands, and desert (181).

A correlation has been made between the volume of the interphase chromosomes in the nucleus of the cell of a species and its susceptibility to radiation damage. The larger this volume, the lower the radiation exposure required to kill the organism (180, 183).

DETECTION AND MONITORING

A number of radioactivity monitoring networks have been established by federal, state, and foreign agencies. The data obtained are generally published as reports to the agencies involved, but much of the information is also correlated and published by EPA (122), UNSCEAR (13), and similar organizations. Environmental radiation measurements, including requirements for measurement and surveillance programs, have been considered systematically by NCRP (184).

EPA and many, more localized agencies operate extensive water quality sampling and analysis programs; most of them include determinations of gross beta and alpha activity and, in some cases, of specific radionuclides among the analyses of surface, ground, and treated water samples (122a). The Radiation Alert Network gathers atmospheric samples at 68 stations throughout the 50 states as well as in Puerto Rico, Panama, and Guam. An elaborate air and water sampling network (NERC-LV) has been established in connection with the

Table 8

Estimated radiation exposures required to damage major ecosystems (in roentgens)

	Level of damage		
Major ecosystems	Minor	Intermediate	Severe
City	200	200	—
Agriculture	200	200	—
Coniferous forest	200	200– 2000	>2000
Deciduous forest	200	200–10000	>10000
Grassland	2000	2000–20000	>20000
Herbaceous successional	4000	4000–70000	>70000

Source: Woodwell, G.M., Sparrow, A.H., "Effects of Ionizing Radiation on Ecological Systems," in "Ecological Effects of Nuclear War," U.S. Atomic Energy Commission, Report BNL-917 (C-43), 1965.

Nevada nuclear testing site. Radioactivity in milk is the most useful indicator of the general population's intake of radionuclides from the environment, and a Pasteurized Milk Network has been established that comprises 61 sampling stations in the U.S., one in Puerto Rico, and one in the Canal Zone. The Canadians have an air and precipitation monitoring network, and a Pan American Air Sampling Program exists at 12 locations throughout South America. The Environmental Measurements Laboratory of the Department of Energy operates a network for determining monthly deposition rates for ^{90}Sr at 33 locations in this country and at 90 sites abroad. Most of this routine monitoring is on the basis of gross beta or alpha activity, but specific isotopes, such as ^{90}Sr, are measured in some cases. Sampling and measurement techniques are not completely standardized among all the surveillance groups, so intercomparisons are sometimes difficult. A tremendous mass of data is nevertheless available.

EPA has produced a guide (185) for establishing surveillance networks around various nuclear facilities in an effort to insure that the data obtained are uniform and can be used to calculate the exposure dose to any of the population at risk. The guide recommends that radiation surveillance be carried out for at least a year around a new facility before it begins operating. During this time, the "probable critical exposure pathways" are identified. This is done on the premise that for a particular facility in a particular location, certain types of possible radioactivity releases must be monitored more carefully than others. (A radioiodine release from a reactor in the center of a desert, for example, would be less serious than one from the same reactor next to a dairy farm because of the possible grass-cow-milk-man pathway in the latter case.) The critical population groups — those most likely to be affected by a release — are also identified during the preoperating surveillance. With this information in hand, decisions can be made on the types of samples to be taken, measurements to be made, and location of sample sites. A discussion of monitoring techniques and philosophy similar to the EPA guide has also appeared (2p). In spite of such guides, however, it is not uncommon to find considerable disagreement in data based presumably on samplings taken under similar circumstances and in similar locations by different groups. This problem has been intensified as private companies have entered the field of providing monitoring services for a fee. A need evidently exists to achieve uniformity in monitoring methods, to upgrade the proficiency of their application, and to improve the quality control of the resulting data.

Recommendation R12: *Federal or state certification of all laboratories offering monitoring services should be considered as a means of standardizing the methods used and improving quality control for the data produced.*

Despite the problems with standardization, the detection and monitoring of radionuclides in the environment have improved enormously since the period of intense weapons-testing fallout in the 1950's and early 1960's. At that time, careful and tedious chemical separation was required, followed by counting of alpha or beta-gamma rays emitted by the separated element. Now that we have much better alpha and gamma spectrometers, simpler separations and in some cases direct measurements are adequate. This improvement is obtained only at considerable cost in equipment and counting time.

Radionuclide releases from reactors or reprocessing plants can be detected best in the plant ventilation system before dilution in the stack or other point of release. Routine releases should be measured quantitatively in terms of the significant isotopes. Total activity measurements are entirely inadequate except as an alarm system and apparently will not be accepted in the future by regulatory authorities. An exception is the measurement of external gamma dose.

Measurements in the environment are made after the fact and should not be considered part of the release detection system. Instead, they should be designed specifically to evaluate potential hazards to man. The most direct sample is the airborne material before it becomes involved with the soil or with plant or animal life. Such samples are most free of interferences and lend themselves well to direct spectrometry. Samples in the transmission chain closer to man may require lesser extrapolation to assess hazard, but at the expense of greater complexity in analysis.

The generalities of radiochemical analysis for the important radionuclides have been covered earlier. A number of excellent laboratory instrumental systems are now available for measuring environmental levels of radionuclides (Table 9). The ICRU has also prepared a summary of radiation protection instrumentation and its application (186). Because of its extremely soft (low-energy) beta radiation, tritium is a particularly difficult radioisotope to measure accurately. Flow-through instruments have been developed for tritium measurements in the field, but they are subject to a number of interferences that make the data obtained of questionable value in low-level activity situations. A more satisfactory method is to remove the tritium from the air by freeze-drying,

Table 9

Limits of detection for environmental radioactivity (400 minute counting time[a])

Procedure	Nuclide	Limit of detection[b]
Chemical separation Alpha spectrometry	Am-241, 243	0.04 pCi
	Pu-239	0.05
Chemical separation Alpha scintillation counting	Ra-226	0.7
Radon emanation Alpha counting	Ra-226	0.1
Chemical separation Low-level beta counting	Ba-140	0.2 pCi
	Ce-141	0.5
	Ce-144	0.3
	Cs-137	0.3
	I-131	0.3
	Sr-90	0.2
Gamma spectrometry ~50 cm^3 Ge(Li) detector	Ba-140	2 pCi
	Ce-141	1
	Ce-144	5
	Cs-137	2
	I-131	2

a. Adequate sensitivities may sometime be obtained with shorter counting times, e.g., the limit of detection would be about double the values shown.

b. Defined as the smallest activity for which there is 95% confidence that activity is present.

silica-gel adsorption, etc., and transfer the sample to the laboratory for measurement by liquid scintillation counting.

Field measurements in the environment are limited largely to measurements of gamma emitters. A gamma level of 1 microroentgen per hour can be measured to better than ±25%. Some additional information is available from *in situ* spectrometry, preferably with a germanium-lithium diode. With such instruments, less than 10 mCi/km^2 of ^{137}Cs or other gamma emitter can be specifically quantified in a 30- to 40-minute count.

Recommendation R13: *Existing radioactivity monitoring programs around possible emitters of radioactivity to the environment should be maintained and in some cases expanded.*

Recommendation R14: *Efforts should continue to develop highly sensitive, reasonably accurate instruments for measuring low-level activity in the field.*

LITERATURE CITED

1. *Fed. Regist.*, **42,** 8382, Feb. 10, 1977; 10836, Feb. 24, 1977.
2. Eisenbud, Merrill, "Environmental Radioactivity," 2nd. ed., Academic Press, New York, N.Y., 1973, p. 210. Page numbers: (a) 210, (b) 212, (c) 230, (d) 424, (e) 292, (f) 428, (g) 289, (h) Chap. 9, (i) 411, (j) 137, (k) 118, (l) 411, (m) 90, (n) 153–155, (o) 140, (p) Chap. 17, (q) 24.
3. "Understanding the Atom Series," U.S. Atomic Energy Commission, Information can be obtained from U.S. Department of Energy, Technical Information Center, P.O. Box 62, Oak Ridge, Tenn. 37830.
4. International Atomic Energy Agency, Safety and Technical Reports Series. Information can be obtained from National Agency for International Publications, Inc., 317 East 35th Street, New York, N.Y. 10016; or International Atomic Energy Agency, Kartner Ring 11, A 1010, Vienna I, Austria.
5. Klement, A.W., Jr., et al., "Estimates of Ionizing Radiation Doses in the United States, 1960–2000," U.S. Environmental Protection Agency, ORP/CSD 72-1, August 1972. Page numbers: (a) 7, (b) 9, (c) 89, (d) 90, (e) 92, (f) 95, (g) 96, (h) 157, (i) 159, (j) 160, (k) 22, (l) 13, (m) 52, (n) 50, (o) 55, (p) 44, (q) 51, (r) 49, (s) 168, (t) 31, (u) 42, (v) 47, (w) 48, (x) 60, (y) 99.
6. Oakley, D.T., "Natural Radiation Exposure in the United States," U.S. Environmental Protection Agency, ORP/SID 72-1, Washington, D.C., June 1972.
7. Kahn, B., et al., "Radiological Surveillance Studies at a Pressurized Water Nuclear Power Reactor," U.S. Environmental Protection Agency Report 61-1, Washington, D.C., August 1971.
8. Kahn, B., et al., "Radiological Surveillance Studies at a Boiling Water Nuclear Power Reactor," Public Health Service Report BRH/DER 70-1, Washington, D.C., 1970.
9. Cochran, J.A., Griffin, W.R., Jr., Troianello, E.J., "Observation of Airborne Tritium Waste Discharge from a Nuclear Fuel Reprocessing Plant," U.S. Environmental Protection Agency, EPA/ORP 73-1, Washington, D.C., February 1973.
10. Magno, P.J., Reavey, T.C., Apidianakis, J.C., "Iodine-129 in the Environment Around a Nuclear Fuel Reprocessing Plant," U.S. Environmental Protection Agency, ORP/SID 72-5, Washington, D.C., October 1972.
11. "Assessment of Environmental Radioactivity in the Vicinity of Shippingport Atomic Power Station," Environmental Protection Agency, November 1973, 38 pp., National Technical Information Service, PB 277 704/5WP, Springfield, Va.

12. "Environmental Radiation Dose Commitment: An Application to the Nuclear Power Industry," U.S. Environmental Protection Agency, Office of Radiation Programs, EPA-520/4-73-002, Washington, D.C., February 1974.
13. "Ionizing Radiation, Levels and Effects," Report of the United Nations Scientific Committee on the Effects of Atomic Radiation, UN General Assembly, New York, N.Y., Official Records, 27th Session. Suppl. No. 25, UN Document A/8725, 1972. Vol. 1, "Levels"; Vol. 2, "Effects." Page numbers (a) 152, 153, 159, (b) 145, (c) 148, (d) 150, (e) 149, (f) 192, (g) 188, (h) 190, (i) 179, (j) 95, (k) 57, (l) 97, (m) 62, 96, (n) 83.
14. "The Effects on Populations of Exposure to Low Levels of Ionizing Radiation," Report of the Advisory Committee on the Biological Effects of Ionizing Radiations, National Academy of Sciences — National Research Council, Washington, D.C., November 1972. Page numbers: (a) 123, (b) 12, (c) 147.
15. Alexander, P., "Atomic Radiation and Life," 2nd ed., Penguin Books, Baltimore, Md., 1965.
16. Cohen, B.L., "Nuclear Science and Society," Anchor Press, Garden City, N.Y., 1974.
17. Friedlander, G., Kennedy, J.W., Miller, J.M., "Nuclear and Radiochemistry," 2nd ed., John Wiley & Sons, New York, N.Y., 1964.
18. McKay, H.A.C., "Principles of Radiochemistry," Butterworths, London, Eng., 1971.
19. "Radiation Quantities and Units," International Commission on Radiological Units and Measurements, Report 19, Washington, D.C., 1971.
20. "Dose Equivalent," International Commission on Radiological Units and Measurements, Suppl. to ICRU Report 19, Washington, D.C., Sept. 1, 1973.
21. "Nuclear Terms, A Glossary," 2nd ed., "Understanding the Atom Series," U.S. Atomic Energy Commission, Washington, D.C., 1974.
22. Glasstone, S., "Sourcebook on Atomic Energy," 3rd ed., D. Van Nostrand Co., Inc., Princeton, N.J., 1967.
23. Shapiro, J., "Radiation Protection," Harvard University Press, Cambridge, Mass., 1972, pp. 253–257.
24. Stewart, D.C., "Techniques of Handling Highly Active Beta- and Gamma-Emitting Material," in "Techniques of Inorganic Chemistry," Vol. III, H.B. Jonassen, A. Weissberg, eds., John Wiley & Sons, New York, N.Y., 1972, pp. 169–170.
25. Recommendations of the International Commission on Radiological Protection, Report of Committee II on Permissible Dose for Internal Radiation, 1959, "Radiation Protection," ICRP Publ. 2, Pergamon Press, Oxford, 1959; *Health Physics*, **3**, 1–380 (1960).
26. Recommendations of the International Commission on Radiological Protection (as amended 1959 and revised 1962), "Radiation Protection," ICRP Publ. 6, Pergamon Press, Oxford, 1964.
27. Recommendations of the International Commission on Radiological Protection (adopted Sept. 17, 1965), ICRP Publ. 9, Pergamon Press, Oxford, 1966.
28. "Protection Against Ionizing Radiation from External Sources," International Commission on Radiological Protection, Report by Committee III, ICRP Publ. 15, Pergamon Press, Oxford, 1970.
29. "The Metabolism of Compounds of Plutonium, and Other Actinides, A Report," International Commission on Radiological Protection, Committee II on Permissible Dose for Internal Radiation, ICRP Publ. 19, Pergamon Press, Oxford, 1972.
30. "Alkaline Earth Metabolism in Adult Man," International Commission on Radiological Protection, Committee II on Permissible Dose for Internal Radiation, ICRP Report 20, Pergamon Press, Oxford, 1973.
31. "Review of the Current State of Radiation Protection Philosophy," National Council on Radiation Protection and Measurements, Report No. 43, U.S. Government Printing Office, Washington, D.C., 1975.

32. "Standards for Protection Against Radiation," Code of Federal Regulations, Title 10 — Atomic Energy, Part 20, U.S. Government Printing Office, Washington, D.C.

33. "Hazardous Materials Regulating Board," Code of Federal Regulations, Title 49, Parts 170 to 199, Department of Transportation, Washington, D.C., 1970 (revision).

34. "Licensing of Production and Utilization Facilities," Code of Federal Regulations, Title 10 — Atomic Energy, Part 50, *Fed. Regist.*, **35,** 18388 (1970).

35. Miller, L.A., "Report of State and Local Radiological Health Programs, Fiscal Year 1972," U.S. Public Health Service, Bureau of Radiological Health, DHEW/FDA-74-8004, Rockville, Md., August 1973.

36. "Packaging of Radioactive Material for Transport," Code of Federal Regulations, Title 10, Part 71, USAEC, Washington, D.C., 1967 (and subsequent revisions).

37. "Disposal of Radioactive Wastes on Land," National Academy of Sciences — National Research Council, Publ. 519, Washington, D.C., 1957.

38. "Disposal of Solid Radioactive Wastes in Bedded Salt Deposits," Committee on Radioactive Waste Management, National Academy of Sciences — National Research Council, Washington, D.C., November 1970.

39. "An Evaluation of the Concept of Storing Radioactive Wastes in Bedrock Below the Savannah River Site," Committee on Radioactive Waste Management, National Academy of Sciences — National Research Council, Washington, D.C., 1972.

40. "Nuclear Fuel Cycle Controversy," *Environ. Report,* **V(2),** 9–13, May 24, 1974.

41. Train, Administrator Environmental Protection Agency, et al., vs. Colorado Public Interest Research Group, Inc., et al., U.S. Supreme Court No. 74-1270, June 1, 1976.

42. Jacobs, D.G., "Sources of Tritium and Its Behavior Upon Release to the Environment," U.S. Atomic Energy Commission, Report TID-24635, 1968.

43. Elwood, J.W., "Ecological Aspects of Tritium Behavior in the Environment," *Nucl. Saf.*, **12,** 326 (1971).

44. Cronkite, E.P., Robertson, J.S., Feinendegen, L.E., "Somatic and Teratogenic Effects of Tritium," in "Tritium," A.A. Moghissi, M.W. Carter, eds., Messenger Graphics, Phoenix, Ariz., May 1973.

45. Feinendegen, L.E., Bond, V.P., "Transmutation Versus Beta Irradiation in the Pathological Effects of Tritium Decay," in "Tritium," A.A. Moghissi, M.W. Carter, eds., Messenger Graphics, Phoenix, Ariz., May 1973.

46. Schroder, J., Munnich, K.O., Ehhalt, D.H., "Krypton-85 in the Troposphere," *Nature*, **233,** 614 (1971).

47. Mellemgaard, K., Lassen, N.A., Georg, J., "Right-to-Left Shunts in Normal Man Determined by the Use of Tritium and Krypton-85," *J. Appl. Physiol.*, **17,** 778 (1962).

48. Muehlbacher, C.A., DeBon, F.L., Featherstone, R.M., "The Solubilities of Xenon and Cyclopropane in Blood and Protein Solutions," *Mol. Pharmacol.*, **2,** 86 (1966).

49. Whipple, G.H., "Possible Effects of Noble Gas Effluents from Power Reactors and Fuel Reprocessing Plants," in "Noble Gases," R.E. Stanley, A.A. Moghissi, eds., U.S. Government Printing Office, CONF-730915, Washington, D.C., 1975, pp. 488–491.

50. Hendrickson, M.M., "Eventual Whole Body Exposure Rate from ^{85}Kr Released to the Atmosphere," U.S. Atomic Energy Commission Report BNWL-SA-3233, July 13, 1970.

51. "Instrumentation for Environmental Monitoring," Vol. 3, "Radiation," Lawrence Berkeley Laboratory, U.S. Atomic Energy Commission Report LBL-1, May 1972.

52. Pelletier, C.A., Cline, J.E., Keller, J.H., "Measurement of Sources of Iodine-131 Releases to the Atmosphere from Nuclear Power Plants," *IEEE Trans.*, **NS-21** (4), 478, February 1974.

53. Wayne, E.J., Koutras, D.A., Alexander, W.D., "Chemical Aspects of Iodine Metabolism," F.H. Davis Co., Philadelphia, Pa., 1964.

54. Grayson, R.R., "Factors Which Influence the Radioactive-Iodine Thyroidal-Uptake Test," *Am. J. Med.*, **28,** 397 (1960).

55. Black, A., Hounam, R.F., "Penetration of Iodine Through the Nose and Mouth and the Clearance and Metabolism of the Deposited Iodine," *Ann. Occup. Hyg.*, **11**, 209 (1968).

56. Morgan, D.J., Morgan, A., "Studies on the Retention and Metabolism of Inhaled Methyl Iodide. I. Retention of Inhaled Methyl Iodide," *Health Phys.*, **13**, 1055 (1967).

57. "Background Material for the Development of Radiation Protection Standards," U.S. Federal Radiation Council, Report No. 5, Washington, D.C., July 1964.

58. Russell, J.L., Hahn, P.B., "Public Health Aspects of I-129 from Nuclear Power Industry," *Radiol. Health Data Rep.*, **12** (4), 189 (1971).

59. U.S. Atomic Energy Commission, Final Environmental Statement for General Electric Midwest Fuel Recovery Plant, Docket No. 50-268, December 1972.

60. Kastner, J., "The Natural Radiation Environment," in "Understanding the Atom Series," U.S. Atomic Energy Commission, Div. Tech. Inf., 1968.

61. Finkel, A.J., Miller, C.E., Hasterlik, R.J., "Radium-Induced Malignant Tumors in Man," in "Delayed Effects of Bone-Seeking Radionuclides," C.W. Mays, W.S.S. Jee, R.D. Lloyd, eds., University of Utah Press, Salt Lake City, Utah, 1969, pp. 195–225.

62. Radiological and Environmental Research Division Annual Report, Center for Human Radiobiology, June 1972–June 1973; U.S. Atomic Energy Commission Report ANL-8060, Pt. II.

63. Kirby, H.W., Salutsky, M.L., "The Radiochemistry of Radium," National Academy of Sciences — National Research Council, Nuclear Science Series, Report NAS-NS-3057, Washington, D.C., December 1964.

64. Beeley, R.J., Buttrey, K., Golan, S., "Fuel Management for LMFBR Plants," *Proc. Am. Power Conf.*, **31**, 180 (1969).

65. "Radioactivity and Human Diet," R.S. Russell, ed., Pergamon Press, Oxford, England, 1966.

66. Harley, J.H., in Proc. Environmental Plutonium Symposium, E.B. Fowler, R.W. Henderson, M.F. Milligan, eds., Los Alamos, N.M., Aug. 4–5, 1971, U.S. Atomic Energy Commission Report LA-4756, December 1971.

67. Ballou, J.E., "Effects of Age and Mode of Ingestion on the Absorption of Plutonium," *Proc. Soc. Exp. Biol. Med.*, **98**, 726 (1958).

68. Bustad, L.K., et al., "Preliminary Observations on Metabolism and Toxicity of Plutonium in Miniature Swine," *Health Phys.*, **8**, 615 (1962).

69. Thompson, S.E., et al., "Concentration Factors of Chemical Elements in Edible Aquatic Organisms," U.S. Atomic Energy Commission Report UCRL-50564, Rev. I, October 1972.

70. Burney, G.A., Dukes, E.K., Groh, H.J., "Analytical Chemistry of Neptunium," in "Progress in Nuclear Energy, Series IX," D.C. Stewart and H.A. Elion, eds., Vol. 6, Pergamon Press, Elmsford, N.Y., 1966, pp.183–209.

71. Keller, C., "The Chemistry of the Transuranium Elements," Weinheim/Bergstr., Germany, Verlag Chemie, 1971.

72. Metz, C.F., Waterbury, G.R., "The Transuranium Actinide Elements," in "Treatise on Analytical Chemistry," I.M. Kolthoff and P.J. Elving, eds., Part II, Vol. 9, Interscience, New York, N.Y., 1962, pp. 189–391.

73. Seaborg, G.T., "Man-Made Transuranium Elements," Prentice-Hall, Englewood Cliffs, N.J., 1963.

74. "Gmelin's Handbuch der Anorganischen Chemie," Band 8, Transuranes, Verlag Chemie, Weinheim/Berstrasse, West Germany.

75. Taylor, D.M., "Chemical and Physical Properties of the Transplutonium Elements," in "Uranium, Plutonium, Transplutonium Elements," H.C. Hodge, ed., Springer-Verlag, New York, N.Y., 1973, pp. 717–738.

76. Magno, P.J., Nelson, C.B., Ellet, W.H., "A Consideration of the Significance of Carbon-14 Discharges from the Nuclear Power Industry," Proc. 13th Atomic Energy Commission Air Cleaning Conference, M.W. First, ed., CONF 740807, Report U-70, Vol. 2, March 1975, pp. 1047–1055.

77. "Natural Background Radiation in the United States," National Council on Radiation Protection and Measurements, Report No. 45, Washington, D.C., November 1975.

78. Hammond, R.P., "Nuclear Power Risks," *Am. Sci.*, **62,** 155 (1974).

79. "Nuclear Power and the Environment, Questions and Answers," American Nuclear Society, Hinsdale, Ill., 1973.

80. United Nations' Scientific Committee on the Effects of Atomic Radiation, Report to the General Assembly, U.N. General Assembly, Official Records, 17th Session, Suppl. No. 16, U.N. Document A/5216, 1962.

81. Gitlin, J.N., Lawrence, P.S., "Population Exposure to X-Rays, U.S., 1964," Report of the Public Health Service X-Ray Exposure Study, U.S. Department of Health, Education and Welfare, PHS Publication No. 1519, Washington, D.C., 1966.

82. "Population Dose from X-Rays, U.S., 1964," Estimates of Gonad Dose and Genetically Significant Dose from the Public Health Service X-Ray Exposure Study, Public Health Service, U.S. Department of Health, Education and Welfare, PHS Publication No. 2001, Washington, D.C., 1969.

83. Penfil, R.L., Brown, M.L., "Genetically Significant Dose to the United States Population from Diagnostic Medical Roentgenology," *Radiology,* **90,** 209 (1968).

84. "Gonad Doses and Genetically Significant Dose from Diagnostic Radiology, U.S., 1964 and 1970," U.S. Department of Health, Education and Welfare, Public Health Service, Bureau of Radiological Health, Rockville, Md., April 1976, pp. 17, 36.

85. Jester, W.A., et al., "Nuclear Engineering Applications in Medical Science," *Iso. Rad. Tech.,* **9,** 485 (1972).

86. Fowler, E.E., "Recent Advances in Applications of Isotopes and Radiation in the United States," *Iso. Rad. Tech.,* **9,** 253 (1972).

87. Hoffer, P.B., "Fluorescent Thyroid Scanning," *Am. J. Roentgenol., Radium Ther. Nucl. Med.,* **105,** 721 (1969).

88. "Survey of the Use of Radionuclides in Medicine," (prepared by the Stanford Research Institute), U.S. Public Health Service, Bureau of Radiological Health, Report BRH/DMRE 70-1, Rockville, Md., 1970.

89. "A Summary of the Washington, D.C., Metropolitan Area Survey of Color Television Receivers," National Center for Radiological Health, *Rad. Health Data Reports,* **12,** 1 (1971).

90. "Performance Standards for Electronic Products," *Fed. Regist.,* **34,** 20271 (1969).

91. Deonigi, D.E., McKee, R.W., Haffner, D.R., "Isotope Production and Availability from Power Reactors," U.S. Atomic Energy Commission Report BNWL-716, July 1968.

92. Rohrmann, C.A., "Values in Spent Fuel from Power Reactors," *Iso. Rad. Tech.,* **6,** 19 (1968).

93. Deonigi, D.E., "Formation of Transuranium Isotopes in Power Reactors," U.S. Atomic Energy Commission Report BNWL-140, Rev. 1, January 1966.

94. Wood, P.M., "Isotope Production in Commercial Power Reactors," *Iso. Rad. Tech.,* **9,** 129 (1971–72).

95. Comar, C.L., "Fallout from Nuclear Tests," U.S. Atomic Energy Commission Div. Tech. Inf. "Understanding the Atom Series," Rev., 1966.

96. Klement, A.W., Jr., "Radioactive Fallout Phenomena and Mechanisms," *Health Phys.,* **11,** 1265 (1965).

97. Keilholtz, G.W., "Krypton-Xenon Removal Systems," *Nucl. Saf.,* **12,** 591 (1971).

98. Hardy, E.P., Jr., "Fallout Program, New York Operations Office (AEC), New York Health and Safety Lab.," HASL-278, 173 pp., Jan. 1, 1974.

99. Solon, L.R., et al., "Further Investigations of Natural Environmental Radiation," U.S. Atomic Energy Commission Report HASL-73, 1959.
100. Beck, H.L., et al., "Further Studies of External Environmental Radiation," U.S. Atomic Energy Commission Report HASL-170, 1966.
101. Holoday, D.A., Archer, V.E., Lundin, F., "A Summary of United States Exposure Experiences in the Uranium Mining Industry," p. 451, Proc. Symp. Diagnosis Treatment Deposited Radionuclides, Excerpta Med. Found., Amsterdam, 1968.
102. U.S. Government, Part 57 of Public Law 89-577, Chap. 57.24–7, *Fed. Regist.*, May 25, 1971.
103. Stein, L., "Noble Gas Compounds: New Methods for Collecting Radon and Xenon," *Chemistry*, **47,** 15 (1974).
104. "Evaluation of Radon-222 near Uranium Tailings Piles," Public Health Service, Bureau of Radiological Health, Report DER 69-1, Rockville, Md., 1969.
105. "Environmental Analysis of the Uranium Fuel Cycle," U.S. Environmental Protection Agency, Office of Radiation Programs, Washington, D.C., 1973. (a) Part I — Fuel Supply, EPA Report EPA-520/9-73-003-B. (b) Part II — Nuclear Power Reactors, EPA Report EPA-520/9-73-003-C. (c) Part III — Nuclear Fuel Reprocessing, EPA Report EPA-520/9-73-003-D.
106. Siek, R.D., Barid, J.B., Jr., "Uranium Mill Tailings Problems in Grand Junction, Colorado," *HPS Newsletter*, **1** (3), 12 (1972).
107. "Use of Uranium Mill Tailings for Construction Purposes," Hearings before the Subcommittee on Raw Materials of the Joint Committee on Atomic Energy, Congress of the United States, 1971.
108. "Hot Town," *Time*, Dec. 20, 1971, p.33.
109. Barton, C.J., "Glove Box Techniques," in "Technique of Inorganic Chemistry," H.B. Jonassen and A. Weissberger, eds., Vol. III, John Wiley & Sons, New York, N.Y., 1963, pp. 259–333.
110. "Industry Report 1977–78," *Nuclear News*, Buyers Guide, mid-February 1978, American Nuclear Society, La Grange Park, Ill.
111. "AEC Is Predicting, 1,200,000 . . . of Nuclear Generating Capacity . . . ," *Nucl. Week*, **14** (10), 2 (1973).
112. Hottel, H.C., Howard, J.B., "New Energy Technology, Some Facts and Assessments," MIT Press, Cambridge, Mass., 1971.
113. *Monthly Energy Review*, April 1978, U.S. Department of Energy, Washington, D.C.
114. Benedict, M., "Electric Power from Nuclear Fission," *Proc. Nat'l. Acad. Sci. U.S.*, **68,** 1923 (1971).
115. "Nuclear Power and the Environment," International Atomic Energy Agency, Vienna, Austria, 1973. Page numbers: (a) 21–47, (b) 52.
116. Seaborg, G.T., Bloom, J.L., "Fast Breeder Reactors," *Sci. Amer.*, 13–21, November 1970.
117. "The Safety of Nuclear Power Reactors and Related Facilities," U.S. Atomic Energy Commission Report WASH-1250, Final Draft, July 1973. Page numbers: (a) Chap. 1.0, (b) 7-20, 1-15, 1-22, (c) 4-14, (d) 4-13, (e) 1-27, (f) 1-35, (g) 4-17, (h) 4-19, (i) 8-1, (j) 5-5, (k) 1-14, 1-20, (l) 1-42, (m) 7-28, (n) 4-31, (o) 4-71, (p) 4-78, (q) 4-77, (r) 4-77, (s) 1-51, (t) 1-36.
118. Eisenbud, M., Petrow, H.G., "Radioactivity in the Atmospheric Effluents of Power Plants that Use Fossil Fuels," *Science*, **144,** 288 (1964).
119. Martin, J.E., Harward, E.D., Oakley, D.T., "Radiation Doses from Fossil-Fuel and Nuclear Power Plants," Chap. 9, "Power Generation and Environmental Change," AAAS, MIT Press, Cambridge, Mass., 1971.

120. Martin, J.E., "Comparative Population Radiation Dose Commitments of Nuclear and Fossil Fuel Electric Power Cycles," Proc., Eighth Midyear Topical Symposium, Health Physics Society, October 1974, CONF-741018, National Technical Information Service, Springfield, Va.

121. Burger, L.L., Ryan, J.L., "The Technology of Tritium Fixation and Storage," Battelle Pacific Northwest Laboratories, BNWL-1807, Richland, Wash., Jan. 18, 1974.

122. U.S. Environmental Protection Agency, Office of Radiation Programs, "Radiation Data and Reports," **14** (8) 463, August 1973, U.S. Government Printing Office, Washington, D.C.

123. "The Potential Radiological Implications of Nuclear Facilities in Upper Mississippi Basin in the Year 2000," U.S. Atomic Energy Commission Report WASH-1209, January 1973.

124. Ybarrando, L.J., Solbrig, C.W., Isbin, H.S., "The 'Calculated' Loss-of-Coolant Accident: A Review," *Am. Inst. Chem. Eng. Monograph*, Series #7, American Institute of Chemical Engineers, New York, N.Y., 1972.

125. Ford, D.F., et al., "The Nuclear Fuel Cycle, A Survey of the Public Health, Environmental and National Security Effects of Nuclear Power," Union of Concerned Scientists, Cambridge, Mass., 1973.

126. Rose, D.J., "Nuclear Electric Power," *Science,* **184,** 351 (1974).

127. "New Idaho ECCS Tests Show Core Is Again Bypassed by Emergency Coolant," *Nucl. Week,* **14** (42), 4, Oct. 18, 1973.

128. "San Onofre-1 Had an ECCS Injection Last Month, Proving the System's Efficiency," *Nucl. Week,* **14** (48), 6, Nov. 29, 1973.

129. "Fire Knocks Out Brown's Ferry 1-2 in Probably Worst Incident Yet," *Nucl. Week,* **16** (13), 1 (1975).

130. "Reactor Safety Study, An Assessment of Accident Risk in U.S. Commercial Nuclear Power Plants," U.S. Nuclear Regulatory Commission, WASH 1400 (NUREG 75/014), Washington, D.C., 1975. (Consists of a main report and 11 appendices published individually.)

131. Gera, F., Jacobs, G.D., "Considerations in the Long-Term Management of High-Level Radioactive Wastes," U.S. Atomic Energy Commission, ORNL-4762, February 1972.

132. "Aqueous Processing of LMFBR Fuels — Technical Assessment and Experimental Program Definition," J.H. Goode, S.D. Clinton, eds., U.S. Atomic Energy Commission ORNL 4436, Suppl. 1, June 1972.

133. Pence, D.T., Duce, F.A., Maeck, W.J., "Application of Metal Zeolites to Nuclear Fuel Reprocessing Plant, Off-Gas Treatment," *Trans. Am. Nuc. Soc.,* **15,** 96 (June 1972).

134. "Chemical Technology Annual Progress Report for Period Ending March 31, 1971," Oak Ridge National Laboratory, U.S. Atomic Energy Commission ORNL-4682, 61, July 1971.

135. Bendixsen, C.L., Rohde, K.L., "Operation Performance and Safety of a Cryogenic System for Krypton Recovery," *Trans. Am. Nuc. Soc.,* **15,** 97 (June 1972).

136. Merriman, J.R., Stephenson, J.J., Dunthorn, D.I., "Recent Developments in Controlling the Release of Noble Gases by Absorption in Fluorocarbons," *Trans Am. Nuc. Soc.,* **15,** 95 (June 1972).

137. "Annual Report to Congress of the Atomic Energy Commission for 1970," U.S. Government Printing Office, Washington, D.C., p. 55, January 1971.

138. "Regulations for the Safe Transport of Radioactive Materials," International Atomic Energy Agency, IAEA Safety Series No. 6, 1973 Rev. Ed., Vienna, Austria, 1973.

139. McCluggage, W.C., "The AEC Accident Record and Recent Changes in AEC Manual, Chap. 0529," U.S. Atomic Energy Commission, BNWL-SA-3906, 1971.

140. "Directorate of Regulatory Standards, Environmental Survey of Transportation of Radioactive Materials to and from Nuclear Power Plants," U.S. Atomic Energy Commission USAEC WASH-1238, December 1972.

141. Code of Federal Regulations, Title 10, Part 20.301-304, "Waste Disposal," U.S. Atomic Energy Commission, Washington, D.C.

142. "The Shallow Land Burial of Low-Level Radioactively Contaminated Solid Waste," Committee on Radioactive Waste Management, National Academy of Sciences — National Research Council, Washington, D.C., 1976.

143. Fox, C.H., "Radioactive Wastes," U.S. Atomic Energy Commission "Understanding the Atom Series," Department of Energy, Office Tech. Info., Oak Ridge, Tenn. (Rev. 1969).

144. "Siting of Fuel Reprocessing Plants and Waste Management Facilities," Oak Ridge National Laboratory, Chemical Technology Division, USAEC ORNL-4451, July 1970. Page numbers: (a) 4-18, 3-66, (b) 3-67, (c) 4-21, (d) 4-22, (e) 4-27.

145. "Radioactive Waste Processing and Disposal," Department of Energy, Technical Information Center, USAEC TID-3311-S5/WP, Oak Ridge, Tenn., September 1973.

146. Schneider, K.J., "Status of Technology in the United States for the Solidification of Highly Radioactive Wastes," U.S. Atomic Energy Commission, BNWL-820, October 1968.

147. "NRDC Sees New Hanford Spill as Worse than Big Leak Last Summer," *Nucl. Week*, **14** (47), 7, Nov. 22, 1973.

148. "Hanford Radiation Leaks May Lead to Suit by Environmental Group," *Nucl. Week*, **14** (30), 4, July 26, 1973.

149. "Interim Storage of Solidified High-Level Radioactive Wastes," Committee on Radioactive Waste Management, National Academy of Sciences — National Research Council, Washington, D.C., 1975.

150. Kubo, A.S., Rose, D.J., "Disposal of Nuclear Wastes," *Science*, **182,** 1205 (1973).

151. Mackay, J.S., "An Evaluation of Some Special Techniques for Nuclear Waste Disposal in Space," NASA-TM-X-72272, 25 pp., National Aeronautics and Space Administration, Washington, D.C., August 1973.

152. Claiborne, H.C., "Neutron-Induced Transmutation of High-Level Radioactive Waste," U.S. Atomic Energy Commission Report ORNL-TM-3964, 1963.

153. Cohen, B.L., "Radioactive Waste Disposal," *Science*, **184,** 746 (1974).

154. "Alternatives for Managing Wastes from Reactors and Post-Fission Operations in the LWR Fuel Cycle," Energy Research and Development Administration, ERDA 76-43, Washington, D.C., May 1976.

155. "Project Rio Blanco: Radioactivity and the Environment," CER Geonuclear Corp., National Technical Information Service, PNE-RB-63 and Addendum, Springfield, Va.

156. Shapley, D., "Credibility Gap Widens on Plutonium Plant Safety," *Science*, **174,** 569 (1971).

157. "Biological Aspects of the Nuclear Age: Selected Readings in Radiation Ecology," V. Schultz and F.W. Whicker, eds., U.S. Atomic Energy Commission Report TID 25978, 1972.

158. "Environmental Behavior of Radionuclides Released in the Nuclear Industry," Proc. 1973 Aix-en-Provence Symp., International Atomic Energy Agency, Unipub, New York, N.Y., 1973, 749 pp.

159. "Physical Behavior of Radioactive Contaminants in the Atmosphere," International Atomic Energy Agency, Unipub, New York, N.Y., 1974.

160. "Tracking Radioactivity by Remote Computer," *Ind. Res.*, September 1973, p. 31.

161. Hammond, A.L., Maugh, T.H., II, "Stratospheric Pollution: Multiple Threats to Earth's Ozone," *Science*, **186,** 335 (1974).

162. Folsom, T.R., Vine, A.C., "Tagged Water Masses for Studying the Oceans," National Academy of Sciences — National Research Council Rpt. 551, 1957, quoted in Ref. 2, p. 143.

163. Pritchard, D.W., "The Application of Existing Oceanographic Knowledge to the Problem of Radioactive Waste Disposal in the Sea," Proc. Disposal of Radioactive Wastes, International Atomic Energy Agency, Vienna, Austria, 1960, quoted in Ref. 2, p. 146.

164. "Columbia River Estuary and Adjacent Ocean Waters. Bioenvironmental Studies," A.T. Pruter, D.L. Alverson, eds., University of Washington Press, Seattle, Wash., 1972.

165. "Radioactivity in the Marine Environment," Panel on Radioactivity in the Marine Environment, National Academy of Sciences — National Research Council, Washington, D.C., 1971.

166. "Radioactivity and Human Diet," R.C. Russell, ed., Pergamon Press, Oxford, Eng., 1966. Page numbers: (a) 61, (b) 48, (c) 92, (d) 97.

167. "Radionuclides in Food," National Academy of Sciences, Washington, D.C., 1973.

168. "Isotopes and Radiation in Soil Organic-Matter Studies," Proc. Symp., July 15–19, 1968, International Atomic Energy Agency, Vienna, Austria, 1968.

169. "Isotopes and Radiation in Soil-Plant Relationships Including Forestry," Proc. Symp., Dec. 13–17, 1971, International Atomic Energy Agency, Vienna, Austria, 1972.

170. "Isotope and Radiation Techniques in Soil Physics and Irrigation Studies," Proc. Symp., Oct. 1–5, 1973, International Atomic Energy Agency, Vienna, Austria, 1974.

171. "Survival of Food Crops and Livestock in the Event of Nuclear War," D.W. Bensen, A.H. Sparrow, eds., Proc. Symposium at Brookhaven National Laboratory, New York, Sept. 15–18, 1970, U.S. Atomic Energy Commission Office of Information Services, December 1971.

172. Martell, E.A., "Tobacco Radioactivity and Cancer in Smokers," *Am. Sci.*, **63**, 404 (1975).

173. "Alpha Emitting Particles in the Lungs," National Council on Radiation Protection and Measurements, Report No. 46, Washington, D.C., 1975.

174. Casarett, A.P., "Radiation Biology," Prentice-Hall, Englewood Cliffs, N.J., 1968.

175. Osborn, W.S., Jr., "Forecasting Long-Range Recovery from Nuclear Attack," Proc. Symp. on Post-Attack Recovery from Nuclear War," National Academy of Sciences — National Academy of Engineering — National Research Council, 1968, pp. 107–135; quoted in Ref. 157, pp. 486–513.

176. Palumbo, R.F., "Recovery of the Land Plants at Eniwetok Atoll Following a Nuclear Detonation," *Radiat. Bot.*, **1**, 182 (1962); quoted in Ref. 157, pp. 302–315.

177. Rhoads, W.A., Ragsdale, H.L., "Artemesia Shrub Size and Radiation Damage from Local Fallout from Project Schooner," Ref. 182, pp. 953–960.

178. Reference 182, Part IV, "Thermal Effects Studies," 22 papers, pp. 513–876.

179. Gibbons, J.W., Sharitz, R.R., "Thermal Alteration of Aquatic Ecosystems," *Am. Sci.*, **62**, 660 (1974).

180. Woodwell, G.M., Sparrow, A.H., "Effects of Ionizing Radiation on Ecological Systems," in "Ecological Effects of Nuclear War," U.S. Atomic Energy Commission Report BNL-917 (c-43), pp. 20–38, 1965; quoted in Ref. 157, pp. 271–289.

181. Platt, R.B., "Ionizing Radiation and Homeostasis of Ecosystems," in Ecological Effects of Nuclear War," U.S. Atomic Energy Commission Report BNL-917 (c-43), pp. 39–60, 1965; quoted in Ref. 157, pp. 249–270.

182. "Radionuclides in Ecosystems," D.J. Nelson, ed., Proc. Third National Symposium on Radioecology, Oak Ridge, Tenn., May 10–12, 1971, Conf. 710501-P2, National Technical Information Service, Springfield, Va.

183. Sparrow, A.H., "The Role of the Cell Nucleus in Determining Radiosensitivity," Brookhaven Lecture Series No. 17, U.S. Atomic Energy Commission Report BNL-766 (T-287), 1962.

184. "Environmental Radiation Measurements," National Council on Radiation Protection and Measurements, Report No. 50, Washington, D.C., 1976.

185. "Environmental Radioactivity Surveillance Guide," U.S. Environmental Protection Agency, Report ORP/SID 72-2, Washington, D.C., 1972.

186. "Radiation Protection Instrumentation and Its Application," International Commission on Radiological Units and Measurements, Report No. 20, Washington, D.C., Oct. 1, 1971.

187. Lederer, C.M., Hollander, J.M., Perlman, I., "Table of Isotopes," 6th ed., John Wiley & Sons, New York, N.Y., 1967.

188. "Handbook of Chemistry and Physics," Weast, R.C., ed., CRC Press, Cleveland, Ohio, 1976.

189. Sloth, E.N., et al., "Tritium in the Thermal Neutron Fission of Uranium-235," *J. Inorg. Nuc. Chem.*, **24,** 337 (1962).

190. Fluss, M.J., Dudey, N.D., Malewicke, R.L., "Tritium and Alpha-Particle Yields in Fast and Thermal Neutron Fission of ^{235}U," *Phys. Rev.*, **6C,** 2252 (1972).

191. Flynn, K.F., Glendenin, L.E., "Yields of Fission Products for Several Fissionable Nuclides at Various Incident Neutron Energies," U.S. Atomic Energy Commission Report ANL-7749, Washington, D.C., December 1970.

192. U.S. Atomic Energy Commission, AEC Manual, Part 0500, Health and Safety, Appendix 0524, Aug. 12, 1963.

INDEX

A

B

C

D

E

F

G

H

I

L

M

N

O

P

R

S

T

U

V

W

X

NOTES